ANNALS OF
THE NEW YORK ACADEMY
OF SCIENCES

Volume 984

EDITORIAL STAFF

Acting Executive Editor
JUSTINE CULLINAN

Associate Editors
JOHN W. KENNEDY
STEFAN MALMOLI

The New York Academy of Sciences
2 East 63rd Street
New York, New York 10021

THE NEW YORK ACADEMY OF SCIENCES
(Founded in 1817)

BOARD OF GOVERNORS, September 2002–September 2003

TORSTEN N. WIESEL, *Chairman of the Board*
JOHN F. NIBLACK, *Vice Chairman*
JOHN T. MORGAN, *Treasurer*
ELLIS RUBINSTEIN, *Chief Executive Officer* [ex officio]

Honorary Life Governors
WILLIAM T. GOLDEN   JOSHUA LEDERBERG

Governors

ELEANOR BAUM	KAREN E. BURKE	PRAVEEN CHAUDHARI
BRIAN FERGUSON	GERALD D. FISCHBACH	RONALD L. GRAHAM
MARNIE IMHOFF	JACQUELINE LEO	BRUCE McEWEN
PAUL MARKS	RONAY MENSCHEL	SANDRA PANEM
PETER RINGROSE	LEE VANCE	DEBORAH WILEY

HELENE L. KAPLAN, *Counsel* [ex officio]

ADVANCED MEMBRANE TECHNOLOGY

ANNALS OF THE NEW YORK ACADEMY OF SCIENCES
Volume 984

ADVANCED MEMBRANE TECHNOLOGY

Edited by
Norman N. Li, Enrico Drioli,
W.S. Winston Ho, and Glenn G. Lipscomb

The New York Academy of Sciences
New York, New York
2003

Copyright © 2003 by the New York Academy of Sciences. All rights reserved. Under the provisions of the United States Copyright Act of 1976, individual readers of the Annals *are permitted to make fair use of the material in them for teaching and research. Permission is granted to quote from the* Annals *provided that the customary acknowledgment is made of the source. Material in the* Annals *may be republished only by permission of the Academy. Address inquiries to the Permissions Department (permissions@nyas.org) at the New York Academy of Sciences.*

Copying fees: *For each copy of an article made beyond the free copying permitted under Section 107 or 108 of the 1976 Copyright Act, a fee should be paid through the Copyright Clearance Center, Inc., 222 Rosewood Drive, Danvers, MA 01923 (www.copyright.com).*

∞ *The paper used in this publication meets the minimum requirements of American National Standard for Information Sciences—Permanence of Paper for Printed Library Materials. ANSI Z39.48-1984.*

Library of Congress Cataloging-in-Publication Data

Advanced membrane technology / edited by Norman N. Li ... [*et al.*].
   p. cm. — (Annals of the New York Academy of Sciences ; v. 984)
Includes bibliographical references and index.
  ISBN 1-57331-426-9 (alk. paper) — ISBN 1-57331-427-7 (pbk. : alk. paper)
  1. Membranes (Technology)—Congresses. I. Li, Norman N. II. Series.
Q11.N5 vol. 984
[TP159.M4]
500 s—dc21
[660'.28424]

2003000359
CIP

K-M Research/CCP
Printed in the United States of America
ISBN 1-57331-426-9 (cloth)
ISBN 1-57331-427-7 (paper)
ISSN 0077-8923

ANNALS OF THE NEW YORK ACADEMY OF SCIENCES

Volume 984
March 2003

ADVANCED MEMBRANE TECHNOLOGY

Editors
NORMAN N. LI, ENRICO DRIOLI,
W.S. WINSTON HO, AND GLENN G. LIPSCOMB

This volume is the result of a conference entitled **Advanced Membrane Technology**, sponsored by the United Engineering Foundation and held October 14–19, 2001, in Castelvecchio Pascoli, Italy.

CONTENTS

Preface. *By* NORMAN N. LI... ix

About the Editors.. x

Part I. Membrane Contactors

Membrane Contactors in the Beverage Industry for Controlling the Water Gas Composition. *By* ALESSANDRA CRISCUOLI, ENRICO DRIOLI, AND UGO MORETTI................................. 1

Application of Hollow Fiber Membrane Contactors for Catalyst Recovery in the WPO Process. *By* INMACULADA ORTIZ, ANE URTIAGA, M. JOSÉ ABELLÁN, AND FRESNEDO SAN ROMÁN........................ 17

Membrane Contactors for Textile Wastewater Ozonation. *By* GIANLUCA CIARDELLI, INGRID CIABATTI, LAURA RANIERI, GUSTAVO CAPANNELLI, AND ALDO BOTTINO........................... 29

Part II. Water Treatment

Closing Pulp and Paper Mill Water Circuits with Membrane Filtration. *By* JUTTA NUORTILA-JOKINEN, TIINA HUUHILO, AND MARIANNE NYSTRÖM. 39

Membrane Technologies Applied to Textile Wastewater Treatment. *By* MANUELE MARCUCCI, INGRID CIABATTI, ALESSANDRO MATTEUCCI, AND GUIDO VERNAGLIONE... 53

Integrating Photocatalysis and Membrane Technologies for Water Treatment. *By* DAVID F. OLLIS... 65

Water Quality Monitoring in Membrane Filtration Systems.
  By ELHADI M. ABOGREAN, SIOBHAN F.E. BOERLAGE, MARIA D. KENNEDY,
  IBRAHIM M. EL-AZIZI, GILBERT GALJAARD, AND JAN S. SCHIPPERS....... 85
Removal and Recovery of Metals and Other Materials by Supported Liquid
  Membranes with Strip Dispersion. By W.S. WINSTON HO............... 97

Part III. Nanofiltration/Ultrafiltration

Membrane Separation in Green Chemical Processing: Solvent Nanofiltration
  in Liquid Phase Organic Synthesis Reactions. By ANDREW LIVINGSTON,
  LUDMILA PEEVA, SHEJIAO HAN, DINESH NAIR, SATINDER SINGH LUTHRA,
  LLOYD S. WHITE, AND LUISA M. FREITAS DOS SANTOS 123
Computer-Aided Simulation and Design of Nanofiltration Processes.
  By MOHAN NORONHA, VALKO MAVROV, AND HORST CHMIEL 142
Advances in Solvent-Resistant Nanofiltration Membranes:
  Experimental Observations and Applications.
  By D. BHANUSHALI AND D. BHATTACHARYYA...................... 159
Characteristics and Application of Ceramic Nanofiltration Membranes.
  By RALPH WEBER, HORST CHMIEL, AND VALKO MAVROV 178

Part IV. Charged Membranes

Nanocomposite Lithium Ion Conducting Membranes.
  By FAUSTO CROCE AND BRUNO SCROSATI 194
Preparation of Nano-Structured Polymeric Proton Conducting Membranes
  for Use in Fuel Cells. By GIULIO ALBERTI, MARIO CASCIOLA,
  MONICA PICA, AND GIUSI DI CESARE............................. 208
High Performance Perfluoropolymer Films and Membranes.
  By VINCENZO ARCELLA, ALESSANDRO GHIELMI, AND GIULIO TOMMASI.... 226
Novel Charge-Mosaic Membranes. By BENHUI SUN AND SHUYING CHENG...... 245
Material Transport through Charged Mosaic Membrane.
  By AKIRA YAMAUCHI AND TAKASHI FUKUDA 256

Part V. Membrane Formation and New Materials

Effect of Compatibility of PVC/P_2 Alloy System on Membrane Structure
  and Performance. By PATRICIA B. SUN AND BENHUI SUN 267
Stable Liquid Membranes: Recent Developments and Future Directions.
  By A. SARMA KOVVALI AND KAMALESH K. SIRKAR 279
The Use of Conducting Polymers in Membrane-Based Separations: A Review and
  Recent Developments. By JOHN PELLEGRINO 289

Part VI. Gas Separations

Natural Gas Cleanup by Means of Membranes.
  By KLAUS OHLROGGE AND TORSTEN BRINKMANN..................... 306

Polysulfone Hollow Fiber Gas Separation Membranes Filled with
Submicron Particles. *By* V. BHARDWAJ, A. MACINTOSH,
I.D. SHARPE, S.A. GORDEYEV, AND S.J. SHILTON 318

Carbon Molecular Sieve Membranes: A Promising Alternative for
Selected Industrial Applications. *By* MAY-BRITT HÄGG,
JON A. LIE, AND ARNE LINDBRÅTHEN 329

Thin Composite Palladium and Palladium/Alloy Membranes for Hydrogen
Separation. *By* YI HUA MA, IVAN P. MARDILOVICH, AND ERIK E. ENGWALL   346

Mixed Matrix Membrane Development. *By* SANTI KULPRATHIPANJA 361

Part VII. Pervaporation

Nonlinear Parameter Estimation for Solution–Diffusion Models of Membrane
Pervaporation. *By* BING CAO AND MICHAEL A. HENSON 370

Poly(Vinyl Alcohol)–Based Polyelectrolyte Pervaporation Membranes.
By BENHUI SUN AND JIAN ZOU 386

Hydrophobic Pervaporation: Toward a Shortcut Method for the Pervaporation-
Decanter System. *By* ROBERT W. FIELD AND VANESSA LOBO 401

Part VIII. Bioprocessing

Membrane Bioreactors for Treating Waste Streams.
By J.A. HOWELL, T.C. ARNOT, AND W. LIU 411

Why and How Membrane Bioreactors with Unsteady Filtration Conditions Can
Improve the Efficiency of Biological Processes. *By* ISABELLE DAUBERT,
MURIEL MERCIER-BONIN, CLAUDE MARANGES, GÉRARD GOMA,
CHRISTIAN FONADE, AND CHRISTINE LAFFORGUE 420

Effect of Immobilization Site and Membrane Materials on
Multiphasic Enantiocatalytic Enzyme Membrane Reactors.
By NA LI, LIDIETTA GIORNO, AND ENRICO DRIOLI 436

CO_2 Capture by Means of an Enzyme-Based Reactor. *By* R.M. COWAN,
J.-J. GE, Y.-J. QIN, M.L. MCGREGOR, AND M.C. TRACHTENBERG 453

Affinity Membranes as a Tool for Life Science Applications.
By HEIKE BORCHERDING, HANS-GEORG HICKE, DIERK JORCKE,
AND MATHIAS ULBRICHT 470

Enzyme Transmission during Crossflow Filtration of Yeast Suspensions
Using Gas/Liquid Two-Phase Flows. *By* MURIEL MERCIER-BONIN
AND CHRISTIAN FONADE 480

Turnup Turndown of Membrane Operation of Membrane Bioreactors.
By J.A. HOWELL, T.C. ARNOT, AND H.C. CHUA 492

Designing Blood Oxygenators. *By* S.R. WICKRAMASINGHE, A.R. GOERKE,
J.D. GARCIA, AND BINBING HAN 502

Part IX. Membrane Evaporative Cooling

Membrane Evaporative Cooling to 30°C or Less: 1. Membrane Evaporative Cooling of Contained Water. *By* SIDNEY LOEB.................... 515

Membrane Evaporative Cooling to 30°C or Less: 2. Membrane Evaporative Air Cooling. *By* SIDNEY LOEB 528

Index of Contributors 539

The New York Academy of Sciences believes it has a responsibility to provide an open forum for discussion of scientific questions. The positions taken by the participants in the reported conferences are their own and not necessarily those of the Academy. The Academy has no intent to influence legislation by providing such forums.

Preface

This book is devoted to membrane science and technology. It has thirty-eight papers that discuss important types of membranes, such as nanofiltration membranes, ultrafiltration membranes, microfiltration membranes, membranes for separations of gases, inorganic membranes, facilitated or liquid membranes, catalytic membranes, and conducting membranes, and their applications and processes, such as wastewater purification and bioprocessing. Virtually all of my colleagues in the membrane community agree that we have needed a book that discusses comprehensively membrane science and technology. This book is meant to fill that role.

The origin of the book is a new international conference on Advanced Membrane Technology, sponsored by the United Engineering Foundation in New York. Dr. Herman Bieber of the United Engineering Foundation invited me to organize the conference. I agreed to undertake the task because we both felt strongly that there was a timely need for an international forum to discuss the important types of membranes and their processes. The conference was held in the beautiful Tuscany region of Italy, specifically, in Barga, Italy, during the week of October 14, 2001. Dr. Barbara Goldman of the New York Academy of Sciences subsequently suggested publishing an *Annals of the New York Academy of Sciences* volume on advanced membrane technology based on the presentations made at the Barga Conference. Although most of the papers in the volume are from the conference speakers, who updated their conference presentations, a few papers in the book were by special invitation.

I would like to express my gratitude to Ms. Barbara K. Hickernell and Dr. Herman Bieber of the United Engineering Foundation for managing the conference, which was indeed a great success, and to acknowledge financial support for the conference from Ausimont SpA, De Nora Technologie Elettrochimiche SpA, DuPont/Air Liquide, the National Science Foundation (U.S.A.), the Italia Research Center of Procter & Gamble Italia SpA, and the Center of Membrane Science of the University of Kentucky. I also thank my three coeditors, Dr. Enrico Drioli, Dr. Winston Ho, and Dr. Glenn Lipscomb—it was indeed a pleasure to work with them—and all the authors who contributed their papers to this book.

NORMAN N. LI
NL Chemical Technology, Inc.,
Mount Prospect, Illinois

About the Editors

Dr. Norman N. Li is a world-renowned industrial scientist and executive. He has been in the chemical and petroleum industries for about 40 years. He worked for Exxon, UOP, and AlliedSignal (now Honeywell). Presently, he is the president of a research, engineering, and consulting company, NL Chemical Technology, Inc., located in Mount Prospect, Illinois.

Dr. Li has about 100 technical papers, 45 U.S. patents, and 13 edited books. He received the Perkin Medal, the highest honor in American chemical industry, and several awards from the American Chemical Society, the American Institute of Chemical Engineers, and the World Congress of Chemical Engineering. He is a member of the National Academy of Engineering of the U.S.A.

Dr. Enrico Drioli is a Professor of Chemistry at the University of Calabria, Italy. He is also the Director of the Institute of Membranes and Chemical Reactors of the National Research Council of Italy, as well as Chairman of the Working Party on Membranes of the European Federation of Chemical Engineering. He served as President of the European Society of Membrane Science and Technology for about 16 years and was given the honorary title, "Founding President". He is an honorary member of the Institute of Petrochemical Synthesis of the Russian Academy of Sciences and an honorary professor at the China Northwest University in Xi'an, China.

Dr. Drioli has authored more than 320 scientific papers and holds 14 patents, all in the field of membrane science and technology.

Dr. W. S. Winston Ho is a Chair Professor of Chemical Engineering at Ohio State University. Before joining Ohio State in 2002, he was a professor at the University of Kentucky. Dr. Ho, however, spent most of his career in industry. He worked at Allied Chemical Corporation, Xerox Corporation, Exxon, and Commodore Separation Technologies for a total of 28 years. He was elected to the National Academy of Engineering, U.S.A., in 2002.

Dr. Ho has more than 50 publications and 42 patents. He is the coauthor of the *Membrane Handbook*, for which he received the Professional and Scholarly Publishing Award. He also received the Industrial Research 100 Award for Exxon's FLEXSORB gas treatment process. He was the Inventor of the Year for 1991, awarded by the New Jersey Inventors Congress and Hall of Fame. Dr. Ho served as the Director of the Separations Division of the American Institute of Chemical Engineers.

Dr. Glenn G. Lipscomb is a Professor of Chemical Engineering at the University of Toledo in Toledo, Ohio. He was on the faculty at the University of Cincinnati for five years and worked at Dow Chemical Company for three years after receiving his Ph.D. degree from the University of California at Berkeley in 1987.

Dr. Lipscomb has 35 publications and 3 patents. He has given 23 invited seminars at universities and industrial laboratories. He served as the President of the North American Membrane Society and is presently Editor of the *Membrane Quarterly*, published by the North American Membrane Society.

Membrane Contactors in the Beverage Industry for Controlling the Water Gas Composition

ALESSANDRA CRISCUOLI,[a] ENRICO DRIOLI,[b] AND UGO MORETTI[c]

[a]*Research Institute on Membrane Technology (ITM-CNR), Via Pietro Bucci Cubo 17/C, Rende (CS) 87030 Italy*

[b]*Department of Chemical Engineering and Materials, University of Calabria, Via Pietro Bucci Cubo 17/C, Rende (CS) 87030 Italy*

[c]*Tecnoproject Industriale s.r.l., Via E. Fermi 40, Curno (BG) 24035 Italy*

ABSTRACT: In the work described here, membrane contactors are used for coupling the removal of species (oxygen and hydrogen sulfide) present in the water with the water carbonation process. We include both experiments and a theoretical study devoted to the analysis of the transport phenomena that occur in the membrane contactor. The main resistance to mass transport was located at the liquid side. Correlations between Sherwood and Reynolds numbers on the shell side that are suitable for the membrane contactor used to carry out our experiments have been determined. In particular, for $Re > 1.6$, the expression proposed by Yang and Cussler in 1986: $Sh = 0.90 Re^{0.40} Sc^{0.33}$ describes the behavior of the system; whereas, for Re between 0.03 and 0.3, a new expression is proposed: $Sh = 0.435 Re^{1.2} Sc^{0.33}$. A comparison with traditional equipment is also furnished. Membrane contactors offer reduced size, CO_2 consumption, and capital costs.

KEYWORDS: membrane contactors; water gas composition; beverage industry

INTRODUCTION

Interest in membrane contactors as new systems for carrying out mass transfer between phases was recently reviewed by Gabelman and Hwang[1] who reported their application in fermentation, pharmaceuticals, wastewater treatment, VOC removal from waste gas, osmotic distillation, and other systems. Commercial applications of membrane contactors have been cited for the carbonation line of a Pepsicola plant in West Virginia, CO_2 removal in beer production, and ultrapure water production for the semiconductor industry. A huge amount literature refers to the use of membrane contactors for carrying out gas absorption or removal into/from a liquid phase.[2-11] Advantages found over traditional systems include independent control of the phases flow rates, the prevention of flooding and loading problems typical of packed towers, and the possibility of achieving better yields and removal with respect to conventional

---

Address for correspondence: Alessandra Criscuoli, Research Institute on Membrane Technology (ITM-CNR), Via Pietro Bucci Cubo 17/C, Rende (CS) 87030 Italy. Voice: +39-0984-492014; fax: +39-0984-402103.
a.criscuoli@itm.cnr.it

equipment. Another interesting characteristic of membrane contactors is the high surface area/volume ratio, which makes them good candidates for reaching the objectives of *process intensification* theory so as to design plants with high productivity/size ratio. The beverage industry differs in the treatments used to obtain mineral water with the desired characteristics. The well water sometimes contains species, such as oxygen or hydrogen sulfide, that have to be removed. In the case of sparkling water production, carbon dioxide has to be added to the water (water carbonation). When the water temperature exceeds 14°C (the temperature at which water is usually carbonated) a refrigeration step needs to be considered. In conventional processes removal and carbonation are carried out separately in different devices, and the desired removal is not often achieved. In the present study, membrane contactors are used to coupling the removal of species present in water to the water carbonation process. Both an experimental and theoretical study has been carried out in order to compare the potential of membrane contactors with respect to traditional equipment in terms of efficiency and cost. The comparison was made by using a model developed for the scaleup of the system in order to determine the size required for treating industrial quantities.

EXPERIMENTAL SECTION

Experiments were carried out at laboratory scale with a commercial hollow fiber module, working in continuous and in recycle mode, for water carbonation coupled with oxygen removal, using carbon dioxide as the strip gas. The system flowsheet for the continuous flow mode is shown in FIGURE 1. In the recycle flow mode, only water was recycled; the gas stream was continuously supplied to the module. Water containing oxygen was fed through a cooling/heating device used to vary its temperature into the shell side of the module; the CO_2 stream was fed countercurrently into the lumen of the membrane fibers. The flow rates, pressures, and temperatures of all streams involved were monitored during each run. The influence of several parameters, such as the water flow rate, the gas flow rate, the gas pressure, and the inlet oxygen concentration, on the efficiency of the process was investigated. In all tests, the water pressure was maintained at 0.5 bar higher than the gas pressure in order to avoid the introduction of CO_2 bubbles in the liquid stream and to better *stabilize* the interfacial area between the two phases. Water was analyzed by both a mass balance on the gas stream and titration for CO_2 determination, and by a microprocessor dissolved oxygen meter (Hanna Instruments) for O_2 determination. The lowest oxygen concentration the microprocessor could register was 0.1 ppm. A 2.5"×8" LiquiCel (Hoechst Celanese) module was used for the experiments. The main characteristics of the module and of the membranes are reported in TABLE 1.

THEORETICAL SECTION

A simulation program was written to describe the experimental system used. In particular, a differential mass balance along the module was formulated for both phases:

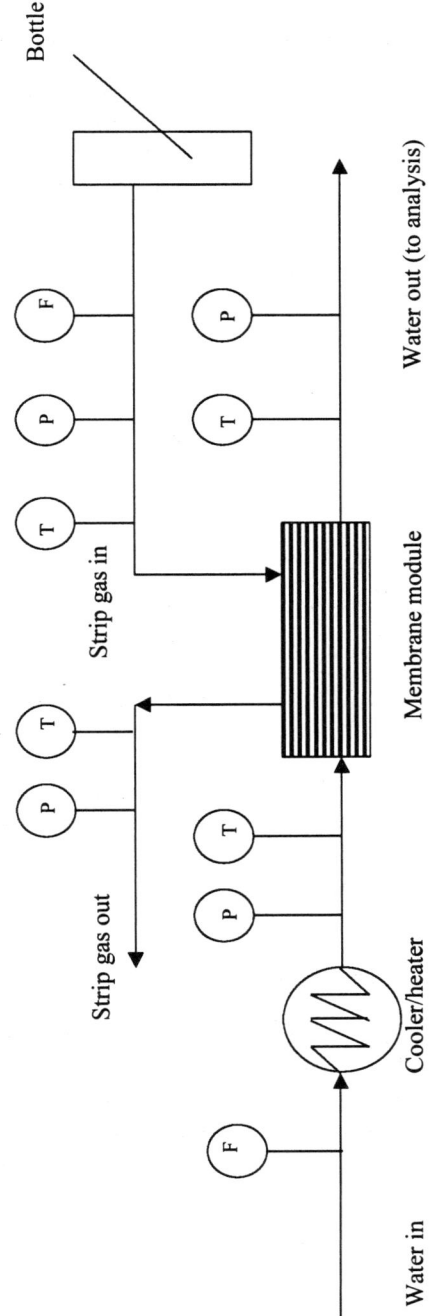

FIGURE 1. Flowsheet of the system used—continuous flow mode.

TABLE 1. Characteristics of the commercial module (2.5″×8″ LiquiCel—Hoechst Celanese)

Main Characteristics of the Module Used		Main Characteristics of the Membranes	
average cartridge i.d. (m)	0.022	inside diameter (m)	240×10^{-6}
average cartridge o.d. (m)	0.055	outside diameter (m)	300×10^{-6}
average number of fibers	10,200	porosity	0.40
effective fiber length (m)	0.16	pore diameter (m)	0.03×10^{-6}
fiber packing fraction	0.435	tortuosity	2.5
total contact area (m^2)	1.4	material	polypropylene

Liquid phase:

$$\frac{dx_i}{dz} = Flu_i \cdot 2\pi r_f \frac{N_f}{F_{H_2O}} \tag{1}$$

Gas phase:

$$\frac{dy_i}{dz} = Flu_i \cdot 2\pi r_f \frac{N_f}{F_{gas}}, \tag{2}$$

where r_f denotes the outer fiber radius; N_f is the number of fibers; F_{H_2O} is the water flowrate; F_{gas} is the gas flowrate; Flu_i is the flux of i through the membrane; x_i and y_i are the concentrations of i in the liquid and gas phase, respectively; and z is the module local length.

There are three resistances to mass transport between phases: (1) resistance on the liquid side, (2) resistance on the gas side, and (3) resistance through the membrane. Thus, the flux can be expressed as

$$Flu_i = K_L \cdot (x_i - y_i). \tag{3}$$

The overall mass transfer resistance is given by

$$\frac{1}{K_L} = \frac{1}{k_l} + \frac{1}{k_m \cdot H} + \frac{1}{k_g \cdot H}, \tag{4}$$

where K_L is the overall mass transfer coefficient (m/sec), k_l is the liquid mass transfer coefficient (m/sec), k_g is the gas mass transfer coefficient (m/sec), k_m is the membrane mass transfer coefficient (m/sec), and H is Henry's constant.

To calculate the flux it is necessary to obtain expressions for the three mass transfer coefficients. For the gas phase, the following expression, valid for a fluid flowing into fibers, can be used:[10]

$$\frac{1}{k_g H} = \frac{0.617}{H} \left[\frac{L d_i}{v_g D_i^2} \right]^{0.33}, \tag{5}$$

where d_i is the inner diameter of the fiber, L is the length of the fiber, v_g is the gas velocity inside the fiber, and D_i is the diffusion coefficient of i in the gas phase.

The membrane mass transfer coefficient can be calculated from[12]

$$\frac{1}{k_m H} = \frac{\delta \tau}{D_{eff} \varepsilon H}, \qquad (6)$$

where $D_{eff} = (1/D_i + 1/D_{kn})^{-1}$ and D_{kn} is the Knudsen diffusion coefficient.

No generalized expressions are available to describe the mass transfer that occurs on the shell side of the module. Several expressions are available in the literature that depend on the particular system analyzed, the Reynolds range used, and the type of flow (parallel or crossflow). The module used to carry out the experiments is characterized by a flow that is mainly perpendicular to the fibers (there is a parallel flow for only a small part of the module). For this reason, several expressions proposed in literature for this type of flow were tested in our theoretical analysis:

For loosely packed fibers and $1 < Re < 25$[13]

$$k_l = \frac{D_i}{2r_f} 0.90 Re^{0.40} Sc^{0.33}. \qquad (7)$$

For closely packed fibers and $1 < Re < 25$[13]

$$k_l = \frac{D_i}{2r_f} 1.38 Re^{0.34} Sc^{0.33}. \qquad (8)$$

For closely packed fibers[14]

$$k_l = \frac{D_i}{2r_f} 0.39 Re^{0.59} Sc^{0.33}. \qquad (9)$$

For $0.01 < Re < 1$[8]

$$k_l = \frac{D_i}{2r_f} 0.57 Re^{0.31} Sc^{0.33}. \qquad (10)$$

All of these expressions refer to the outer diameter of the fibers. By solving the differential mass balances it was possible to determine theoretically how the membrane contactor behaves in terms of oxygen removal and water carbonation.

RESULTS AND DISCUSSION

To determine the liquid mass transfer coefficient expression best able to describe the behavior of the experimental device, a comparison between the experimental results obtained as function of Reynolds number and the theoretical predictions was made. FIGURE 2 shows the oxygen removal achieved experimentally (points) and theoretically (lines). The theoretical results are relative to the liquid mass transfer coefficients calculated from Equations (9) and (10). Equations (7) and (8) were not used because the experimental range of Reynolds number was lower than the value they relate to. It can be seen that, experimentally, there is an increase in oxygen removal with Reynolds, due to the reduction of the resistance at the liquid side, whereas the trend obtained by considering both expressions is opposite. The expression that was found to describe the experimental trend well is as follows:

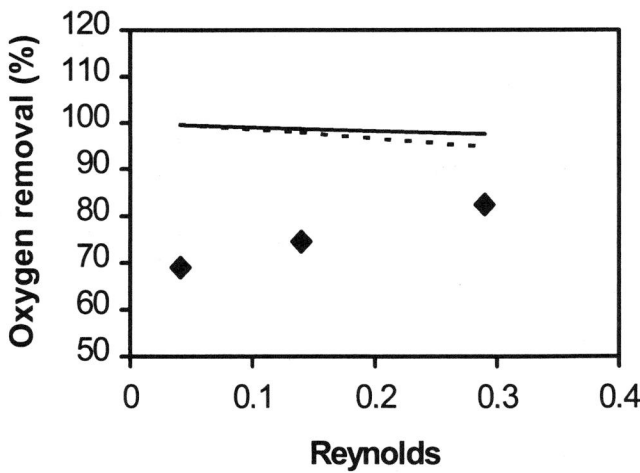

FIGURE 2. Influence of the Reynolds number on oxygen removal. Comparison between experimental data and theoretical predictions (T, 20°C; P, 1 atm; gas flowrate, 290 mL/min; $O_{2,in}$, 7 ppm). - - -, 3; ———, 4; ♦, Exp.

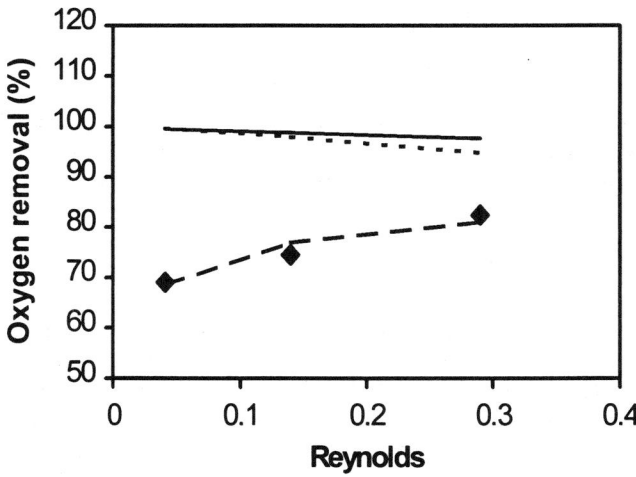

FIGURE 3. Influence of the Reynolds number on oxygen removal. Comparison between experimental data and theoretical predictions (T, 20°C; P, 1 atm; gas flowrate, 290 mL/min; $O_{2,in}$, 7 ppm). - - -, 3; ———, 4; ♦, Exp; — —, 5.

$$k_l = \frac{D_i}{2r_f} 0.435 Re^{1.2} Sc^{0.33}, \tag{11}$$

as is shown in FIGURE 3. The multiplicative term 0.435 is the fiber packing fraction of the module, as furnished by the producer. Also in this expression, the outer diameter of the fibers is considered. To define the range of validity of the new proposed expression, a comparison between the results obtained by the Hoechst Celanese Company working at higher Reynolds values (1.6–9.6) and the theoretical predictions was made. FIGURE 4 shows the results obtained by using **(7)**, **(8)**, **(9)**, and **(11)**. In this case, all literature expressions follow the trend reported by the Company, whereas the proposed new expression is not able to predict this trend well. Furthermore, among literature expressions **(7)**, relating to loosely packing fractions, gives the best fit. An average Reynolds exponent equal to 0.41 was reported by the producers,[7] in agreement with what was found here. On the basis of these observations we conclude that, for Re ranging between 0.03 and 0.3, the expression that should be used to describe the system is **(11)**, whereas for $Re > 1.6$, Equation **(7)** should be considered. The larger exponent value found for lower Reynolds number could be attributed to the possible presence of polydisperse channels between fibers, due to the very low flow rates used during the experimental tests (35–350 mL/min). By increasing the Reynolds number contact between the membrane surface and the liquid is enhanced. For this reason, the mass transfer coefficient exhibits a stronger dependence on Reynolds number at low flow rates than at higher rates. The fact that, depending on Reynolds values, different expressions should be used to calculate the liquid mass transfer coefficients, was also reported by Wickramasinghe *et al.*[15] The effect of the gas flow rate on oxygen removal is shown in FIGURE 5. Both experimentally and theoretically, the gas flow rate does not influence the result; the mass

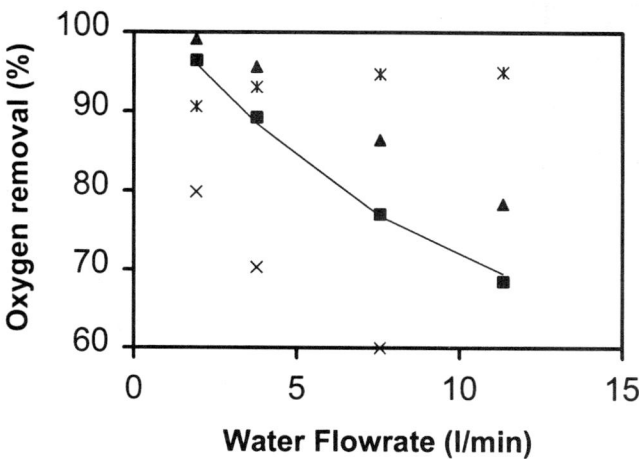

FIGURE 4. Oxygen removal as function of the water flowrate. Comparison between data from the producer and theoretical predictions (2.5″×8″ module; T, 20°C; P, 1 atm; excess of gas flowrate). —, producer; ■, (1); ▲, (2); ×, (3); ✶, (5).

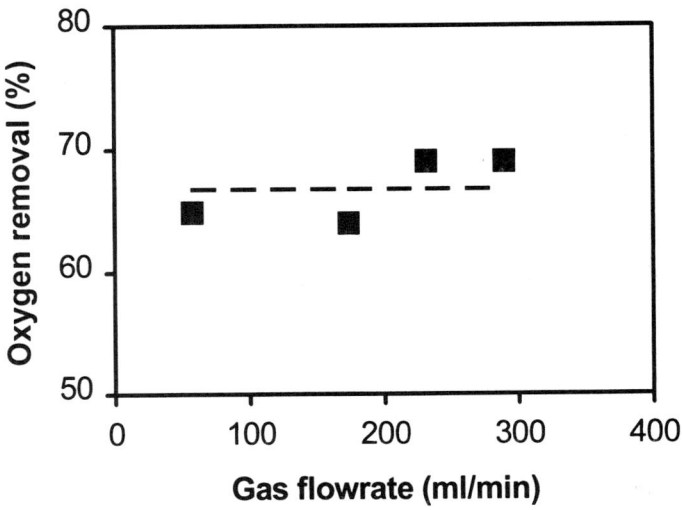

FIGURE 5. Influence of the gas flowrate on oxygen removal (T, 20°C; P, 1 atm; water flowrate, 40 mL/min; $O_{2,in}$, 7 ppm). — —, 5; ■, Exp.

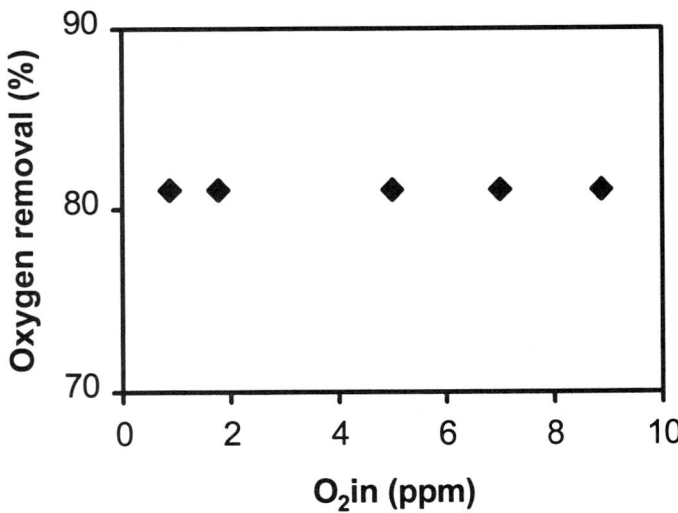

FIGURE 6. Influence of the oxygen concentration on the oxygen removal (T, 20°C; P, 1 atm; water flowrate, 350 mL/min; gas flowrate, 290 mL/min).

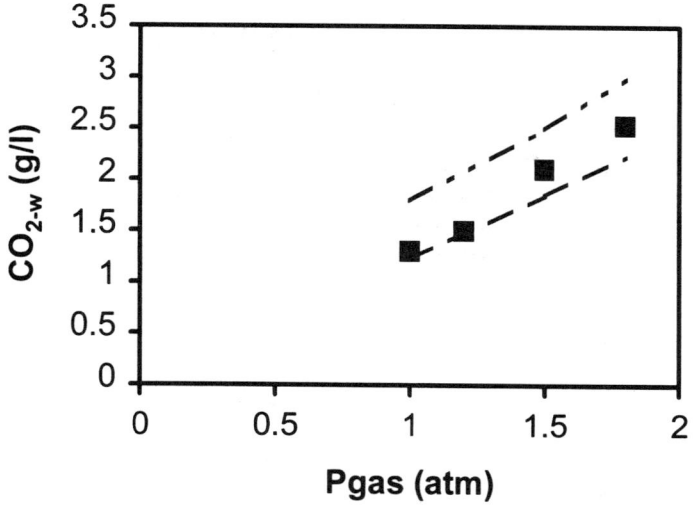

FIGURE 7. Influence of gas pressure on CO_2 concentration in water (CO_2–w) (T, 20°C; water flowrate, 200 mL/min; gas flowrate, mL/min). — —, 5; ■, Exp; — - -, solubility.

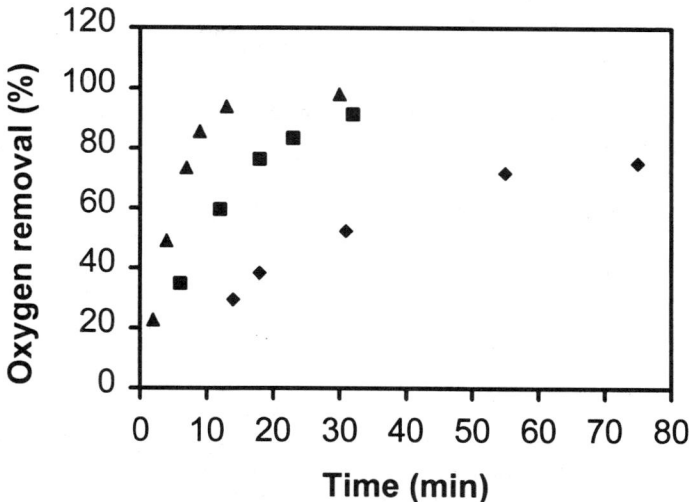

FIGURE 8. Dependence of oxygen removal on the water flow rate—recycling tests (T, 20°C; P, 1 atm; gas flowrate, 290 mL/min; $O_{2,in}$, 7 ppm; volume of liquid in the entire recycle loop, 1.5 liters). Q1 = 35 (◆), 180 (■), and 350 (▲) mL/min.

transfer resistance at the liquid side controls, in fact, the performance of the system. By varying the inlet oxygen concentration we obtained the same extent of oxygen removal, an index of the higher flexibility of the system (see FIGURE 6). FIGURE 7 shows the CO_2 concentration in water as function of gas pressure. The theoretical values agree with the experimental data. By increasing the gas pressure, a higher degree of water carbonation can be achieved according to thermodynamic considerations. The same type of tests were made in the recycle flow mode. FIGURE 8 shows the oxygen removal with time for three different liquid flow rates. By increasing the water flow rate there is a reduction in the mass resistance on the liquid side and, thus, 99% removal is achieved in less time. The final water was saturated in CO_2.

SCALEUP

In beverage industries the water carbonation is generally carried out by using two units in series (see FIGURE 9):
1. A deareation stripper (height 5.5 m) in which 60 kg/h of CO_2 (99.96% purity, pressure 7 bar) is fed to remove oxygen (the residual oxygen content in water after stripping is 0.05 ppm) from 30 m^3/h of water (pressure 2 bar, T = 15–20°C). The CO_2 at the exit (with O_2) is vented to the atmosphere.
2. A saturation column (approximately the same dimensions as the stripper) in which cold water (15°C) is fed and saturated with pressurized CO_2 (less than 7 bar) in a short time (this can be considered a continuous process).

In the present study the mathematical model developed was used to scaleup the system. In particular, we calculated the membrane area and the strip gas flow rate required to treat (deareate and carbonate) 30 m^3/h of water. A larger module, the 4"×28"—commercialized by Hoechst Celanese—was considered for this calculation. Its characteristics are reported in TABLE 2. Because of the modularity of the membrane systems, the same expression for the liquid mass transfer coefficient calculation, found to be valid at Re > 1.6, for the data reported by the Company for the 2.5"×8" module, well fits the data reported by the Company for the larger module operating at 2.9 < Re < 20 (see FIGURE 10). This expression has been used for the larger module. The CO_2 content in carbonated water usually ranges from 1 to 4.4 g/L. Because the solubility of CO_2 at 20°C and 2.5 atm is 4.5 g/L, the operating pressure of the CO_2 stream was 2.5. FIGURE 11 shows the scheme for the proposed membrane system based on the simulation results. The water is fed to two blocks in series, each made with 20 modules in parallel. The outlet water contains 4.3 g/L of CO_2 and 0.05 ppm of O_2. TABLE 3 reports a comparison between traditional and proposed systems. From this table it can be seen that the membrane system presents lower capital costs and substantially lower volumes, but higher CO_2 consumption and operating costs due to membrane replacement. To improve the performance of the membrane devices, a membrane system with three membrane contactor units (units 1, 2, and 3) in series was analyzed (the units consisted of, respectively, twenty, ten, and fifteen modules in parallel) for treating water and a mixer (unit 4) employed to dilute the oxygen in the strip gas stream (carbon dioxide, oxygen, nitrogen, and water vapor) in order to recycling it to the three membrane contactors (see FIGURE 12). A comparison between traditional and the *improved proposed system* is shown in TABLE 4.

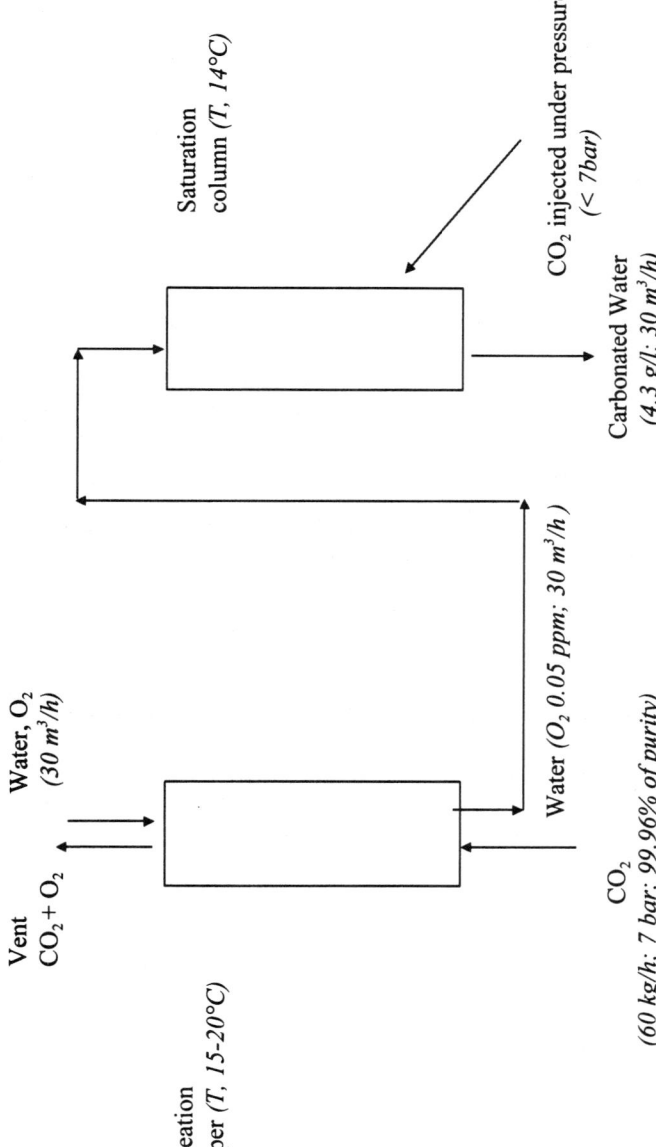

FIGURE 9. Scheme of the system employed in the beverage industries for water carbonation.

TABLE 2. Characteristics of the commercial module (4″ × 28″ LiquiCel—Hoechst Celanese)

Main Characteristics of the Module Used		Main Characteristics of the Membranes	
average cartridge i.d. (m)	0.032	inside diameter (m)	240×10^{-6}
average cartridge o.d. (m)	0.088	outside diameter (m)	300×10^{-6}
average number of fibers	31,800	porosity	0.40
effective fiber length (m)	0.62	pore diameter (m)	0.03×10^{-6}
fiber packing fraction	0.435	tortuosity	2.5
total contact area (m^2)	18.6	material	polypropylene

Membrane contactors present lower carbon dioxide consumption and a release to the atmosphere comparable with that related to the traditional system. The 20 kg/h of carbon dioxide at 99.20% coming from the mixer might be sold. With respect to membrane cost, it is noted that the 92.96 Euro/m^2 used is the commercial value of one of the modules considered in the analysis. This cost can be drastically reduced if the number of modules required increases, as in our case, allowing a significant reduction in replacement costs.

From the initial calculations for hydrogen sulfide removal, it is estimated that 85% removal can be achieved by using 682 kg/h of carbon dioxide and 1,116 m^2 membrane area.

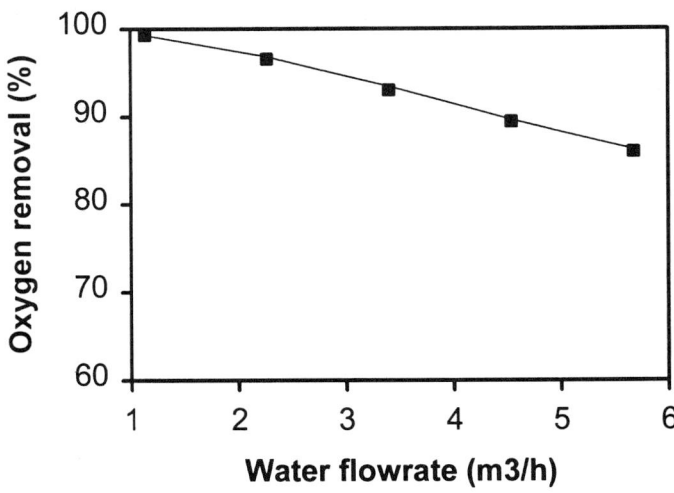

FIGURE 10. Oxygen removal as function of the water flowrate. Comparison between data from the producer and theoretical predictions (4″×28″ module; T, 20°C; P, 1 atm; excess gas flowrate). ——, producer; ■, (1).

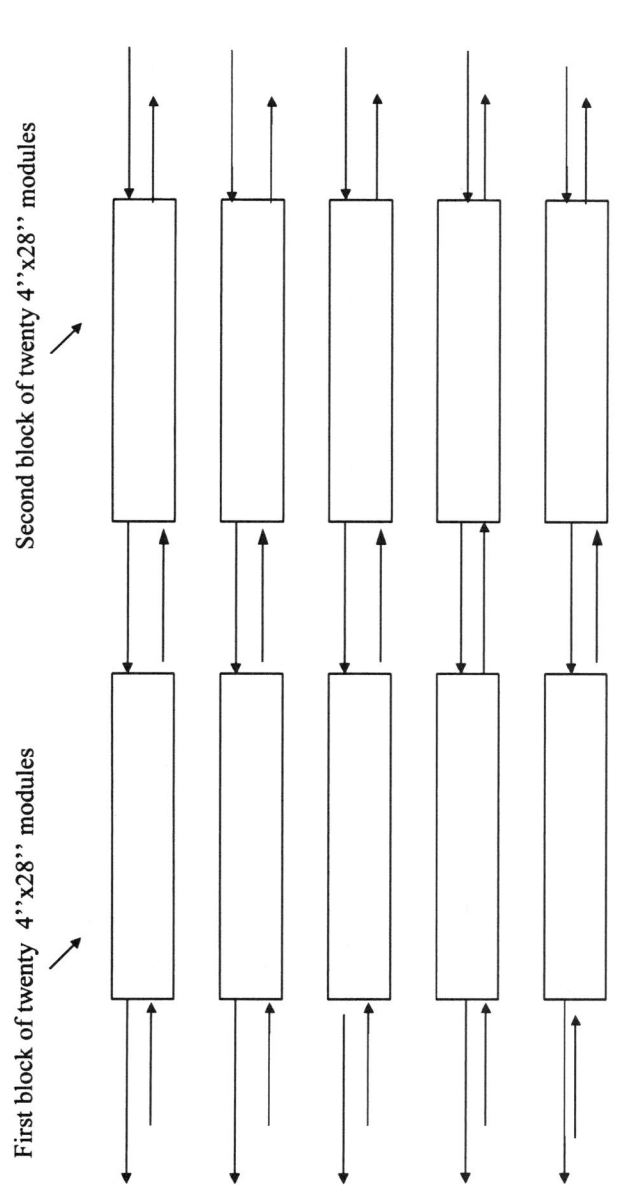

FIGURE 11. Scheme for the first proposed membrane system. Feed: water (30 m^3/h, O$_2$ 7 ppm); CO$_2$ (659 kg/h, 2.5 bar, 99.96% purity). Exit: carbonized water (30m^3/h, O$_2$ 0.05 ppm, CO$_2$ 4.3 g/L). (T 20°C, membrane area required 744 m^2, oxygen removal 99.28%).

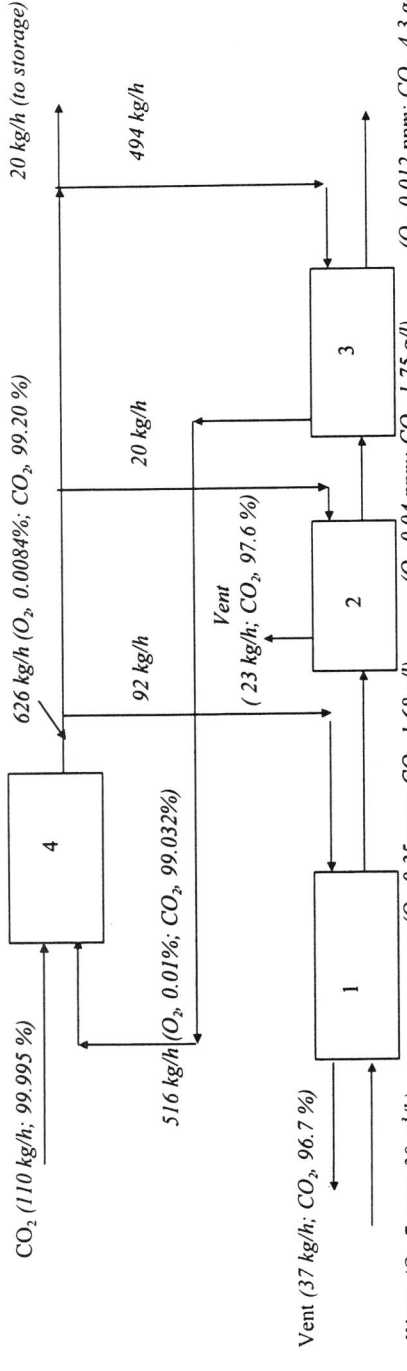

FIGURE 12. Scheme for the second proposed membrane system. Unit 1: block of 20 4″×28″ modules; operating conditions 20°C, 1 bar; oxygen removal 95%; membrane area 372 m². Unit 2: block of 10 4″×28″ modules; operating conditions 20°C, 1 bar; oxygen removal 88.57%; membrane area 186 m². Unit 3: block of 15 4″×28″ modules; operating conditions 20°C, 2.5 bar; oxygen removal 70%; membrane area 279 m². Unit 4: mixer; operating conditions 20°C, 2.5 bar. Total membrane area 837 m². Overall oxygen removal 99.83%.

TABLE 3. Comparison between traditional and first proposed systems

		Traditional	Proposed
Equipment cost (Euro)	Deareation unit	71,013	
	Saturation unit	103,291	
	Total	174,304	69,160[a]
CO_2 consumption (kg/h)	Deareation unit	60	
	Saturation unit	130[b]	
	Total	190	659
CO_2 cost (Euro/h)[c]		30	102
Membrane replacement (Euro/y)[d]		—	23,000
Volume (m^3)		3	0.22

[a] A membrane cost of 92.96 Euro/m^2 has been considered.
[b] For saturating at 4.3 g/L, 30 m^3/h of water, 130 kg/h of CO_2 need to be used.
[c] The CO_2 cost is 0.155 Euro/kg.
[d] Membrane lifetime, three years.

CONCLUSIONS

Our work has been useful to define appropriate mass transfer correlations at shell side of the commercial module used. From the comparison between the proposed membrane system and traditional equipment used for the water carbonation process several benefits have been found:

TABLE 4. Comparison between traditional and proposed systems

		Traditional	Proposed
Equipment cost (Euro)	Deareation unit	71,013	
	Saturation unit	103,291	
	Total	174,304	77,800[a]
CO_2 consumption (kg/h)	Deareation unit	60	
	Saturation unit	130[b]	
	Total	190	110
CO_2 cost (Euro/h)[c]		30	17
CO_2 in the atmosphere (kg/h)		60	60
Membrane replacement (Euro/y)[d]		—	25,900
Volume (m^3)		3	0.25

[a] A membrane cost of 92.96 Euro/m^2 has been considered.
[b] For saturating at 4.3 g/L, 30 m^3/h of water, 130 kg/h of CO_2 need to be used.
[c] The CO_2 cost is 0.155 Euro/kg.
[d] Membrane lifetime, three years.

1. It is possible to operate at ambient temperature without cooling the water;
2. It is possible to remove up to 99.83% oxygen;
3. Lower CO_2 consumption is required;
4. Lower investment costs are needed;
5. The independence of the efficiency of the process on the inlet concentration of oxygen makes the system flexible and able to handle the inevitable variations in water composition;
6. Hydrogen sulfide can also be removed;
7. Lower volumes (due to the high surface/volume ratios of the membrane modules) are necessary, as required by the *process intensification* strategy and the rationalization of industrial production.

REFERENCES

1. GABELMAN, A. & S.-T. HWANG. 1999. Hollow fiber membrane contactors. J. Membr. Sci. **159:** 61–106.
2. LI, K., M.S.L. TAI & W.K. TEO. 1994. Design of a CO_2 scrubber for self-contained breathing systems using a microporous membrane. J. Membr. Sci. **86:** 119–125.
3. GHOSH, A.C., S. BORTHAKUR & N.N. DUTTA. 1994. Absorption of carbon monoxide in hollow fiber membranes. J. Membr. Sci. **96:** 183–192.
4. SCHNEIDER, M., F. REYMOND, I.W. MARISON, et al. 1995. Bubble-free oxygenation by means of hydrophobic porous membranes. Enzyme Microb. Technol. September **17:** 839–847.
5. RANGWALA, H.A. 1996. Absorption of carbon dioxide into aqueous solutions using hollow fiber membrane contactors. J. Membr. Sci. **112:** 229–240.
6. AL-SAFFAR, H.B., B. OZTURK & R. HUGHES. 1997. A comparison of porous and non-porous gas–liquid membrane contactors for gas separation. Trans. IChemE **75**(A): 685–692.
7. SEGUNPTA, A., P.A. PETERSON, B.D. MILLER, et al. 1998. Large-scale application of membrane contactors for gas transfer from or to ultrapure water. Sep. Pur. Technol. **14:** 189–200.
8. BHAUMIK, D., S. MAJUMDAR & K.K. SIRKAR. 1998. Absorption of CO_2 in a transverse flow hollow fiber membrane module having a few wraps of the fiber mat. J. Membr. Sci. **138:** 77–82.
9. VLADISAVLJEVIC, G.T. 1999. Use of polysulfone hollow fibers for bubbleless membrane oxygenation/deoxygenation of water. Sep. Pur. Technol. **17:** 1–10.
10. MAHMUD, H., A. KUMAR, R.M. NARBAITZ, et al. 2000. A study of mass transfer in the membrane air-stripping process using microporous polypropylene hollow fibers. J. Membr. Sci. **179:** 29–41.
11. LEIKNES, T. & M.J. SEMMENS. 2001. Vacuum degassing using microporous hollow fiber membranes. Sep. Pur. Technol. **22:** 287–294.
12. QI, Z. & E.L. CUSSLER. 1985. Microporous hollow fibers for gas absorption. II. Mass transfer across the membrane. J. Membr. Sci. **23:** 333–345.
13. YANG, M.-C. & E.L. CUSSLER. 1986. Designing hollow-fiber contactors. AIChE J. **32:** 1910–1916.
14. KREITH, F. & W.Z. BLACK. 1980. Basic Heat Transfer. Harper & Row, New York.
15. WICKRAMASINGHE, S.R., M.J. SEMMENS & E.L. CUSSLER. 1992. Mass transfer in various hollow fiber geometries. J. Membr. Sci. **84:** 1–14.

Application of Hollow Fiber Membrane Contactors for Catalyst Recovery in the WPO Process

INMACULADA ORTIZ, ANE URTIAGA,
M. JOSÉ ABELLÁN, AND FRESNEDO SAN ROMÁN

*Department of Chemical Engineering, University of Cantabria,
Avda. de los Castros, Santander, Spain*

ABSTRACT: In this work the use of a membrane based liquid extraction process for recovery of the homogeneous catalyst employed in the wet peroxide oxidation process (WPO) is studied. In the WPO process the oxidation agent is the hydroxyl radical that is obtained by using a combination of hydrogen peroxide and a mixture of Fe(II), Cu(II), and Mn(II) in aqueous solution. The mixture of metallic cations permits the almost total degradation of the refractory organic compounds, but the use of metallic salts as catalysts induces additional pollution. To recover the homogeneous catalyst of the WPO process by means of non-dispersive solvent extraction (NDSX) two hollow fiber membrane contactors are employed, one for the extraction step and the second for the back-extraction step. From the initial assays, the extractant LIX 622N was selected for Cu(II) recovery and Cyanex 272 for Fe(II) and Mn(II) recovery. Selective separation of Fe(II) and Mn(II) can be obtained by adjusting the pH of the feed aqueous phase. The three metals are stripped using sulfuric acid to give concentrated solutions of $CuSO_4$, $FeSO_4$, and $MnSO_4$ that can be recycled to the formulation of the catalyst solution of the WPO process. A mathematical model has been proposed to describe the recovery of Cu. Two design parameters are required: the membrane mass transport coefficient of the extraction and stripping modules ($k_m = 3.07 \times 10^{-7}$ m/sec) and the equilibrium parameter of the extraction reaction ($K_{Ex} = 0.0832$).

KEYWORDS: copper; LIX 622N; wet peroxide oxidation; membrane-based solvent extraction; hollow fiber

NOMENCLATURE:

C	concentration, mol/m^3
D	diffusion coefficient, m^2/sec
d_{lm}	logarithmic mean diameter of the fiber, m
d_o	outer diameter of the fiber, m
F	volumetric flowrate, m^3/sec
K_{EX}	chemical equilibrium constant
k_m	membrane mass transfer coefficient, m/sec
L	length of the membrane contactor, m
l	membrane thickness, m
n_f	number of fibers
r_i	inner diameter of the fibers, m

Address for correspondence: Inmaculada Ortiz, Department of Chemical Engineering, University of Cantabria, Avda. de los Castros s/n, 39005 Santander, Spain. Voice: 34 942 201587; fax: 34 942 201591.
ortizi@unican.es

r_o	outer diameter of the fibers, m
t	time
V	volume, m³
z	axial coordinate

Greek Letters

ε	porosity of the membrane
τ	tortuosity factor of the membrane

Subscripts

E	aqueous feed phase
S	aqueous stripping phase
O	organic extractant phase

Superscripts

i	initial
mE	extraction module
mS	stripping module
T	tank
$*$	interfacial

INTRODUCTION

Advanced oxidation technologies (AOPs) offer an alternative to the treatment of industrial effluents characterized by the presence of refractory organic substances. In AOPs the primary oxidant is the hydroxyl radical (·OH) that can be obtained by conversion of hydrogen peroxide in a variety of mechanisms: solutions of hydrogen peroxide exposed to UV light (H_2O_2/UV) or mixed with ozone (H_2O_2/O_3) or iron salts (H_2O_2/Fe).[1] In the wet peroxide oxidation (WPO) process hydroxyl radical generation is obtained by using a combination of hydrogen peroxide and a mixture of Fe(II), Cu(II), and Mn(II) as a homogeneous catalyst.[2] The WPO process is adapted from the classical Fenton's reagent, allowing high oxidation efficiencies under mild conditions (90–130°C, 0.1–0.5 MPa). Fenton's reaction leads, in an acidic medium, to elimination of the initial organic compounds and to a significant

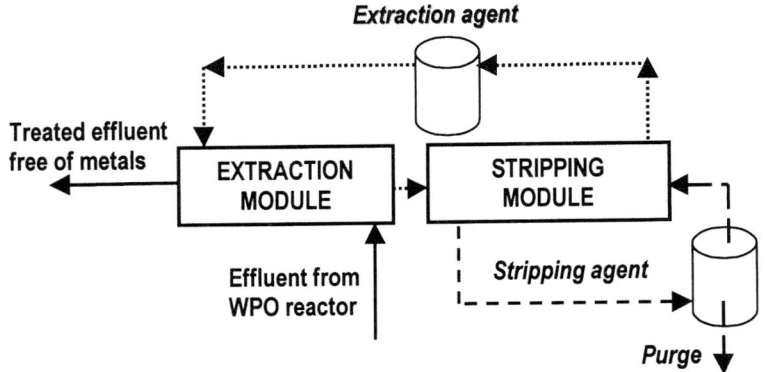

FIGURE 1. Flowsheet for the non-dispersive solvent extraction process.

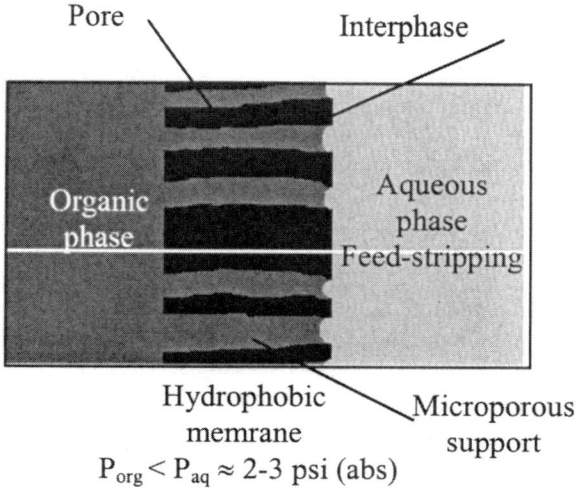

FIGURE 2. Interface immobilization in the non-dispersive extraction process.

reduction in COD and TOC. During the reaction, species refractory to the oxidation accumulate.[3] These are mainly volatile fatty acids (oxalic, acetic, formic, etc.). In the WPO process the mixture of metallic cations permits the almost total degradation of refractory organic acids that are generated during the oxidation of the initial organic compounds.

The use of metallic salts as catalysts, however, induces additional pollution that needs to be treated.[4] When oxidation of the wastewater is complete, the metal catalysts are precipitated to form a sludge that must be managed as a toxic waste.

In this work we analyze the recovery of the homogeneous catalyst of the WPO process by means of non-dispersive solvent extraction (NDSX), as an alternative to the metal precipitation step. Two hollow fiber membrane contactors are employed, one for the extraction step and the second for the backextraction step, following the scheme shown in FIGURE 1. In the extraction step the metals are transferred from the aqueous phase to the organic phase by selective reaction with the extraction agent. In the stripping step the extraction agent is regenerated and the metals are concentrated in the stripping liquor. NDSX is characterized by stabilization of the aqueous–organic interface at a porous material, avoiding dispersion of the phases and, thus, eliminating emulsion formation and phase entrainment (see FIGURE 2). Previous experience in the application of the NDSX technology to metal recovery is referenced.[5–7]

EXPERIMENTAL

The hollow fiber membrane contactors used for NDSX in this work are commercialized by Hoechst Celanese (see FIGURE 3). The main characteristics are 10,000

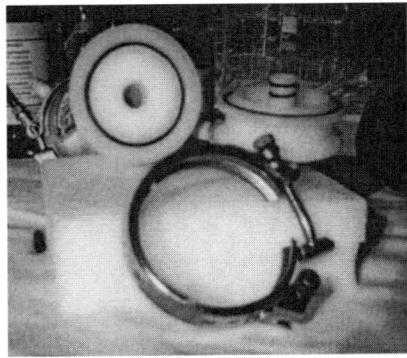
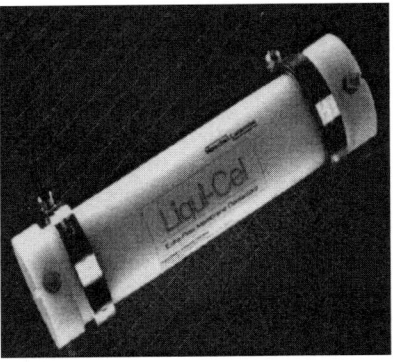

FIGURE 3. Commercial hollow fiber membrane contactors.

porous polypropylene fibers (X-30), nominal porosity 40%, internal diameter 2.4×10^{-4} m, wall thickness 3.0×10^{-5} m; and mass transfer area $1.4\,\text{m}^2$. An experimental setup according to the scheme shown in FIGURE 1 was installed. The aqueous feed phase and the aqueous stripping phase were fed through the inner of the fibers and the organic phase flowed through the shell side.

The pressures of the aqueous and organic phases were monitored at the inlet and outlet ports of the extraction and stripping modules in order to control the pressure drop across the membrane wall. Aqueous inlet gauge pressures were 0.7 bar and the organic inlet gauge pressure was 0.5 bar. The pressure drop across the module length was less than 0.1 bar in all phases. The flow rates used in the copper recovery experiments were: $F_E = 400\,\text{mL/min}$, $F_S = 460\,\text{mL/min}$, and $F_O = 430\,\text{mL/min}$. Preliminary experiments were performed without pH control of the feed aqueous phase, thus leading to acidification of the aqueous feed phase during the extraction test. In the copper recovery experiments a pH control system was used to maintain constant the pH in the feed tank. Extraction and stripping experiments were performed at room temperature, $19 \pm 2°C$.

Synthetic aqueous solutions of $Fe_2(SO_4)_3$, $CuSO_4$, and $MnSO_4$ were used as feed phase; initial metals concentration simulated the average values in the aqueous stream after the wet peroxide oxidation process; that is, Cu 250 mg/L, Fe 170 mg/L, and Mn 170 mg/L. The performance of two extractants was studied: LIX 622N (Cognis) and Cyanex 272 (Cytec). The extractants were used as supplied. The active substance of LIX 622N is 5-nonylsalicylaldoxime, which also contained tridecanol as a modifier. Sulfuric acid solutions were used as stripping phase. The volumes of the three phases were 1 liter each.

EXPERIMENTAL RESULTS

Preliminary Experiments

The work started with the selection of extractants for the removal and recovery of Fe, Cu, and Mn from the oxidized aqueous stream. A good review of the possibilities of solvent extraction in achieving metal separation from solutions containing iron is given by Ritcey.[8] The extraction of Fe(III) has been widely investigated since the leaching solutions of metal ores frequently contain iron as impurity. Cyanex 272 is a selective extracting agent for iron and manganese. It is a phosphinic acid that extracts metals by a cationic exchange mechanism. Depending on the extraction pH, iron and manganese can be separated selectively. The solvent extraction of copper is a well-established hydrometallurgic process and being the hydroxy oximes group of chelation type extractants of general use in commercial solvent extraction–electrowinning plants. Chelating extractants are also referenced as selective for copper over Fe^{3+}.[9]

The equilibrium isotherms (see FIGURE 4), obtained from the manufacturer information (Cyanex 272) and the results of preliminary experiments (LIX 622N), indicate that selective separation and recovery of a mixture of Cu, Fe, and Mn can be achieved in three consecutive steps. In the first step, the extractant LIX 622N permits the extraction of Cu at a pH of 2.5. In the second step, and using Cyanex 272, Fe is extracted at a pH 2.5. The remaining Mn can be extracted using Cyanex 272 at pH 5.

The viability of the selective removal of copper from an aqueous solution containing Cu, Fe, and Mn was experimentally investigated. In FIGURE 5 it can be observed that the concentration of Cu in the feed tank decreased with time, whereas the concentrations of Fe and Mn remained at their initial values. According to these results, the separation of Cu can be performed in a first step using LIX 622N as selective extractant. Next the behavior of Cyanex 272 was tested. The results are shown in FIGURE 6. Under the experimental conditions, Fe is extracted and recovered in the stripping phase whereas Mn remains in the feed phase. Additional experiments allowed us to check that Cyanex 272 is able to extract Mn at a feed pH of 5.25 and that Mn loaded Cyanex 272 is stripped using sulfuric acid solutions.[10]

Copper Recovery

The NDSX process for the recovery of Cu was studied in detail using LIX 622N as extractant and sulfuric acid solution 1 mol/L as stripping solution. Kinetic exper-

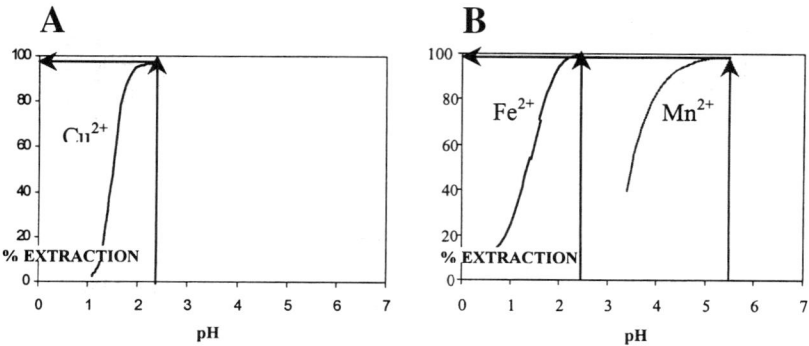

FIGURE 4. Equilibrium isotherms for (**A**) Cu extraction using LIX 622N and (**B**) Fe and Mn extraction using Cyanex 272.

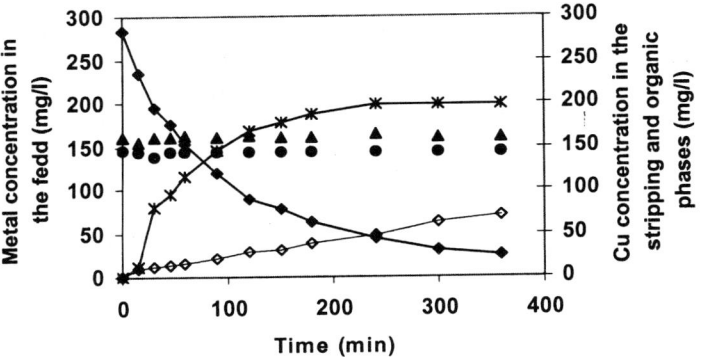

FIGURE 5. NDSX of Cu using LIX622N. Organic phase: 10%v/v of LIX 622N, 10%v/v of TBP in kerosene. Initial feed pH = 2. Stripping phase: H_2SO_4 2M. F_E = 185 mL/min; F_S = 110 mL/min; F_O = 110 mL/min. ◆ Cu, ● Fe, ▲ Mn, ◇ Cu stripping, ∗ Cu organic.

iments were performed working with a synthetic aqueous copper sulfate solution using an initial Cu concentration of 200 mg/L as feed solution. The pH of the feed phase was maintained constant in the feed tank by continuous addition of NaOH. The influence of the pH of the feed phase was investigated. The results are shown in FIGURE 7. In the experiments at feed pH 2.55 and 3.55 both the feed and the stripping solutions were replaced in each successive run. It is observed that Cu is quantitatively stripped, although the stripping process is slower than the extraction process. In order to increase the concentration of Cu in the stripping solution, two sets of experiments (feed pH 3 and 4) were carried out, replacing the feed solution by fresh solu-

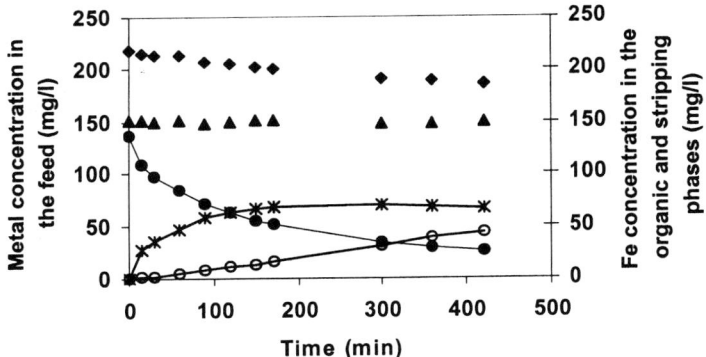

FIGURE 6. NDSX of Fe. Organic phase: 7.5%v/v of Cyanex 272, 7.5%v/v of TBP in kerosene. Initial feed pH = 2.75. Stripping phase: H_2SO_4 2M. ● Fe, ◆ Cu, ▲ Mn, ○ Fe stripping, ∗ Fe organic.

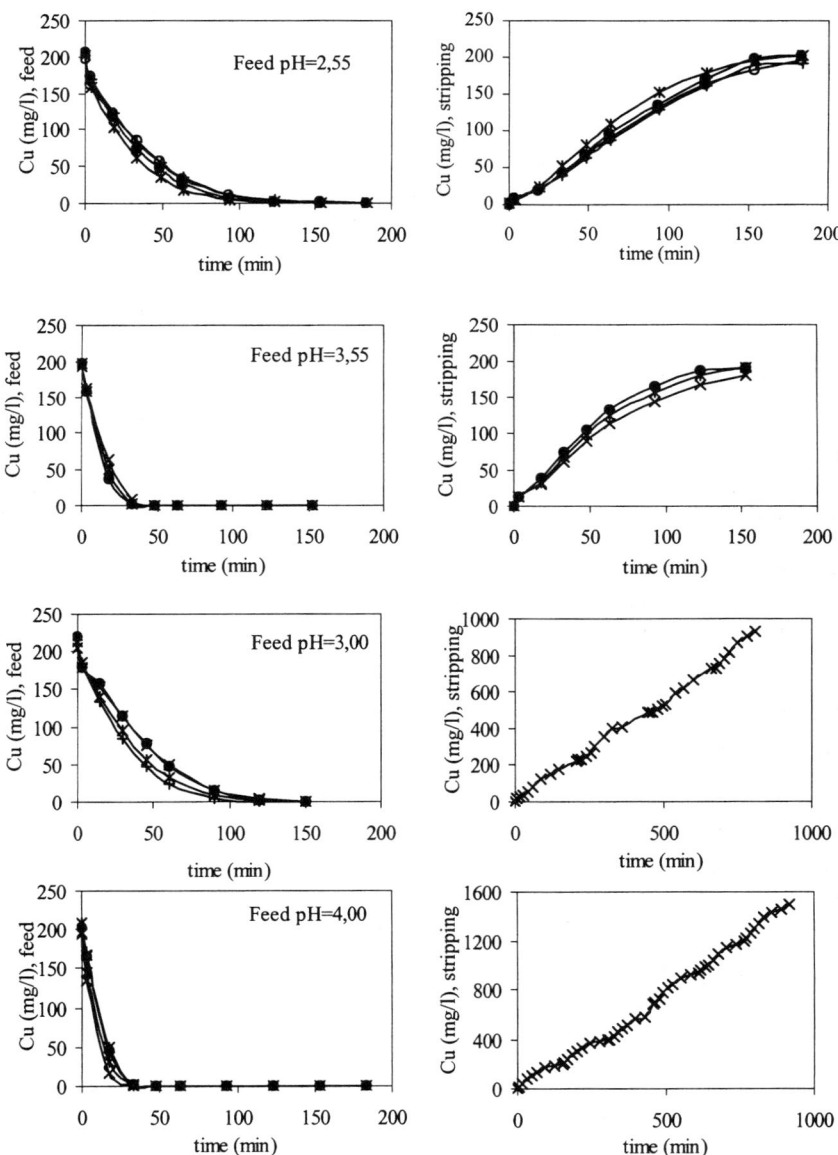

FIGURE 7. Cu extraction and stripping. Influence of the pH of the feed phase. $C^i_{Cu/E}$ = 200 mg/L. Organic phase: 10%v/v of LIX 622N, 10%v/v of TBP in kerosene. Stripping phase: H_2SO_4 1 M.

tion in consecutive runs, whereas the organic and strip solutions were maintained the same. Concentrations of 1,600 mg/L Cu in the stripping phase were reached when working at pH = 4. In the two types of configurations it was observed that the extraction rate is higher at the upper value of the pH.

MATHEMATICAL MODELING

The dynamic response of the system is determined by solving a system of differential equations that describe the evolution of the concentration of Cu in the feed, organic, and stripping phases when flowing along the extraction and stripping modules, together with the mass balances for the three tanks considered as ideal stirred vessels. Several references dealing with the mathematical description of metal separation and recovery using NDSX technology are referenced.[11–15,18]

The assumption of kinetic control of the interfacial reaction shared with mass transfer resistance in the fluid phase circulating through the inner side of the hollow fibers has led to the determination of representative mathematical models for the separation flux and velocity of copper recovery processes.[15,16] However, in this work, as a first attempt, the development of a kinetic model was carried out following the same procedure previously employed by the authors for the recovery of various metallic systems.[5,6,16–18] Thus, under the assumptions of negligible boundary layer resistances in the aqueous and organic bulks and fast interfacial reaction, control of the mass transfer rate mainly lies with diffusion of the Cu–extractant complex through the membrane. The mass balances in a counter-current flowing scheme are as follows:

Aqueous Feed Phase

Extraction module mass balance

$$-\frac{V_{mE}}{F_E L}\frac{\partial C_{Cu/E}^{mE}}{\partial t} = \frac{\partial C_{Cu/E}^{mE}}{\partial z} + \frac{2\pi n_f}{F_E} k_m (C_{Cu/O}^{mE*} - C_{Cu/O}^{mE}) \qquad (1)$$

$$z = 0 \qquad C_{Cu/E}^{mE} = C_{Cu/E_{z=0}}^{i} \qquad (2)$$

$$t = 0 \qquad C_{Cu/E}^{mE} = C_{Cu/E}^{i} \qquad (3)$$

Feed tank mass balance

$$V_E^T \frac{dC_{Cu/E}^T}{dt} = F_E (C_{Cu/E_{z=L}}^{mE} - C_{Cu/E_{z=0}}^{mE}) \qquad (4)$$

$$t = 0 \qquad C_{Cu/E}^T = C_{Cu}^i \qquad (5)$$

Aqueous Stripping Phase

Stripping module mass balance

$$\frac{V_{mS}}{F_S L}\frac{\partial C_{Cu/S}^{mS}}{\partial t} = \frac{\partial C_{Cu/S}^{mS}}{\partial z} + \frac{2\pi n_f}{F_E} k_m (C_{Cu/O}^{mS} - C_{Cu/O}^{mS*}) \qquad (6)$$

$$z = 0 \qquad C_{Cu/S}^{mS} = C_{Cu/S_{z=0}}^{mS} \qquad (7)$$

$$t = 0 \quad C^{mS}_{Cu/S} = C^{i}_{Cu/S} \tag{8}$$

Stripping tank mass balance

$$V^T_S \frac{dC^T_{Cu/S}}{dt} = F_S(C^{mS}_{Cu/S_{z=L}} - C^{mS}_{Cu/S_{z=0}}) \tag{9}$$

$$t = 0 \quad C^T_{Cu/S} = C^{i}_{Cu/S} \tag{10}$$

Organic Extractant Phase

Extraction module mass balance

$$\frac{V_{mO}}{F_O L} \frac{\partial C^{mS}_{Cu/O}}{\partial t} = \frac{\partial C^{mS}_{Cu/O}}{\partial z} + \frac{2\pi n_f}{F_O} k_m (C^{mE^*}_{Cu/O} - C^{mE}_{Cu/O}) \tag{11}$$

$$z = 0 \quad C^{mE}_{Cu/O} = C^{mE}_{Cu/O_{z=0}} \tag{12}$$

$$t = 0 \quad C^{mE}_{Cu/O} = C^{i}_{Cu/O} \tag{13}$$

Stripping module mass balance

$$-\frac{V_{mO}}{F_O L} \frac{\partial C^{mS}_{Cu/O}}{\partial t} = -\frac{\partial C^{mS}_{Cu/O}}{\partial z} + \frac{2\pi n_f}{F_O} k_m (C^{mS}_{Cu/O} - C^{mS^*}_{Cu/O}) \tag{14}$$

$$z = 0 \quad C^{mS}_{Cu/O} = C^{mS}_{Cu/O_{z=0}} \tag{15}$$

$$t = 0 \quad C^{mS}_{Cu/O} = C^{i}_{Cu/O} \tag{16}$$

Organic tank mass balance

$$V^T_S \frac{dC^T_{Cu/O}}{dt} = F_O(C^{mS}_{Cu/O_{z=0}} - C^{mE}_{Cu/O_{z=L}}) \tag{17}$$

$$t = 0 \quad C^T_{Cu/O} = C^{i}_{Cu/O} \tag{18}$$

Equilibrium expression of the reversible extraction equilibrium of Cu(II) with LIX 622N permits the copper concentration in the aqueous phase ($C^{mE}_{Cu/E}$) to be related to the concentration of the Cu–extractant complex at the aqueous–organic interface ($C^{mE^*}_{Cu/O}$):

$$Cu^{2+} + 2\overline{RH} \leftrightarrow \overline{CuR_2} + 2H^+$$

$$K_{EX} = \frac{C^{mE^*}_{Cu/O}(C^{TE}_{H^+} + 2C^{mE^*}_{Cu/O})^2}{C^{mE}_{Cu/E}(C_T - 2C^{mE^*}_{Cu/O})^2}. \tag{19}$$

The equilibrium constant of the back-extraction process is considered to be the inverse of the extraction constant, $K_S = 1/K_{EX}$. According to the previous model the mass transfer parameters required to describe the mass Cu transport in the NDSX system are the membrane mass transfer coefficient k_m and the equilibrium constant of the reaction of extraction K_{EX}.

The value of the equilibrium constant was obtained from the experimental data shown in FIGURE 4, that is, the LIX 622N-Cu extraction isotherm, $K_{EX} = 0.083$. The value of the membrane mass transfer coefficient was estimated by comparison of the

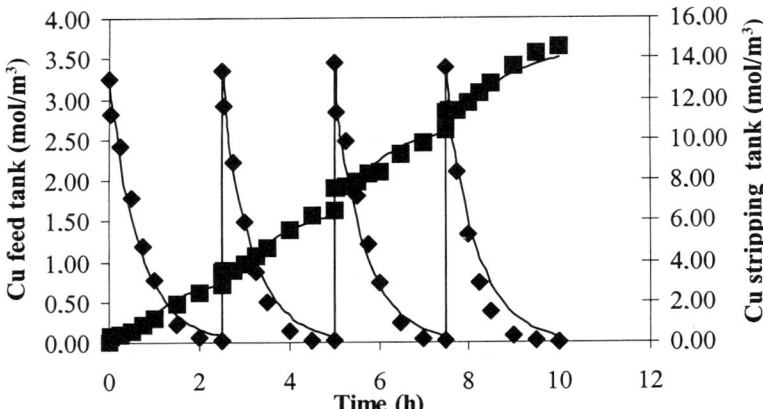

FIGURE 8. Comparison of simulated and experimental results: (♦) experimental feed; (■) experimental stripping, simulated. Conditions: feed tank pH = 3. $C^i_{Cu/E}$ = 3 mol/m³. Organic phase: 10%v/v LIX 622N, 10%v/v TBP in kerosene. Stripping phase: H_2SO_4 initial concentration 1 M.

experimental results with simulated results obtained using the proposed model. In the estimation, the extraction and stripping results obtained in four successive runs at a feed tank pH = 3 were used. The resulting value is $k_m = 3.07 \times 10^{-7}$ m/sec. FIGURE 8 shows the best match between simulated and experimental results. Other reported values of k_m in hollow fiber contactors for complexes of Cu with hydroxy oxime extractants are: Cu–LIX84 in tridecanol[19] $k_m = 6.69 \times 10^{-7}$ m/sec; Cu–LIX 65N[21] $k_m = 5 \times 10^{-7}$ m/sec; Cu–SME529[21] $k_m = 3 \times 10^{-7}$ m/sec.

The value of the mass transfer coefficient in the membrane for Cu–LIX 622N complexes can also be approximated[22] by

$$k_m = \frac{\varepsilon D d_{lm}}{\tau l \, d_o}. \qquad (20)$$

The diffusivity of the complex in the bulk organic phase D was estimated by the Wilke–Chang equation.[23] The viscosity value of the organic phase (80%v/v kerosene, 10%v/v LIX 622N, 10%v/v TBP) was measured, $\mu = 6$ cP. The molar volume of the complex R_2Cu was calculated to be 637.1 cm³/mol by the method of additive group contributions.[23] The molecular weight of kerosene, LIX 622N, and TBP are 184, 263, and 266.2 g/mol, respectively. The membrane tortuosity value 2.6 was taken from Lin and Juang.[15] The calculated diffusion coefficient is $D = 9.87 \times 10^{-11}$ m²/sec. For the hollow fiber employed in this work the logarithmic mean diameter of the fibers d_{lm} was equal to 2.689×10^{-4} m. Application of Equation (20) leads to the value $k_m = 3.80 \times 10^{-7}$ m/sec. This last value is reasonably close to the value of k_m obtained by estimation according to the mass transfer model. Thus, we conclude that the mathematical model and parameters are adequate for the description of the experimental results reported in this work.

Future work will deal with mathematical modeling of the discrimination between the different mathematical models reported in the literature, as well as the recovery of the remaining metals, iron and manganese, in successive nondispersive solvent extraction steps.

CONCLUSIONS

This work focuses on the recovery of the metallic catalyst used in wet peroxide oxidation processes constituted by mixtures of Fe(III), Cu(II), and Mn(II) by means of non-dispersive solvent extraction. A three-step process is proposed. In a first step, Cu can be extracted using LIX 622N as a selective Cu extractant. In the second and third steps, Fe and Mn can be separated using Cyanex 272, by adjustment of the pH of the aqueous phase.

The non-dispersive extraction of Cu(II) from sulfate solution using LIX 622N and simultaneous back-extraction to a stripping sulfuric acid solution was experimentally studied. The extraction rate increased with increasing feed pH. A mass transfer model considering membrane diffusion as the rate limiting step permits the description of the separation system. A simulation procedure permitted us to obtain an estimated value for the mass transfer parameter $k_m = 3.07 \times 10^{-7}$ m/sec that was in good agreement with the calculated value using the system geometric and physical–chemical parameters, $k_{m,\text{calc}} = 3.80 \times 10^{-7}$ m/sec.

ACKNOWLEDGMENTS

Financial support from the Spanish CICYT (MEC) and the European Commission under project 1FD97-1189 is gratefully acknowledged.

REFERENCES

1. BIGDA, R.J. 1998. Fenton's chemistry in wastewater treatment. *In* Encyclopedia of Environmental Analysis and Remediation, Vol. 3. R.A. Meyers, Ed.: 1661-1674. John Wiley & Sons, New York.
2. FALCON, M., *et al.* 1995. Wet oxidation of carboxylic acids with hydrogen peroxide. Wet peroxide oxidation (WPO®) process. Optimal ratios and role of Fe:Cu:Mn metals. Environ. Technol. **16**: 501–513.
3. LUCK, F. 1996. A review of industrial catalytic wet air oxidation processes. Catalysis Today **27**: 195–202.
4. OVEJERO, G., *et al.* 2001. Wet peroxide oxidation of phenolic solutions over different iron containing zeolitic materials. Ind. Eng. Chem. Res. **40**: 3921–3928.
5. ALONSO, A.I., A.M. URTIAGA, A. IRABIEN & I. ORTIZ. 1994. Extraction of Cr(VI) with Aliquat 336 in hollow fiber contactors: mass transfer analysis and modeling. Chem. Eng. Sci. **49**: 901–909.
6. ALONSO, A.I., A.M. URTIAGA, S. ZAMACONA, *et al.* 1997. Kinetic modelling of cadmium removal from phosphoric acid by nondispersive solvent extraction. J. Membr. Sci. **130**: 193–203.
7. URTIAGA, A.M., A. ALONSO, I. ORTIZ, *et al.* 2000. Comparison of liquid membrane processes for the removal of cadmium from wet phosphoric acid. J. Membr. Sci. **164**: 229–240.

8. RITCEY, G.M. 1986. Iron—an overview of its control in solvent extraction of metals. *In* Iron Control in Hydrometallurgy. J.E. Dutrizac & A.J. Monhemius, Eds. Ellis Horwood.
9. SZYMANOWSKI, J. 1993. Hydroxyoximes and Copper Hydrometallurgy. CRC Press, Boca Raton.
10. URTIAGA, A.M. & I. ORTIZ. 2002. Catalyst recovery of the wet peroxide oxidation process by means of non-dispersive solvent extraction. Proceedings of the International Solvent Extraction Conference, ISEC 2002, Cape Town, South Africa, March 17–21, 2002. Vol. 2, 846–851. Chris van Rensburg Pub., Melville, South Africa.
11. SATO, Y., K. KONDO & F. NAKASHIO. 1990. A novel membrane extractor using hollow fibers for separation and enrichment of metal. J. Chem. Eng. Japn. **23:** 23–29.
12. PRASAD, R. & K.K. SIRKAR. 1992. Membrane-based solvent extraction. *In* Membrane Handbook. W.S.W. Ho & K.K. Sirkar, Eds. Chapman & Hall, New York.
13. YANG, Z.-F., A. GUHA & K.K. SIRKAR. 1996. Novel membrane-based synergestic metal extraction and recovery processes. Ind. Eng. Chem. Res. **35:** 1383–1394.
14. YANG, C. & E.L. CUSSLER. 2000. Reaction dependent extraction of copper and nickel using hollow fiber. J. Membr. Sci. **166:** 229–238.
15. LIN, S.-H. & R.-S. JUANG. 2001. Mass-transfer in hollow-fiber modules for extraction and backextraction of copper(II) with LIX 64N carriers. J. Membr. Sci. **188:** 251–262.
16. ORTIZ, I., B. GALAN & A. IRABIEN. 1996. Membrane mass transport coefficient for the recovery of Cr(VI) in hollow fiber extraction and back-extraction modules. J. Membr. Sci. **118:** 213.
17. ALONSO, A., B. GALAN, M. GONZALEZ & I. ORTIZ. 1999. Experimental and theoretical analysis of a non-dispersive solvent extraction pilot plant for the removal of Cr(VI) from a galvanic process wastewater. Ind. Eng. Chem. Res. **38:** 1666–1675.
18. ORTIZ, I., B. GALAN, F. SAN ROMAN & R. IBAÑEZ. 2001. Kinetic of separating multicomponent mixtures by non-dispersive solvent extraction: Ni and Cd. AIChE J. **47:** 895–905.
19. HU, S.-Y. & J.M. WIENCEK. 1998. Emulsion liquid membrane extraction of copper using a hollow fiber contactor. AIChE J. **44:** 570–581.
20. DIMITROV, K., S. ALEXANDROVA & M. BURGARD. 1997. Recovery of copper from solutions by rotating film pertraction. Sep. Purif. Technol. **12:** 165–173.
21. TERAMOTO, M. & H. TANIMOTO. 1983. Mechanism of copper permeation through hollow fiber liquid membranes. Sep. Sci. Technol. **18:** 871–892.
22. HO, W. & T.K. PODDAR. 2001. New membrane technology for removal and recovery of chromium from wastewaters. Environ. Prog. **20:** 4452.
23. POLING, B.E., J.M. PRAUSNITZ & J.P. O'CONELL. 2000. The Properties of Gases and Liquids, 5th edit. McGraw-Hill, New York.

Membrane Contactors for Textile Wastewater Ozonation

GIANLUCA CIARDELLI,[a] INGRID CIABATTI,[a] LAURA RANIERI,[a] GUSTAVO CAPANNELLI,[b] AND ALDO BOTTINO[b]

[a]Tecnotessile S.r.l., via del Gelso 13, Prato, Italy

[b]Dipartimento di Chimica e Chimica Industriale, Università di Genova, Via Dodecaneso 31, Genoa, Italy

ABSTRACT: This paper deals with the application of a membrane contactor for the ozone treatment of textile wastewater. Ceramic (α-Al_2O_3) membranes were chosen because of their ozone resistance. A thin metal oxide (TiO_2 and γ-Al_2O_3) layer was deposited on the membrane surface to eliminate large defects. Membranes were characterized by bubble pressure and gas permeability tests. Mass transfer coefficients were calculated by using the double-film theory. Decolorization kinetics were studied with model dye solutions. Decolorization experiments with a real exhausted dyebath (untreated and after biological treatment) were also carried out. The potential advantages of membrane contactors for the treatment of these types of effluents are demonstrated.

KEYWORDS: membrane contactors; ozonation; reuse; textile wastewater

INTRODUCTION

Water is used extensively throughout textile processing operations and its consumption varies widely depending on the type of unit process and the type of equipment employed. Textile wastewater may include a large variety of dyes, detergents, insecticides, pesticides, fungicides, grease and oils, sulfide compounds, solvents, heavy metals, inorganic salts, and fibers in amounts that vary from industry to industry.[1]

Dyeing is among the most water-intensive processes in textile production, which in turn produces waste streams with high environmental impact, mainly because of their color (due to soluble and insoluble dyes). It has been estimated that the dye loss in the discharged effluent is around 5–20% for acid dyes, 10% for disperse dyes, and 5–30% for direct dyes. Even if dyes are generally non-toxic, the dark color in water streams reduces light penetration that affects plant growth and wildlife, among other environmental concerns. Furthermore, color can cause æsthetic problems to the surroundings.[2]

Textile wastewaters are often purified by means of biological processes but because dyestuffs are highly structured, complex polymers that are very difficult to biologically decompose,[3] they only partially remove color. For example, it is known

---

Address for correspondence: Gianluca Ciardelli, Tecnotessile S.r.l., via del Gelso 13, I-59100 Prato, Italy. Voice: +39-0574-634040; fax: +39-0574-634045.
chemtech@tecnotex.it

that 90% of the reactive dyes entering activated sludge sewage treatment plants will pass through unchanged.[4] Other methods that were found to successfully evaluate the removal of color from textile effluents have been studied. These include activated carbon adsorption,[5] chemical precipitation,[6] flocculation,[7] and oxidation with hydrogen peroxide (Fenton's reaction)[8] even coupled with UV light.[9] Membrane processes, such as micro-, ultra-, and nanofiltration and reverse osmosis have also been applied to dyehouse effluents with the main objective of recovering dyes and water.[10] Although effective, these processes are still far from large-scale industrial application, because of the high costs involved.

A less expensive way to decolorize and reuse textile dyeing effluents has been found by inserting ozonation after an ærobic biological plant.[11] This process removes biodegradable matter from the water, thus reducing the amount of pollutant present. Therefore, ozone consumption is reduced since the ozone serves only to break down the straight, unsaturated bonds in the dyeing molecules.[12]

Conventional methods of gas–liquid contact for the ozonation of wastewater, such as bubble columns and packed beds, are limited by low mass transfer of ozone into the aqueous phase.[13] In order to increase the number of industrial sites performing wastewater purification by ozonation it is important to attain a high degree of ozone utilization in the contactors and to minimize the costs related to ozone production and ozone destruction in the outlet gas.

Membrane-based equipment for bringing gas and liquid phases into contact could be suitable for industrial wastewater ozonation. Membrane contactors are systems in which porous membranes are used to promote gas–liquid or liquid–liquid mass transfer without causing the dispersion of one phase into the other. This is accomplished by flowing the fluids on opposite sides of a porous membrane. By properly controlling the pressure difference between the fluids, a fluid–fluid interface is created in the pores. Details on membrane contactors and their advantages and benefits are reviewed by Gabelman and Hwang.[14]

The use of membrane-based ozonators in water treatment has not been fully developed. Wikol *et al.* used polytetrafluoroethylene membrane contactors to investigate the ozonization of tap water.[15] Shanbhag *et al.*[13] tested a silicone capillary membrane-based ozonator for destroying phenol, acrylonitrile, and nitrobenzene in wastewater. However, the durability of the silicone membranes appeared to present a limiting factor. Janknecht *et al.*[16–19] studied the more inert ceramic membranes for ozonization in a wet-oxidative treatment and investigated the effect of a hydrophobic coating of their surface on the ozone mass transfer.

In this paper a variety of ceramic membranes with various pore sizes and configurations are examined as membrane contactors for the treatment of textile wastewater. The effect of the membrane characteristic on the ozone mass transfer and decolorization yield is shown.

EXPERIMENTAL

Membranes

TABLE 1 lists the membranes used and their relevant properties. All membranes were 15 cm long but their porous permeable area was smaller since about 2 cm at

TABLE 1. Membranes used and their main properties

Manufacturer	Configuration	Outer Diameter (cm)	Channel Diameter (cm)	Selective Layer Properties		Membrane Area (cm²)
				Pore Size	Material[a]	
SCT (France)	single channel	1.1	0.7	200nm	$\alpha\text{-}Al_2O_3$	21
SCT (France)		1.1	0.7	800nm	$\alpha\text{-}Al_2O_3$	22
ATECH Innovations (Germany)		1	0.6	200nm	$\alpha\text{-}Al_2O_3$	16
TAMI Industries (France)	three channel	1	0.36[b]	450nm	ZrO_2 $\alpha\text{-}Al_2O_3$	39
TAMI Industries (France)		1	0.36[b]	140nm	ZrO_2 $\alpha\text{-}Al_2O_3$	40

[a]In the TAMI membrane the ZrO_2 represents the selective layer.
[b]Hydraulic diameter.

each end of the membrane was sealed using a vitrification process. For this reason the values of the membrane area reported in TABLE 1 are lower than those that can be calculated from channel diameter and membrane length. One membrane (TAMI, 140nm, 3 channels) among those listed was modified by coating a thin layer of inorganic oxides (TiO_2 and γ-Al_2O_3) on the surface channels.[20]

Pilot Plant

A schematic diagram of the membrane contactor laboratory scale plant is shown in FIGURE 1. The liquid is fed (at flow rate $0.1\,m^3/h$) to the membrane lumen by means of an electromagnetic pump and then recirculated to the reactor. The liquid pressure is regulated by a gate valve and is read on a manometer placed at the outlet of the stainless steel membrane module. The gas (a mixture of air and pure ozone generated by the ozonizer) flows first along the shell side of the membrane module and then through the KI solution contained in the trap prior to being discharged. A KI trap is also connected to the liquid reservoir in order to avoid any possible ozone discharge into the atmosphere. The gas pressure is regulated by a needle valve and read using a manometer.

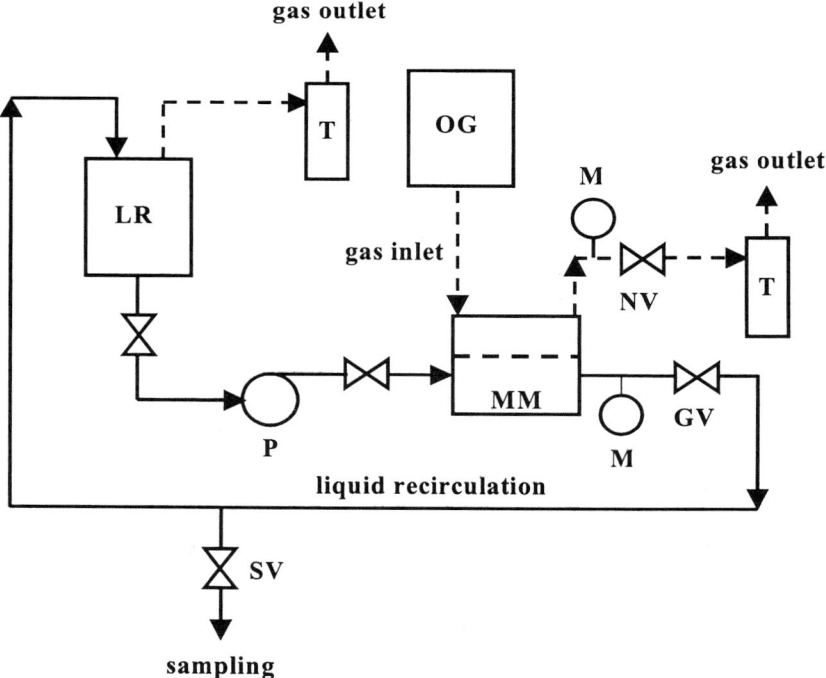

FIGURE 1. Schematic of the laboratory scale pilot plant used for the ozonation tests: GV, gate valve; LR, liquid reservoir (2 L capacity); M, manometers; MM, membrane and stainless steel vessel; NV, needle valve; OG, ozone generator; P, electromagnetic pump; T, traps; SV, sampling valve.

Membrane Characterization

Membranes were characterized through bubble pressure and gas permeability measurements. The former were carried on the same plant used for the textile wastewater treatment (FIG. 1), but employing deionized water as the liquid and air as the gas. The air pressure was gradually increased until reaching a value high enough to displace the water from the (largest) pore of the membrane pores. A glass membrane vessel was also used to directly observe the air bubble stream passed through this pore.

Nitrogen permeability was determined at 20°C by measuring the pressure drop through the membrane due to the permeation of a controlled nitrogen flux.[20]

Ozone Mass Transfer

The well-known two-film theory was used to calculate the overall liquid phase mass transfer coefficients,

$$K_L = \frac{N_A}{A(C_L^* - C_L)}. \tag{1}$$

The ozone flux N_A through the gas–liquid interface was determined from the know amount of ozone transferred (at 25°C) to the liquid phase over time. This amount was evaluated by feeding the laboratory plant with a 2,000 mL solution (1,750 mL potassium iodide and 250 mL starch indicator) and titrating the iodine developed from the reaction between KI and O_3, with a solution of sodium thiosulphate. The interfacial area A was assumed to be equal to the membrane wet surface. The concentration of ozone in the liquid phase C_L was assumed to be zero, since the ozone was immediately consumed at the gas liquid interface. The equilibrium concentration of ozone in the water phase C_L^*, was calculated using Henry's law by assuming the ozone–air mixture to behave as an ideal gas.

Decolorization Experiments

The simplified equation proposed by Chu and Ma[21]

$$\ln\left(\frac{D_0}{D}\right) = kt \tag{2}$$

was used to study the decolorization kinetics ($T = 25°C$) of model aqueous solution of Blue 19 reactive dye. The initial dye concentration (D_0) and at a given reaction time (D) were determined spectrophotometrically. The decolorization of exhausted dyebaths (untreated and after biological treatment) were also investigated at the same temperature. Color was determined by measuring the absorbance at 420 nm. The decolorization percentage was calculated from the ratio between the absorbance after a given reaction time and the initial absorbance of the exhausted solution.

RESULTS AND DISCUSSION

The bubble pressure of all the membranes listed in TABLE 1 was found to be on the order of 1 bar or less and, hence, much lower than the theoretical value (P) that may be calculated by the Young–Laplace equation,

$$P = 2\sigma\cos\frac{\theta}{r}, \qquad (3)$$

where σ is the surface tension (72 dynes/cm for the air–water system), θ is the contact angle (usually assumed to be zero), and r is the pore radius.

This fact, connected to the presence of defects on the membrane surface, may cause severe problems for the industrial implementation of an application using these types of membranes as contact devices. In fact, when the bubble pressure is too low, to control the gas–liquid interface inside the membrane pores becomes more difficult, especially when long industrial membrane elements are employed. This is because the pressure of the liquid that flows through the lumen-side of the membrane progressively decreases due to friction loss. Coating the membrane surface with a thin layer of metal oxides eliminates the defects, thus increasing the bubble pressure. However, it obviously reduces the overall membrane porosity and consequently lowers the gas membrane permeability.

FIGURE 2 shows the mass transfer coefficients for various types of uncoated membranes. The single channels SCT and ATECH uncoated membrane exhibits very high K_L values due to their high pore size and gas permeability. The low K_L value of the TAMI multichannel membrane is due to the fact that the gas–liquid transfer occurs almost entirely in the outer part of the channels that first come into contact with ozone.

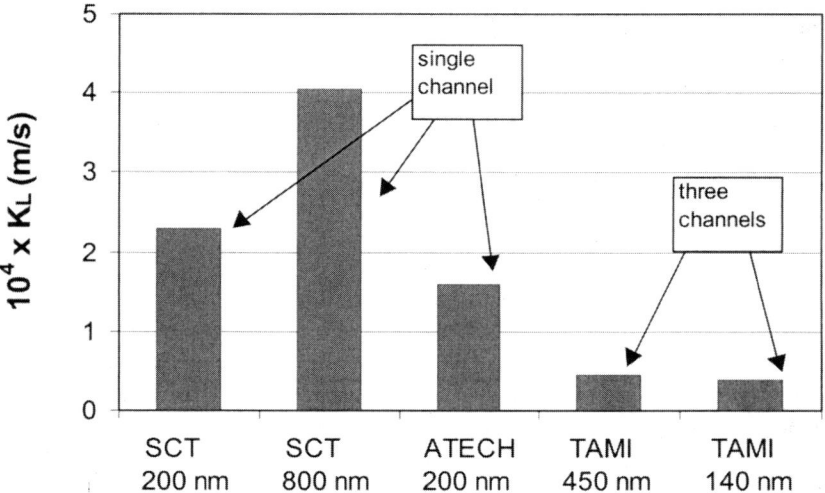

FIGURE 2. Overall liquid phase mass transfer coefficient for various types of uncoated membranes. $P_{gas} = 0.5$ bar; $P_{liquid} = 0.1$ bar.

It is worth observing that the reported K_L values are evaluated at a very low gas pressure ($P_{gas} = 0.5$ bar) to avoid the liquid being forced out of the pores. The deposition on the membrane of a thin and less porous metal oxide (TiO_2 and $\gamma\text{-}Al_2O_3$) layer allows for operations to occur at higher pressure, thus improving the ozone transfer (see FIGURE 3). Consequently (see FIGURE 4) K_L at 1.75 bar reaches a much higher value than that found for the uncoated membrane at 0.5 bar (FIG. 2). The importance of this finding is obvious in view of the need to have sufficiently high bubble pressures at the beginning of this section.

Decolorization Experiments

Coated TAMI membranes were used for decolorization experiments because of the feasibility to better control the gas–liquid interface and, hence, the ozone transfer. The results of decolorization kinetics studies carried out on three types of model solutions with different initial dye (Blue 19) concentrations are shown in FIGURE 5. A straightforward relation is observed according to Equation (2) and, as TABLE 2 clearly indicates, the higher the initial dyestuff concentration the lower the kinetic constant k.

The results of decolorization tests carried out on real solutions derived from untreated and biologically treated dyebaths are shown in TABLE 3. As can be seen the decolorization became much easier for the biologically treated dyebath. After two hours operating time the amount of dissolved ozone is about 50 mg and the decolorization percentage is high enough to allow the reuse of the treated stream in textile technological processes.

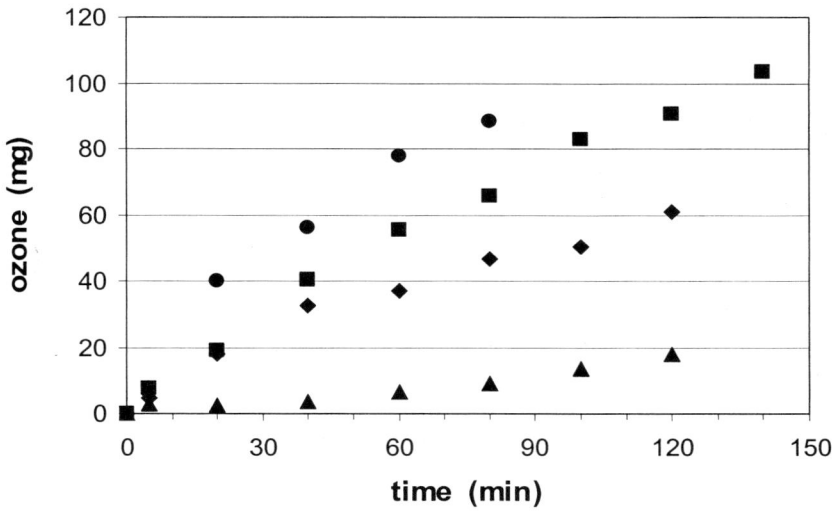

FIGURE 3. Ozone transferred to the liquid phase ($P_{liquid} = 0.1$ bar) at various gas pressures for a TAMI 140 nm 3C membrane coated with a thin layer of metal oxides. $P_g = 1.75$ (●), 1.5 (■), 1 (◆), and 0.5 (▲) bar.

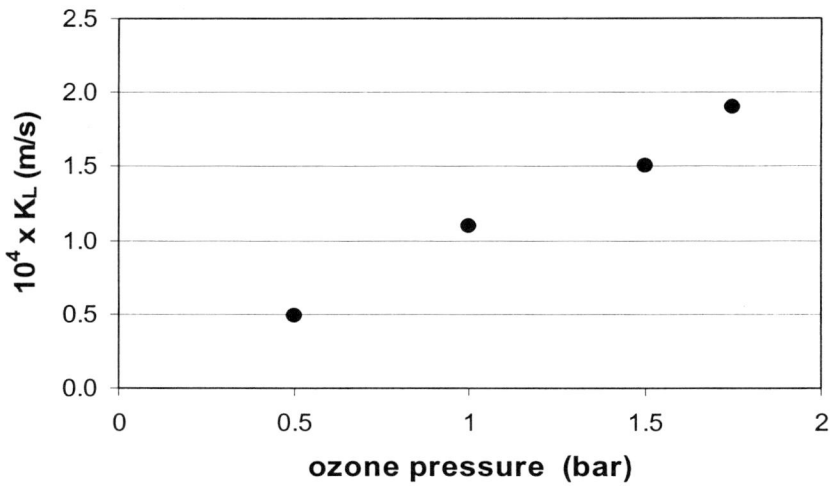

FIGURE 4. Overall liquid phase mass transfer coefficient for a TAMI 140 nm 3C membrane coated with a thin layer of metal oxides. $P_{liquid} = 0.1$ bar.

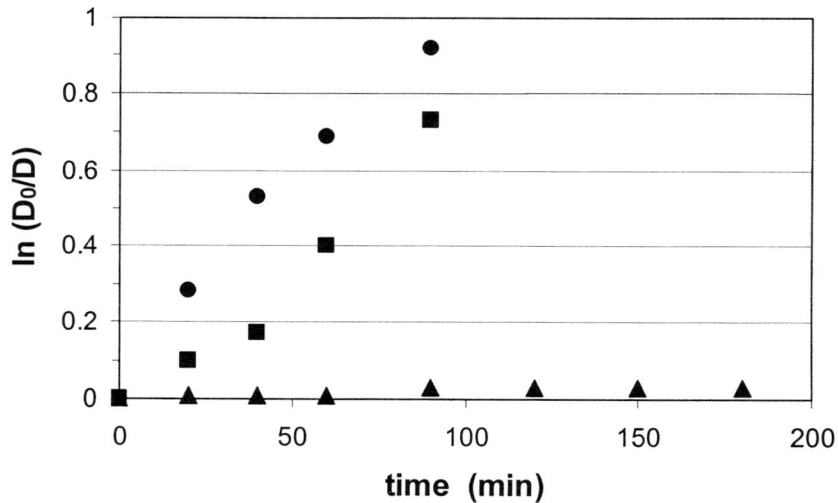

FIGURE 5. Variation of $\ln(D_0/D)$ versus operating time during decolorization of Blue 19 dye solutions of various initial concentrations. $D_0 = 0.072$ (●), 0.036 (■), and 0.018 (▲) mmol/L.

TABLE 2. Kinetic constant k as a function of the initial dyestuff concentration

Solution Concentration (mmol/L)	k (sec^{-1})
1.8×10^{-2}	1.8×10^{-4}
3.6×10^{-2}	1.2×10^{-4}
7.2×10^{-2}	0.03×10^{-4}

TABLE 3. Amount of dissolved ozone and decolorization percentage (see EXPERIMENTAL section) over the operating time

Time (min)	Dissolved Ozone (mg)	Decolorization (%)	
		Untreated Dyebath	Biologically Treated Dyebath
30	12.936	83	75
60	25.04	65	52
90	37.57	46	38
120	50.09	38	20
150	62.61	35	15

CONCLUSIONS

It was demonstrated that ceramic membranes may be effectively used for textile wastewater ozonization. Coating the membrane surface with thin metal oxides layers eliminates defects and the resulting increase of the bubble pressure permits operation at higher gas pressure, with a substantial improvement of the ozone transfer.

ACKNOWLEDGMENTS

The authors wish to acknowledge Dr. R. Prudente for ozone transfer and decolorization experiments.

REFERENCES

1. LOPEZ, A., G. RICCO, R. CIANNARELLA, et al. 1999. Textile wastewater reuse: ozonation of membrane concentrated secondary effluent. Water Sci. Tech. **40:** 99–105.
2. GONÇALVES, I.M.C., A. GOMES, R. BRAS, et al. 2000. Biological treatment of effluent containing textile dyes. J. Soc. Dyers Colour **116:** 393–397.
3. LIN, S.H. & C.L. LAI. 2000. Kinetic characteristics of textile wastewater ozonation in fluidized and fixed activated carbon beds. Water Res. **34:** 763–772.

4. ABADULLA, E., K.H. ROBRA, G.M. GÜBITZ, et al. 2000. Enzymatic decolorization of textile dyeing effluents. Text. Res. J. **5:** 409–414.
5. BARLAS, H. & T. AKGUN. 2000. Colour removal from the textile wastewaters by adsorption techniques. Fresenius Env. Bull. **9:** 597–602.
6. TAN, B.H., T.T. TENG & A.K.M. OMAR. 2000. Removal of dyes and industrial dye wastes by magnesium chloride. Water Res. **34:** 597–601.
7. PAPIC, S., N. KOPRIVANAC & A. METES. 2000. Optimizing polymer-induced flocculation process to remove reactive dyes from wastewater. Env. Tech. **21:** 97–105.
8. INCE, N.H. & G. TEZCANL. 1999. Treatability of textile dye-bath effluents by advanced oxidation: preparation for reuse. Water Sci. Tech. **40:** 183–190.
9. LIAO, C.H., S.F. KANG & H.P. HUNG. 1999. Simultaneous removal of COD and color from dye manufacturing process wastewater using photo-Fenton oxidation process. J. Env. Sci. Health, Part A: Toxic/Hazard. Subst. Env. Eng. **34:** 989–1012.
10. BUCKLEY, C.A. 1992. Membrane technology for the treatment of dyehouse effluents. Water Sci. Technol. **25:** 203–209.
11. CIABATTI, I., G. CIARDELLI & M. MARCUCCI. 2001. Il riutilizzo degli effluenti tessili mediante ozono. Il caso del Lanificio Pecci. Innovare **1:** 50–54.
12. KROISS, H. & H. MÜLLER. 1999. Development of design criteria for highly efficient biological treatment of textile wastewater. Water Sci. Tech. **40:** 399–407.
13. SHANBHAG, P.V., A.K. GUHA & K.K. SIRKAR. 1998. Membrane-based ozonation of organic compounds. Ind. Eng. Chem. Res. **37:** 4388–4398.
14. GABELMAN, A. & S.T. HWANG. 1999. Hollow fiber membrane contactors. J. Membr. Sci. **159:** 61–106.
15. WIKOL, M.J., M. KOBAYASHI & S.J. HARDWICK. 1998. Application of PTFE membrane contactors to the infusion of ozone into ultra-high purity water. ICCS 14th Int. Symp. on Contamination Control, 44th Annual Technical Meeting, Phoeniz AZ, 26 April–1 May 1988.
16. JANKNECHT, P., P.A. WILDERER, C. PICARD, et al. 2000. Bubble-free ozone contacting with ceramic membranes for wet oxidative treatment. Chem. Eng. Tech. **23:** 674–677.
17. JANKNECHT, P., P.A. WILDERER, C. PICARD, et al. 2000. Bubble-free ozone transfer through ceramic membranes by wet-oxidative waste-water treatment. Chem. Ing. Tech. **72:** 122–126.
18. JANKNECHT, P., P.A. WILDERER, C. PICARD, et al. 2000. Investigations on ozone contacting by ceramic membranes. Ozone Sci. Eng. **22:** 379–392.
19. JANKNECHT, P., P.A. WILDERER, C. PICARD & A. LARBOT. 2001. Ozone–water contacting by ceramic membranes. Sep. Purif. Tech. **25:** 341–346.
20. BOTTINO, A., G. CAPANNELLI & A. COMITE. 2002. Catalytic membrane reactors for the oxidehydrogenation of propane: experimental and modelling study. J. Membr. Sci. **197:** 75–88.
21. CHU, W. & C.W. MA. 2000. Quantitative prediction of direct and indirect dye ozonation kinetics. Water Res. **34:** 3153–3160.

Closing Pulp and Paper Mill Water Circuits with Membrane Filtration

JUTTA NUORTILA-JOKINEN,[a] TIINA HUUHILO,[b] AND MARIANNE NYSTRÖM[b]

[a]*Laboratory of Membrane Technology and Technical Polymer Chemistry, Lappeenranta University of Technology, Lappeenranta, Finland*

[b]*Lappeenranta University of Technology, Lappeenranta, Finland*

ABSTRACT: In this study membrane filtration, ultrafiltration, and nanofiltration alone and as part of hybrid processes are considered as means to purify pulp and paper mill process waters suitable for reuse. Thermophilic aerobic biological treatment, pH adjustment, flocculation, and ozonation were tested as pretreatment methods on pilot or on laboratory scale. The aim was to increase flux and reduce fouling by various pretreatment steps and, thus, increase the competitiveness of the membrane process. The results were also evaluated by comparing the benefits obtained against the costs. It was discovered that benefits could be obtained with all the pretreatments tried. Thermophilic aerobic biology assisted in the removal of organic material and increased flux significantly, but the costs were the highest. The most cost-effective processes, however, seem to be pH-adjusted nanofiltration and flocculation nanofiltration hybrid processes, which is understandable because of their significantly lower investment costs compared to, for example, those of biological process. The pH adjustment increased the electrostatic repulsion between negatively charged solutes and membrane, thereby increasing the flux. Flocculation removed the foulants effectively from the feed and it both increased flux and reduced fouling. Yet, many noteworthy benefits were obtained also with ultrafiltration and ozonation. All of the hybrid processes tested could be applied at various points of the water circuit of an integrated pulp and paper mill for purification purposes. The eventual superiority and cost-effectiveness of the applied process remains to be determined case by case.

KEYWORDS: ultrafiltration; nanofiltration; CR-ultrafilter; VSEP; hybrid processes; thermophilic aerobic biology; ozonation; flocculation; costs

INTRODUCTION

Fresh water consumption in the pulp and paper industry varies between $10\,m^3$/ton up to $40\,m^3$/ton paper produced depending, for example, on the paper grade produced and the technical age of the mill. As a result of legislation (e.g., the IPPC-directive and the cluster rules), lack of water resources, and customer demands, the pulp and paper industry all over the world must find competitive ways to reduce its specific water consumption.

Address for correspondence: Jutta Nuortila-Jokinen, Laboratory of Membrane Technology and Technical Polymer Chemistry, Lappeenranta University of Technology, P.O. Box 20, FIN-53851 Lappeenranta, Finland. Voice: +358 5 621 2173; fax: +358 5 621 2199.
jnj@lut.fi

Process water recycling is required in an integrated pulp and paper mill when the specific water consumption is reduced below $10 m^3$/ton paper, produced with the help of water segregation, for example. At this stage various kinds of disturbing substances begin to enrich in the water circuits causing runnability and quality problems, such as corrosion, slime, and color. Membrane filtration offers a competitive way to purify process waters before recycling.

Membrane processes are used in many applications in the pulp and paper industry.[1–7] However, one of the factors reducing the attractiveness of a membrane process is fouling. Yet, there are many ways to improve the membrane filtration process and reduce fouling. First, shear enhanced membrane modules apply high shear forces and, thus, reduce the concentration polarization layer. Moreover, fouling can be reduced by using pretreatment methods, such as biological[8–10] and chemical processes,[11,12] oxidation,[13] and electric field.[14,15]

In this study, groundwood mill circulation water (GMW) was treated with ultra- and nanofiltration. For ultrafiltration shear enhanced modules, a cross rotational (CR)-filter (Valmet-Raisio Oy) and a vibrating shear enhanced processing (VSEP)-filter (New Logic International Inc.) were used. A laboratory flat sheet crossflow module or the VSEP-filter were used in nanofiltration experiments. Thermophilic (+55°C) aerobic biological processing, ozonation, and chemical treatment, such as pH adjustment and flocculation, were used as pretreatment methods to enhance the membrane filtration process. The results are discussed with respect to flux, fouling, and costs.

MATERIALS AND METHODS

Groundwood Mill Circulation Water (GMW)

Modern integrated pulp and paper mills now run according to the counter current principle. This means that the fresh (or recycled) water is brought into the mill only as paper machine shower water. All possible other water fractions are brought countercurrently toward the pulp plant. The effluent is then removed only from the pulp plant. This effluent, which is the most concentrated, is either sent to an external purification plant (e.g., activated sludge plant) or to further treatment, that is, to an internal "kidney". If a closed water circuit system is the target, membrane filtration, which produces water pure enough for recycling purposes, is an excellent choice. By choosing micro- (MF), ultra- (UF), or nanofiltration (NF), the water can be cleaned to a desired degree of purity. In this study, circulation water from the grinding room of a mechanical pulp plant (later groundwood mill water, GMW) was used as a *model* effluent for the closed water system. Groundwood mill water contains the colloidal and dissolved substances originating from the wood itself (e.g., cellulose, hemicelluloses, lignin, lipophilic extractives, and salts). Moreover, the paper machine clear filtrate that is used to pick the pulp from the pulp plant to the paper machine, is mixed with it. Thus, residuals of all chemicals used in the paper making process end up in the GMW. Examples of such chemicals retention aids and antifoaming agents.

Membranes and Modules

Due to the huge amounts of water to be treated in pulp and paper applications the membrane processes need to produce high fluxes and occupy small floor space, especially if installed in an existing mill where no extra space is available. Thus, shear enhanced membrane modules were mainly used in this study. The CR-ultrafilter from Valmet-Raisio Oy uses rotors on the membrane surface to produce high shear forces. The VSEP-filter (New Logic International Inc.) produces high shear forces by means of vibration. The VSEP was mainly used in nanofiltration experiments.

In the ultrafiltration experiments CR1000/10 (pilot), CR550/3 (pilot), and CR250/1 (laboratory) modules were used. The first number refers to the diameter of the membrane and the second to the number of cells in the module. One cell consists of two membrane sheets. The VSEP-filter is the laboratory model (VSEP L) using a single sheet of membrane in one cell (area 465cm^2).

In most of the ultrafiltration experiments regenerated cellulose membranes C30G and C30F (MWCO 30kD) from Nadir Filtration GmbH were used. In addition, the PES 50H (polyethersulfone, 50kD), the PA 50H (polyaramide), 50kD) ultrafiltration membranes (both from Nadir Filtration GmbH) and the polyimide G50 membrane (MWCO 8kD, Osmonics Ltd.) were used in some experiments. Desal-5 (DS-5, Osmonics Ltd.) membrane made from aromatic polyamide and polysulfone was used in the nanofiltration experiments. NaCl-rejection of the DS-5 membrane is 50% according to the manufacturer. FilmTec NF45 polyaramide membrane (NaCl rejection 45%) was also used in some experiments.

Biological Treatment

The experiments were made at the mill premises with a pilot unit. The pilot unit for the suspended carrier biofilm process (SCBP) was an automated floobed unit. In this process biomass grows on carrier elements moved by aeration. In the unit there are two components in series and the total volume of the reactors is 2m^3. The process was run under thermophilic conditions ($T = 45-55°C$). Technical urea and phosphorus acid were added as nutrients so that the COD:N:P ratio was 100:5:1. A $1,000 \mu\text{m}$ bow screen was placed before the pilot unit to remove the fibres. Before the pilot unit was also a heat exchanger, which adjusted the water temperature to 60°C. Aeration decreased the temperature in the reactors to 55°C. The thermophilic SCBP was operated applying increasing loading rates from 3-4 and 6-8 kg $\text{SCOD}/(\text{m}^3\text{d})$ to 13.5-15 and 27-30 kg $\text{SCOD}/(\text{m}^3\text{d})$ in reactors R1 and R2, respectively. The hydraulic retention time (HRT) decreased in the reactors from 8-9 to 2-3 h and from 4-4.5 to 1-1.5 h, respectively. The temperature in the reactors varied between 50 to 59°C.

The ultrafiltration unit was a CR550/3 pilot filter and the membrane was the regenerated cellulose membrane C30F (Nadir Filtration GmbH). The pressure in the filtrations was 0.8-1 bar and the rotor speed was 470 rpm. A $200 \mu\text{m}$ bow screen was set up between the biological process and the ultrafiltration unit to remove most of the biological sludge and fibers.

Ozonation

Ground wood mill circulation water was treated with continuous flow ozonation with various ozone doses delivered to the water. Adjusting the water feed rate and the ozone concentration in the gas inlet changed the ozone dose. The gas flow rate was kept constant at $5\,dm^3$/min in all experiments. The calculated ozone dose in the reactor was 68, 120, or $147\,mg\ O_3/dm^3$. Filtration tests were made with a CR200/1 filter equipped with the C30F membrane. Samples were taken from the reactor and absorbance (UV/VIS-spectra) and color (°PtCo) were analyzed.

Flocculation

Three commercial flocculation agents from Kemira Chemicals were chosen for further investigation, screened according to their ability to remove turbidity. The chosen agents were FRC + PEO (resin + polyethylene oxide), K594 (cationic polyelectrolyte), and Altonit SF (bentonite, clay) + Fennopol A321 (anionic polyelectrolyte). The filtration tests were run with two different dosages for each flocculation agent. The doses for K594 were 15 ppm and 25 ppm. For (bentonite + A321) the doses were (143 ppm + 13 ppm) and (176 ppm + 16 ppm). For FRC and PEO the doses were 20 ppm + 5 ppm and 25 ppm + 10 ppm. The filtration tests were run with a laboratory scale crossflow nanofiltration equipment. The membrane was Desal 5 DK and the filtration temperature was 50°C. The water was first treated with the flocculation agent and then the flocs were removed with a 150 µm sieve. The water was nanofiltered at 11 bar.

RESULTS AND DISCUSSION

The shear enhanced membrane modules have proved to be very efficient in the pulp and paper applications. For the groundwood mill water (GMW) very high ultrafiltration fluxes (in excess of $400\,L/m^2h$) with very little fouling were obtained with the CR1000/10-filter (see FIGURE 1). The membrane stack was separated internally into three stages and it was seen that, as the feed concentrated, the flux decreased from about $450\,L/m^2h$ to about $270\,L/m^2h$ with the C30F regenerated cellulose membrane. It is also seen that more hydrophobic membranes, such as polyethersulfone (PES50H) and aromatic polyamide (PA50H) membranes, gave a steady, but much too low flux—about $25\,L/m^2h$—regardless of the stage they were placed in. In general, it has been observed, that the more hydrophilic the membrane the less hydrophobic foulants, such as lipophilic extractives, adsorb on its surface.[11] In the CR-filter the membranes are also kept very clean because the operating pressure (0.8 bar) is beyond the critical flux of the membrane. The VSEP-filter was also tried in the ultrafiltration of the pulp and paper mill effluents. However, the fluxes obtained were relatively low and the membranes were subjected to quick and severe fouling. However, nanofiltration fluxes close to $200\,L/m^2h$ were obtained with the VSEP-filter. It was observed that, in ultrafiltration, the pressures used in the VSEP (minimum 2.5 bar) well exceeded the critical flux and, thus, ultrafiltration membranes fouled readily. It is, however, also possible that the maximum flux possible was obtained with the C30G membrane as shown in FIGURE 2, when the plateau was reached. In

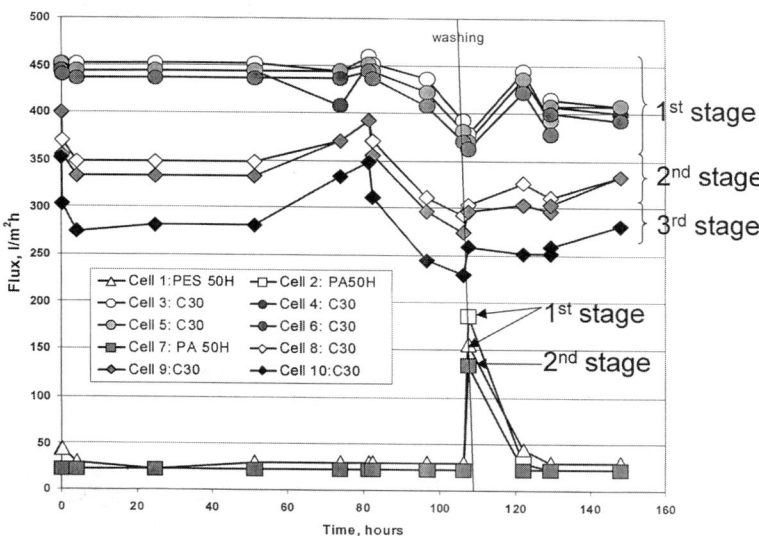

FIGURE 1. CR1000/10 ultrafiltration pilot runs at mill site. Grinding room water (GMW), pH 5, 0.8 bar, 56–70 °C, 13.5 m^2. The membrane stack was internally separated into three stages.

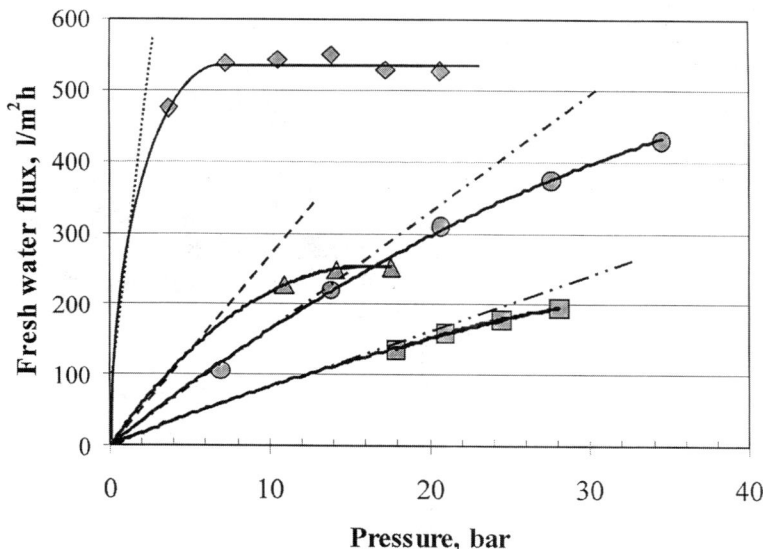

FIGURE 2. Ultra- and nanofiltration fresh water fluxes recalculated to 25 °C of C30G (30 kD), G-50 (8 kD), NF45 (NaCl-rejection 45%), and Desal-5 (NaCl-rejection 50%) membranes. VSEP L, 465 cm^2, $T = 21$–30 °C. ■, DS-5; ●, NF45; ▲, G-50; ◆, C30G.

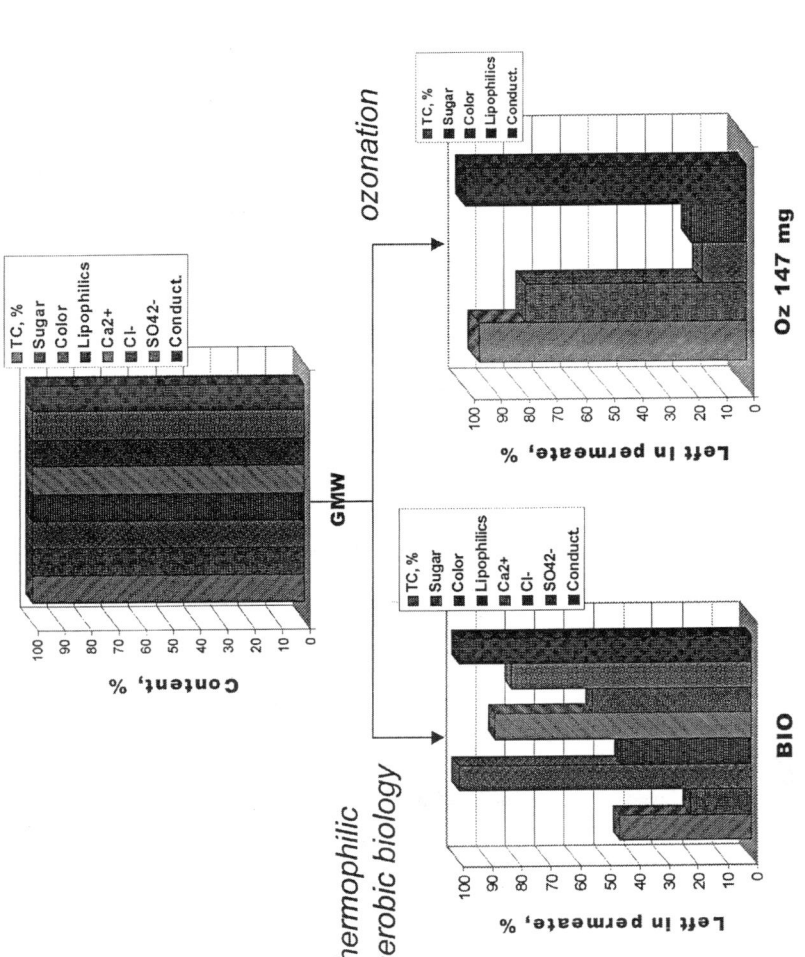

FIGURE 3. Comparison of the removed substances from the groundwood mill water (GMW) by thermophilic aerobic biology and ozonation.

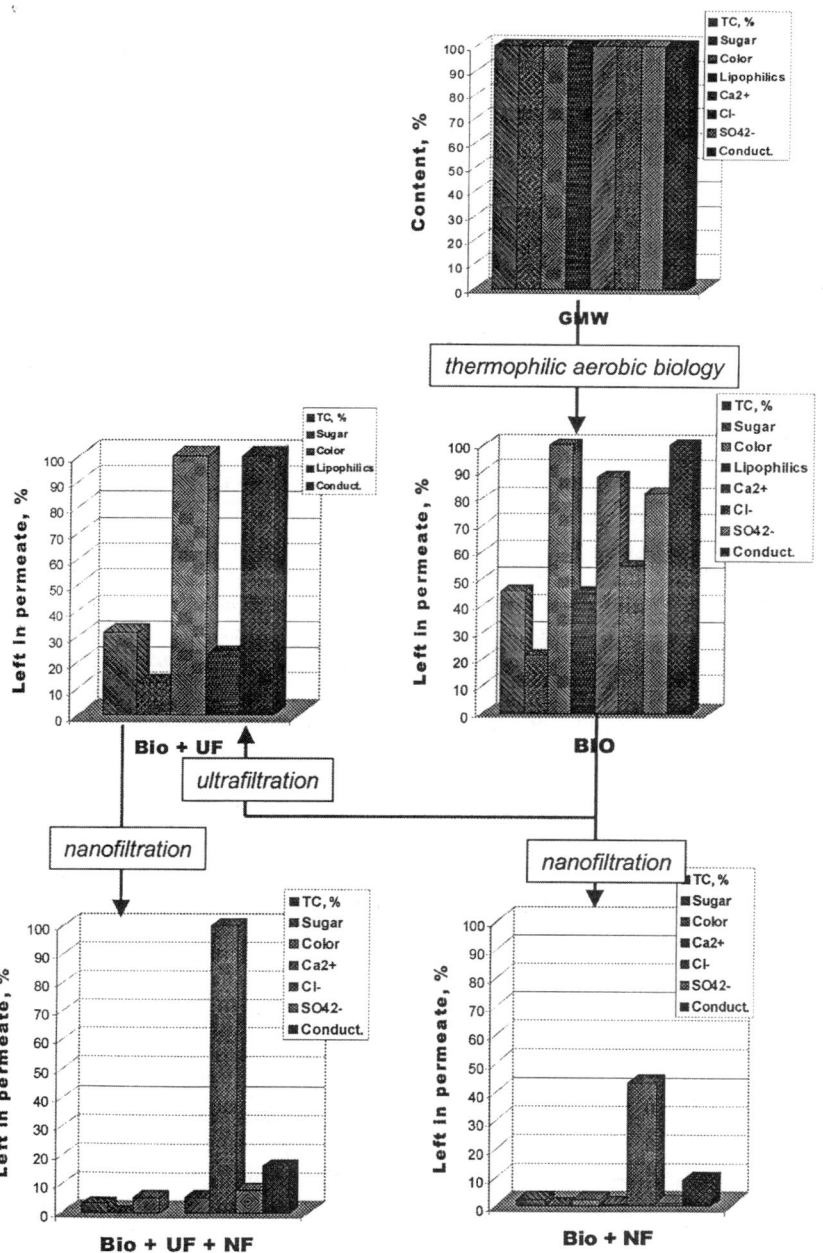

FIGURE 4. The effect of hybrid processes on the removal of organic and inorganic compounds from the groundwood mill water (GMW). Hybrid processes: thermophilic ærobic biology—ultrafiltration—nanofiltration and thermophilic ærobic biology—nanofiltration.

nanofiltration, however, the operation can be performed below critical flux and thus the fluxes are high and hardly any fouling is observed (FIG. 2).

To enhance the ultrafiltration process various hybrid processes were tried out. The aim was to increase flux and reduce membrane fouling. Thermophilic ærobic biological treatment was used to reduce the amount of the low molar mass compounds in the feed water, as can be seen in FIGURE 3. Ozonation, on the other hand, can specifically destroy compounds, such as lipophilic extractives, that have been shown to act as foulants (FIG. 3). The results from the hybrid process test runs showed that the biological treatment and ozonation increased the ultrafiltration flux somewhat. However, the fouling, measured as a reduction of the pure water flux before and after filtration, was significantly decreased only when ozonation was used.

Thermophilic ærobic biological pretreatment produced approximately 45% higher flux for nanofiltration at 45–50°C of GMW compared to the untreated water. This can be explained by the effect of the pH adjustment. For the biological process the pH is increased from its original pH 5 to pH 7. At pH 7 electrostatic repulsion is created between the negatively charged membrane and the solutes in the feed.

For the nanofiltration permeate quality it was observed that a cleaner permeate was produced when only thermophilic ærobic biology was used as pretreatment compared to when biology was followed by ultrafiltration (see FIGURE 4). This can be explained by the fact that ultrafiltration removes long–chain molecules that act as a filtering aid in nanofiltration thus permitting small ions and molecules to pass the membrane more freely. The same phenomenon was observed when ultrafiltration was used as a pretreatment method for nanofiltration compared to when nanofiltration alone was used.

Flocculation with commercial flocculation chemicals also proved to be an effective pretreatment method for nanofiltration (see FIGURE 5). The nanofiltration flux doubled with a relatively low dosage of flocculation chemical (15 ppm, K594) and at the same time the flux became more stable (flux reduction only 10% during filtration). This shows that the chemical was able to effectively form flocs with the foulants thus preventing the membrane from fouling.

Within the hybrid processes tried, the highest flux and biggest organic matter removal was achieved when thermophilic ærobic biology was applied as a first stage (see FIGURE 6). As previously mentioned, flocculation was also shown to be more effective than nanofiltration alone. The placement of ultrafiltration as a pretreatment method in FIGURE 6 shows clearly how the removal of long-chain organics improves flux but decreases retention in the nanofiltration stage.

The cost-effectiveness of the tested unit and the hybrid processes in organic matter removal is evaluated in FIGURE 7. An S-shaped curve is observed and a plateau is reached when the removal rate exceeds 90%. The most economic process, as judged by the organic removal, was nanofiltration. All the benefits achieved by the pretreatments are not seen if only the organic removal is examined. For example, even though the organic compound removal rate obtained with ozonation is rather low—only 10%—the membrane foulants, such as lipophilic extractives, are specifically destroyed. Thus, the benefit and the cost-effectiveness of the ozonation-ultrafiltration hybrid process is actually higher than that shown in FIGURE 7.

If the flux increase due to a pretreatment method is plotted against the cost (see FIGURE 8), it can be seen that the nanofiltration flux is already doubled with a simple

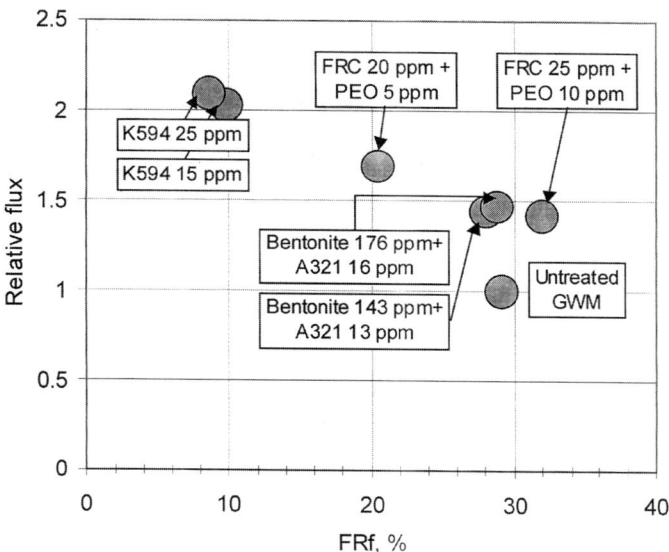

FIGURE 5. The effectiveness of chemical flocculation as a pretreatment method for nanofiltration. Desal-5 DK, pressure 11 bar, $T = 50°C$. Formed flocs were removed by 150 μm sieve before nanofiltration.

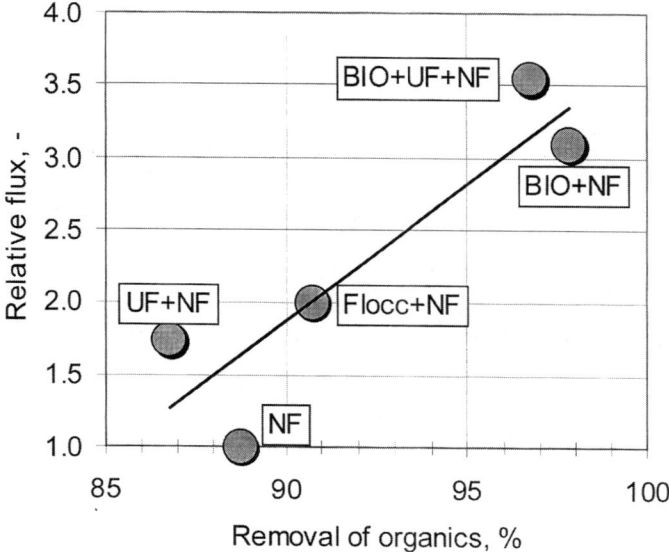

FIGURE 6. The effect of some hybrid processes on the obtained flux and removal of organics in the membrane stage. Groundwood mill water. In UF and NF the C30F and the Desal-5 DK membranes were used, respectively.

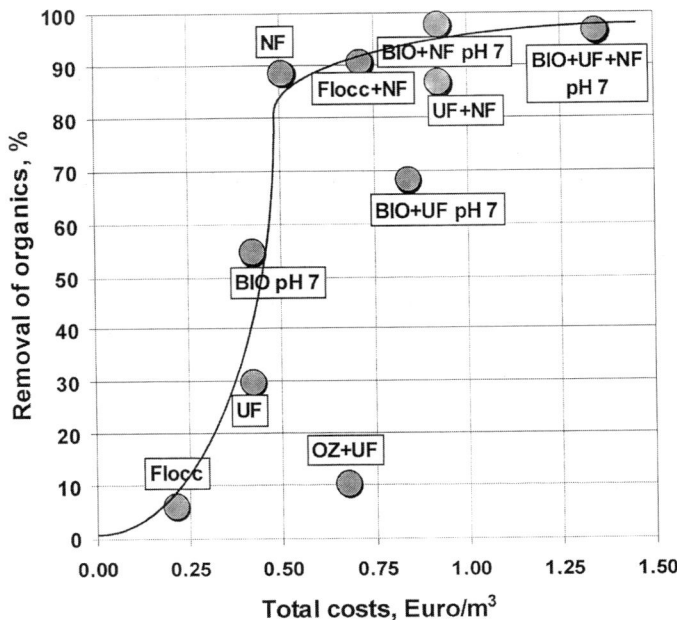

FIGURE 7. The cost-effectiveness of unit and hybrid processes in removal of organic matter from the groundwood mill water.

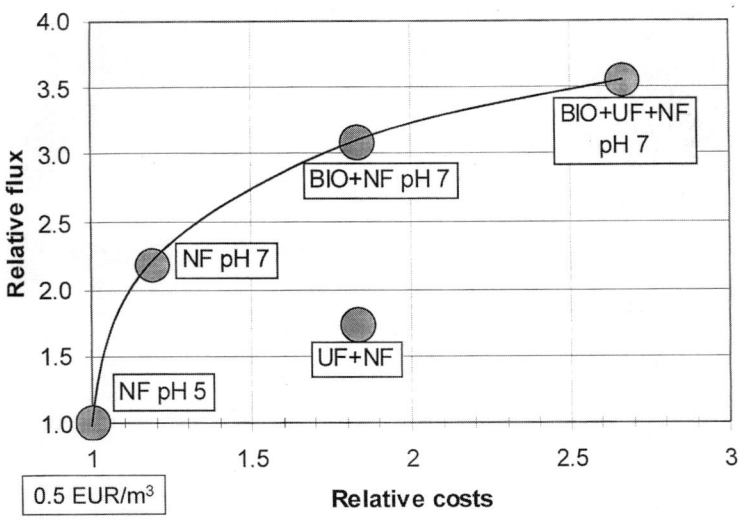

FIGURE 8. The costs of flux increase obtained with hybrid processes. Groundwood mill water. In UF and NF the C30F and the Desal-5 DK membranes were used, respectively.

procedure like pH adjustment (from natural pH 5 to pH 7), and the cost increase is only 20%. Flocculation also doubles the nanofiltration flux with a 30% increase in cost. Trebling the flux (biology-nanofiltration) demands 75% higher cost compared to nanofiltration alone. However, if flux recovery is considered, thermophilic aerobic biology gives the highest flux recovery, about 80%, whereas pH adjustment allows only 50% flux recovery. Without any pretreatment the flux recovery in NF is about 20%. Biological pretreatment also increases the flux during the filtration, thus prolonging the washing intervals and indirectly also the expected membrane lifetime.

When the results of this and previous[16–19] studies are combined, the resulting closed water circuit of an integrated pulp and paper mill using membrane filtration in various locations could be featured, as shown in FIGURE 9. Taking into account the recent simulation results,[20] water circuits should be segregated between the mechanical pulp plant and the paper mill to optimize the energy and water use. This indicates that the pulp should be transferred from the pulp plant to the paper plant at as high a consistency as possible to minimize the water amount mixed between the two plants. Moreover, effluents should be taken out from where they are formed and treated instantly. This kind of arrangement also makes the use of membrane filtration as an internal purification method easier; the water volumes to be treated are reduced significantly and the effluent quality is more uniform than when total mill effluent is treated.

On this basis it is suggested that nanofiltration or biology-nanofiltration or ozonation-ultrafiltration hybrid processes are implied in the mechanical pulp plant, because they produce water with very low lipophilic extractive content. The use of pretreatment produces an extra benefit for the membrane process by increasing flux and reducing fouling resulting in a more cost-effective process.

The quality demands for the paper machine shower water suggest that the recycled water should be free from suspended solids and possess no biological activity. Ultrafiltration easily fulfils these requirements and it has also been shown in practice that the use of ultrafiltered process water prevents slime forming in the paper machine shower nozzles and pipes.[21] If the paper machine water circuit could be further purified, nanofiltration could be used to treat a specific fraction of the ultrafiltered water. Water with low organic material and ionic content would be required for transporting the pulp from the pulp plant to the paper plant. Nanofiltered water would be an obvious choice for this purpose, but so far the nanofiltration fluxes are still insufficient to meet the flux requirements.

Even though the ultrafiltration–nanofiltration hybrid process does not seem very cost-effective according to FIGURES 7 and 8, it is justified if spiral wound nanofiltration membranes are used. Ultrafiltration as a pretreatment effectively removes the suspended solids that are very apt to plug the spiral wound modules. The total cost of this hybrid process is reduced due to the fact that the spiral wound modules are the cheapest on the market.

It could also be justifiable to take advantage of the existing external biological treatment plant, usually the activated sludge (AS) plant, as a pretreatment stage for membrane filtration. However, the AS plants are normally located quite far from the paper mill and, thus, the construction of new pipelines from the AS plant back to the mill is considered too expensive. Furthermore, the water needs heating from about 35°C to process temperature which, in a closed water circuit, can rise above 60°C.

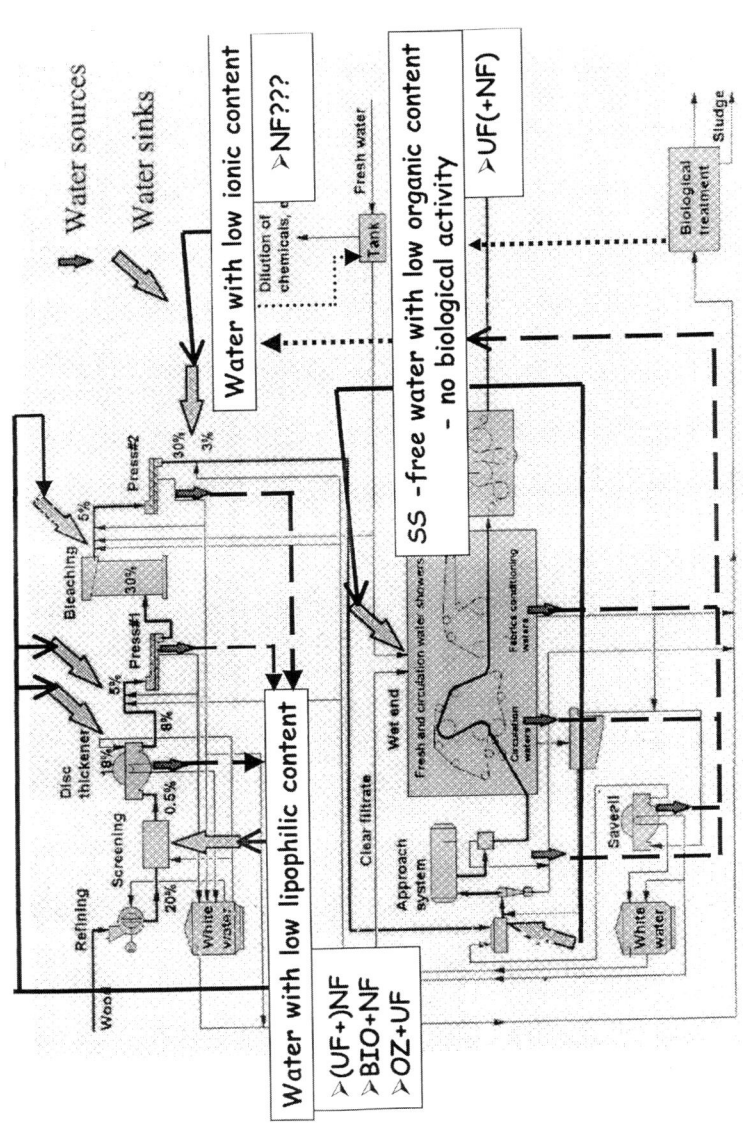

FIGURE 9. A suggestion for a closed water circuit of an integrated pulp and paper mill using membrane filtration or hybrid processes including membrane stage.

Moreover, our experience from the mesophilic ($T < 40°C$) biological treatment[22] combined with membrane filtration showed that more membrane fouling was obtained the lower the temperature in the biological treatment. On the other hand, an existing example of utilizing an AS plant as a part of a closed water circuit is found in the McKinley paper mill in New Mexico, USA.[23] In the McKinley case the arrangement is well-grounded because of the restricted fresh water sources and the lack of receiving water ways.

CONCLUSION

In this study it has been shown that membrane filtration offers a very competitive option as a "kidney" process in the closed water circuits of integrated pulp and paper mills. Shear enhanced membrane modules offer high fluxes and the compact structure demanded from an internal purification process; the CR-filter for the ultrafiltration and the VSEP-filter for nanofiltration.

Thermophilic ærobic biology assisted the removal of organic material and increased flux significantly. The most cost-effective processes, however, seem to be a pH adjusted nanofiltration and a flocculation–nanofiltration hybrid process, which is understandable due to their significantly lower investment costs. Yet, many noteworthy benefits were obtained also with ultrafiltration and ozonation.

Thus, it seems obvious that an unambiguous approach for the most cost-effective internal purification method cannot be found, but that it will always be a compromise between the benefits and costs.

ACKNOWLEDGMENTS

The Technology Development Centre (TEKES) and the Academy of Finland are acknowledged for their financial support. All the Cactus-project partners are warmly thanked for their valuable contribution to this study.

REFERENCES

1. JÖNSSON, A.-S. 1990. Membranteknik—tillämpningar inom skogsindustrin. Svensk Papperstidn. **93**(13): 35–38.
2. GAVELIN, G. 1991. Membranfilter för rening och återvändning av processvatten. Svensk Papperstidn. **94**(14): 35–37.
3. GAVELIN, G. 1992. ABB Flootek: Membranfiltrering med CR-filter. Svensk Papperstidn. **95**(13): 26–29.
4. LIEN, L. & D. SIMONIS. 1995. Case histories of two large nanofiltration systems reclaiming effluent from pulp and paper mills for reuse. In Poc. TAPPI 1995 Int. Environmental Conference. Book 2. 1023–1027. TAPPI Press, Atlanta.
5. MERRY, A., R. GREAVES & G. DANIELSSON. 1995. Bleach waste COD reduction: a case study. In Proc. Topical Conference: Recent Developments and Future Opportunities in Separations Tecnology. Vol I: 492–495. Miami Beach, USA.
6. TEPPLER, M., P. NURMINEN, H. DAMEN, et al. 1999. PM white water treatment at Metsa-Serla Kirkniemi mill in Finland. In 6th International Conference on New Available Technologies, Stockholm, Sweden, 1–4 June 1999. 332–342. SPCI Swedish Association of Pulp and Paper Engineers, Stockholm, Sweden.

7. ALHO, J., I. ROITTO, S. NYGÅRD & S. HIETANEN. 1998. A review on coating effluent treatment by ultrafiltration. *In* Second EcoPaperTech Conference. 219–231. Jyväskylä, Finland.
8. HALL, E.R., K.A. ONYSKO & W.J. PARKER. 1995. Enhancement of bleached Kraft organochlorine removal by coupling membrane filtration and anaerobic treatment. Env. Tech. **16**(2): 115.
9. BOMAN, B., M. EK, W. HEYMAN & B. FROSTELL. 1991. Membrane filtration combined with biological treatment for purification of bleach plant effluents. Water Sci. Tech. **24**(3–4): 219–228.
10. HUUHILO, T., J. SUVILAMPI, L. PURO, *et al.* 2003. Internal treatment of pulp and paper mill process waters with a high temperature aerobic biofilm process combined with ultrafiltration and/or nanofiltration. Pap. Puu **83**(8). In press.
11. NUORTILA-JOKINEN, J. 1999. Strategy of optimum membrane process selection for efficient and economical circuit water filtration. *In* 1st PTS CTP Symposium Environmental Technologies. I. Demel & H.-J. Öller, Eds.: 9-1–9-10. Munich, PTS. PTS Symposium WU-SY 908.
12. LAGACE, P., P.R. STUART & G.J. KUBES. 1993. Development of a physical-chemical treatment as a closed cycle technology for a TMP-newsprint mill. *In* Int. Environ. Symp. 163–171. Paris, France.
13. HONGDE, Z. & D.W. SMITH. 1997. Process parameter development for ozonation of kraft pulp mill effluents. Water Sci. Tech. **35**(2–3): 251–259.
14. JAGANNADH, S.N. & H.S. MURALIDHARA. 1996. Electrokinetics methods to control membrane fouling. Ind. Eng. Chem. Res. **35**(4): 1133–1140.
15. HUOTARI, H. & M. NYSTRÖM. 2000. Electrofiltration in industrial wastewater applications. Trans. Filtration Soc. **1**(1): 17–22.
16. NUORTILA-JOKINEN, J. 2001. Fresh water minimization by membrane filtration in the pulp and paper industry. *In* Our Fragile World: Challenges and Opportunities for Sustainable Development, Forerunner to the Encyclopedia of Life Support Systems (EOLSS), Vol. II, Section 3. Knowledge, Technology, and Management. M.K. Tolba, Ed.: EOLSS Publishers, Oxford. CD-ROM.
17. HUUHILO, T., P. VÄISÄNEN, J. NUORTILA-JOKINEN & M. NYSTRÖM. 2001. Influence of shear enhanced membrane filtration of an integrated pulp & paper mill circulation water. Desalination **141**(3): 245–258.
18. NUORTILA-JOKINEN, J., T. HUUHILO, M. MÄNTTÄRI & M. NYSTRÖM. 2001. Hybrid processes for process water recycling in the pulp and paper industry. *In* Proc. Engineering with Membranes, June 3–6, 2001, Granada, Spain. Vol. 1: I-268–I-273.
19. NUORTILA-JOKINEN, J. 2000. Reducing paper mill fresh water consumption by membrane filtration. *In* Proc. DITP 27th Int. Annual Symposium: Novelties in Technique, Technology, Researching and Ecology in Papermaking at the Transition to the 21st Century. 29–33. 15.-17.11.2000, Bled, Slovenia.
20. KAIJALUOTO, S. 2001. Simulation of closed water circuits. *In* Water Treatment Methods in the Closed Water Circuits of Pulp and Paper Mills 15.-19.1.2001, LUT, Lappeenranta, Finland.
21. SUTELA, T. 2001. CR-Filter Presentation Material. Valmet-Raisio Oyj, Raisio, Finland.
22. NUORTILA-JOKINEN, J., J. HEIKKINEN & C.-O. WASENIUS. 1999. A study on biologically enhanced membrane filtration in a pulp and paper application. Abstract. *In* ICOM 1999: 80. June 12–18, 1999, Toronto, Canada. Extended abstract on CD-ROM.
23. WEBB, L. 1999. IPPC steps up to BAT. Pulp Paper Int. **4**: 29–32.

Membrane Technologies Applied to Textile Wastewater Treatment

MANUELE MARCUCCI, INGRID CIABATTI,
ALESSANDRO MATTEUCCI, AND GUIDO VERNAGLIONE

Tecnotessile S.r.l., via del Gelso 13, Prato, Italy

ABSTRACT: This paper describes the experimental results of a pilot-scale application of membrane technologies to textile wastewater advanced treatment, downstream of a biological activated sludge process, aimed at water reuse in textile technology processes. The chosen approach consisted of sand filtration as a pretreatment, a microfiltration (MF) or ultrafiltration (UF) membrane process, and a final separation treatment performed by means of a nanofiltration (NF) or a reverse osmosis (RO) membrane. An experimental study to compare spiral wound membranes, operating under pressure, to flat membranes, operating under vacuum was conducted. The technical results and a preliminary economic analysis indicate the possibility of technological transfer of the membrane technologies to an industrial scale for textile wastewater reclamation.

KEYWORDS: water reuse; textile wastewater; membrane technologies; flat membrane ultrafiltration

INTRODUCTION

Textile Wastewater Characteristics

In the past few years pollution control of industrial effluents has become more stringent in Europe and the demand for more efficient wastewater treatment systems has increased, in accord with the European Union legislation in force.[1] The textile industry plays a critical role in this context because: (1) it is a large user of water, typically 0.2–0.5 m^3 are needed to produce 1 kg of finished product;[2,3] (2) it is well represented in Europe—in 1997, the European Union output was 1.4 times and 1.8 times larger than the respective production levels of the USA and Japan.[4]

Textile effluents contain several types of pollutants, such as dispersants, levelling agents, salts, carriers, acids, alkali, and various dyes;[5] wastewater quality is variable and depends on the kind of process that generates the effluent. In general, three main types of textile wastewater are recognized:

- cooling waters, with medium–high temperature, but low pollutant content;

- process waters, generated by the processes of dyeing, bleaching, scouring, sizing, desizing, and finishing; they do not have a high flow rate but often present a high pollutant content; and

Address for correspondence: Manuele Marcucci, Tecnotessile S.r.l., via del Gelso 13, I-59100 Prato, Italy. Voice: +39-0574-634040; fax: +39-0574-634045.
chemtech@tecnotex.it

- washing waters, which have high flow rate and a pollutant content reduced with reference to process waters, but not negligible.[6]

Most environmental concern relates to the effluents of the dyeing and finishing processes that contain a variety of components of varying concentration; that is, 50–5,000 mg/L of chemical oxygen demand (COD); 200–300 mg/L biological oxygen demand (BOD); 50–500 mg/L suspended solids; 18–39 mg/L organic nitrogen; 0.3–15 mg/L total phosphorus; and 0.2–0.5 mg/L total chromium. Color is usually noticeable at dye concentrations above 1 mg/L and has been reported in effluent from textile manufacturing at concentrations exceeding 300 mg/L,[7] mainly because 10–15% of the dye is lost into wastewater during the dyeing processes.

Textile wastewater is usually treated in a chemical–physical, or most commonly in an activated sludge plant, to comply with the limits imposed by legislation for discharge; water needs further treatments, called tertiary or advanced treatments, in order to produce a final effluent suitable for reuse in the textile factories.

The interest in the advanced treatments aimed at textile wastewater reuse has grown at a rapid pace, especially to deal with problems of water shortages that are of concern in the South of Europe where the textile sector is widespread. Various treatment processes were studied and applied to investigate the possibility of textile wastewater reuse, such as ozonation, Fenton's reagent oxidation, electrolysis, and flotation.[8,9] Within this research field, the authors performed several experimental campaigns and, according to their experience, membrane processes are the most promising methods.

Membrane Technologies Aimed at Textile Wastewater Reuse

A number of pressure-driven membrane processes have been investigated for advanced textile wastewater treatment, including microfiltration, ultrafiltration, nanofiltration, and reverse osmosis.

Several studies have been conducted relating to the application of reverse osmosis to the treatment and reuse of textile effluents.[10–12] RO membranes are suitable for removing ions and larger species from dyebath effluents; the permeate produced is usually colorless and low in total salinity.[13] Nanofiltration was also studied as treatment for secondary textile effluents after microfiltration: the NF permeate quality was satisfactory and acceptable for water reuse.[14]

Even if the application of membrane processes to textile wastewater treatment and reuse has been proved effective from a technical point of view, to date this kind of application has been restricted to studies on a pilot scale because of the high investment and management costs involved.

Membrane shape is a critical factor, because it determines the membrane susceptibility to clogging and the cleaning efficiency. NF and RO membrane modules are commonly fabricated in a spiral-wound configuration, whereas MF and UF membranes often use a hollow-fiber geometry. MF and UF may also operate as dead-end filtration systems, that is, with no cross-flow. Tubular membrane modules were also found effective for the treatment of dyehouse effluents by ultrafiltration.[15] The flat membrane modules have a novel configuration that has been applied to membrane bioreactors.[16] UF and MF modules with flat membranes operating under vacuum have recently been industrialized by Filterpar S.r.l. (Italy), under the name FLAMEC.

To extend NF/RO membrane life and, consequently, decrease the costs related to membrane module replacement, the use of effective pre-treatments, such as flotation or other membrane processes (e.g., microfiltration or ultrafiltration) is fundamental. Critical evaluation of the systems developed to date suggests that preliminary removal of fine suspended solids and colloids from effluents is necessary to prevent fouling and module damage up to clogging. This is particularly important when spiral-wound membrane modules are applied to biologically treated water.[17]

This paper deals with the *in situ* application of membrane technologies to textile wastewater treatment, downstream of a biological activated sludge process, with the aim of water reuse. Two different approaches were investigated on a pilot scale:

- in the first case-study, a flat UF membrane module, operating under vacuum, was placed upstream of a spiral-wound RO module; and

- in the second case-study, a MF module was placed upstream of a NF module, containing a spiral-wound membrane. The choice of the treatments depended on the different quality of the effluents to be purified, which differed mainly in terms of salt content (higher for the first case study).

The objective of the study was to gain information on the most effective approach from technical and economic points of view; that is to collect data on:

- the most suitable kind of membrane to ensure the production of the desired quality of filtrate;

- the optimum conditions for the operation of the plant and the recovery rate that can be obtained; and

- the influence of the periodic membrane washing.

EXPERIMENTAL

Industrial Effluents Tested

Treated wastewaters from two dyeing and finishing mills were used to dye hanks, skeins, and flocks of various natural and synthetic fibers and blends. Textile effluents were pretreated by means of a biological activated sludge plant on an industrial scale, prior to the advanced treatment in the pilot unit.

Membranes

First Case Study

A Filterpar FLAMEC filter 150.1 system with flat polyvinylidenefluoride (PVDF) membranes, operating under vacuum, was used for the UF step. The total filtration surface of the module was $47 m^2$. Membrane molecular weight cut-off (MWCO) was 70,000 Da. RO was realized by means of a FilmTec TW30-LE (low energy)-4040 spiral-wound membrane.

Second Case Study

Two NADIR P150F membranes (configuration: flat sheet in rolls), produced by Nadir Filtration (formerly Celgard), were used for MF. Each membrane was 101.6 mm in diameter and 1,016 mm in length. Membrane MWCO was 150,000 Da and the filtration surface 11 m^2. An Osmonics Desal DL4040F spiral-wound membrane was used for NF. This membrane has a MWCO of 150300 Da and a filtration surface of 8.4 m^2.

Pilot Plants

First Case Study

The pilot plant consisted of three separation units: sand filtration, UF, and RO. Part of the effluent of a biological activated sludge plant was first filtered through an industrial-scale sand filter and then sent to the pilot-scale sand filter (1.8 bar pressure), which had an output of 1,200 L/h. Water from sand filtration was treated in the UF module. Water permeation in this module was started by the extraction pump of the permeate working under vacuum (average value: −0.4 bar); the liquid to filter was recirculated among the membranes by means of a low head and large capacity pump, to ensure turbulent motion. The flat membranes contained in the module were back-washed with the permeate every 20 minutes for 90 seconds, at a maximum pressure of +0.4 bar. The average flow of the UF permeate was 1,000 L/h. The UF membranes were also chemically washed when hydraulic performance worsened (usually after 100 h of working). Acid and alkaline chemicals were used; sometimes hypochlorite solutions (with a maximum free chlorine content of 50 mg/L) were used for biofouling prevention. The UF effluent was stored and then sent to a RO module working at 13 bar pressure. The RO permeate to be reused was 60–65% of the feed. The 35–40% of the feed was discharged as concentrate.

Second Case Study

In this case, the pilot plant consisted of sand filtration, MF, and NF. Part of the effluent of a biological activated sludge plant was sent to the pilot scale sand filter (2.8 bar pressure), which had an output of 1,000 L/h. Water from sand filtration was treated in the MF module (average operating pressure ranging from 3.5 to 3.8 bar). The MF effluent was finally sent to the NF module (average operating pressure ranging from 6.5 to 7.0 bar). The NF permeate to be reused was 65–70% of the feed; 30–35% of the feed was discharged as concentrate. Chemical washings of both MF and NF membranes were usually performed after 100 h of working; that is, once a week (the plant worked 20 h/day for five days per week) to guarantee a high efficiency of permeation. For MF membranes, a hot aqueous solution (40–50°C temperature) containing 3–4% of alkaline detergent was periodically used for membrane cleaning. For NF membrane, two washing cycles were performed in series:

- acid washing, by means of a solution containing 1–2% of acid detergent for removal of the inorganic polluting agents (scaling); and

- alkaline washing, by means of a solution containing 1–2% of alkaline detergent for removal of organic polluting agents (fouling).

Membrane washing was performed at low pressures and at high flows.

Effluent Analysis

The most important parameters for water reuse in textile processes were determined: pH, chemical oxygen demand (COD), total suspended solids (TSS), turbidity, conductivity, total hardness, chlorides, nonionic and anionic surfactants, and color content.

Dyeing Tests

In the first case study, RO permeate was used in dyeing experiments on an industrial scale by means of a machine that dyes 10 kg of yarn. Several fibers and fiber blends were used in the tests: cotton, mercerized cotton, nylon–cotton, viscose, viscose–wool, wool, and other mixed fibers.

In the second case study, tests with NF permeate were carried out in dark, medium, and light coloration on a laboratory scale. All tests were carried out on 100% wool yarns. In both cases, samples dyed with recycled water were compared to samples dyed with freshwater by means of spectrophotometric color measurements.

RESULTS AND DISCUSSION

Hydraulic Performance of the Membrane Units

Although crossflow filtration (performed by all the membranes tested) shears the membrane surface and helps minimize accumulation of solids, the application of membrane technologies to wastewater treatment systems still encounters problems of handling flow and pressure variations because of membrane fouling. To perform a cost-effective treatment, it is necessary to use a membrane cleaning method at regular intervals to break up the deposit and to limit the membrane replacement. Consequently, during both investigations, special emphasis was given to the monitoring of the hydraulic performance of the membrane systems. This made it possible to evaluate the effectiveness of chemical membrane washing in restoring the initial working conditions in terms of pressure and permeate flows.

In the first case study, testing of flat UF membranes working under vacuum was of a special interest, because this membrane system has a different conception compared to traditional spiral-wound membrane processes. The working cycle of the UF was monitored by measuring the operating pressure regularly, before and after back-washing with UF permeate. FIGURE 1 shows the pressure progress over time in the UF module. The negative pressure values correspond to the membrane working phase, whereas the positive values correspond to the back-washing phase that was started every 20 minutes. During the working phase, the absolute value of the operating pressure increased because of membrane fouling. The experiment demonstrated that back-washing is effective in cleaning the flat membranes contained in the UF module. In fact, as shown in FIGURE 1, periodic back-washing permitted restoration of the low value of the UF operating pressure, and, therefore, the low energy consumption.

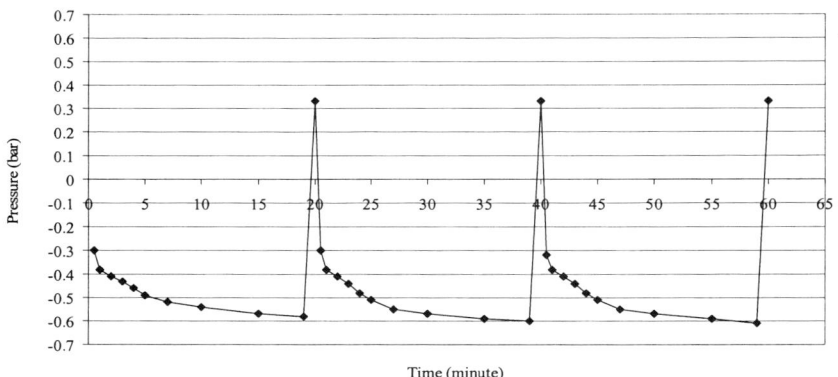

FIGURE 1. First case study: typical pressure progress over time in the UF module with a flat membrane operating under vacuum.

In the second case study, a working cycle of two membranes connected in series, MF and NF, was monitored for 530h, measuring the permeate flows before and after chemical membrane cleaning performed every 100h. FIGURE 2 illustrates a typical evolution of the MF and NF permeate flow over time: a decrease in flow was observed because of fouling, but chemical washing was effective in restoring the initial flow values of both MF and NF permeates. FIGURE 3 illustrates the pressure progress in the two membrane modules over time. The diagram shows that, despite membrane fouling, which caused a decrease in the permeate flux, the pressure of neither systems varied significantly over time.

In accord with the results of the experiments, UF technology with flat membranes, compared to a conventional MF process, guaranteed:

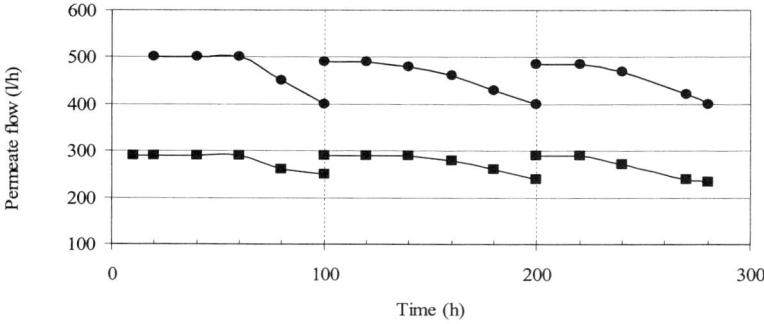

FIGURE 2. Second case study: typical permeate flow over time for MF (●) and NF (■) membranes. Membrane washing was performed every 100h.

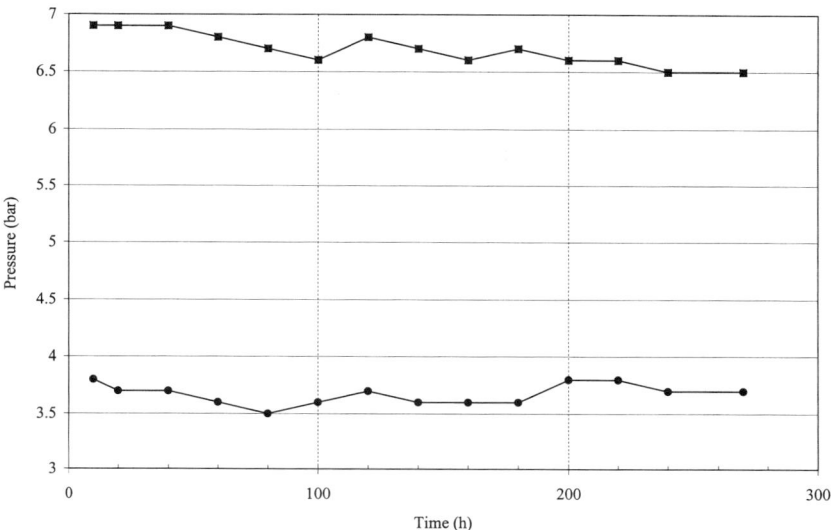

FIGURE 3. Second case study: typical pressure progress over time for MF (■) and NF (●) membranes.

- a fairly constant permeate flow rate; for MF membranes the observed flux indicated a decline with time (permeate flow decreased of 20% in 100h), emphasizing the significance of periodic chemical cleaning to maintain a constant flux;
- an energy saving, since the operating pressures of the UF module were very low during both working and back-washing phases; and
- a reduction in the frequency of chemical washing, because back-washing is effective in automatically cleaning the UF membranes.

Consequently, it seems that the novel UF technology with flat membranes operating under vacuum offers a cost-effective solution compared to traditional pressure-driven, spiral-wound membranes.

Waste Removal

Water quality parameters of both pilot units were controlled by means of sampling of the sand filtration inlet and of the permeates of the two membrane modules connected in series. TABLES 1 and 2 show the main characteristics of the analyzed samples for case studies 1 and 2, respectively. FIGURES 4 and 5 show the removal efficiency of the first two separation processes (sand filtration+UF and sand filtration+MF, respectively) and of the last purification step (RO and NF, respectively) for case studies 1 and 2, respectively. The percentages of removal always refer to the sand filter inlet.

TABLE 1. First case study: water quality parameters for sand filtration inlet, UF permeate, and RO permeate

Parameter	Sampling Point		
	Sand Filter Inlet	UF Permeate	RO Permeate
pH	7.8	8.1	7.3[a]
COD (mg/L O_2)	96	70	5
TSS (mg/L)	21	< 5	< 5
Turbidity (NTU)	3.2	0.7	0.6
Conductivity (µS/cm)	3,950	3,665	192
Total hardness (°F)	10	10	< 1
Chlorides (mg/L)	954	950	29
BIAS[b] (mg/L)	1.3	1.3	0.3
MBAS[c] (mg/L)	0.8	0.8	0.6
Color removal[d] (%)	Reference	5	> 95

[a] After correction with sodium bicarbonate.
[b] Non-ionic surfactants as bismuth–iodine active substances (BIAS) according to IRSA method 5160-ISO 7875/2 First Edition, 1984.
[c] Anionic surfactants as methyl blue active substances (MBAS) according to IRSA method 5150-ISO 7875/1 First Edition, 1984.
[d] Integral of the absorbance curve through the entire visible range (400–800 nm).

TABLE 2. Second case study: water quality parameters for sand filtration inlet, MF permeate, and NF permeate

Parameter	Sampling Point		
	Sand Filter Inlet	UF Permeate	RO Permeate
pH	6.7	6.4	6.1
COD (mg/L O_2)	66	42	12
TSS (mg/L)	21	< 5	< 5
Turbidity (NTU)	4.1	0.9	0.5
Conductivity (µS/cm)	1,150	1,145	582
Total hardness (°F)	9	8	2
Chlorides (mg/L)	100	88	78
BIAS (mg/L)	1.1	1.1	0.4
MBAS (mg/L)	0.9	0.6	0.3
Color removal (%)	Reference	13	94

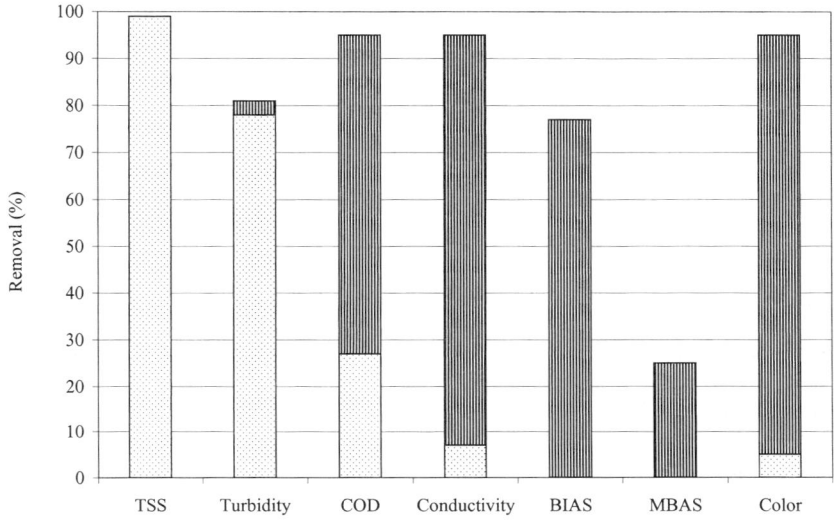

FIGURE 4. First case study: contribution of: ☐ sand filtration and UF and ▥ RO to the total removal of the parameters shown. The percentages refer to the sand filtration inlet.

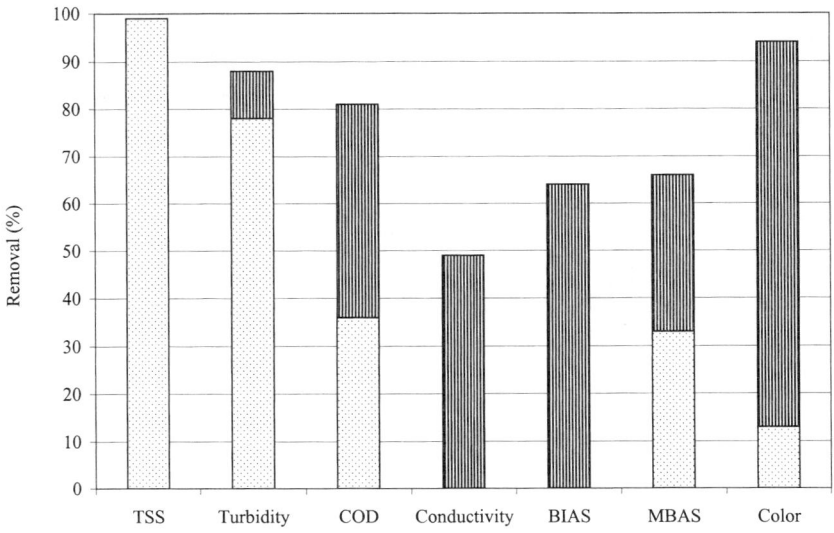

FIGURE 5. Second case study: contribution of: ☐ sand filtration and MF and ▥ NF to the total removal of the parameters shown. The percentages refer to the sand filtration inlet.

First Case Study

Sand filtration and ultrafiltration made it possible to achieve practically turbidity free effluent, as required to minimize RO membrane fouling. Sand filtration and UF partially removed COD (less than 30%) and color (about 5%); their effect in removing the other pollutants was negligible. The RO permeate was of excellent quality. Reverse osmosis significantly achieved the removal of COD (the overall tertiary treatment achieved a COD removal of 95%) and also strongly contributed to removing water salt content and color—two of the most important parameters for water to be used in textile processes.

Second Case Study

Even in this case, sand filtration and MF were fundamental in removing suspended solids and turbidity (removal in excess of 99% and about 80%, respectively). COD was partially removed by sand filtration and MF (36%) and completely removed after NF. The overall advanced treatment allowed a color removal of 94%, mainly thanks to nanofiltration, since the pretreatment stages had only a slight effect in water decolorization (color removal 13%).

Results of the Dyeing Tests

In the first case study, industrial-scale dyeing tests with 100% RO permeate showed excellent results for all the fibers and fiber blends under evaluation, even when dyed with light shades. Spectrophotometric measurements of color were performed to compare yarns dyed with softened freshwater and yarns dyed with recycled water. The results indicated a very high quality of the dyeings with RO permeate, because the samples dyed with the purified water and the samples dyed with well water showed color differences largely within the limits required by the market.

In the second case study, laboratory-scale wool dyeing tests with 100% NF permeate were very satisfactory, even if the water quality in terms of salt content and chlorides was not as good as the RO permeate produced by the first pilot unit. Spectrophotometric analyses of color showed negligible differences between yarns dyed with NF permeate and yarns dyed with freshwater in 95% of the cases.

CONCLUSIONS

Both studies indicated that the final permeate from the two-stage membrane system (UF+RO or MF+NF) has a high quality and can be used as process water for the textile industry. Using this kind of advanced treatment system, about 60–65% of the mill effluents can be recycled.

Some differences emerged concerning the hydraulic performance of the first membrane process in the two pilot units. In the first plant, an UF process with flat membranes operating under vacuum was proposed, whereas in the second plant a MF process with conventional membranes was used. The novel UF module operating under vacuum exhibited very good hydraulic performance because it maintained low operating pressure, and it produced a constant permeate flow; hence it can replace the traditional spiral-wound membrane module in textile wastewater advanced treatment.

The various membrane treatments chosen as the final purification step, RO and NF, were both suitable for the specific cases. The RO membrane produced a permeate of very good quality at an average pressure of 13 bar, that is, working with a low energy consumption with reference to other RO treatments. This makes it possible to minimize the operating costs. It is noted that the use of a low-pressure RO membrane was possible, thanks to the presence of the UF step, whose permeate was of a good quality.

The NF membrane did not reach the retention behavior of the RO membrane; nevertheless, as demonstrated by the dyeing tests, the treated water can replace freshwater in textile processes. Moreover, since NF works under less demanding conditions (pressure 6.5–7.0 bar) than RO, it allows a further reduction in cost.

ACKNOWLEDGMENTS

The authors are grateful to Filterpar (Bergamo, Italy), Fildrop (Florence, Italy), Laura Ranieri, Mauro Russo, and Silvia Tarocchi for technical support.

REFERENCES

1. Directive 2000/60/EC of the European Parliament and of the Council of 23 October 2000 establishing a framework for Community action in the field of water policy. Official J. Eur. Commun. No. L 327, 1.
2. TOMANELLI, R. 1997. I reflui dell'industria tessile. Acqua Aria **9:** 24–34.
3. MARCUCCI, M., G. NOSENZO, G. CAPANNELLI, et al. 2001. Treatment and reuse of textile effluents based on new ultrafiltration and other membrane technologies. Desalination **138:** 75–82.
4. EUROSTAT & DIRECTORATE-GENERAL FOR ENTERPRISE. 2000. Panorama of European Business 1999, data 1998–1999, Theme 4: Industry, Trade and Services. Office for Official Publications of the European Communities, Luxembourg.
5. COOPER, P. 1995. Colour in Dyehouse Effluent. Society of Dyers & Colourists, Bradford.
6. CIABATTI, I., G. CIARDELLI & M. MARCUCCI. 2001. Il riutilizzo degli effluenti tessili mediante ozono. Il caso del Lanificio Pecci. Innovare **1:** 50–54.
7. LAING, I.G. 1991. The impact of effluent regulations on the dyeing industry. Rev. Prog. Coloration Related Top. **21:** 56–71.
8. VANDEVIVERE, P.C., R. BIANCHI & W. VERSTRAETE. 1998. Treatment and reuse of wastewater from the textile wet-processing industry: review of emerging technologies. J. Chem. Tech. Biotech. **72**(4): 289–302.
9. ROTT, U. & R. MINKE. 1999. Overview of wastewater treatment and recycling in the textile processing industry. Water Sci. Tech. **40**(1): 137–144.
10. WENZEL, H., H.H. KNUDSEN, G.H. KRISTENSEN & J. HANSEN. 1996. Reclamation and reuse of process water from reactive dyeing of cotton. Desalination **106:** 195–203.
11. ROZZI, A., F. MALPEI, R. BIANCHI & D. MATTIOLI. 2000. Pilot-scale membrane bioreactor and reverse osmosis studies for direct reuse of secondary textile effluents. Water Sci. Tech. **41**(10–11): 189–195.
12. BOTTINO, A., G. CAPANNELLI, G. TOCCHI, et al. 2001. Membrane separation processes tackle waste-water treatment. Membrane Tech. **130:** 9–11.
13. BUCKLEY, C.A. 1992. Membrane technology for the treatment of dyehouse effluents. Water Sci. Tech. **25**(10): 203–209.

14. ROZZI, A., M. ANTONELLI & M. ARCARI. 1999. Membrane treatment of secondary textile effluents for direct reuse. Water Sci. Tech. **40**(4–5): 409–416.
15. MAJEWSKA-NOWAK, K. 1989. Synthesis and properties of polysulfone membranes. Desalination **71**(2): 83–95.
16. PETERS, T.A., R. GÜNTHER & K. VOSSENKAUL. 2000. Reuse of filter backwash water using ultrafiltration technology. Filtration Separation **37**(1): 18.
17. BAL, A.S., C.G. MALEWAR & A.N. VAIDYA. 1991. Development of non-cellulosic membranes for wastewater treatment. Desalination **83**(1–3): 325–330.

Integrating Photocatalysis and Membrane Technologies for Water Treatment

DAVID F. OLLIS

Chemical Engineering Department, North Carolina State University, Raleigh, North Carolina, USA

ABSTRACT: Removal of organic contaminants in water may sometimes be more easily achieved with a pair rather than with a single unit operation. We explore here the combination of an ambient temperature, chemical oxidation photocatalysis (PC) process with physical separation via a membrane operation, microfiltration (MF), ultrafiltration (UF), or reverse osmosis (RO) using both a conceptual statement and a literature review format. Four configurations are noted and discussed here: (1) PC+MF for catalyst slurry recycle, (2) PC+UF for catalyst slurry and (polymer) reactant recycle, (3) immobilized PC and UF/RO for reactant recycle, and (4) immobilized PC on UF/RO membrane for membrane self-cleaning. Although the literature review is encouraging with respect to plausibility, the paucity of examples indicates a need for substantial effort to fully exploit the suggested possibilities for process development.

KEYWORDS: photocatalysis; membrane technology; water treatment

INTRODUCTION

Water treatment must often rely on the process integration of two or more unit operations, rather than a single unit alone in order to produce the desired water quality. This paper reviews recent efforts involving the integration of ambient temperature, photocatalyzed oxidation (PCO) with membrane separation; microfiltration (MF), ultrafiltration (UF), and/or reverse osmosis (RO). To appreciate this process integration potential, we briefly review the defining characteristics of each type of operation, then indicate plausible motivations for integrating these operations to achieve a desired process advantage. Finally, we review recent research literature efforts involving the use of both operations for processing and purification of water.

Heterogeneous photo-assisted catalysis, or photocatalysis for short, involves use of near UV light (wavelength less than 360 nm) activated metal oxides, especially titanium dioxide (anatase form, TiO_2) or zinc oxide (ZnO) to create forms of active oxygen (e.g., hydroxyl, $\cdot OH$, or hydroperoxyl, $\cdot OH_2$) radicals at room temperature in the presence of oxygen (O_2) or air. These potent oxidants can attack virtually any organic water pollutant and can eventually mineralize them completely. Importantly, for the present discussion, such photoproduced active species can also cleave polymers into lower molecular weight solutes, kill (and oxidize) microbial cells, and deactivate viruses.[1–5]

Address for correspondence: David F. Ollis, Chemical Engineering Department, North Carolina State University, Raleigh, NC 27695-7905, USA. Voice: 919-515-2329.
ollis@eos.ncsu.edu

Photocatalysis is the only advanced chemical oxidation process that uses dissolved oxygen and a heterogeneous catalyst to carry out an ambient temperature, photo-driven oxidation. The dominant catalyst investigated in the literature is the semiconductor oxide, anatase TiO_2, with ZnO a modest second choice. Most early reports on photocatalysis used nanometer sized metal oxides, often aggregated into micron-size clusters during synthesis; for example, Degussa P-25 TiO_2, d_p = 20–30 nm, a primary particle with d(aggregate) = 1–3 microns.[6,7] These ambient temperature photocatalysts are now know to oxidize virtually all organic water contaminants, given a dissolved oxygen supply, including both common molecular solutes and microbial cells, viruses, biopolymers, and oils.[8–10]

Membrane processes are many. For our purposes, it suffices to distinguish among the following broad categories:[11,12] *microfiltration* (removal/retention of micron sized particles), *nano-* or *ultrafiltration* (removal of 1–10 nm diameter particles, such as proteins, other polymers, and biopolymers), and *reverse osmosis* (removal of 5 Å or smaller ions and solutes).

Also important for our discussion is the use of membranes from ceramic (i.e, non-oxidizable) or organic (oxidizable) components, inasmuch as the former are clearly compatible with the ceramic photocatalyst metal oxides TiO_2 and ZnO, whereas the latter may be chemically unstable in the presence of the active oxygen radicals (·OH and ·OH_2) commonly produced by these illuminated photocatalysts.

The metal oxide photocatalysts of common interest have dimensions that depend on their method of formation. Common among these are: (1) vapor phase hydrolysis of $TiCl_4(v)$[13] to yield 20–30 nm primary particles that, in the same process, are agglomerated into 1–5 micron aggregates (e.g., Degussa P-25 titania), (2) sol–gel techniques in which metal-alkoxide precursors[14] are hydrolyzed in liquid water and the resulting sol is then gelled and may be coated and dried on various surfaces to produce a porous *ceramic membrane* layer.

INTEGRATION OF CHEMICAL OXIDATION AND MEMBRANE SEPARATION

Four distinct motivations for photocatalyst–membrane combinations are noted here, as illustrated schematically in FIGURE 1.

A. Filtration for Suspended Catalyst Recycle: A microfilter is used to recycle micron-size photocatalyst slurry particles. Reaction occurs in a photoreactor that is separate from the microfilter; the filter passes all (soluble) reactants and products with no selectivity.

B. Filtration for Reactant and Suspended Catalyst Recycle: Both particulate catalyst suspension and (macromolecular) organic reactant are recycled via UF or RO. The membrane aids in catalyst recycle. Photocatalyzed destruction of reactant may diminish or eliminate (some causes of) membrane fouling.

C. Immobilized Photocatalyst + Reactant Recycle: Catalyst is a fixed bed; no catalyst recovery is necessary. The UF/RO membrane functions only to recycle the (macromolecular) organic reactant.

D. Photocatalyst Immobilized on Crossflow Membrane: The photocatalyst is deposited on the upstream face of a membrane in order to diminish or eliminate membrane fouling by destruction of polymers or microbial film growth.

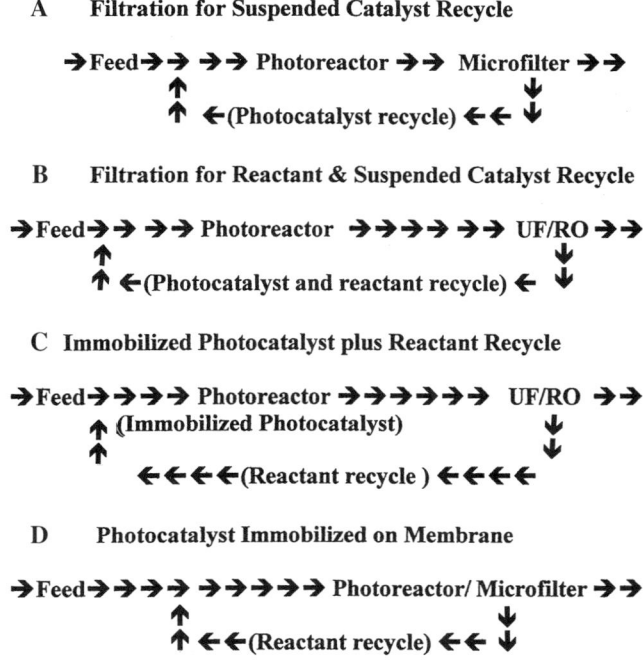

FIGURE 1. Photocatalysis reactor–membrane configurations.

Each example configuration indicates an opportunity for process enhancement via process integration. We now survey the literature to report examples to date in these domains and to conclude with opportunities for further exploration.

Filtration for Suspended Catalyst Recycle

The earliest reports of photocatalytic activity for the destruction of water pollutants concerned use of slurries of microparticulate TiO_2, typically using a Degussa P-25 form with $50 nm^2/g$ surface area, a primary particle size of 20–30 nm, and an agglomerated size, as synthesized, of the order of 1 micron.[6,7] These slurries were often used in a recirculating batch system (see FIGURE 2) allowing for reaction monitoring and strong fluid movement to maintain catalyst in suspension.[15–20] Although 0.1 wt% suspensions in laboratory reactors with a volume 100–1,000 mL were optically opaque and, hence, efficient in photon capture, the slurry itself was immediately problematic in its potential for use in a continuous flow process.

Two approaches to this slurry problem were (1) to devise a means of catalyst recovery and recycle (e.g., membranes in the present paper) or (2) to immobilize the photocatalyst particles on various solid supports (meshes, monoliths, beads, wires, optical fibers, tube walls, etc.). Procedure (1) we discuss in the present and following section; procedure (2) appears later in this paper.

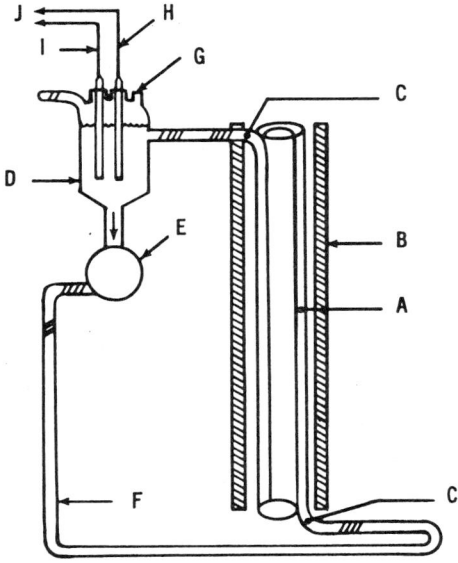

FIGURE 2. Recirculating, differential conversion photoreactor. (A) Quartz annular photoreactor; (B) black lights parallel to reactor axis (7BLB GE, 15 W each); (C) thermocouple; (D) pyrex sampling vessel; (E) centrifugal recirculation pump; (F) teflon tubing; (G) sampling port with teflon-faced septum; (H) chloride ion electrode; (I) reference electrode; and (J) collimator. (Reprinted by permission of the *Journal of Catalysis*.)

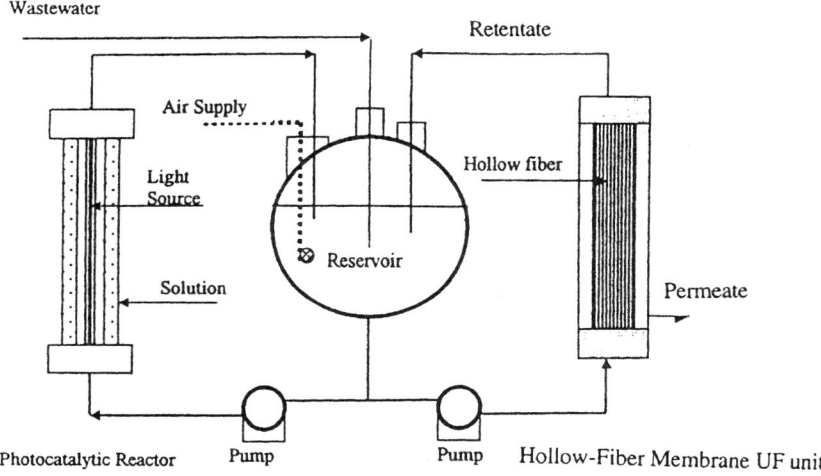

FIGURE 3. Schematic diagram of (integrated) photoreactor–membrane ultrafiltration (UF) system. (Reference 24; reprinted by permission of Kluwer Academic Publishers.)

The earliest example of photocatalyst recycle appears in a patent of Cooper.[21] A related version operates in a commercial system of Matrix Photocatalytic, Ontario, Canada.[22] A recent complete study from an engineering view is that of Sopajaree et al., who examined quantitatively the coupling of polysulphone membrane (molecular weight cutoff [MWCO], 105 Da), UF, and PC.[23] Using an apparatus designated as an integrated photoreactor–membrane UF system (see FIGURE 3), they demonstrated a two-step batch process: (1) operation of an annular photoreactor and the recirculation (mixed) reservoir for pollutant (methylene blue, MB) degradation, and (2) operation of a reservoir charged with photocatalyst suspension and reacted medium in a recirculation loop with the ultrafilter. Thus, continuous operation was not explored; rather the individual batch operating results were tested against appropriate kinetic models.

Their batch recirculation system consisted of a plug-flow photoreactor (PFR) in series with a continuous stirred tank (CST) reservoir. They operated at low concentration, so that first order, rather than a Langmuir–Hinshelwood, rate form applies; the batch kinetics of reactant disappearance were described by

$$-\frac{dC}{dt} = \frac{(1-e-k\theta_2)C}{\theta_1} = k_{app}C, \quad (1)$$

where $\theta_1 = V_1/Q$ and $\theta_2 = V_2/Q$.

In ultrafiltration mode, a second report by these authors examined the influence of crossflow velocity, transmembrane pressure, and photocatalytic slurry solids concentration on the flux of permeate.[24] Upon batch start-up, an eventual pseudo-steady flux during batch ultrafiltration was fitted to a gel–layer model with the expression

$$\frac{1}{J} = \frac{1}{J_{max}} + \frac{\mu R}{\Delta P}, \quad (2)$$

where J is the flux, μ is the viscosity, and R is the total (membrane+photocatalyst gel) resistance.

Titania slurry particles are 10–1,000 times larger that the radius of the membrane pores, as estimated from the Hagan–Poiseuille equation,

$$Rp = \left(\frac{8J\mu L}{e\Delta P}\right)^{0.5}, \quad (3)$$

where L is the membrane thickness and e the membrane porosity.

The reciprocal flux versus reciprocal pressure plots indicated linearity with variable crossflow velocities, the changing slope indicating progressive decrease in R with increasing crossflow (shear rate) (see FIGURE 4A) and a constant slope, variable intercept when velocity is fixed and titania concentration is varied (constant viscosity with R but declining J_{max} as particle concentration increases, see FIG. 4B).

Repeated batch cycling of their apparatus between reaction and ultrafiltration modes (with reactant recharged before each reaction time) created a substantial eventual diminution in pollutant MB conversion at fixed operating times. A substantial shift toward larger particle size was observed in the particle size distribution of the titania slurry (see FIGURE 5). Such pH driven particle agglomeration could have created 5–10 micron aggregates in suspension, producing some shadowing of the photocatalyst primary particles and the observed decrease in the photochemical rate.

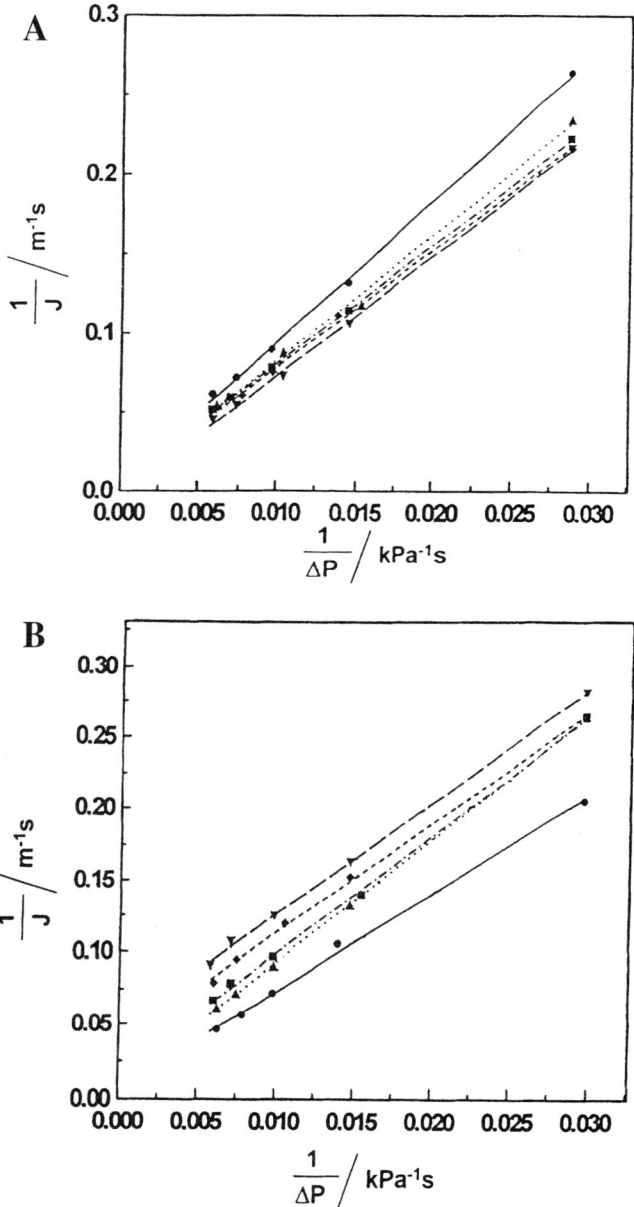

FIGURE 4. (A) Reciprocal plot of $1/J$ against $1/\Delta P$ with a crossflow velocity (in m/sec) as a parameter: (●) 0.54, (▲) 0.97, (■) 1.50, (★) 2.03, and (▼) 2.92. Lines are least-squares fits. Data are for 1 g/dm^3 TiO$_2$ slurry. (B) Reciprocal plot of $1/J$ against $1/\Delta P$ with TiO$_2$ dose (in g/dm^3) as a parameter: (▼) 0.5, (◆) 1.0, (■) 1.5, (▲) 2.0, and (●) 3.0. (Reference 25; reprinted by permission of Kluwer Academic Publishers.)

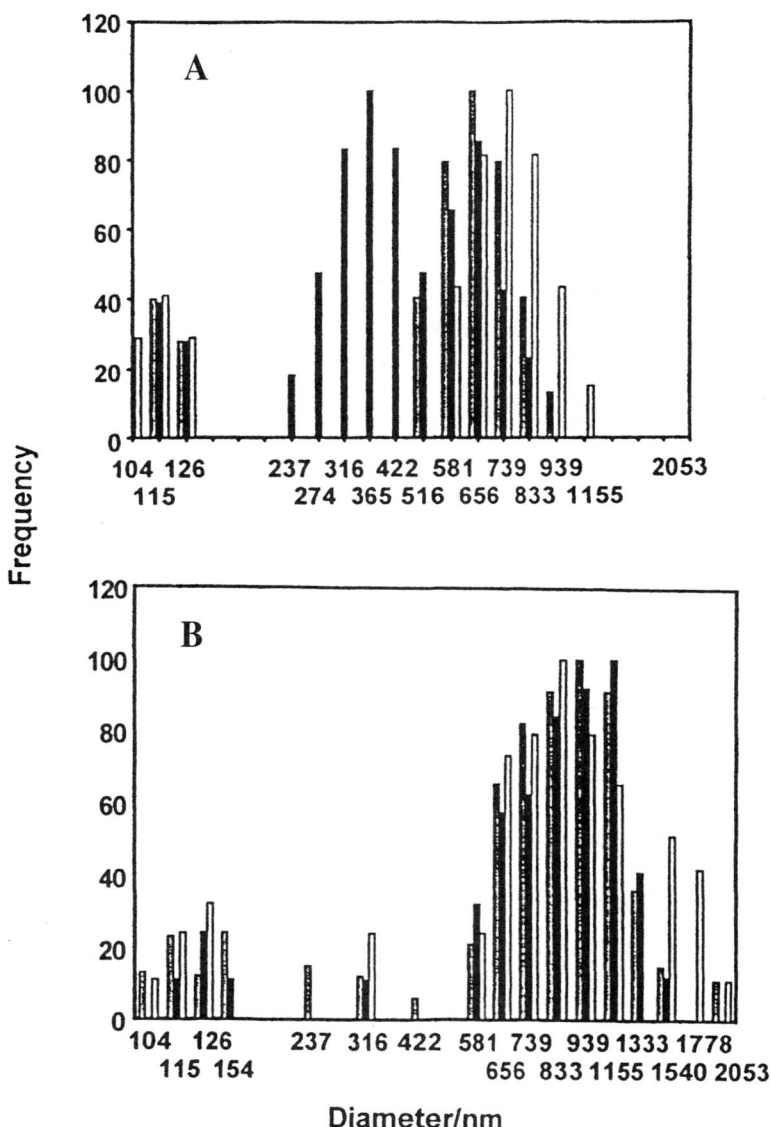

FIGURE 5. Dynamic laser light scattering data on a TiO_2 suspension before (**A**) and after (**B**) use in the combined system In the latter case, the suspension was recovered after the tenth cycle of operation. Three replicate data sets are shown in each case in histogram format for the particle size distribution. (Reference 25; reprinted by permission of Kluwer Academic Publishers.)

Photocatalytic oxidation of organics routinely produces carboxylic acid intermediates. Formic acid in particular has been demonstrated to cause suspension agglomeration, via acidification leading to pH near the point of zero charge (pzc) of titania (about 4.5).[25]

The clear advantage of this configuration is the catalyst recycle. Steady continuous flow operation of such a system is an evident next step. In this case, the model equation for system performance would clearly involve both the catalyst kinetics and the membrane hydraulic properties.

Filtration for Reactant and Suspended Catalyst Recycle

Replacement of the microfilter in Cooper's photocatalyst recycle scheme[21] by an ultrafilter allows for photocatalyst and partial or complete recycle of pollutant (reactant). This conceptual approach was claimed by Molinari *et al.*[26] using the apparatus illustrated in FIGURE 6. This piping configuration allows for continuous flow operation.

The authors first examined the reaction kinetics for the reactant (model pollutant) using a suspended slurry TiO_2 photocatalyst in an illuminated CSTR.[27] Their final

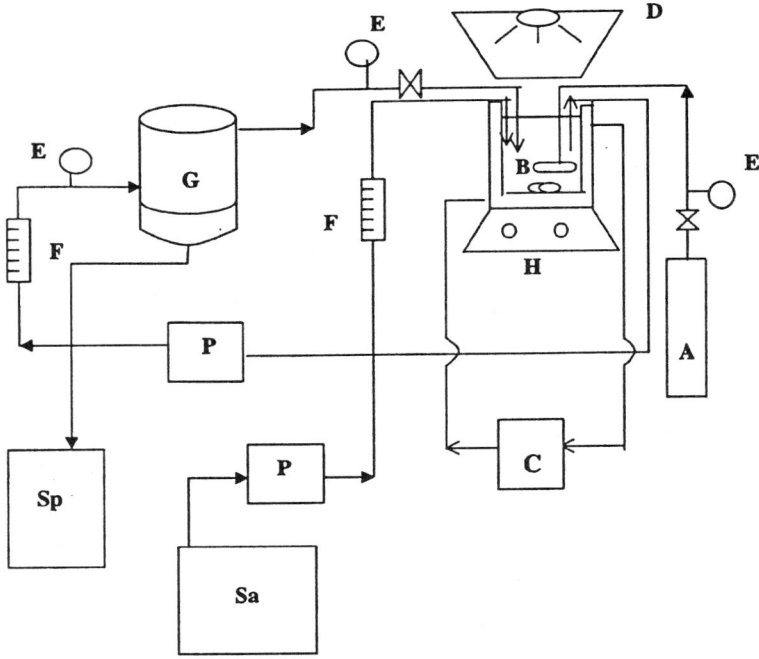

FIGURE 6. Scheme for a continuous membrane photoreactor system with suspended catalyst: A, oxygen reservoir; B, recirculatioin reservoir (reactor); C, thermostatting water; D, UV lamp; E, manometer; F, flowmeter; G, membrane cell; H, magnetic stirrer; P, peristaltic pump; Sa, feed reservoir; and Sp, permeate reservoir. (Reference 26; reprinted by permission of *Catalysis Today*.)

rate equation below indicates the influences of reactant concentration (nitrophenol, C_{4NP}), dissolved oxygen (O_2), intensity I, and photocatalyst concentration T. This empirical equation reflects a linear dependence on I (often observed elsewhere for I less than about $1\,mw/cm^2$), optical opacity at $T = 9.138\,mg/cm^2$, and a negative influence of the initial nitrophenol level (probably due to an internal filter effect). The rate form suggests that 4NP and O_2 adsorb on different sites, as also often reported for other liquid phase organics:

$$-r_{4NP} = -\frac{dC_{4NP}}{dt} = \frac{T[O_2]I(1.36 \times 10^{-3} + 14.8/(C_{4NP})_0^2)C_{4NP}}{(9.138 + T)(0.646 + [O_2])}. \quad (4)$$

When the photocatalytic reactor was coupled to a membrane recycle and used in a continuous flow system (FIG. 6), the apparent ultrafiltration of 4NP vanished over time (see FIGURE 7). Although the authors suggested retentate 4NP decline was due to photoreaction, and the 4NP rise in the permeate level was due to both membrane reaction and adsorption, it appears possible that there is no membrane rejection (4NP has a small diameter), and that when adsorption of 4NP into the membrane is nearly "complete", no further selective removal of 4NP occurs and the retentate and permeate concentrations approach each other. If this is true for all membranes tested here, the data of FIGURE 7 primarily reflects variation in membrane adsorption.

Performance of the integrated systems is shown in FIGURE 8. If no 4NP rejection is occurring, then this joint recycle of reactant and photocatalyst has yet to be demonstrated. A total mass balance was not reported, so we do not know the final partitioning of 4NP among the retentate, permeate, and adsorption into the membrane.

Interactions between a pollutant solute, humic acid, and the particulate titania photocatalyst were explored more fundamentally by Lee et al.[27] These authors measured properties of filter cakes formed by humic acid solutions, titania suspensions,

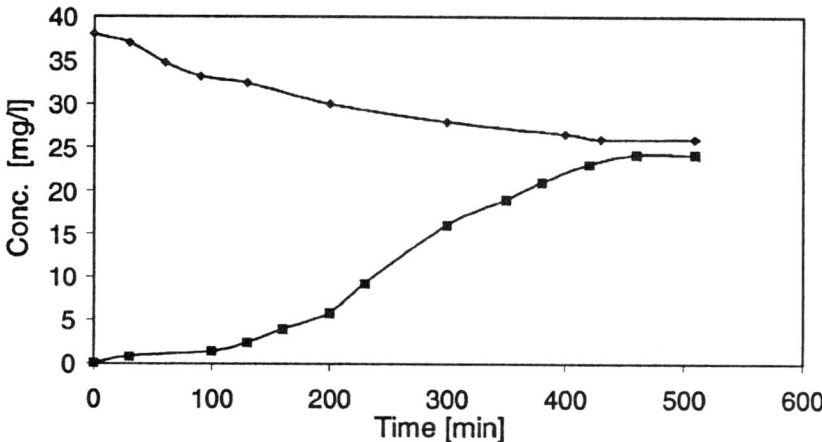

FIGURE 7. 4-Nitrophenol (4NP) concentration versus time using an N30F membrane: $T = 30°C$; $P = 4\,bar$; steady-state flux, $10.21\,hLm^{-2}$. ●, C, ret.; ■, C. perm. (Reference 26; reprinted by permission of *Catalysis Today*.)

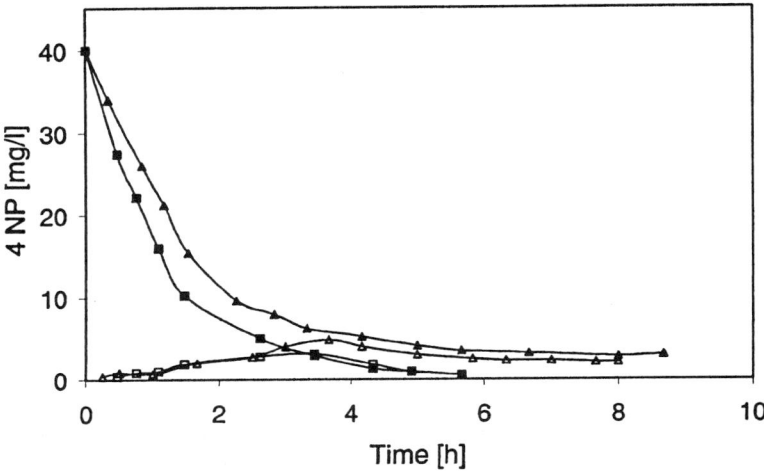

FIGURE 8. 4NP concentration in the retentate and permeate versus time for continuous and recycle system configuration: $T = 30°C$; $P = 3.5$ bar; $[TiO_2] = 1$ g/L; $[O_2] = 22$ ppm; $I = 3.6$ mW/cm^2; tangential flowrate, 500 mL/min; permeate flux, 10 Lh^{-1}m^{-2}. 4NP concentration in the retentate versus time varying the total volume of the recycle system in comparison to the system without recycle. Retentate: continuous system, ▲; recycle, ■. Permeate: continuous system, △; recycle, □. (Molinari et al.[26]; reprinted by permission of *Catalysis Today*.)

and their mixtures, finding that (1) humic solutions or titania suspensions alone produced essentially a constant membrane flux of 160–170 Lm^{-2}h^{-1}, and (2) the mixture yielded fluxes of 145, decreasing to 101, Lm^{-2}h^{-1}, a thirty percent reduction in the first 30 minutes of operation. This result the authors connected to humic acid binding to TiO$_2$ particles to create a more tightly bound layer. Furthermore, during ultrafiltration, the enhanced humic acid level produced additional adsorption within the TiO$_2$ cake, further restricting permeate flow. SEM photographs showed apparently denser cakes for titania–humic acid solution that those created via titania suspensions alone.

To rationalize individual titania or humic acid deposition, assuming particle and molecular diameters of 0.1 to 10 microns and 5×10^{-4} to 3×10^{-2} microns, respectively, the authors estimated particle velocity back transport from the cake surface (see FIGURE 9). Arguing that only Brownian diffusion and scouring due to turbulent flow (crossflow velocity of 1 m/sec) were important, they concluded that particle back transport velocities less than about 4×10^{-3} cm/sec would produce deposition on the cake surface. This purely particle transport approach did not rationalize the dense nature of the mixed cake situation, as the authors noted. The authors did propose that a humic acid–photocatalyst interaction could create a denser cake (FIG. 9). We add that the cake properties may also be affected by the method of formation: (1) prefiltration of titania followed by humic solution addition (as done, forms cake, then compresses via crosslinking during adsorption and pore filling) or (2) premixing of titania and humic acids followed by filtration of the resulting larger agglomerates may give a more open cake than obtained in the present research.

Specific cake resistance measurements at 0.7 bar showed that the titania and mixture cake values were 8.7×10^{12} m/kg and 3.8×10^{13} m/kg, respectively. The mixture cakes were also more compressible than the titania layers (compressibility coefficient, 0.94 vs. 0.39).

Ultrafiltration of humic acid through the membrane was characterized by UV_{254} measures, inasmuch as negligible TiO_2 permeation was noted. Increasing crossflow velocity provided a reduced humic level in the permeate, which the authors attributed to a reduction of concentration polarization and, hence, a reduction in permeate concentration driving force. Humic acid adsorption equilibration was shown to be relatively rapid, and no change in particle size was noted; thus, only the concentration polarization argument was involved according to the authors.

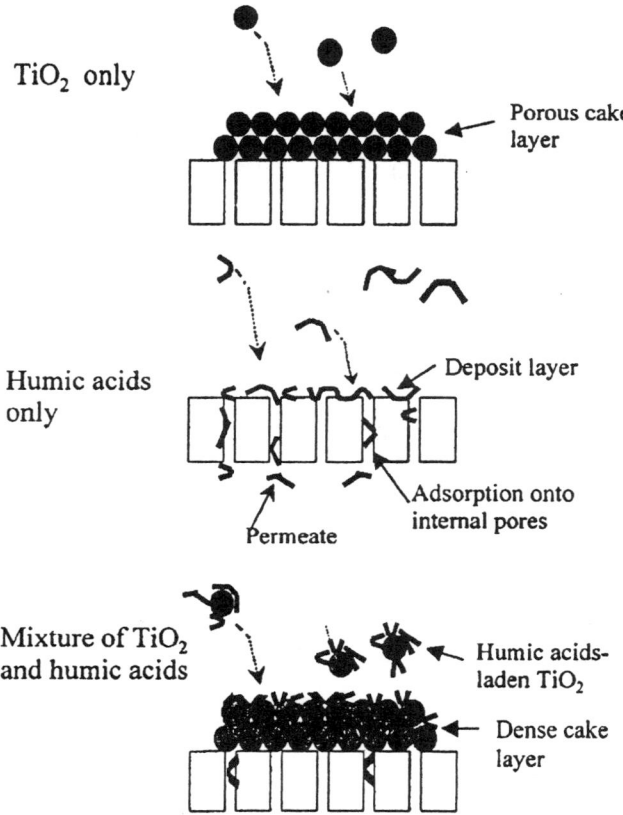

FIGURE 9. Illustration of possible mechanisms for the formation of deposit/cake layers at the membrane surface during UF of different combinations of solutes (including photocatalyst slurry). (Reference 28; reprinted by permission of *Industrial and Engineering Research*.)

When the photoreactor (see FIGURE 10) was illuminated, a constant flux ($Lm^{-2}h^{-1}$) (see FIGURE 11) was measured during six hours, consistent with destruction of humic acid (see FIGURE 12A). Permeate UV_{254} was less than $0.009 cm^{-1}$. The authors noted that the percent UV_{254} nm reduction was larger than the decrease in DOC (FIG. 12B). The photocatalyzed conversion of humic acid into smaller and/or less absorptive species was proposed. Importantly, the authors concluded that "photocatalytic reactions appear to be attractive for the control of membrane fouling and removal of humic acids during the UF of TiO_2 particles in the presence of fouling materials such as natural organic matter". They concluded that "A UF process is potentially an attractive method for the separation of photocatalyst from drinking water without any concern about membrane fouling as long as photocatalytic reactions are occurring", an optimistic statement achievable only if fouling by (non-oxidizable) inorganic solutes and particles is simultaneously absent.

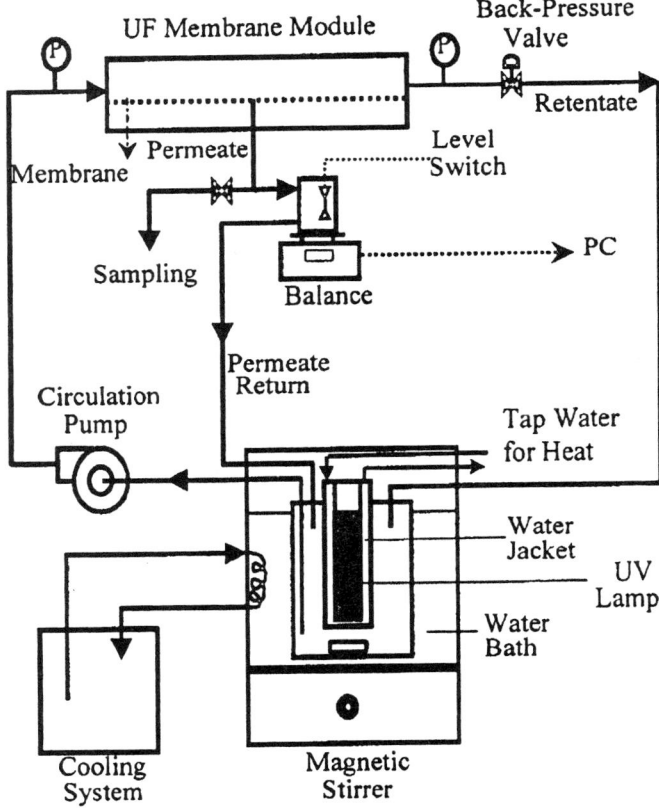

FIGURE 10. Schematic of a laboratory-scale photocatalysis/ultrafiltration system. (Reference 28; reprinted by permission of *Industrial and Engineering Research*.)

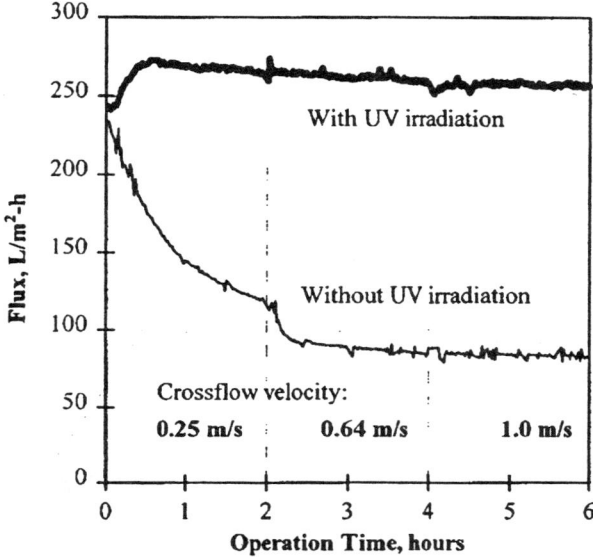

FIGURE 11. Effect of UV irradiation (photocatalysis) on UF flux. (Reference 28; reprinted by permission of *Industrial and Engineering Research*.)

Immobilized Photocatalyst + Reactant Recycle

We are unaware of any literature examples to date. This configuration has at least one potential advantage in situations where a large MW initial pollutant is hazardous, but its smaller partial oxidation products are not, then a UF or RO process could

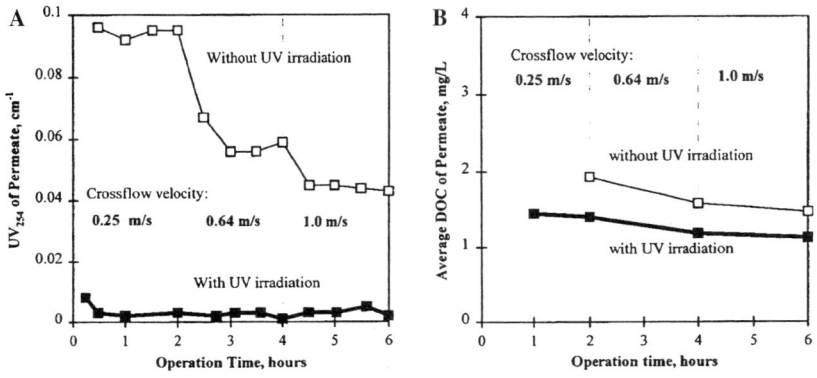

FIGURE 12. Effect of UV irradiation on **(A)** permeate UV_{254} and **(B)** permeate DOC removal. (Reference 28; reprinted by permission of *Industrial and Engineering Research*.)

recycle target reactant to the reactor, while removing partial oxidation (harmless) products which would otherwise remain and compete with the target reactant for photogenerated active centers, for example, hydroxyl and perhydroxyl species.

A second process improvement would involve retaining the larger, non-biodegradable starting material, and passing, via RO or UF, the intermediates (biodegradable) on to a biological oxidation system. We discuss integrated (photo)chemical and biological oxidation elsewhere.[29–31]

Photocatalyst Immobilized on Crossflow Membranes

In this configuration, the catalyst is either deposited on the upstream surface of the membrane, or is incorporated into a light-transmitting membrane. This configuration has been most extensively considered, with a half dozen or so studies appearing in the last two years.

The present discussion considers only flow-through membranes. Photocatalysts have been deposited as ceramic membranes on walls, on or within polymer films, and on opaque or optically transparent electrodes or optical fibers; however, we do not consider such cases here since the separation selectivity of the membrane does not operate; that is, these membranes are used only for support, not separation during flow.

Molinari et al.[27] explored a range of commercial membranes composed, respectively, of polylsulphone (PS), polyamide (PA), PEEK, fluoride+polyupropylene

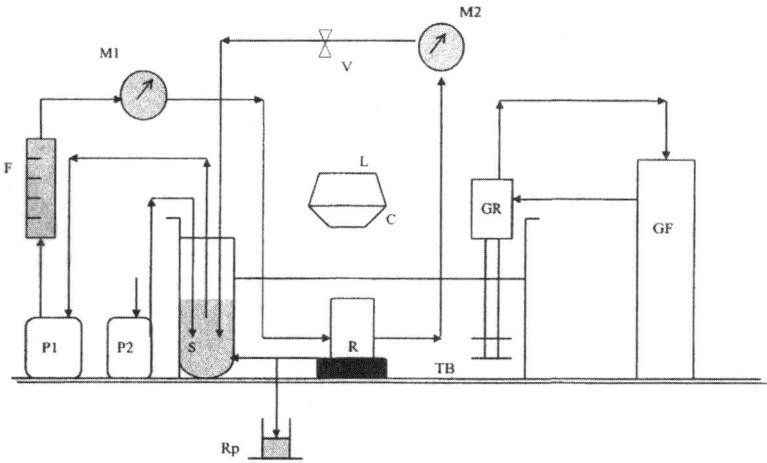

FIGURE 13. Scheme of the setup for the photoreactivity experiments carried out with immobilized TiO_2. F, flowmeter; M1 and M2, pressure gauges; P1, peristaltic pump; P2, air pump or oxygen cylinder; S, recirculation tank; Rp, permeate reservoir; R, flat sheet membrane photoreactor; L, UV lamp; C, ray concentration cone; v, pressure control valve; GR, thermostatic group; TB, thermostatic water bath; and GF, refrigerator group. (Reference 27; reprinted by permission of *Catalysis Today*.)

(PP), polyethersulphone (PES), polyvinylfluoride (PVF) (PP+PS), polyacrylonitrile (PAN), and cellulose acetate (CA)+PP. A preliminary screening of membrane photostability showed PAN, fluoride+PP, and PP+PS to be photostable to near UV (300–400 nm) illumination. In contrast, other membranes exhibited an increase in permeate flux after illumination, attributed to membrane photolysis (polylsulfone, PEEK, and polyamide) and/or a release of soluble carbon, reflected in increased levels of dissolved total organic carbon (TOC), CA+PP, FS+PP, PES, PS+PP, PVF, and PAN, extractable from the light-treated membranes.

Photocatalysis studies with the apparatus shown in FIGURE 13, using the PAN (highest flux, constant in time) membrane were executed by ultrafiltering on to the upper membrane surface various amounts (0.76 to 6.12 mg/cm^2) of TiO$_2$. Flux decline due to TiO$_2$ cake formation was of the order of 80%, and nearly independent of pressure, in contrast to bare surface membrane fluxes which were linear in ΔP. An intermediate cake loading of 4.08 mg TiO$_2$/cm^2 appeared to give the best uniform coverage (lower values gave incomplete surface coverage, higher values produced uneven films.) In continuous systems, these two systems failed to show any long term performance of either a photocatalyst or membrane type.

The use of a titania layer deposited on an alumina microfilter ($d_{p,ave}$ = 1 micron) was explored by Tsuru et al.[32] by coating TiO$_2$ colloidal sols on the microfilter, firing at 450°C to yield a 1-micron titania coat. Two potential photocatalytic actions were proposed as (see FIGURE 14). Here PCO can oxidize a target pollutant (trichloroethylene, TCE) and, simultaneously, oxidize to smaller fragments any model organic

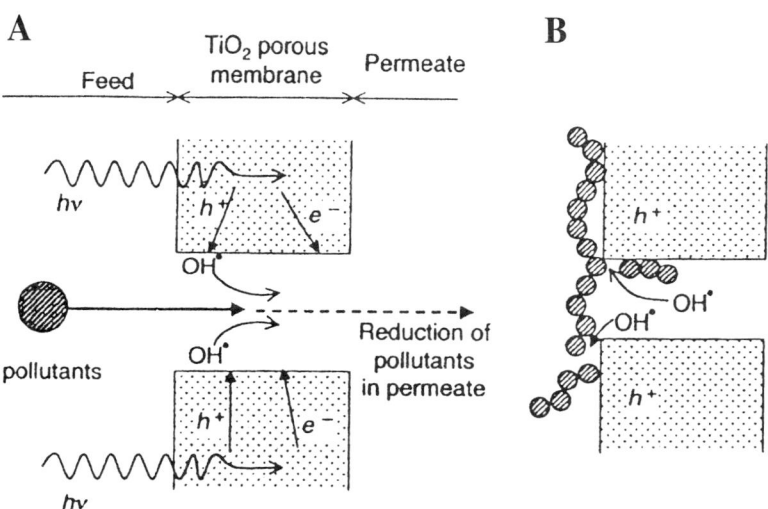

FIGURE 14. Schematic concept of photocatalytic reaction in a filtration system: **(A)** organic pollutants are decomposed while permeating TiO$_2$ pores; **(B)** foulants on the surface of TiO$_2$ membranes are decomposed, and permeating volume flux is increased. (Reference 32; reprinted by permission of *Journal of Chemical Engineering of Japan*.)

foulant (e.g., polyeethyleneimine, PEI). The modest temperature sintering of nanometer TiO_2 deposits creates an ideal situation in which the diffusional time to the titania pore walls is small, because the titania layer itself is both the ultrafilter and photocatalyst.

The photocatalytic activity in single pass operation was shown by the substantial decrease in permeate TCE level (see FIGURE 15) and the anti-fouling activity demonstrated to hold under steady operation by achievement of substantial flux increase (see FIGURE 16) following UV illumination of a titania/alumina membrane (MWCO, 500) after introduction in the dark of PEI (MW, 1,800) solution.

It would be interesting to repeat this experiment in the opposite order of illumination followed by PEI introduction, to see if preillumination could further reduce fouling. The authors, therefore, demonstrate a novel PCO system, "in which filtration and a photocatalytic reaction occur simultaneously" and in which "membrane fouling was also found to be reduced by photocatalytic reaction."

In RO, fouling by biofilms (organic films containing, and often formed by, microbial life) occurs in water treatment plants and may contribute substantially to the total operating costs for the facility. Microbial biofilms are particularly problematic for aromatic polyamide thin-film-composite (TFC) membranes, a common RO material. To address this problem, researchers have explored the use of titania modified membranes for anti-bacterial action. For example, Kwak et al.[33] used the direct immobilization of the sol–gel titania nanoparticles achieved through a "self-assembly" recipe, using polyamide carboxylate groups (–COOH) and hydrogen

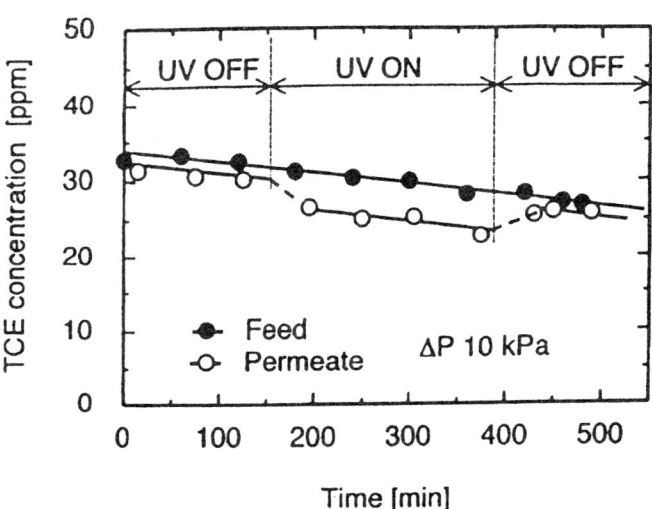

FIGURE 15. Trichloroethylene (TCE) concentration as a function of elapsed time: $\Delta P = 10\,kPa$; MWCO = 30,000. (Reference 32; reprinted by permission of *Journal of Chemical Engineering of Japan*.)

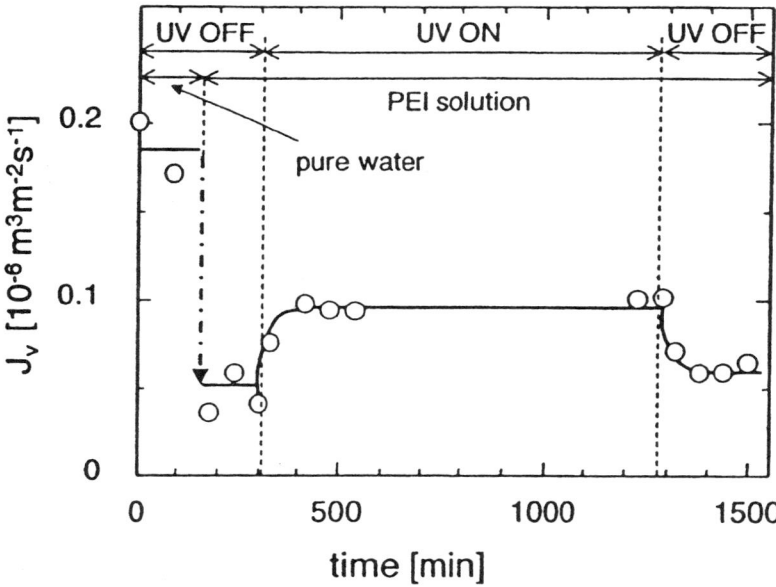

FIGURE 16. Permeate flux, J_v, as a function of elapsed time. The experimental sequences were as follows: pure water (UV off) → PEI solution (UV off) → PEI (UV on) → PEI (UV off). PEI concentration, 100 ppm, $\Delta P = 20$ kPa, pH 10.5, MWCO = 500. (Reference 32; reprinted by permission of *Journal of Chemical Engineering of Japan*.)

bonding to bind the titania particles. Long-term laboratory experiments showed substantial retention of nanometer titania particles on the membrane after seven days in crossflow operation.

Illumination of the hybrid titania-alumina membrane system following deposition of 150 microliters (10^4 cells) onto a titania surface, illuminating (8w near UV), removed by pipetting and plated on to Luria–Bertrani (LB) agar medium. The decline in cell viability of *E. coli* bacteria (DH5alpha strain) was faster in the presence than in the absence of illumination, and clearly involved achievement of total cell destruction in four hours, in contrast with the non-illuminated results (FIGURE 17). Although this antibacterial result is encouraging, the cell-killing activity under UF and RO flow operations should be greater, inasmuch as the filtration flux should assist in bringing the live cells directly up to the illuminated photocatalyst. It is likely that a (nearly) direct contact of cell and catalyst is needed for maximum efficiency.

This section has demonstrated (1) creation of titania–ceramic and titania–organic modifications of existing ceramic (alumina) and organic (polyamide) membranes, and (2) use of photocatalysis to prevent/reduce fouling as evidenced by destruction of organic polymers (polyethyleneimine) and cell killing (*E. coli* bacteria). Only the polymer fouling reduction was demonstrated under actual filtration conditions.

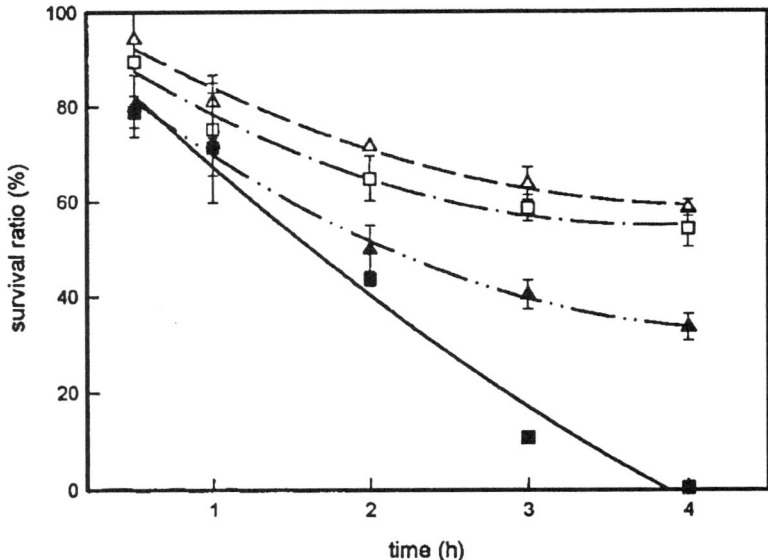

FIGURE 17. Photocatalytic bactericidal effects of the TiO_2 hybrid and neat aromatic polyamide thin film composite (TFC) membranes in the dark and with UV light illumination: △, neat membrane+dark; ☐, TiO_2 hybrid membrane+dark; ☐, neat membrane+UV; ■, TiO_2 membrane+UV. (Reference 33; reprinted by permission of *Environmental Science and Technology*.)

CONCLUSIONS

Four potential combinations of photocatalysis and membrane technologies are considered in this brief review: (a) PC+MF for catalyst slurry recycle, (b) PC+UF for catalyst slurry and (polymer) reactant recycle, (c) immobilized PC and UF/RO for reactant recycle, and (d) immobilized PC on UF/RO membrane for membrane self-cleaning. Of these, only three (a, b, and d) have been demonstrated in literature reports, despite the presumed ability of untested configuration (c) to provide reactant recycle to extinction in a continuous flow or batch reactor.

Both batch and cycled batch versions of (a) have been shown, including development of preliminary models for system performance. However, the organic acid generation characteristic of all photocatalyzed oxidation of organics leads to a changing titania agglomerate size, which in turn alters MF membrane performance in the cycled batch case. No model yet exists for this presumably common phenomenon in photocatalysis.

The simultaneous conversion of polymer reactant and recycle of catalyst with a characteristic surface water natural pollutant, humic acid (configuration b), was also demonstrated. Here again, the configuration was demonstrated, but interactions of humic acid and catalyst particles generated colloidal phenomena, which altered the filter cake and, thus, the permeation flux.

The potential for photocatalyst immobilized on the ceramic membrane upstream side for maintenance of membrane flux by destroying a polymeric contaminant was demonstrated as configuration (d). The single report validates nicely this potential advantage, and suggests a need for similar demonstrations with other common foulants.

In summary, three of the four configurations identified as possible have been demonstrated, but unanticipated colloidal influences in (b) and (d) remind us of the need for more fundamentally designed experiments. Furthermore, the need for fundamental engineering science models for all cases is obvious, so that some more strongly predictive approaches will be available in this promising area of integrating a room temperature total oxidation photocatalyst with established membrane technologies.

REFERENCES

1. OLLIS, D.F., & H. AL-EKABI, Eds. 1993. Photocatalytic Treatment and Purification of Water and Air. Elsevier, Amsterdam.
2. OLLIS, D.F., E. PELIZZETTI & N. SERPONE. 1991. Environmental Science and Technology.
3. HOFFMANN, M., et al. 1995. Environmental applications of semiconductor photocatalysis. Chem. Rev. 69–96.
4. BLAKE, D.M., et al. 1999. Application of the photocatalytic chemistry of titanium dioxide to disinfection and the killing of cancer cells. Separ. Purific. Meth. 25: 1–50.
5. OLLIS, D.F. 2000. Photocatalytic purification and remediation of contaminated air and water. Comp. Rend. Acad. Sci. II(C3): 405–411.
6. NARGIELLO, M. & T. HERZ. 1993. Physical-chemical characteristics of P-25 making it extremely suitable as the catalyst for the photodegradation of organic compounds. In Photocatalytic Treatment and Purification of Water and Air. D.F. Ollis & H. Al-Ekabi, Eds.: 801–808. Elsevier, Amsterdam.
7. SUZUKI, et al. 2000. Chem. Lett. 130–132.
8. JACOBY, W.A., et al. 1998. Mineralization of bacterial cell mass on a photocatalytic surface in air. Environ. Sci. Tech. 32: 2650–2653.
9. SITKIEWITZ, S. & A. HELLER. 1996. Photocatalytic oxidation of benzene and stearic on sol–gel derived TiO_2 thin films attached to glass. New J. Chem. 20: 233–241.
10. ROMEAS, V., et al. 1999. Degradation of palmitic acid deposited on TiO_2-coated self cleaning glass: kinetics of disappearance, intermediates products, and degradation pathways. New J. Chem. 23: 365–373.
11. BITTER, J.G.A. 1991. Transport Mechanisms in Membrane Separation Processes. Plenum, New York.
12. SCOTT, K. & R. HUGHES. 1996. Industrial Membrane Separation Technology.
13. FORMENTI, M., et al. 1970. Heterogeneous Photocatalysis. Chem. Tech. 1: 680.
14. CANDAL, R.J., et al. 1999. Titanium-supported titania photoelectrodes made by sol–gel process. J. Environ. Eng. 125: 906–912.
15. TURCHI, C.S. & D.F. OLLIS. 1989, Mixed reactant photocatalysis: intermediates and mutual rate inhibition. J. Catalysis 119: 483–496.
16. OLLIS, D.F. 1985. Contaminant degradation in water. Envir. Sci. Technol. 19: 480–484.
17. NGUYEN, T. & D.F. OLLIS. 1984. Complete heterogeneously photocatalyszed transformation of 1,1-dibromoethane and 1,2-dibromoethane to CO_2 and HBr. J. Phys. Chem. 88: 3386–3388.
18. OLLIS, D.F., C.Y. HSIAO & L. BUDIMAN. 1984, Heterogeneously photoassisted catalysis-conversions of perchloroethylene, dichloroethane, chloroacetic acids, and chlorobenzenes. J. Catalysis 88: 89–96.
19. PRUDEN, A.L. & D.F. OLLIS. 1983. Degradation of chloroform by photoassisted heterogeneous catalysis in dilute aqueous suspensions of titianium dioxide. Envir. Sci. Tech. 17: 628-631.

20. PRUDEN, A.L. & D.F. OLLIS. 1983. Photoassisted heterogeneous catalysis—the degradation of trichloroethylene in water. J. Catalysis **82:** 404–417.
21. COOPER, G. & M.A. RATLIFF. 1992. Photocatalytic treatment of water. U.S. Patent 5,118,422.
22. ENZWEILER, R., et al. 1993. AIChE Meeting, Seattle.
23. SOPAJAREE, K., et al. 1999. An integrated flow reactor-membrane filtration system for heterogeneous photocatalysis. Part I: experiments and modeling of a batch-recirculated photoreactor. J. Appl. Electrochem. **29:** 533–539.
24. SOPAJAREE, K., et al. 1999. An integrated flow reactor-membrane filtration system for heterogeneous photocatalysis. Part II: experiments on the ultrafiltration unit and combined operation. J. Appl. Electrochem. **29:** 1111–1118.
25. BUECHLER, K., et al. 2001. Investigation of the mechanism for the controlled periodic illumination effect in TiO_2 photocatalysis. Ind. Eng. Chem. Res. **40:** 1097–1102.
26. MOLINARI, R., et al. 2000. Study on a photocatalytic membrane reactor for water purification. Catalysis Today **55:** 71–78.
27. LEE, S.A., et al. 2001. Use of ultrafiltration membranes for the separation of TiO_2 photocatalysis in drinking water treatment. Ind. Eng. Chem. Res. **40:** 1712–1719.
28. SCOTT, J. & D.F. OLLIS. 1995. Integration of chemical and biological oxidation processes for water treatment: review and recommendations. Envir. Prog. **14:** 88–102.
29. SCOTT, J. & D.F. OLLIS. 1996. Engineering models of combined chemical and biological processes. J. Envir. Eng. ASCE **122:** 1110–1114.
30. OLLIS, D.F. 2001. On the need for engineering models of integrated chemical and biological oxidation of wastewaters. Water Sci. Tech. **44:** 117–123.
31. MOLINARI, R., et al. 2001. Photocatalytic membrane reactors for degradation of organic pollutants in water. Catalysis Today **67:** 273–279.
32. TSURU, T., et al. 2001. Photocatalytic reactions in a filtration system through porous titanium dioxide membranes. **34:** 844–847.
33. KWAK, S.Y., et al. 2001, Hybrid organic/inorganic reverse osmosis (RO) membrane for bactericidal anti-fouling. 1. Preparation and characterization of TiO_2 nanoparticle self-assembled aromatic polyamide thin-film-composite (TFC) membrane. **35:** 2388–2394.

Water Quality Monitoring in Membrane Filtration Systems

ELHADI M. ABOGREAN,[a] SIOBHAN F.E. BOERLAGE,[b] MARIA D. KENNEDY,[b] IBRAHIM M. EL-AZIZI,[a] GILBERT GALJAARD,[c] AND JAN S. SCHIPPERS[a,c]

[a]*Tajura Research Center, Tajura, Tripoli – Libya*

[b]*International Institute for Infrastructural, Hydraulic, and Environmental Engineering (IHE), Delft, the Netherlands*

[c]*Kiwa Research and Consultancy, Nieuwegein, the Netherlands*

ABSTRACT: We report on an experimental study of UF membrane fouling by colloidal particles. Deposition colloidal particles during membrane filtration causes a decline in permeate flux. Membrane flux is monitored on a laboratory scale, crossflow employing UF membranes. The existing modified fouling index (MFI) uses a microfilter membrane as a quick test of feed water quality. The MFI is based on cake filtration, and thus, a model can be developed for flux decline predication. However, this MFI is not sensitive to the presence of smaller particles. Therefore, more recently MFI using ultrafiltration membranes (MFI-UF) was developed. This research investigates various critical aspects of the MFI-UF test for use as a water quality indicator; stability of the MFI-UF over time, linearity of the index with particulate concentration, and reproducibility (1) of the test (reusability of a UF module) and (2) module manufacture. Pressure dependence of the MFI-UF was also examined. The aforementioned criteria were examined using a polyacrylonitrile module with 13,000 molecular weight cutoff for low fouling (tap and process water). The MFI-UF was stable over time and directly related to colloidal concentration. The MFI-UF test was reproducible for one module with repeated testing; reproducible module manufacture was found for 80% of the test modules.

KEYWORDS: modified fouling index; colloidal fouling; cake compression

NOMENCLATURE:
- A membrane surface area, m^2
- A_0 reference surface area of a 0.45 μm membrane filter, $13.8 \times 10^{-4} m^2$
- C_b concentration of particles in feed water, g/m^3
- CWF specific clean water flux, $L/m^2 h\, bar$
- d_s particle diameter, m
- I propensity index for particles to form a hydraulic resistant layer, L/m^2
- J permeate water flux, $m^3/m^2 sec$
- n compressibility coefficient
- Δp_2 reference applied transmembrane pressure, 2 bar
- Q flow at temperature t, L/h
- R retention by membrane filter
- R_b resistance due to blocking, L/m
- R_c resistance of the cake, L/m

Address for correspondence: Elhadi M. Abogrean, International Institute for Infrastructural, Hydraulic, and Environmental Engineering (IHE), P.O. Box 3015, 2601 DA Delft, the Netherlands.

abogrean@yahoo.co.uk

R_m	membrane filter resistance, L/m
t	filtration time, sec
TMP	applied transmembrane pressure, bar
V	filtrate volume, m^3
T	temperature, °C
α	specific cake resistance, m/g
α_0	constant, m/g
ε	cake porosity
η_{20}	water viscosity at 20°C, Nsec/m^2
η_t	water viscosity at temperature t, Nsec/m^2
ρ_s	density of solute-material forming the cake, g/m^3

INTRODUCTION

The era of pressure-driven membrane processes, ultra-, nano-, and hyperfiltration, is relatively recent. It is only within the last 30 years that these processes have emerged in large-scale operation in drinking and industrial water treatment. Recent innovation in membrane technology and process design, for example, ultra low pressure membranes have led to a dramatic decrease in energy costs. Consequently, in industrial and potable water production these processes are increasingly applied to all water. Similarly, in recycling and water reuse, membrane technology is playing a role that increase in importance year by year.

However, one of the limits to future growth in this technology is membrane fouling—that is, the deposition of undesirable material on the membrane or on the spacer in the feed brine channel. Particulate fouling—that is, colloids, suspended solids, and microbial cells—is a persistent problem in potable water applications, especially colloidal fouling. Its complete removal by pretreatment of the feed water is not always easy or feasible. Membrane fouling causes a reduction in product water flux or increasing operational pressure to maintain flux, which translates to increased operational costs. Membrane cleaning to remove foulants results in increased down time, energy and chemical use, and the production of waste water adds further costs attributable to fouling.

No quality parameter exists to accurately measure the particulate content of feed waters and to predict particulate fouling. Therefore, pilot plant operation is commonly used for the prediction of fouling to design full-scale systems. However, although this method generally provides reasonably good reproducibility, it is time consuming and expensive.

Existing indices to measure the particulate fouling potential of feed water are the silt density index (SDI) and modified fouling index (MFI). Both are designed as tests to simulate membrane fouling by feeding water through a 0.45µm microfilter in dead-end flow under constant pressure filtration. Disadvantages of the SDI are that it is not based on a distinction between filtration mechanisms occurring during the test and, therefore, cannot model flux decline in membrane systems. Moreover, since there is no cake filtration, it can be used to model flux decline in membrane systems. In addition, the MFI index is linear with the concentration of particles in feed water.[1] However, it does not satisfactorily correlate with colloidal fouling observed in membrane installations. This is because many particles, less than 0.45µm, that are responsible for membrane fouling are not measured in the test.[2]

Recent research focused on applying ultafiltration (UF) membranes to incorporate these particles in the existing MFI. More accurate flux decline prediction was indeed found for the pilot plant examined. Although cake filtration was assumed to occur in the MFI-UF test, previous research proved inconclusive.[3] In order for the MFI-UF test to be operated, several criteria must be satisfied. First, the MFI-UF value must be stable over time. Second, the MFI-UF index must be linear with the concentration of particles in the feed water (which is also a proof of cake filtration). Third, the MFI-UF test must be reproducible, thus membrane module manufacture needs to be consistent for a module. In addition, because UF modules are expensive, for the test to be commercially viable the modules must be reusable after cleaning and offer reproducible measurement.

Furthermore, the MFI-UF test is carried out under a constant applied reference pressure of 2 bar. If the cake is composed of compressible particles, then an increase in the pressure causes the formation of a denser cake with a higher resistance.[4] In previous research the existing MFI was found to depend on pressure and the extent of cake compression was determined using the compressibility index, which relates the specific resistance of the cake to applied pressure.[5] The pressure dependency of the MFI-UF is unknown. If cake compression occurs, it may interfere with the MFI-UF measurement, particularly in the case of water with a higher colloidal content. To minimize this effect, the measurement should be carried out at a lower pressure and, thus, requires pressure correction. In addition, for application in the MFI flux decline model, the effect of cake compression needs to be considered since the applied pressure in, for example, hyperfiltration installations using ultralow-pressure membranes, is about 10–15 bar. Thus, the dependence of the MFI-UF on pressure needs to be ascertained and a method for pressure correction of the MFI-UF test must be developed.

This research aims at further exploring the MFI-UF test for monitoring particulate fouling. The above criteria will be investigated, using a selected polyacrylonitrile UF module from previous research for varying types of feed water. A preliminary investigation will also be carried out UF. If cake compression is found to occur, pressure correction incorporating the compressibility index will be examined in order that the MFI-UF can be corrected for different applied test pressures.

BACKGROUND

Various filtration mechanisms can occur when feed water is filtered through a membrane; that is, blocking filtration, cake filtration, and cake filtration with compression. Particles retained by the membrane during filtration add additional resistance to the membrane and reduction in flux can be described by the resistance in series model, Equation (1):[6]

$$J = \frac{\Delta p}{(R_m + R_c + R_b)\eta}, \qquad (1)$$

where J is the permeate flux through a membrane, p the applied transmembrane pressure, and η is the water viscosity. Initially only the membrane provides a resistance to water flux, R_m, and membrane permeability is function of properties such as membrane thickness a and porosity. Particles deposited inside pores or blocking pore entry restrict flow and reduce the flux. This added resistance is defined as R_b. When the membrane is fully or partially blocked, particles may build up on the membrane as a *filter cake* described by R_c). If the cake becomes compressed or clogged the resistance increases further.

Cake Filtration

The MFI is based on cake filtration theory whereby particles are retained on the membrane during filtration by a mechanism of surface deposition. Equations for cake filtration are based on the assumption of viscous or streamline flow. The R_m of the membrane is assumed to be constant, and once the cake has formed it has time independent permeability and a uniform cake.[4] The filtration rate through a membrane on which a cake has formed is described as follows:

$$\frac{dV}{Adt} = \frac{\Delta p}{\eta} \times \frac{1}{R_m + R_c}, \qquad (2)$$

where V is the filtrate volume, t the filtration time, and A is the membrane surface area.

The resistance of the cake is expressed by taking into account its continuous growth as thickness of the cake increase, assuming no cake compression occurs, and the retention R is constant. Cake resistance is described as the specific resistance of the cake per unit α, the total volume of filtrate collected V, and the concentration of particles per unit volume of filtrate, C_b,

$$R_c = \frac{VR}{A} \times \alpha C_b. \qquad (3)$$

The specific cake resistance α is constant for incompressible cakes under constant pressure filtration and can be calculated according to the Carman relationship,[7]

$$\alpha = \frac{180(1-\varepsilon)}{\rho_s d_s^2 \varepsilon^3}. \qquad (4)$$

According to the Carman relationship, a reduction in the porosity of the cake ε or decrease in particle diameter size d_s increases the specific resistance of the deposited cake.

In the MFI, the I index is taken to be the product of the specific resistance of the cake and concentration of particles in the feed water, and is assumed to be independent of pressure. I is a function of the dimension and nature of the particles, and is directly correlated with concentration in feed water,[1]

$$I = \alpha C_b. \qquad (5)$$

Substituting for I in Equation (3), combining the result with Equation (2), and integrating at constant p from $t = 0$ to $t = t$ (assuming $R = 1$), results in expressions for filtration time and filtrate volume:

$$\frac{1}{v} = \frac{\eta R_m}{\Delta p A} + \frac{\eta V I}{2 \Delta p A^2} \qquad (6)$$

$$\frac{\eta I}{2 \Delta p A^2}, \qquad (7)$$

where the term of (7) serves as the MFI index for the fouling potential of a feed water containing particles, when fixed reference values are used for Δp, η, and A. The MFI is determined from the plot of t/v versus V. Typically this plot shows three regions corresponding to blocking, cake filtration, and cake filtration with compression (see FIGURE 1). The first sharp increase in slope is attributed to blocking, followed by cake filtration, which is a linear region of minimum slope. The MFI is defined as the slope (tangent) of this line.[1]

Cake Compression

If the shape of the particles or their physical strength is such the packing arrangement in the cake formed on the UF membrane can sustain the viscous drag of the filtrate flow over the particle surface without deformation, the cake is regarded as incompressible.[4] The porosity of the cake and its specific resistance are independent of the imposed transmembrane pressure, and (5) can applied. However, many cakes are composed of clays and microbial cells that are highly compressible, resulting in a decrease in cake porosity and hence an increase in specific resistance as the transmembrane pressure or the rate of flow is increased.[4,8] Then (5) does not apply and the effect of pressure on the specific resistance and, hence, the I value needs to be considered. An empirical relationship, often applied to account for pressure, is provided by Equation (8), which relates the specific resistance to pressure, to the power of compressibility coefficient n, and a constant α.[9] For incompressible cake, n is zero and the higher the compressibility coefficient the more compressible the cake:

$$\alpha = \alpha_0 P^n. \qquad (8)$$

I is then defined to incorporate the compressibility factor as follows:

$$I = \alpha_0 P^n C_b. \qquad (9)$$

The increase in flux due to a higher applied pressure may also cause cake compression. Ruth[10] indicated that the velocity with which particles are arrested when they impinge on the filter medium affects the initial void volume resulting in a higher initial α at higher rate of flow.

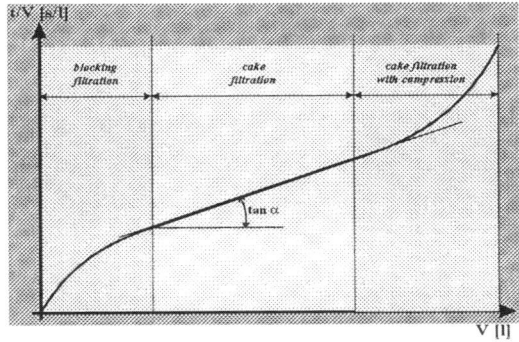

FIGURE 1. Ratio of time and filtrate volume as a function of total filtrate volume (based on applied transmembrane pressure).

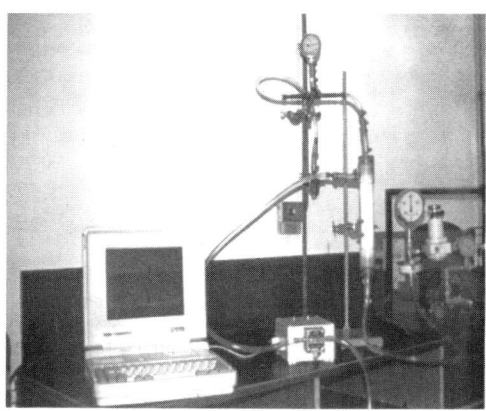

FIGURE 2. MFI-UF equipment.

EXPERIMENTAL

A polyacrylonitrile model with a molecular weight cutoff of 13,000 Dalton (abbreviated as PAN 13) was chosen for investigation based on previous research.[3] In the reproducibility study six modules were employed, three manufactured from one batch (A), three from other batches (B) and (C), where a batch refers to the same month of module manufacture. The specific clean water flux (CWF) was measured in the MFI-UF test equipment (see FIGURE 2) using hyperfiltered water and calculated according to Equation **(10)** corrected to 20°C (see TABLE 1):

$$CWF = \frac{\eta_t}{\eta_{20}} \times \frac{Q}{A I \Delta P}, \quad (10)$$

where Q is the clean water flow (L/h) at temperature t and A is the membrane surface area (PAN 13, $0.2\,\mathrm{m}^2$).

Feed waters tested were (1) a process water pretreated by coagulation, sedimentation, and filtration (WRK), (2) tap water, and (3) canal water. The canal water has

TABLE 1. Initial specific clean water flux of PAN 13 ultra-filtration module

Membrane Module	Initial Clean Water Flux (20°C, L/m²h bar)
A1	170
A2	165
A3	165
B1	155
C1	165
C2	165

quite a high turbidity (10–20 NTU) and was prefiltered through a 60-μ nylon mesh before use. For the MFI-UF test, the feed water was pumped to the UF module inlet at a constant transmembrane pressure (0.5–1.6 bar), using a pressure reducing valve. The reference pressure used in previous research was not obtained in this study due to limitations in pump capacity. The module was operated in dead-end flow. Flow was measured by a microoval flowmeter (flowmate LSN41) with convertor (DGHD1101). Measured data, time t, flow, and total volume V of filtered water, were recorded by computer with a specially designed software program at 1- or 5-min intervals. The MFI-UF was calculated, according to the following formula by the software program:

$$MFI - UF = \frac{\eta_{20}}{\eta_t} \frac{\Delta P}{\Delta P_2} \frac{A^2}{A_0^2} \frac{d(t/v)}{dt}. \quad (11)$$

The MFI-UF value (and standard deviation) reported represents the average value calculated from the stable region of the MFI-UF over time. The MFI-UF was corrected to a temperature of 20°C, a reference pressure of 2 bar (ΔP_2) and a reference surface area of the MFI (0.45 μm) microfilter (A_0) in Equation (11).

After a MFI-UF test the modules were back-washed (1.5 bar) with hyperfiltered water (ambient temperature) for 15 min followed by a chemical cleaning using a 200 ppm sodium hypochlorite solution, which was recirculated for one hour at 1 bar. Sodium hypochlorite was prepared from analytic grade reagent and hyperfiltered water. To remove any chemicals the modules were back-washed again before the specific CWF restoration was measured.

For the linearity and pressure experiments one module (A3) was employed and after a test the CWF was restored to 90% before subsequent testing. Linearity experiments were carried out with the MFI-UF determined for the most concentrated feed water first, after which the feed water was serially diluted and tested. The feed waters tested, tap water, WRK, and canal water, were collected and allowed to equilibrate to ambient temperature. The diluents, tap water (for canal water), were added to the calculated height of the 1 m³ test tank for the desired dilution, stirred, and allowed temperature equilibrate. Serial dilutions were prepared taking into account the previous dilution. Experiments for the feed waters were carried out at an applied transmembrane pressure (TMP) of 1 bar. In addition, for tap water, linearity experiments at applied TMP in the range of 0.5–10^6 bar were also carried out.

In the MFI-UF reproducibility experiments the module C2 was employed using tap water as the feed water in four repeated measurements. One test was carried out per day, and the module was cleaned to 90% CWF between tests. The tap water in a 1 m³ tank allowed to temperature equilibrate and was continuously stirred during testing. For the module reproducibility, tap water was collected and treated in the same way as for C2. Six modules with 100% CWF were tested using the collected tap water. In the reproducibility studies the applied TMP was 1.0 bar.

RESULTS AND DISCUSSION

Cake Filtration

The existing MFI and the MFI-UF under development are based on the assumption that cake part of the filtration time during measurement; that is, linearity is

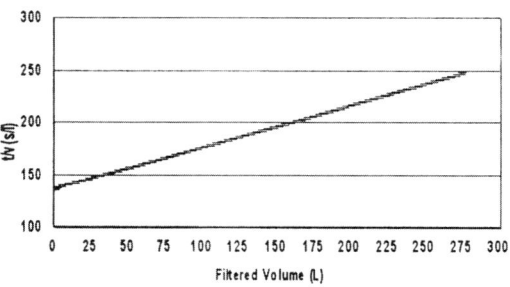

FIGURE 3. Filtration curve t/v versus V for Pan13 (UF) at 1.0 bar.

achieved in t/V versus V filtration curve. In previous research, cake filtration was shown to occur within about 1–2 h depending on the water quality, after which compression of the cake occurred.[3,5] An example of the behavior generally found for the PAN 13 (UF) filtration curve is shown in FIGURE 3. Linearity was found in the filtration curve for the PAN 13 modules, indicating that cake filtration occurs during the MFI-UF test.

MFI-UF Monitor Requirements

Stable MFI-UF Value

Examination of the behavior of the MFI-UF over time for WRK (see FIGURE 4), tap water, and diluted canal water (see FIGURE 5) showed that the MFI-UF was stable after about three hours and for the entire experiment duration. The more fouling water (30%) had a higher MFI-UF, $17,800 \pm 212 \,\text{sec/L}^2$, whereas the lower fouling waters had an MFI-UF of $1,860 \pm 75 \,\text{sec/L}^2$ (MAM) and $4,200 \pm 63 \,\text{sec/L}^2$ (tap water).

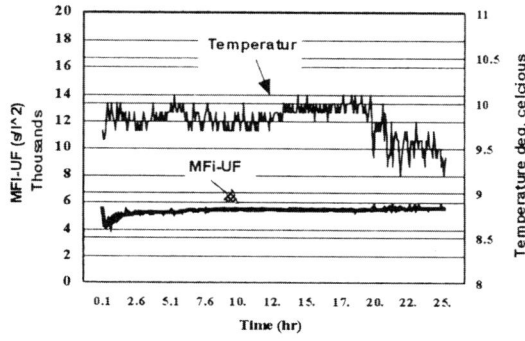

FIGURE 4. MFI-UF measured over time for tap-water at 1.0 bar.

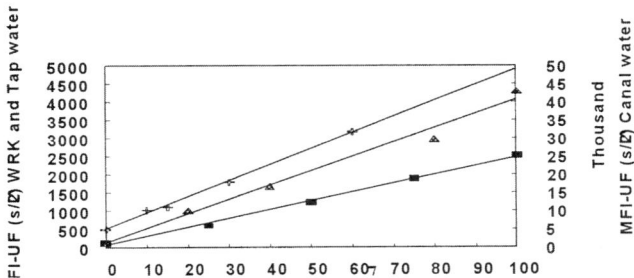

FIGURE 5. MFI-UF versus dilution of feed water for WRK ■, tap ▲, and canal + water at 1.0 bar.

Linearity of MFI-UF with Colloid Concentration

Additional evidence that cake filtration occurred during the MFI-UF test can be observed in the result of the MFI-UF value as a function of particulate concentration. This premise is based on the I value being directly related to the concentration of particles C_b. Linearity was found for all the test waters, the regression coefficients calculated for tap, WRK, and canal water were 0.998, 0.993, and 0.998, respectively. The MFI-UF for WRK water was found to be directly related to the dilution factor. Thus, the MFI-UF value at 25% dilution is $590 \pm 27 \sec/L^2$ and approximately double at 50%, $1,230 \pm 22 \sec/L^2$. In contrast, the MFI-UF of tap water and canal water were higher than expected at lower dilution; however, this may be due to an dilution error.

Another test was carried out with diluted tap water to investigate if linearity could be observed at various pressures. In this case, the MFI-UF values of tap water were found to be related to the dilution factor at three pressures and the linear regression coefficients calculated for the data were all greater than 0.995 (see TABLE 2). Thus, the MFI-UF was found to be linear with particulate content for all the feed waters tested and throughout the pressure range 0.5–1.6 bar, which further supports cake filtration as the operative filtration mechanism in the MFI-UF test.

TABLE 2. MFI-UF ± (standard deviation) versus dilution of tap water (with hyperfiltrated water) at varying TMP of feed water

Percent DTW	TMP of Feed Water (bar)		
	0.5	1.5	1.6
25	698 (18)	970 (12)	1,598 (8)
50	1,403 (28)	2,682 (37)	2,996 (29)
75	2,101 (54)	4,080 (39)	4,766 (96)
100	2,837 (65)	5,322 (61)	6,169 (27)
R^2	0.999	0.996	0.999

R^2, linear regression coefficient.

Reproducibility

The MFI-UF in four repeated measurements for the same feed water (tap water) and module show that the MFI-UF value is reproducible for the first two measurements. The value increases slightly (2.3%) in the third measurement and again in the fourth (5%). The standard deviation calculated from the MFI-UF values for the stable region of each test shows an increase, indicating that the MFI-UF is not quite as stable as the first two measurements.

Previously, the MFI-UF was found to increase in the first few tests for the same module when only back washing was employed between measurements. It was observed to become reproducible after several measurements. This was most likely due to the module stabilizing at an *effective* molecular weight cutoff value attributable to pore constriction through blocking.[3] However, in this study the modules were chemically cleaned to the same clean water flux and, therefore, no change in the *effective* MWCO value is expected. Despite the small increase in the MFI-UF value, the MFI-UF test is reproducible in the four repeated tests, as all values fall within the 95% confidence interval of the mean calculated for the four tests.

The average MFI-UF calculated for the six modules in the module manufacture reproducibility experiment was 2,968 ± 170. The standard deviation calculated was higher than for repeated testing of one module as can be expected considering the variation that can occur during membrane manufacture. Differences in the CWF of the modules (see TABLE 3) were found, which may be attributable to a change in porosity during manufacture and hence the R_m, Equation (**1**). However, no trend was observed in the MFI-UF value related to the module CWF. Similarly, the MFI-UF values measured for modules within a batch and between batches showed no clear trend. The C1 module shows a markedly lower MFI-UF values that the C2 modules from the same batch—about 10% difference. However, five modules fall within 95% confidence interval calculated for the mean of the six modules, indicating that membrane manufacture can be reproducible.

TABLE 3. Reproducibility of the MFI-UF test (module C2) in repeated tests with tap water and reproducibility of PAN 13 module manufacture in the MFI-UF test (six modules)

C2 Module (Test Number)	MFI-UF sec/L^2 (± σ)	Module	MFI-UF sec/L^2 (± σ)
1	2,534 (61)	A1	3,089 (69)
2	2,543 (85)	A2	3,165 (122)
3	2,602 (128)	A3	2,980 (17)
4	2,741 (138)	B1	2,960 (52)
—	—	C1	2,667 (22)
—	—	C2	2,948 (21)
Mean	2,605 (96)		2,948 (170)
95% CI	149		178

TABLE 4. I index value measured over time for tap water at various applied transmembrane pressures

Pressure (bar)	0.5	1.0	1.5	1.6
Fouling Index (I) (L/m^2)	1,800	2,800	4,000	4,750

Pressure Dependence of MFI-UF

A difference in pressure or the rate of flow is known to cause a difference in the specific cake resistance if the cake composed of compressible particles.[4,10] Treated surface water in the Netherlands may contain a relatively large amount of considerable hydrated colloids.[1] Thus, at higher applied pressure these colloids compress, forming a denser cake due to a reduction in porosity. The specific resistance of the cake increase from Equation (8) and, hence, MFI-UF (see TABLE 4) the I index values over time are obtained for undiluted tap water at various applied pressures.

No compression was found to occur within a test, indicated by the stability of the I value over time. However, in a subsequent test when the pressure is increased from 0.5 to 1.0 bar, the I value is circa 1.6 times higher, but remains constant over time. C_b can be assumed to be constant for all the tests, and the same module was employed to prevent any experimental artifacts occurring during the test. Therefore, if the cake were composed of incompressible particles, the same I value would have been found irrespective of the applied operating pressure. Cake compression observed in the MFI-UF test may also be a result of the increased flux to membrane as a result of the increased pressure. The structural arrangement and the extent of cake compression could therefore be the combined result of a pressure and flux increase.

CONCLUSIONS

The selected polyacrylonitrile module gave a stable and reliable value over time for high and low fouling feed waters. The MFI-UF value was demonstrated to be directly proportional to the concentration of particles in the feed waters for the range of applied test pressures. The linear region of the filtration curve t/V versus V and the linearity of the MFI-UF value with colloidal concentration proved that cake filtration occurs in the MFI-UF test. The module tested was found to be reusable with chemical cleaning and the measurement reproducible with repeated testing. Module manufacture within and between batches was found to be reproducible for 80% of the modules tested.

The fouling index I of the MFI-UF depends on pressure. A higher applied test pressure gave a higher I value. Cake compression most likely occurred due to a porosity reduction in the cake, which increased the specific resistance of the cake formed on the UF membrane.

REFERENCES

1. J.C. SCHIPPERS & J. VERDOUW. 1980. Desalination **32**: 137.

2. J.C. SHIPPERS, J.H. HANEMAAYER, C.A. SMOLDERS & A. KOSTERISE. 1981. Desalination **38:** 339.
3. S.F. BOERLAGE, M.D. KENNEDY, P.A.C. BONNE, *et al.* 1997. Proc. AWWA Membrane Technology Conference, New Orleans.
4. R.J. WAKEMAN. 1979. Second World Filtration Congress.
5. J.C. SCHIPPERS. 1989. KIWA Research and Advice. Ph.D. Thesis. N.V. the Netherlands. ISBN 90-9003055-7.
6. P.M. HEERTJES. Trans. Instn. Chem. Engineers.
7. P.C. CARMAN. 1938. Fundamental Principles of Industrial Filtration. Transaction Institution of Chemical Engineers.
8. H. MALLUBHOTLA & G. BELFORT. 1997. Science **125:** 75.
9. D.B. PURCHAS. 1977. Solid/Liquid Separation.
10. B.F. RUTH. 1935. Ind. Eng. Chem. **27:** 708.

Removal and Recovery of Metals and Other Materials by Supported Liquid Membranes with Strip Dispersion

W.S. WINSTON HO

Department of Chemical Engineering, Department of Materials Science and Engineering, Center for Materials Research, The Ohio State University, Columbus, Ohio, USA

ABSTRACT: This paper reviews recent advances in supported liquid membranes (SLMs) with strip dispersion for removal and recovery of metals including chromium, copper, zinc, and strontium; it also discusses potential applications of SLMs for removal and recovery of other materials, including cobalt and penicillin G. The technology for chromium that we developed, not only removes the Cr(VI) from about 100–1,000 ppm to less than 0.05 ppm in the treated effluent allowable for discharge or recycle, but also recovers the chromium product at a high concentration of about 20% Cr(VI) (62.3% Na_2CrO_4) suitable for resale or reuse. In other words, we have achieved the goals of zero discharge and no sludge. The stability of the SLM is ensured by a modified SLM with strip dispersion, where the aqueous strip solution is dispersed in the organic membrane solution in a mixer. The strip dispersion formed is circulated from the mixer to the membrane module to provide a constant supply of the organic solution to the membrane pores. The copper SLM system that we have identified, not only removed the copper from 150 ppm in the inlet feed to less than 0.15 ppm in the treated feed, but also recovered the copper at a high concentration of greater than 10,000 ppm in the strip solution. For the zinc SLM system identified, zinc at an inlet feed concentration of 550 ppm was removed to less than 0.3 ppm in the treated feed, whereas a high zinc concentration of more than 17,000 ppm was recovered in the strip solution. For strontium removal, we synthesized a family of new extractants, alkyl phenylphosphonic acids. The SLM removed radioactive ^{90}Sr to the target of 8 pCi/L or lower from feed solutions of 300–1,000 pCi/L. The SLM removed cobalt from about 525 ppm to 0.7 ppm in the treated feed solution, concentrating it to at least 30,000 ppm in the aqueous strip solution. Concerning penicillin G recovery, the SLM removed penicillin G from a feed of 8,840 ppm and concentrated it to a high concentration of 41,011 ppm in the aqueous strip solution with a high recovery of about 93%.

KEYWORDS: supported liquid membrane; strip dispersion; chromium; cobalt; copper; zinc; strontium; penicillin G

Address for correspondence: W.S. Winston Ho, Department of Chemical Engineering, Department of Materials Science and Engineering, Center for Materials Research, The Ohio State University, 291 Watts Hall, 2041 College Road, Columbus, OH 43210-1178, USA. Voice: 614-292-9970; fax: 614-292-3769.
ho@che.eng.ohio-state.edu

INTRODUCTION

Liquid membranes combine into a single step extraction and stripping processes; these are generally carried out in two separate steps in conventional processes, such as solvent extraction. A one-step liquid membrane process provides the maximum driving force for the separation of a target species, leading to the best cleanup and recovery of the species.[1]

There are two types of liquid membranes: (1) emulsion liquid membranes (ELMs) and (2) supported liquid membranes (SLMs). In ELMs, first invented by Li,[2] an emulsion acts as a liquid membrane for the separation of the target species from a feed solution. An ELM is created by forming a stable emulsion, such as a water–in–oil emulsion, between two immiscible phases, followed by dispersion of the emulsion into a third, continuous phase by agitation for extraction. The membrane phase is the oil phase that separates the encapsulated, internal aqueous droplets in the emulsion from the external, continuous phase.[1–3] The organic membrane phase contains a species-extracting agent, a diluent, which is usually an inert organic solvent, and sometimes a modifier. The internal aqueous phase contains a stripping agent. Emulsions formed from these two phases are generally stabilized by use of a surfactant in the membrane phase. The external, continuous phase is the aqueous feed solution containing the target species. The target species is extracted from the aqueous feed solution into the membrane phase and then stripped into aqueous droplets in the emulsion. The target species can then be recovered from the internal aqueous phase by breaking the emulsion, typically via electrostatic coalescence, followed by electroplating or precipitation.

One disadvantage of ELMs is that the emulsion swells on prolonged contact with the feed stream.[1,4] This swelling causes a reduction in the stripping reagent concentration in the aqueous droplets and, thus, reduces stripping efficiency. It also results in dilution of the target species that has been concentrated in the aqueous droplets, resulting in lower separation efficiency of the membrane. The swelling further results in a reduction in membrane stability by making the membrane thinner. Finally, swelling of the emulsion increases the viscosity of the spent emulsion, making it more difficult to demulsify. A second disadvantage of ELMs is membrane rupture, resulting in leakage of the contents of the aqueous droplets into the feed stream and a concomitant reduction of separation efficiency.[1,4] Wiencek and his coworkers[5–7] have described the use of microporous hollow-fiber contactors as an alternative contacting method to direct dispersion of ELMs thereby minimizing membrane swelling and leakage. This is due to the fact that the hollow-fiber contactors do not have the high shear rates typically encountered with the agitators used in the direct dispersion.

In SLMs, the liquid membrane phase is the organic liquid imbedded in pores of a microporous support; for example, microporous polypropylene hollow fibers.[1] When the organic liquid contacts the microporous support it readily wets the pores of the support and the SLM is formed. For the extraction of a target species from an aqueous feed solution, the organic-based SLM is placed between two aqueous solutions, the feed solution and the strip solution, where the SLM acts as a semipermeable membrane for the transport of the target species from the feed solution to the strip solution. The organic in the SLM is immiscible in the aqueous feed and strip streams and contains an extractant, a diluent which is generally an inert organic solvent, and sometimes a modifier.

In both types of liquid membranes, facilitated transport is the mass transfer mechanism for the target species to go from the feed solution to the strip solution.[1,3,4,8–11] Quinn[8,9] and Li[10,11] pioneered the facilitated transport mechanism and laid a solid foundation for this mechanism. Shown in FIGURE 1 for this mechanism using copper as an example, the extractant (represented by HA, typically in the dimer form of $[HA]_2$) in the organic membrane phase reacts with Cu^{2+} at the interface between the aqueous feed solution and the organic membrane phase to form the copper–extractant complex, $CuA_2[HA]_2$. The complex diffuses from this interface (which has a higher complex concentration than the interface adjacent to the aqueous strip solution), through the organic membrane phase, to the interface between the organic membrane phase and the aqueous strip solution, where the decomplexation/stripping reaction occurs and the stripping agent (a strong acid, e.g., H_2SO_4 or HCl) in the aqueous strip solution strips Cu^{2+} into the aqueous strip solution. The stripping reaction also regenerates the extractant, which diffuses across the membrane phase (the extractant concentration at the interface adjacent to the aqueous strip solution is higher than that at the interface adjacent to the aqueous feed solution) and back to the interface adjacent to the aqueous feed solution, to complete the facilitated transport cycle. Thus, the extractant is also called "carrier" in that it carries Cu^{2+} through the organic membrane phase.

The use of supported liquid membranes (SLMs) for the removal of metals, including chromium,[1,12–19] copper,[1,20–32] zinc,[1,24,29,33–41] strontium,[1,42–49] and cobalt[1,50–56] from aqueous solutions and waste waters has long been pursued by the scientific and industrial community. The use of SLMs for the recovery of benzylpenicillin,[57] penicillin G,[58–60] phenylalanine,[61–63] lactic acid,[64–69] citric acid,[67,70–72] propionic acid,[65,73] and butanoic acid[65,73] from aqueous solutions and fermentation broths has also long been pursued by the scientific and industrial community.

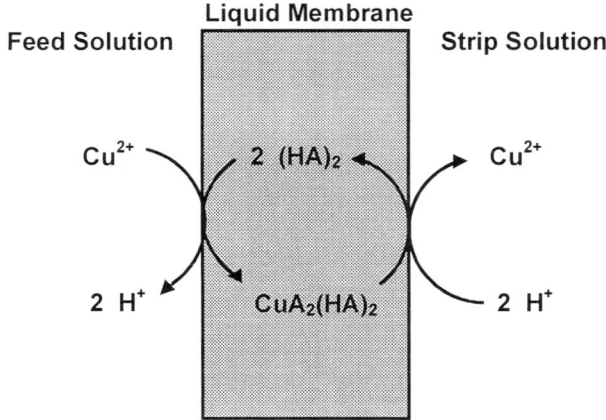

FIGURE 1. Facilitated transport mechanism for copper transfer across a liquid membrane.

Although a SLM process is very effective for the removal of trace contaminants to very low levels due to its ability to circumvent equilibrium limitations,[1] its use has been hampered by its stability. The traditional SLM suffers from a gradual loss of the organic membrane phase to the aqueous feed and strip solutions, due to emulsification (e.g., resulting from lateral shear forces) at the membrane–aqueous interfaces, and to the osmotic pressure difference across the membrane.[74–76] The osmotic pressure difference displaces the organic membrane phase from the micropores of the support. Displacement of the organic membrane phase from the pores can ultimately result in mixing of the feed and strip solutions, leading to complete failure of the separation unit.

This paper reviews and describes recent advances in the supported liquid membranes (SLMs) with strip dispersion that have solved the stability problem and that offer the new SLM technology we have developed recently for removal and recovery of metals, including chromium, copper, zinc, and strontium, from waste waters. This paper also discusses potential applications of the SLMs for removal and recovery of other materials, including cobalt and penicillin G.

SUPPORTED LIQUID MEMBRANE WITH STRIP DISPERSION

The supported liquid membrane (SLM) system we developed uses a *SLM with strip dispersion*,[77–79] which is shown schematically in FIGURE 2. As shown in this figure, an aqueous strip solution is dispersed in an organic membrane solution containing an extractant or extractants in a mixer, and the water–in–oil dispersion formed is then pumped into a membrane module to contact with one side of a microporous support (which is passed through the shell side of a microporous polypropylene hollow-fiber module, for example). The aqueous feed solution containing a target species to be extracted is on the other side of the support (which

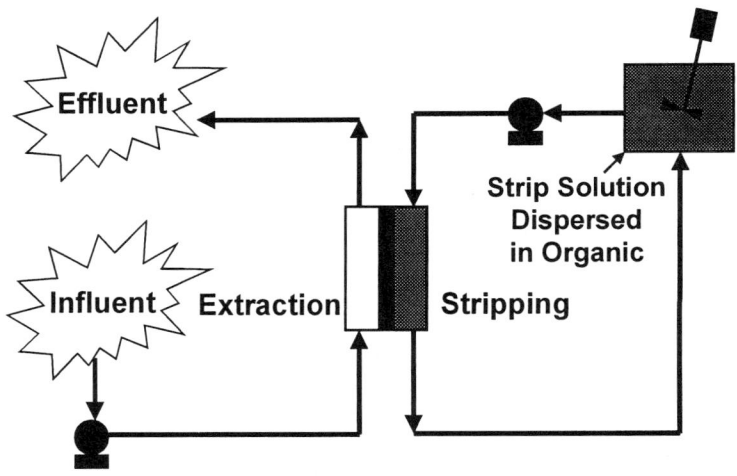

FIGURE 2. A schematic diagram of the supported liquid membrane with strip dispersion.

is passed through the other side of the fibers, the tube side, for example). The continuous organic phase of the dispersion readily wets the pores of a hydrophobic microporous support (e.g., microporous polypropylene hollow fibers in the module), and a stable liquid membrane (the organic phase) supported in the pores of the microporous support is formed.

FIGURE 3 shows an enlarged view of the SLM with strip dispersion. A low pressure differential (about 13.8 kPa [2 psi]) between the aqueous feed solution side (P_a) and the strip dispersion side (P_o) is applied to prevent the organic solution of the strip dispersion from passing through the pores to come into the feed solution side. The dispersed droplets of the aqueous strip solution in a typical size of 80–800 micrometers are orders of magnitude larger than the pore size of the microporous polypropylene support employed for the SLM, which is 0.03 micrometer. Thus, these droplets are retained on the strip dispersion side and cannot pass through the pores to pass into the feed solution side.

This SLM may be considered as a SLM with a constant supply of the organic membrane solution into the pores. This ensures a stable and continuous operation. Following the approach of including organic membrane solution in the strip side, Teramoto et al.[80] stabilized their SLM by adding their organic membrane solution in the strip side. In addition, for the SLM with strip dispersion, the direct contact between the organic and strip phases (with high-shear mixing if necessary) on the strip dispersion side provides an additional mass transfer area between these two phases for stripping, in addition to the hollow-fiber surface area, resulting in very efficient mass transfer for stripping. The stripping is more efficient than that with the conventional SLM.

The SLM supported in a hollow-fiber module has a very large mass transfer area. For example, using microporous polypropylene hollow fibers of 300-micrometer outside diameter, a commercial-size module of 25.4 cm (10 inches) in diameter by 71.1 cm (28 inches) in length has a surface area of 135 m², a pilot plant module of 10.2 cm (4 inches) in diameter by 71.1 cm (28 inches) in length possesses a surface

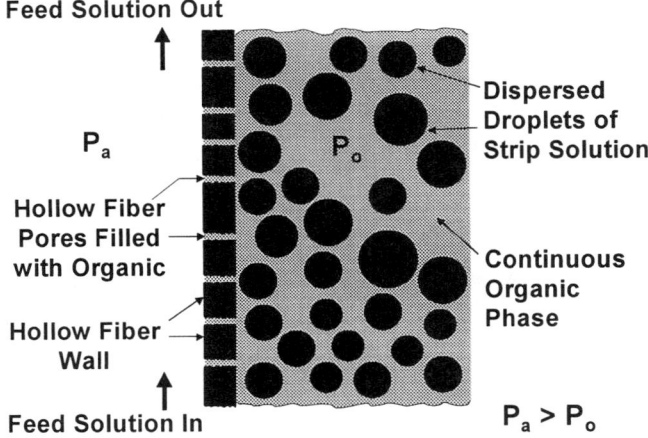

FIGURE 3. An enlarged view of the supported liquid membrane with strip dispersion.

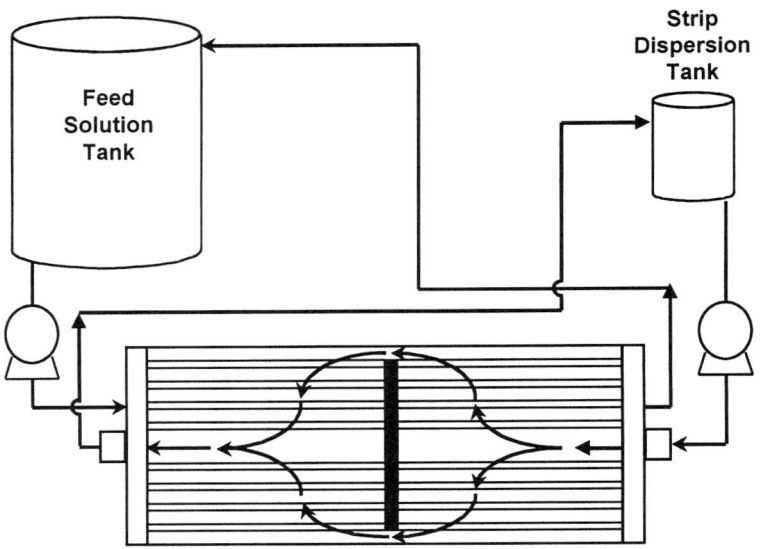

FIGURE 4. A supported liquid membrane system with a recycle feed and strip operation.

area of $19\,m^2$, and a laboratory module of 6.35 cm (2.5 inches) in diameter by 20.3 cm (8 inches) in length provides a surface area of $1.4\,m^2$.

A typical SLM system consists of a SLM hollow-fiber module (or a series of SLM modules), a feed solution vessel, a feed pump, a strip dispersion vessel, a mixer for the strip dispersion vessel (for dispersing an aqueous strip solution in an organic solution; this may be omitted under pumping and flow conditions), and a strip dispersion pump. FIGURE 4 shows a schematic diagram of a SLM system with a recycle mode of operation for the feed solution. The feed solution can also be operated in a single-pass mode. The strip dispersion is typically in the recycle mode. The configuration of the SLM module is similar to that of a shell-and-tube heat exchanger. Upon the completion of the removal of the target species, the mixer for the strip dispersion is turned off, and the dispersion separates into the two phases, the organic solution and the concentrated strip solution, upon standing. Phase separation is very fast (less than about one minute), and there is no formation of an emulsion. The concentrated strip solution can be the product of this process.

CHROMIUM REMOVAL AND RECOVERY

Organic Membrane Solution

For chromium removal and recovery, the organic membrane solution of the SLM used for both laboratory experiments and field demonstration was composed of 10 wt.% (0.213 M) of N-lauryl-N-trialkylmethylamine with a molecular weight of 372 (a total number of 25.3 carbon atoms per amine molecule; e.g., Amberlite

LA-2), 1 wt.% of 1-dodecanol (modifier for the extractant), and 3 wt.% of PLURONIC L31 (a block copolymer of ethylene oxide and propylene oxide, additive to enhance the phase separation of the aqueous strip solution from the organic membrane solution in the absence of agitation for the recovery of the strip solution) in Isopar L (an isoparaffinic hydrocarbon solvent with a flash point of 62°C, a boiling point of 207°C, a viscosity of 1.5 cp at 25°C, and a density of 0.767 g/mL at 15.6°C).[78,81]

Facilitated Transport with Chemical Reactions

The supported liquid membrane (SLM) containing the secondary amine (R_2NH) effectively removes Cr (VI), for example, chromic acid, from waste waters or aqueous solutions under acidic conditions (about pH 1.5) according to the extraction reaction

$$R_2NH + H^+ + HCrO_4^- \rightarrow (R_2NH_2)HCrO_2. \qquad (1)$$

Stripping using sodium hydroxide takes place as follows:

$$(R_2NH_2)HCrO_4 + 2Na^+ + 2OH^- \rightarrow R_2NH + 2Na^+ + CrO_4^{2-} + 2H_2O. \qquad (2)$$

The facilitated transport mechanism for the chromium (VI) transfer across the supported liquid membrane is shown schematically in FIGURE 5.

Chromium Flux

The flux of Cr (VI) through the SLM as a function of Cr (VI) concentration in the feed is shown in FIGURE 6.[78,81] As shown in this figure, a carrier saturation phenomenon with a constant flux for the facilitated transport mechanism becomes evident when the feed Cr (VI) concentration is 1,500 ppm or greater. This flux was about 12.2 g/m²/h, which is quite high. From this flux, the overall mass transfer coefficient for chromium across the SLM was 2.3×10^{-4} cm/sec at a feed Cr (VI) concentration of 1,500 ppm.

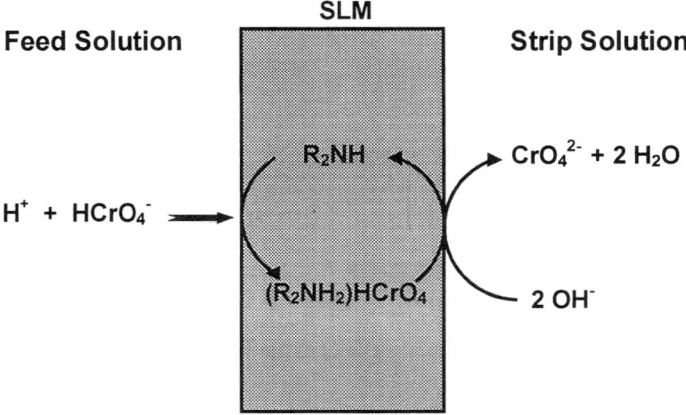

FIGURE 5. Mechanism of chromium (VI) transfer across the supported liquid membrane.

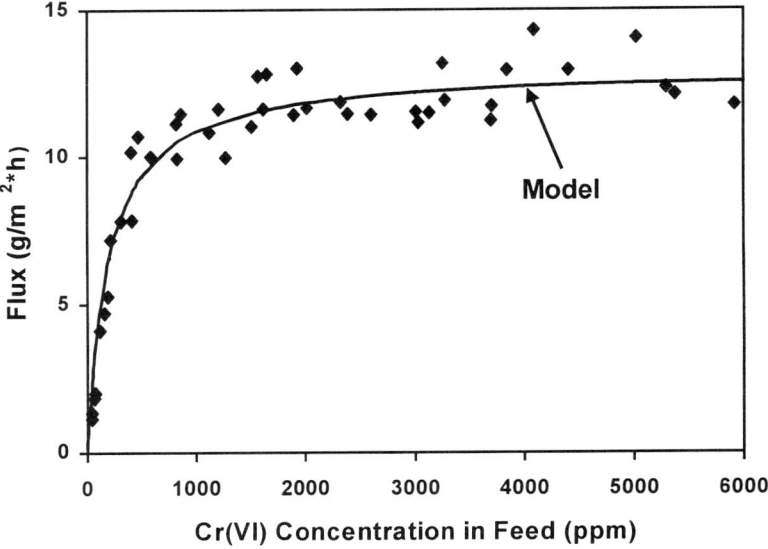

FIGURE 6. Chromium (VI) flux as a function of chromium (VI) concentration in the feed.

Model Prediction of Chromium Flux

The flux predicted from the model developed as a function of the chromium concentration in the aqueous feed solution is shown in FIGURE 6 in comparison with the experimental data.[78,81] As shown, the model prediction is in good agreement with the data for a very wide range of aqueous feed Cr (VI) concentrations, from approximately 20 to 6,000 ppm.

Sulfuric Acid Flux

Sulfuric acid was used for the pH adjustment of the aqueous feed solution (pH 1.5). However, the supported liquid membrane also removes sulfuric acid in the similar mechanism,

$$2R_2NH + 2H^+ + SO_4^{2-} \rightarrow (R_2NH_2)_2SO_4 \qquad (3)$$

$$(R_2NH_2)_2SO_4 + 2Na^+ + 2OH^- \rightarrow 2R_2NH + 2Na^+ + SO_4^{2-} + 2H_2O. \qquad (4)$$

Thus, sulfuric acid competes with chromic acid in the feed solution for complexation with the amine in the membrane phase. The extraction of sulfuric acid into the strip solution is not desirable since it degrades the quality of chromate recovered in the strip solution. However, we unexpectedly found that the flux of sulfuric acid depends on the Cr (VI) concentration in the feed. The concentration of sulfuric acid in the feed was approximately 1,500 ppm at feed pH 1.5. FIGURE 7 shows the sulfuric acid flux as a function of Cr (VI) concentration in the feed at pH 1.5.[78,81] As shown, the sulfuric acid flux is reduced significantly as the Cr (VI) concentration in the feed

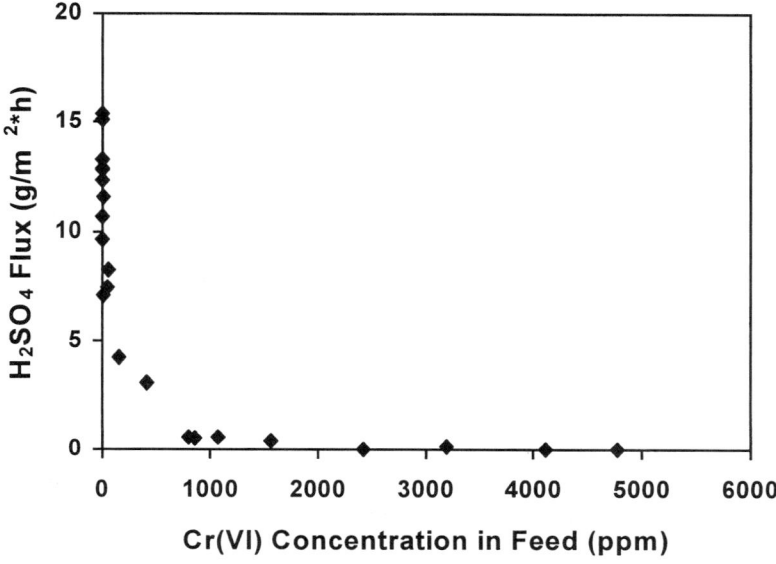

FIGURE 7. Sulfuric acid flux as a function of Cr(VI) concentration in the feed.

increases. The sulfuric acid flux for a feed Cr (VI) concentration of greater than about 100 ppm is much less than that for a lower concentration. In other words, the chromium/sulfuric acid selectivity for a feed Cr (VI) concentration of greater than about 100 ppm is much higher than that for a lower concentration.

The Chromium Removal and Recovery Process

We developed a novel chromium removal and recovery process based on the new unexpected findings described above, where the sulfuric acid flux for a feed Cr (VI) concentration of greater than about 100 ppm is much less than that for a lower concentration; that is, the chromium/sulfuric acid selectivity for a feed Cr (VI) concentration of greater than about 100 ppm is much higher than that for a lower concentration.[78,81] The novel process comprises two SLM steps: (1) a feed solution containing hexavalent chromium, Cr (VI), is treated to decrease the chromium to an acceptable level, 0.05 ppm Cr (VI) or lower, for discharge or recycle with an aqueous strip solution (Strip 1), which results in a moderately concentrated Cr (VI) solution (1,000–6,000 ppm) and (2) the aqueous strip solution, after its phase separation from the continuous, organic phase and pH adjustment to 1.5, is then processed to decrease Cr (VI) to about 100 ppm or greater, a concentration similar to that in the feed solution in Step 1, for recycling back to the feed solution in Step 1, by the use of a new aqueous strip solution (Strip 2) that results in a highly concentrated Cr (VI) solution (about 200,000 ppm). This process is shown schematically in FIGURE 8. This two-step process allows for the production of a highly concentrated Cr (VI) solution with low sulfate concentration from treating a moderately concentrated Cr (VI) solution with a Cr (VI) concentration of about 100 ppm or greater during the second step of this process.

In this process, the highly concentrated Cr (VI) solution with low sulfate concentration is a product that is suitable for reuse or resale. Therefore, this process not only solves the environmental problem, but also recovers the chromium. In each SLM step of this process, the organic phase, after its phase separation from the aqueous strip solution, is reused to make the strip dispersion with the fresh aqueous strip solution.

We demonstrated Step 1 of this process by the use of two commercial-size hollow-fiber modules, each with 25.4 cm (10 inches) in diameter by 71.1 cm (28 inches) in length. Each module (135 m^2) contained 225,000 microporous polypropylene hollow fibers. The feed solution, at a flow rate of 9.46 liters/min (2.5 gallons/min), was passed through the tube side in the once-through, single-pass mode of operation. The strip dispersion, composed of the aqueous sodium hydroxide solution (3 M) dispersed in the organic membrane solution at a volume ratio of 1:2, was circulated at a flow rate of 15.1 liters/min (4 gallons/min) in the shell side. The Cr (VI) concentration in the feed solution was reduced from 100 ppm to less than about 4 ppm from the first module, and it was decreased to less than 0.05 ppm on the effluent from the second module. The Cr (VI) concentration in the aqueous strip solution was concentrated to at least 3,000 ppm (existing as sodium chromate), which still made it possible to meet the specification of 0.05 ppm in the effluent.

After the feed solution for the Step 2 of this process was adjusted to pH 1.5, it was treated in a SLM module in a recycle mode of operation, in which the feed solution was circulated through the tube side of the module. The Cr (VI) concentration in the treated feed solution was reduced to about 100 ppm for recycling to be a part of the feed for the Step 1 of this process, accounting for less than 4% of the total Step 1 feed. The strip dispersion, consisting of an aqueous 12.5 M sodium hydroxide solution dispersed in the same organic membrane solution described previously, with an

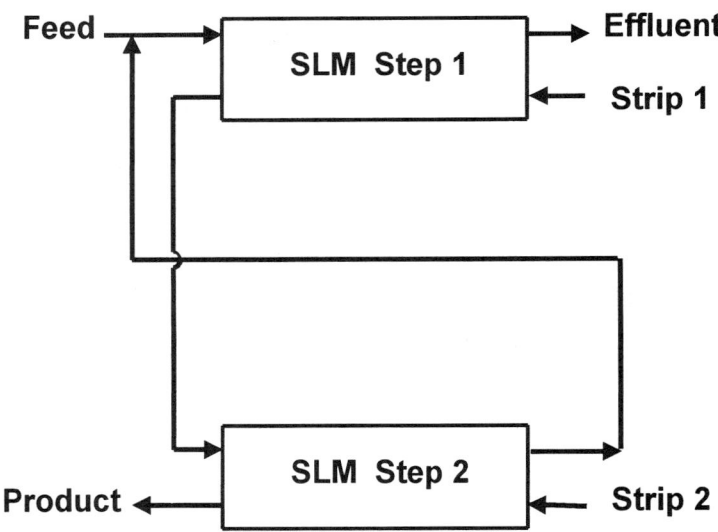

FIGURE 8. Schematic of chromium removal and recovery process comprising two supported liquid membrane steps.

adjustable, low-volume ratio of 1:4, was circulated in the shell side also in the recycle mode of operation. The adjustable, low-volume ratio of 1:4 was achievable to minimize the aqueous strip solution volume and, thus, maximize the chromium concentration in the strip solution and to maximize the strip dispersion volume to fill the holdup volume in the shell side. Cr (VI) concentrations in the aqueous strip solutions were increased to 170,000–200,000 ppm (200,000 ppm Cr (VI) is equivalent to 62.3% sodium chromate) for the Step 2 feed concentrations of 2,500–5,600 ppm Cr(VI). The sulfate concentrations in the aqueous strip solutions were less than 8,000 ppm (much lower than the typical specification of 100,000 ppm). The aqueous strip solutions were suitable for resale or reuse.

COPPER REMOVAL AND RECOVERY

Facilitated Transport of Metal Ions across a Supported Liquid Membrane

The transport of metal ions in a supported liquid membrane (SLM) is shown schematically in FIGURE 1 using copper as an example, where the extractant is represented by HA, typically in the dimmer form $(HA)_2$, in the organic membrane phase. The metal ion (M^{2+}) in the aqueous feed solution forms a complex with the extractant in the organic membrane phase at the interface between these two bulk phases as follows:[1,3,4,8–11]

$$M^{2+} + 2(HA)_2 \rightarrow MA_2(HA)_2 + 2H^+. \qquad (5)$$

At the interface between the organic phase and the aqueous strip solution, the metal ion is stripped into the aqueous strip solution containing a strong acid, for example, sulfuric acid, according to the following stripping reaction:

$$MA_2(HA)_2 + 2H^+ \rightarrow M^{2+} + 2(HA)_2. \qquad (6)$$

Laboratory Results

From Equations (5) and (6), the net result of metal ion removal from the aqueous feed solution into the aqueous strip solution is the transfer of proton ions from the strip solution into the feed solution. Thus, the pH of the feed solution is reduced as metal is removed. This has a significant effect on module utilization efficiency for supported liquid membrane (SLM) systems, and an SLM system operating at a low feed pH of about 2 is highly desirable in order to attain a module utilization efficiency of 100%.[82]

We recently developed a SLM system with strip dispersion effective at the low feed pH of about 2 for copper removal and recovery from waste waters and process streams.[77,82–85] The organic membrane solution in the strip dispersion comprised 15 wt.% LIX 973N (about 46% nonylsalicyl aldoxime, 18% ketoxime, 6% nonylphenol, and 30% diluent), 2 wt.% 1-dodecanol, and 83 wt.% *n*-dodecane. The aqueous strip solution was 3 M sulfuric acid. A high volume ratio of 19 between the organic solution (950 mL) and the strip solution (50 mL) was used to concentrate the copper in the strip solution. The SLM system treated the 5-liter feed solution of the Berkeley Pit water containing both copper and zinc (about 150 ppm copper and 550 ppm zinc after iron removal with pH ≈ 4.3 adjustment) in the recycle mode of operation as shown in FIGURE 4.

As shown in FIGURE 9, the SLM selectively removed the copper to less than 0.1 ppm in the treated feed solution in less than about 90 min in laboratory experiments.[77,82–85] It also concentrated copper in the aqueous strip solution from about 4,800 parts per million by weight (ppm) to about 11,000 ppm. From this figure, a copper flux of about $2 g/m^2/h$ was determined at a copper concentration of about 110 ppm. After the copper removal, the zinc concentration in the treated Berkeley Pit water remained at its original value of about 550 ppm. Zinc removal is discussed below.

Pilot Plant/Scaleup Results

The SLM system was recently scaled up in a pilot plant unit using two 10.2-cm (4-inch) diameter microporous polypropylene hollow-fiber modules in parallel with 16.7 L/min (4.4 gal/min) total feed flow; that is, 8.3 L/min (2.2 gal/min) per module, for the selective removal and recovery of copper from the Berkeley Pit water containing copper, zinc, and the other metals after iron removal with pH (≈4.3) adjustment. In a three-hour pilot plant run with a feed volume of 129.8 L (34.3 gallons), the copper in the feed was removed from 152 ppm to 0.08 ppm for a run time of two hours and to less than 0.07 ppm for the three-hour run time, whereas it was concentrated to 3,700 ppm in the aqueous strip solution for a run time of two hours.

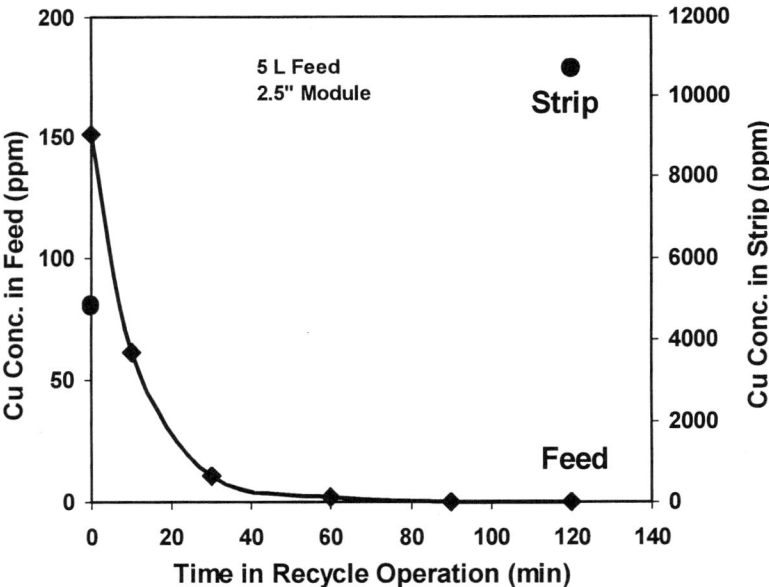

FIGURE 9. Copper concentrations in feed and strip solutions as a function of time in recycle operation from the laboratory experiment.

FIGURE 10 shows the copper concentrations in the aqueous feed and strip solutions as a function of time in recycle operation.[84]

As demonstrated in FIGURES 9 and 10, the scaleup results from pilot plant runs agreed reasonably well with the laboratory results, particularly for the copper concentrations in the treated feed solutions—less than 0.1 ppm achieved from both the laboratory and pilot plant runs. Treated feed solutions with such a low copper concentration would be allowed for discharge in terms of the copper concentration. The copper concentration in the aqueous strip solution recovered from the pilot plant run was in line with that from the laboratory experiment under their operating conditions. The copper concentration from the pilot plant run was about 3,700 ppm, and it was high enough for reuse, including its use in electrowinning for copper metal recovery. Thus, the goals of zero discharge and no sludge are achievable.

From the extraction results shown in FIGURES 9 and 10, the overall mass transfer coefficients for copper across the SLM were calculated.[84] The overall mass transfer coefficient results from the laboratory experiment agreed reasonably well with those from the pilot plant run. The average overall mass transfer coefficient from the laboratory experiment was 5.03×10^{-4} cm/sec, and that from the pilot plant run was 4.12×10^{-4} cm/sec. The pilot plant result was about 18% lower than the laboratory result. This was presumably due to the scale-up factor, the maldistribution of hollow fibers in the larger pilot plant module. The maldistribution is an issue that needs attention in the scaleup to a commercial-size module.

In addition to the analyses of the copper concentrations in both the aqueous feed and strip solutions, the solutions were also analyzed for the other metals, zinc, cobalt, nickel, aluminum, manganese, and cadmium. There were no significant

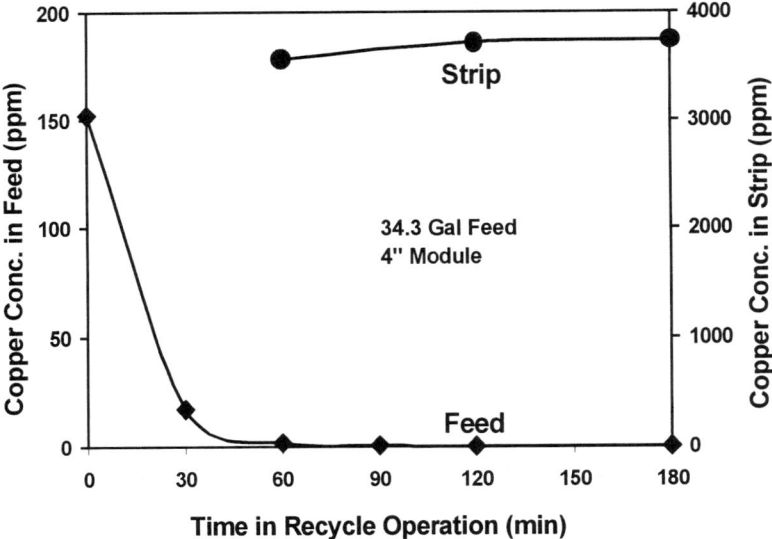

FIGURE 10. Copper concentrations in feed and strip solutions as a function of time in recycle operation from the pilot plant/scaleup run.

changes in the concentrations of the other metals in either the aqueous feed and strip solutions before and after the pilot plant run. In other words, there were no significant contaminants in the concentrated strip solution recovered; that is, the strip solution product had high quality. This also demonstrated the selective removal and recovery of copper from the feed solution containing the contaminant metals, and reinforced the reuse opportunity of the strip solution recovered.

In both the laboratory and pilot runs, the phase separation for the strip dispersion was very satisfactory. Upon the completion of the removal of the target species, copper, the mixer for the strip dispersion was turned off, and the dispersion separated into the two phases, the organic solution and the concentrated strip solution, upon standing. Phase separation was very fast (less than about one minute), and there was no formation of an emulsion. There was no physical loss of the organic membrane solution except its solubility in the aqueous solution that was very low due to the use of the isoparaffinic hydrocarbon solvent. Thus, the stability of the SLM with strip dispersion appeared to be satisfactory for a period of several weeks.

ZINC REMOVAL AND RECOVERY

Laboratory Results

We also developed a SLM system with strip dispersion effective at low feed pH (about 2) for zinc removal and recovery from waste waters and process streams.[77,82–85] The organic membrane solution in the strip dispersion consisted of 8 wt.% di(2,4,4-trimethylpentyl) dithiophosphinic acid (Cyanex 301), 2 wt.% dodecanol, and 90 wt.% Isopar L. The aqueous strip solution was 3 M sulfuric acid. A relatively high volume ratio of about 5.7 between the organic solution and the strip solution was used to concentrate zinc in the strip solution. The SLM system treated the 5-liter feed solution of the Berkeley Pit water after copper removal containing about 550 ppm zinc in the recycle mode of operation as shown in FIGURE 4.

FIGURE 11 shows that SLM removed the zinc to less than 0.3 ppm in the treated feed solution in less than about 120 min.[77,82–85] It also concentrated zinc in the aqueous strip solution from about 4,600 ppm to about 18,000 ppm. From this figure, a zinc flux of about $3.2 \text{g/m}^2/\text{h}$ was determined at a zinc concentration of about 480 ppm.

Pilot Plant/Scaleup Results

The treated Berkeley Pit water remained at a pH value of about 2 after copper removal, and it was used for the removal and recovery of zinc at this pH. The pilot plant runs for zinc used one 10.2-cm (4-inch) diameter module with the feed flow rate of 3.8 L/min (1 gal/min). In a five-hour pilot plant run with a feed volume of 121.1 L (32 gallons), the zinc in the feed was removed from 526 ppm to 0.65 ppm for a run time of four hours and to less than 0.62 ppm for the five-hour run time, whereas it was concentrated to 20,000 ppm in the strip for a run time of four hours. FIGURE 12 shows the zinc concentrations in the aqueous feed and strip solutions as a function of time in recycle operation.[84]

As depicted in FIGURES 11 and 12, the scaleup results from pilot plant runs agreed reasonably well with the laboratory results in terms of the zinc concentrations in the

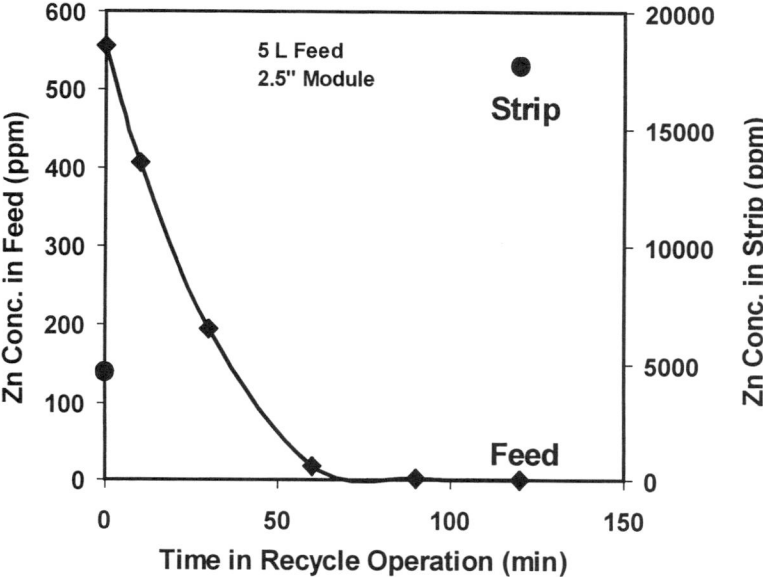

FIGURE 11. Zinc concentrations in feed and strip solutions as a function of time in recycle operation from the laboratory experiment.

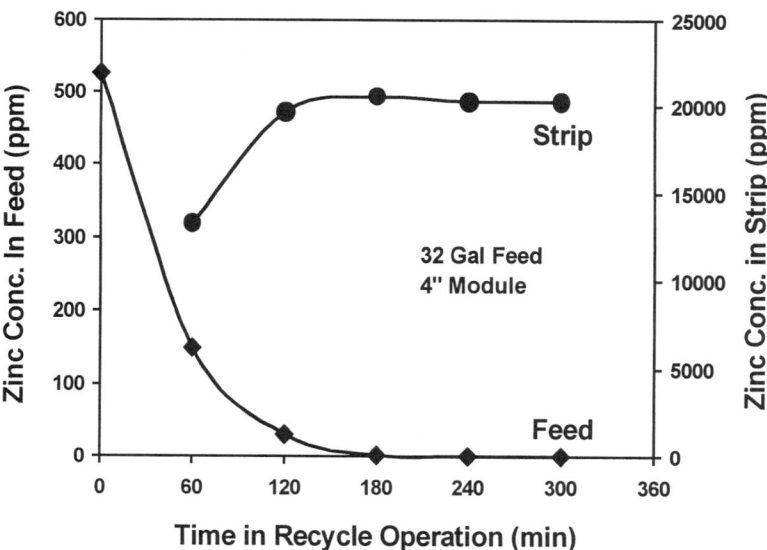

FIGURE 12. Zinc concentrations in feed and strip solutions as a function of time in recycle operation from the pilot plant/scaleup run.

treated feed and concentrated strip solutions. The zinc concentration in the aqueous strip solution recovered from the pilot plant run was in good agreement with that from the laboratory experiment. The zinc concentration was more than 17,000 ppm, and it was high enough for reuse, including its use in electrowinning for zinc metal recovery. The zinc concentration in the treated feed solution from the pilot plant run was in line with that from the laboratory experiment under those operating conditions. The treated feed solutions with such a low zinc concentration (0.3–0.65 ppm) would be allowed for discharge in terms of the zinc concentration. Thus, the goals of zero discharge and no sludge would be achievable.

From the extraction results shown in FIGURES 11 and 12, the overall mass transfer coefficients for zinc across the SLM were determined.[84] The overall mass transfer coefficient results from the laboratory experiment agreed reasonably well with those from the pilot plant run. This was consistent with the copper results described above. The average overall mass transfer coefficient from the laboratory experiment was 3.37×10^{-4} cm/sec, and that from the pilot plant run was 2.79×10^{-4} cm/sec. The pilot plant result was about 17% lower than the laboratory result. Again, this was presumably due to the scaleup factor, the maldistribution of hollow fibers in the larger pilot plant module.

In addition to the analyses of the zinc concentrations in both the aqueous feed and strip solutions, the solutions were also analyzed for the other metals, cobalt, nickel, aluminum, manganese, and cadmium. There were no significant changes in the concentrations of aluminum, manganese, and cadmium in either the aqueous feed or strip solutions before and after the pilot plant run. However, about 35% of the cobalt and 25% of the nickel extracted from the feed solution were transferred to the aqueous strip solution. The cobalt and nickel were about 0.1 wt.% and 0.04 wt.% of the zinc concentration respectively in the aqueous strip solution recovered. Thus, there were no significant contaminants in the concentrated strip solution recovered; that is, the strip solution product had high quality. This also demonstrated the selective removal and recovery of zinc from the feed solution containing the contaminant metals. This also reinforced the reuse opportunity of the strip solution recovered.

The phase separation for the strip dispersion and the stability of the SLM with strip dispersion in both the laboratory and pilot runs for zinc were similar to those for copper described earlier. Both the phase separation and the stability were satisfactory.

STRONTIUM REMOVAL

Non-Radioactive ^{87}Sr Removal

We have synthesized a family of new extractants, alkyl phenylphosphonic acids, for strontium removal by supported liquid membranes (SLMs) with strip dispersion.[77,79] The extractants synthesized were 2-butyloctyl phenylphosphonic acid (C12 BOPPA), 2-hexyldecyl phenylphosphonic acid (C16 HDPPA), 2-octyldecyl/2-hexyldodecyl phenylphosphonic acid (C18 ODPPA/HDPPA), and 2-octyldodecyl phenylphosphonic acid (C20 ODPPA).

Laboratory SLM experiments were carried out in the recycle mode of operation (shown in FIG. 4) for both a feed solution and a strip dispersion. The strip dispersion was prepared by dispersing an aqueous strip solution of 1 M HCl in the organic membrane solution at an organic-to-strip solution volume ratio of 3. Each of the organic membrane solutions of the SLMs used for the experiments comprised 8 wt.% of one of the four alkyl phenylphosphonic acids and 2 wt.% of 1-dodecanol (modifier for the extractant) in n-dodecane.

In the experiments, all the four extractants synthesized were used for strontium removal from aqueous feed solutions containing about 5.5 ppm of non-radioactive ^{87}Sr at pH 3. FIGURE 13 shows the results for all of these four extractants by the use of 1 M HCl as the stripping solutions.[77,79] As shown, all of these four extractants removed strontium from the aqueous feed solutions and recovered it in the aqueous strip solutions. The strontium concentration in the aqueous strip solutions was about 40 ppm, which was nearly the maximum concentration achievable for a feed-to-strip solution volume ratio of 8. Thus, the stripping with hydrochloric acid was very satisfactory and nearly quantitative.

From FIGURE 13, an overall mass transfer coefficient of 1.7×10^{-4} cm/sec was determined for strontium across the SLM. From this mass transfer coefficient, a strontium flux of about $0.03 \text{ g/m}^2/\text{h}$ was obtained at a strontium concentration of 3 ppm. It should be noted that the aqueous feed solutions also contained about 80 ppm calcium, 20 ppm magnesium, and 50 ppm zinc in addition to strontium to simulate ground water.

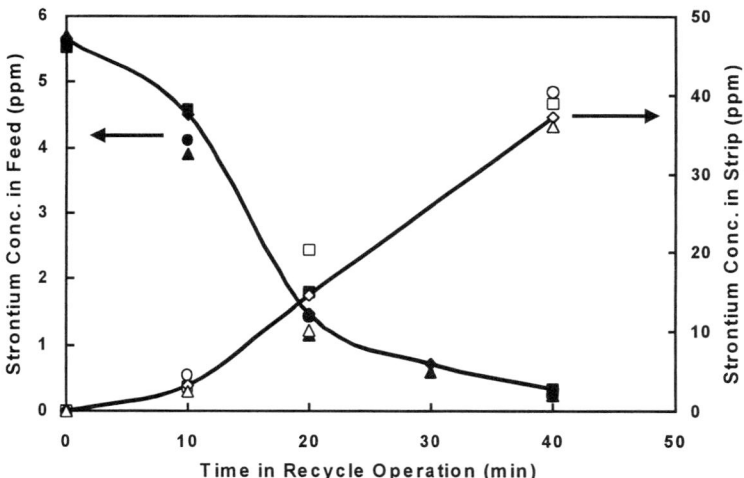

FIGURE 13. Strontium concentrations in feed and strip solutions as a function of time in recycle operation for four extractants. C12 feed (●) concentration in the feed solution for the C12 BOPPA extractant; C12 strip (○) concentration in the strip solution for the C12 BOPPA extractant. Similar for the C16 HDPPA (■ and □), C18 ODPPA/HDPPA (◆ and ◇), and C20 ODPPA (▲ and △) extractants. The aqueous feed solutions also contained about 80 ppm calcium, 20 ppm magnesium, and 50 ppm zinc in addition to ^{87}Sr.

TABLE 1. Radioactive ^{90}Sr removal using the C20 ODPPA extractant

Run	^{90}Sr Concentration in Feed (pCi/L)	Time (min)	^{90}Sr Concentration in Treated Feed (pCi/L)
1	317	120	3.3
2	317	120	3.5
3	317	120	3.3
4	317^a	240	4.0
5	1,000^a	240	5.5
6	1,000^a	360	1.0
7	30,000	60	1,171
		120	352
8	30,000^a	300	84

^{a}The feed solution also contained about 80 ppm calcium, 20 ppm magnesium, and 50 ppm zinc.

Radioactive ^{90}Sr Removal

After we demonstrated that the new extractants was effective for non-radioactive ^{87}Sr removal, we pursued the experiments using radioactive ^{90}Sr. In view of the fact that the four extractants removed ^{87}Sr equally well, as described already, 2-octyl-dodecyl phenylphosphonic acid (C20 ODPPA) (with the longest alkyl chain for the lowest solubility in water) was used for ^{90}Sr removal in order to increase the probability of achieving the goal of 8 pCi/L or lower. TABLE 1 shows the results for the ^{90}Sr removal from aqueous feed solutions at pH 3 with 1M HCl as the stripping solution. As indicated in this table, the supported liquid membrane (SLM) containing C20 ODPPA removed ^{90}Sr to the target concentration of 8 pCi/L (the drinking water standard) or lower from feed solutions containing 317–1,000 pCi/L ^{90}Sr. In particular, this target concentration was also achieved from a feed solution containing 1,000 pCi/L ^{90}Sr with the presence of about 80 ppm calcium, 20 ppm magnesium, and 50 ppm zinc.

This SLM also removed ^{90}Sr to a concentration of 352 pCi/L or lower from feed solutions containing 30,000 pCi/L ^{90}Sr (also shown in TABLE 1) with and without the presence of about 80 ppm calcium, 20 ppm magnesium, and 50 ppm zinc. The strip concentration was concentrated to more than 263,000 pCi/L (with an aqueous strip solution volume of about 220 mL, i.e., a feed-to-strip solution volume ratio of about 9). Based on the results for the feed solutions of 317–1,000 pCi/L, the treated concentration of 352 pCi/L or lower (in the treated solutions), which is lower than 1,000 pCi/L, can be reduced to meet the target concentration of 8 pCi/L or lower by the use of a second SLM step.

REMOVAL AND RECOVERY OF OTHER METALS AND MATERIALS

Supported liquid membranes (SLMs) with strip dispersion have the potential for removal and recovery of other materials including (1) other metals, such as cobalt, nickel, cadmium, selenium, and arsenic and (2) penicillin, phenylalanine, and organic acids such as lactic, citric, propionic, and butanoic acids. We described SLMs for the removal and recovery of cobalt and penicillin G, including their initial results, in the following.

Cobalt Removal and Recovery

We have identified a SLM system with strip dispersion effective at the low feed pH of about 2 for cobalt removal and recovery from waste waters and process streams.[77,85] This SLM system can not have a feed pH reduced significantly due to proton transfer during extraction in a hollow-fiber module in order to use the entire length of the module.[82] The organic membrane solution in the strip dispersion consisted of 8 wt.% di(2,4,4-trimethylpentyl) dithiophosphinic acid (Cyanex 301), 2 wt.% 1-dodecanol, and 90 wt.% Isopar L (described already). The aqueous strip solution was 5 M hydrochloric acid. A relatively high volume ratio of about 13.3 between the organic solution (800 mL) and the strip solution (60 mL) was used to greatly concentrate cobalt in the strip solution. The SLM system treated the 1-liter feed solution containing about 525 ppm cobalt in the recycle mode of operation as shown in FIGURE 4.

FIGURE 14 shows the results.[77,85] The SLM removed the cobalt to 0.7 ppm in the treated feed solution in just 15 min. It also concentrated cobalt to about 30,000 ppm in the aqueous strip solution (from the strip solution of 5 M hydrochloric acid preloaded with about 21,250 ppm cobalt). The cobalt flux at a cobalt concentration of about 380 ppm in the feed solution was about $5 g/m^2/h$, which is very high. The overall mass transfer coefficient at this concentration was about 4×10^{-4} cm/sec.

An SLM using the same organic membrane solution but with an aqueous strip solution of 6.5 M hydrochloric acid was used to remove cobalt from a 40-liter feed solution containing 492 ppm cobalt and concentrate it in the strip solution. To concentrate cobalt in the strip solution from an initial concentration of about 3,100 ppm to a final concentration of about 100,000 ppm, a relatively high volume ratio of about 8.6 between the organic solution and the strip solution was used (a very high volume ratio of about 260 between the feed solution and the strip solution). Again, the SLM system treated the 40-liter feed solution in the recycle mode of operation as shown in FIGURE 4. FIGURE 15 shows the results. As shown, the SLM concentrated the cobalt effectively to about 100,000 ppm in the strip solution in about seven hours in the recycle operation.

Penicillin G Removal and Recovery

We developed a SLM system with strip dispersion effective at the low feed pH of about 3 for penicillin G removal and recovery from aqueous solutions.[86] The organic membrane solution in the strip dispersion consisted of 10 wt.% N-lauryl-N-trialkylmethylamine with a molecular weight of 372 (a total number of 25.3 carbon atoms per amine molecule, e.g., Amberlite LA-2), 1 wt.% o-nitrophenyl octyl ether (o-NPOE), and 89 wt.% Isopar L (described above). The aqueous strip solution was 1.2 M sodium carbonate (Na_2CO_3) solution.

A strip dispersion was prepared by mixing 200 mL of the 1.2 M sodium carbonate (Na_2CO_3) solution and 800 mL of the organic solution. The strip dispersion was fed

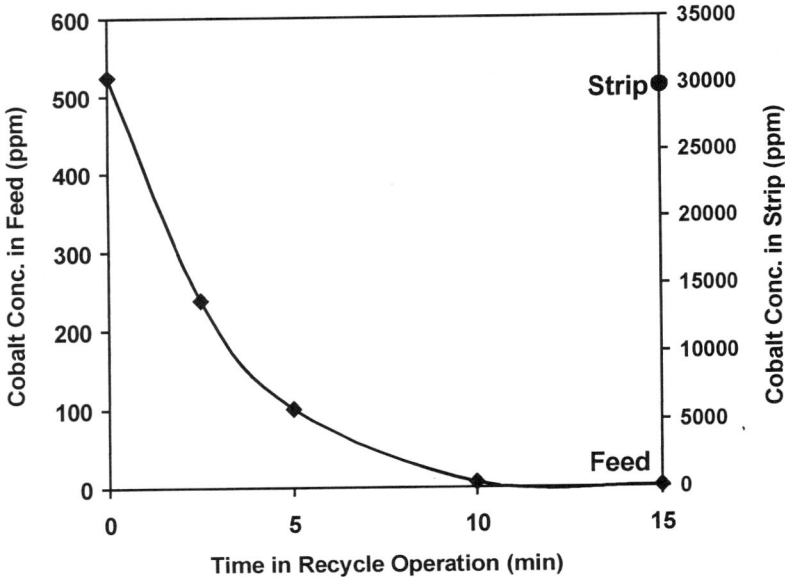

FIGURE 14. Cobalt concentrations in feed and strip solutions as a function of time in recycle operation.

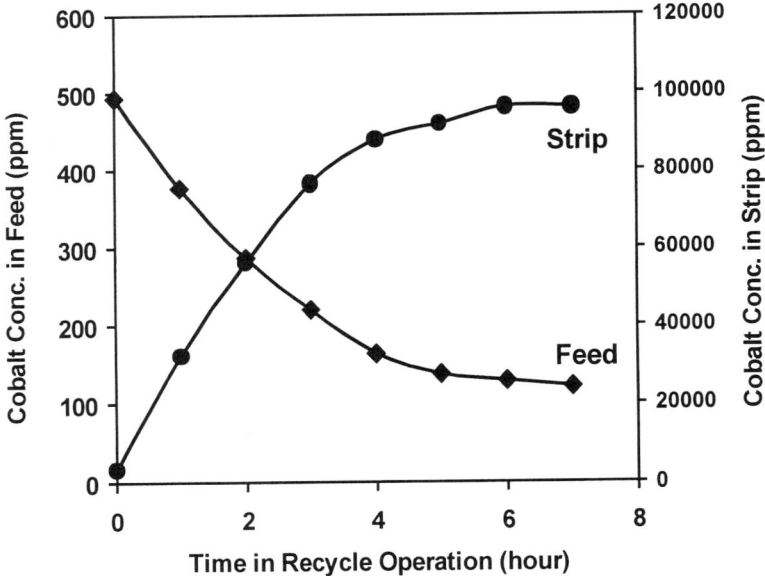

FIGURE 15. Cobalt concentrations in the feed and strip solutions as a function of time in recycle operation for concentrating to about 100,000 ppm in the final strip.

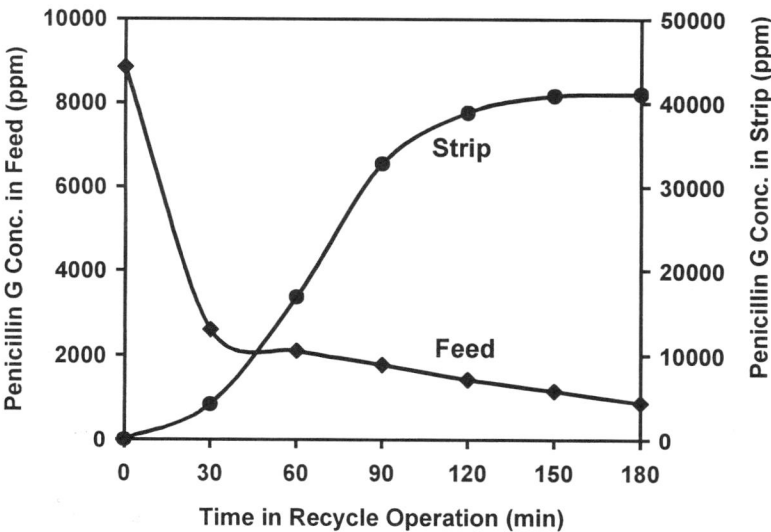

FIGURE 16. Penicillin G concentrations in feed and strip solutions as a function of time in recycle operation.

into the shell side of a 2.5-inch polypropylene hollow fiber module (2.5 inches in diameter by 8 inches in length). A 1-liter feed solution containing a penicillin G concentration of 8,840 ppm was passed into the tube side of the hollow fiber module countercurrently to the flow of the strip dispersion in the recycle mode of operation as shown in FIGURE 4. The pH of the feed solution was maintained at 3±0.1 by adding 3 M sulfuric acid as needed. Samples of the feed and strip solutions were collected at timed intervals and analyzed by UV.

FIGURE 16 shows the results.[86] The SLM removed the penicillin G from a high concentration of 8,840 ppm to a relatively low concentration of 877 ppm in the feed solution in three hours. On the other hand, the penicillin G was recovered and concentrated to a high concentration of 41,011 ppm in the aqueous strip solution at the same time. This represents a high recovery efficiency of about 93%. The penicillin G flux at the penicillin G concentration of about 5,500 ppm in the feed solution was about $9 g/m^2/h$, which is quite high. The overall mass transfer coefficient at this concentration is about 0.5×10^{-4} cm/sec.

CONCLUSIONS

We have developed SLM technology for removal and recovery of metals, including chromium, copper, zinc, and strontium, from waste waters. The technology for chromium not only removes the Cr(VI) from about 100–1,000 ppm to less than 0.05 ppm in the treated effluent allowable for discharge or recycle, but also recovers the chromium product at a high concentration of about 20% Cr (VI) (62.3% Na_2CrO_4) suitable for resale or reuse. In other words, we have achieved the goals of zero discharge and no sludge. This technology is based on our recent findings that a high chromium flux combined with a high chromium/sulfate selectivity can be achieved with a Cr (VI) concentration greater than about 100 ppm. The chromium flux in its entire range, from low Cr (VI) concentrations in the feed solution to high feed concentrations at which the carrier saturation of the facilitated transport mechanism occurs, exhibits good agreement with a film model.

The stability of the SLM is ensured by a modified SLM with strip dispersion, where the aqueous strip solution is dispersed in the organic membrane solution in a mixer. The strip dispersion formed is circulated from the mixer to the membrane module to provide a constant supply of the organic solution to the membrane pores.

We have identified effective SLM systems for the selective removal and recovery of copper and zinc from the Berkeley Pit water containing 150 ppm copper, 550 ppm zinc, and other metals. The copper SLM system using nonylsalicyl aldoxime and ketoxime not only removed copper from 150 ppm in the inlet feed to less than 0.15 ppm in the treated feed, but also recovered the copper at a high concentration of greater than 10,000 ppm in the strip solution. For the zinc SLM system with di(2,4,4-trimethylpentyl) dithiophosphinic acid, the zinc at the inlet feed concentration of 550 ppm was removed to less than 0.3 ppm in the treated feed and a high zinc concentration of more than 17,000 ppm was recovered in the strip solution. Both the copper and zinc concentrations in the treated feed meet the target specifications and both the strip solutions recovered had high purity.

For strontium removal, we synthesized a family of new extractants, alkyl phenylphosphonic acids. The SLM removed radioactive ^{90}Sr to the target of 8 pCi/L or lower from feed solutions of 300–1,000 pCi/L. Stripping with HCl was very satisfactory and nearly quantitative.

SLMs with strip dispersion have the potential for removal and recovery of other materials including (1) other metals such as cobalt and (2) penicillin and organic acids such as penicillin G. The cobalt SLM system identified removed cobalt from about 525 ppm to 0.7 ppm in the treated feed solution, and it also concentrated cobalt to at least 30,000 ppm in the aqueous strip solution. On penicillin G recovery, the SLM removed it from a feed of 8,840 ppm and concentrated it to a high concentration of 41,011 ppm in the aqueous strip solution with a high recovery of about 93%.

ACKNOWLEDGMENTS

The author wishes to thank his colleagues at Commodore Separation Technologies, Inc., particularly, Jim DeAngelis and Mike Kiehnau for their marketing and operation efforts for the supported liquid membrane technology. He also would like to thank Samir Kalini, Travis Neumuller, Tarun Poddar, Tim Pruett, Justin Roller, and Bing Wang for their assistance in some experiments.

REFERENCES

1. Ho, W.S. & K.K. SIRKAR, Eds. 1992. Membrane Handbook. Chapman & Hall, New York.
2. LI, N.N., Inventor; Exxon Research and Engineering Company, Assinee. 1968. Separating hydrocarbons with liquid membranes. U.S. Patent 3,410,794.

3. Ho, W.S. & N.N. Li. 1992. Definitions and theory for emulsion liquid membranes. *In* Membrane Handbook. W.S. Ho & K.K. Sirkar, Eds.: 595–655. Chapman & Hall, New York.
4. Gu, Z.M., W.S. Ho & N.N. Li. 1992. Design considerations for emulsion liquid membranes. *In* Membrane Handbook. W.S. Ho & K.K. Sirkar, Eds.: 656–700. Chapman & Hall, New York.
5. Raghuraman, B. & J.M. Wiencek. 1993. Extraction with emulsion liquid membranes in a hollow-fiber contactor. AIChE J. **39:** 1885–1889.
6. Hu, S.-Y.B. & J.M. Wiencek. 1998. Emulsion-liquid-membrane extraction of copper using a hollow-fiber contactor. AIChE J. **44:** 570–581.
7. Wiencek, J.M. & S.-Y. Hu. 2000. Emulsion liquid membrane extraction in a hollow-fiber contactor. Chem. Eng. Technol. **23:** 551–553.
8. Otto, N.C. & J.A. Quinn. 1971. Facilitated transport of carbon dioxide through bicarbonate solutions. Chem. Eng. Sci. **26:** 949–961.
9. Smith, D.R., R.J. Lander & J.A. Quinn. 1977. Carrier-mediated transport in synthetic membranes. Recent Dev. Sep. Sci. 3(Part B): 225–241.
10. Matuleviclus, E.S. & N.N. Li. 1975. Facilitated transport through liquid membranes. Sep. Purif. Methods **4:** 73–96.
11. Li, N.N. 1978. Facilitated transport through liquid membranes—an extended abstract. J. Membr. Sci. **3:** 265–269.
12. Smith, K.L., W.C. Babcock, R.W. Baker & M.G. Conrod. 1981. Coupled transport membranes for removal of chromium from electroplating rinse solutions. Chem. Water Reuse **1:** 311–324.
13. Molinari, R., E. Drioli & G. Pantano. 1989. Stability and effect of diluents in supported liquid membranes for chromium (III), chromium (VI), and cadmium (II) recovery. Sep. Sci. Technol. **24:** 1015–1032.
14. Chiarizia, R. 1991. Application of supported liquid membranes for removal of nitrate, technetium (VII) and chromium (VI) from groundwater. J. Membr. Sci. **55:** 39–64.
15. Alonso, A.I., M.I. Ortiz & A. Irabien. 1992. Supported liquid membranes for chromate recovery. Experimental study. Process Metall. 7B (Solvent Extr. 1990, Pt. B): 1535–1538.
16. Huang, T.C., C.C. Huang & D.H. Chen. 1998. Transport of chromium (VI) through a supported liquid membrane containing a tri-n-octylphosphine oxide. Sep. Sci. Technol. **33:** 1919–1935.
17. Wang, Y., Y.S. Thio & F.M. Doyle. 1998. Formation of semi-permeable polyamide skin layers on the surface of supported liquid membranes. J. Membr. Sci. **147:** 109–116.
18. Park, S.-W., G.-W. Kim, S.-S. Kim & I.-J. Sohn. 2001. Facilitated transport of Cr (VI) through a supported liquid membrane with trioctylmethylammonium chloride as a carrier. Sep. Sci. Technol. **36:** 2309–2326.
19. Yang, X.J., A.G. Fane & S. MacNaughton. 2001. Removal and recovery of heavy metals from wastewaters by supported liquid membranes. Water Sci. Technol. **43:** 341–348. (First World Water Congress, Part 2: Industrial Wastewater and Environmental Contaminants, 2000.)
20. Largman, T. & S. Sifniades. 1978. Recovery of copper (II) from aqueous solutions by means of supported liquid membranes. Hydrometallurgy **3:** 153–162.
21. Komasawa, I., T. Otake & T. Yamashita. 1983. Mechanism and kinetics of copper permeation through a supported liquid membrane containing a hydroxyoxime as a mobile carrier. Ind. Eng. Chem. Fundam. **22:** 127–131.
22. Danesi, P.R. 1984. A simplified model for the coupled transport of metal ions through hollow-fiber supported liquid membranes. J. Membr. Sci. **20:** 231–248.
23. O'Hara, P.A. & M.P. Bohrer. 1989. Supported liquid membranes for copper transport. J. Membr. Sci. **44:** 273–287.
24. Juang, R.S. 1993. Permeation and separation of zinc and copper by supported liquid membranes using bis(2-ethylhexyl) phosphoric acid as a mobile carrier. Ind. Eng. Chem. Res. **32:** 911–916.
25. Vander Linden, J. & R.F. De Ketelaere. 1998. Selective recuperation of copper by supported liquid membrane (SLM) extraction. J. Membr. Sci. **139:** 125–136.

26. BREEMBROEK, G.R.M., A. VAN STRAALEN, G.J. WITKAMP & G.M. VAN ROSMALEN. 1998. Extraction of cadmium and copper using hollow fiber supported liquid membranes. J. Membr. Sci. **146:** 185–195.
27. VALENZUELA, F., C. BASUALTO, C. TAPIA & J. SAPAG. 1999. Application of hollow-fiber supported liquid membranes technique to the selective recovery of a low-content of copper from a Chilean mine water. J. Membr. Sci. **155:** 163–168.
28. YANG, X.J. & A.G. FANE. 1999. Performance and stability of supported liquid membranes using LIX 984N for copper transport. J. Membr. Sci. **156:** 251–263.
29. GHERROU, A., H. KERDJOUDJ, R. MOLINARI & E. DRIOLI. 2002. Facilitated co-transport of Ag (I), Cu (II), and Zn (II) ions by using a crown ether as carrier: influence of the SLM preparation method on ions flux. Sep. Sci. Technol. **37:** 2317–2336.
30. GHERROU, A., H. KERDJOUDJ, R. MOLINARI & E. DRIOLI. 2002. Removal of silver and copper ions from acidic thiourea solutions with a supported liquid membrane containing D2EHPA as carrier. Sep. Sci. Technol. **28:** 235–244.
31. VALENZUELA, F., M.A. VEGA, M.F. YANEZ & C. BASUALTO. 2002. Application of a mathematical model for copper permeation from a Chilean mine water through a hollow fiber-type supported liquid membrane. J. Membr. Sci. **204:** 385–400.
32. ALGUACIL, F.J., M. ALONSO & A.M. SASTRE. 2002. Copper separation from nitrate/nitric acid media using Acorga M5640 extractant. Part II. Supported liquid membrane study. Chem. Eng. J. **85:** 265–272.
33. DANESI, P.R., R. CHIARIZIA & A. CASTAGNOLA. 1983. Transfer rate and separation of cadmium (II) and zinc (II) chloride species by a trilaurylammonium chloride-triethylbenzene supported liquid membrane. J. Membr. Sci. **14:** 161–174.
34. FERNANDEZ, L., J. APARICIO & M. MUHAMMED. 1987. The role of feed metal concentration in the coupled transport of zinc through a bis-(2-ethylhexyl)phosphoric acid solid supported liquid membrane from aqueous perchlorate media. Sep. Sci. Technol. **22:** 1577–1595.
35. HUANG, T.C. & R.S. JUANG. 1987. Transport of zinc through a supported liquid membrane using di(2-ethylhexyl) phosphoric acid as a mobile carrier. J. Membr. Sci. **31:** 209–226.
36. TANIGAKI, M., T. SHIODE, S. OKUMI & W. EGUCHI. 1988. Facilitated transport of zinc chloride through hollow fiber supported liquid membrane. Part 3. Module operation. Sep. Sci. Technol. **23:** 1171–1181.
37. SAITO, T. 1990. Transportation of zinc (II) ion through a supported liquid membrane. Sep. Sci. Technol. **25:** 581–591.
38. KANUNGO, S.B. & R. MOHAPATRA. 1995. Coupled transport of Zn (II) through a supported liquid membrane containing bis(2,4,4-trimethylpentyl)phosphinic acid in kerosene. II. Experimental evaluation of model equations for rate process under different limiting conditions. J. Membr. Sci. **105:** 227–235.
39. SHAMSIPUR, M., G. AZIMI & S.S. MADAENI. 2000. Selective transport of zinc as Zn(SCN)42-ion through a supported liquid membrane using K^+-dicyclohexyl-18-crown-6 as carrier. J. Membr. Sci. **165:** 217–223.
40. ALGUACIL, F.J. & S. MARTINEZ. 2001. Solvent extraction of Zn (II) by Cyanex 923 and its application to a solid-supported liquid membrane system. J. Chem. Technol. Biotechnol. **76:** 298–302.
41. CANET, L., M. ILPIDE & P. SETA. 2002. Efficient facilitated transport of lead, cadmium, zinc, and silver across a flat-sheet-supported liquid membrane mediated by lasalocid A. Sep. Sci. Technol. **37:** 1851–1860.
42. BUCHALTER, E.M., D.L. HOFMAN, W.M. CRAIG, *et al.* 1987. Supported liquid membrane technology applied to the recovery of useful isotopes from reactor pool water. Chem. Eng. Res. Des. **65:** 381–385.
43. RAMADAN, A. & P.R. DANESI. 1988. Transfer rate and separation of strontium (2+) and cesium (1+) by supported liquid membranes utilizing synergized crown ether carriers. Solvent Extr. Ion Exch. **6:** 157–166.
44. LAMB, J.D., R.L. BRUENING, D.A. LINSLEY, *et al.* 1990. Characterization of a macrocycle-mediated dual module hollow fiber membrane contactor for making cation separations. Sep. Sci. Technol. **25:** 1407–1419.

45. DOZOL, J.F., J. CASAS & A.M. SASTRE. 1994. Influence of the extractant on strontium transport from reprocessing concentrate solutions through flat-sheet supported liquid membranes. Sep. Sci. Technol. **29:** 1999–2018.
46. CHAUDRY, M.A., U.-I. NOOR & I. AHMAD. 1994. Extraction and stripping study of strontium ions across D2EHPA-TBP-kerosine oil-based supported liquid membranes. J. Radioanal. Nucl. Chem. **185:** 369–385.
47. MACKOVA, J. & V. MIKULAJ. 1996. Transport of strontium cation through a hollow fiber supported dichlorobenzene membrane using 18-C-6 crown ether. Nitrate and anion of dinonylnaphthalene sulfonic acid. J. Radioanal. Nucl. Chem. **208:** 111–122.
48. DE GYVES, J. & E. RODRIGUEZ DE SAN MIGUEL. 1999. Metal ion separations by supported liquid membranes. Ind. Eng. Chem. Res. **38:** 2182–2202.
49. LEE, C.W., K.H. HONG, M.H. LEE, et al. 2000. Separation and preconcentration of strontium from calcium in aqueous samples by supported liquid membrane containing crown ether. J. Radioanal. Nucl. Chem. **243:** 767–773.
50. DANESI, P.R., L. REICHLEY-YINGER, C. CIANETTI & P.G. RICKERT. 1984. Separation of cobalt and nickel by liquid-liquid extraction and supported liquid membranes with bis(2,4,4-trimethylpentyl)phosphinic acid [Cyanex 272]. Solvent Extr. Ion Exch. **2:** 781–814.
51. CHAUDRY, M.A., M.T. MALIK & A. ALI. 1990. Transport of cobalt (II) ions through di(2-ethylhexyl) phosphoric acid-carbon tetrachloride supported liquid membranes. Sep. Sci. Technol. **25:** 1161–1174.
52. HUANG, T.C. & T.H. TSAI. 1992. Separation of cobalt and nickel ions with 2-ethylhexylphosphonic acid mono-2-ethylhexyl ester dissolved in kerosene. Process Metall. 7B (Solvent Extr. 1990, Pt. B): 1529–1534.
53. JUANG, R.S. & J.D. JIANG. 1994. Rate-controlling mechanism of cobalt transport through supported liquid membranes containing di(2-ethylhexyl) phosphoric acid. Sep. Sci. Technol. **29:** 223–237.
54. YOUN, I.J., Y. LEE, J. JEONG & W.H. LEE. 1997. Analysis of Co-Ni separation by a supported liquid membrane containing HEH(EHP). J. Membr. Sci. **125:** 231–236.
55. GEGA, J., W. WALKOWIAK & B. GAJDA. 2001. Separation of Co (II) and Ni (II) ions by supported and hybrid liquid membranes. Sep. Purif. Technol. **22–23:** 551-557.
56. ALGUACIL, F.J. 2002. Facilitated transport and separation of manganese and cobalt by a supported liquid membrane using DP-8R as a mobile carrier. Hydrometallurgy **65:** 9–14.
57. MARCHESE, J., J.L. LOPEZ & J.A. QUINN. 1989. Facilitated transport of benzylpenicillin through immobilized liquid membrane. J. Chem. Technol. Biotechnol. **46:** 149–159.
58. TSIKAS, D., E. KALTSIDOU-SCHOTTELIUS & G. BRUNNER. 1992. Hollow fiber-supported liquid membranes for the extraction of penicillins and the synthesis of 6-aminopenicillanic acid. Chem. Ing. Tech. **64:** 545–548.
59. LEE, C.J., H.J. YEH, W.J. YANG & C.R. KAN. 1994. Separation of penicillin G from phenylacetic acid in a supported liquid membrane system. Biotechnol. Bioeng. **43:** 309–313.
60. JUANG, R.-S., S.-H. LEE & R.-C. SHIAU. 1998. Carrier-facilitated liquid membrane extraction of penicillin G from aqueous streams. J. Membr. Sci. **146:** 95–104.
61. BRYJAK, M., J. KOZLOWSKI, P. WIECZOREK & P. KAFARSKI. 1993. Enantioselective transport of amino acid through supported chiral liquid membranes. J. Membr. Sci. **85:** 221–228.
62. CAMPBELL, M.J., R.P. WALTER, R. SINGLETON & C.J. KNOWLES. 1994. Investigation of the stability and selectivity of phenylalanine transport across a supported liquid membrane. J. Chem. Technol. Biotechnol. **60:** 263–273.
63. DZYGIEL, P., P. WIECZOREK, J.A. JONSSON, et al. 1999. Separation of amino acid enantiomers using supported liquid membrane extraction with chiral phosphates and phosphonates. Tetrahedron **55:** 9923–9932.
64. SIRMAN, T., L. PYLE & A.S. GRANDISON. 1991. Extraction of organic acids using a supported liquid membrane. Biochem. Soc. Trans. **19:** 274S.
65. SHEN, Y., L. GROENBERG & J.A. JOENSSON. 1994. Experimental studies on the enrichment of carboxylic acids with tri-n-octylphosphine oxide as extractant in a supported liquid membrane. Anal. Chim. Acta **292:** 31–39.

66. JU, L.-K. & A. VERMA. 1994. Characteristics of lactic acid transport in supported liquid membranes. Sep. Sci. Technol. **29:** 2299–2315.
67. JUANG, R.-S., R.-H. HUANG & R.-T. WU. 1997. Separation of citric and lactic acids in aqueous solutions by solvent extraction and liquid membrane processes. J. Membr. Sci. **136:** 89–99.
68. YAHAYA, G.O., B.J. BRISDON & R. ENGLAND. 2000. Facilitated transport of lactic acid and its ethyl ester by supported liquid membranes containing functionalized polyorganosiloxanes as carriers. J. Membr. Sci. **168:** 187–201.
69. YAHAYA, G.O. 2001. Kinetic studies on organic acid extraction by a supported liquid membrane using functionalized polyorganosiloxanes as mobile and fixed-site carriers. Sep. Sci. Technol. **36:** 3563–3584.
70. FRIESEN, D.T., W.C. BABCOCK, D.J. BROSE & A.R. CHAMBERS. 1991. Recovery of citric acid from fermentation beer using supported-liquid membranes. J. Membr. Sci. **56:** 127–141.
71. JUANG, R.-S. & L.-J. CHEN. 1997. Transport of citric acid across a supported liquid membrane containing various salts of a tertiary amine. J. Membr. Sci. **123:** 81–87.
72. ROCKMAN, J.T., E. KEHAT & R. LAVIE. 1997. Thermally enhanced extraction of citric acid through supported liquid membrane. AIChE J. **43:** 2376–2380.
73. NUCHNOI, P., T. YANO, N. NISHIO & S. NAGAI. 1987. Extraction of volatile fatty acids from diluted aqueous solution using a supported liquid membrane. J. Ferment. Technol. **65:** 301–310.
74. KEMPERMAN, A.J.B., D. BARGEMAN, TH. VAN DEN BOOMGAARD & H. STRATHMANN. 1996. Stability of supported liquid membranes: state of the art. Sep. Sci. Technol. **31:** 2733–2762.
75. HILL, C., J.-F. DOZOL, H. ROUQUETTE, *et al.* 1996. Study of the stability of some supported liquid membranes. J. Membr. Sci. **114:** 73–80.
76. DREHER, T.M. & G.W. STEVENS. 1998. Instability mechanisms of supported liquid membranes. Sep. Sci. Technol. **33:** 835–853.
77. HO, W.S., Inventor; Commodore Separation Technologies Inc., Assinee. 2001. Combined supported liquid membrane/stripping dispersion process for removal and recovery of radionuclides and metals. U.S. Patent 6,328,782.
78. HO, W.S. & T.K. PODDAR. 2001. New membrane technology for removal and recovery of chromium from waste waters. Environ. Prog. **20:** 44–52.
79. HO, W.S. & B. WANG. 2002. Strontium removal by new alkyl phenylphosphonic acids in supported liquid membranes with strip dispersion. Ind. Eng. Chem. Res. **41:** 381–388.
80. TERAMOTO, M., Y. SAKAIDA, S.S. FU, *et al.* 2000. An attempt for the stabilization of supported liquid membrane. Sep. Purif. Technol. **21:** 137–144.
81. HO, W.S., Inventor; Commodore Separation Technologies Inc., Assinee. 2001. Supported liquid membrane process for chromium removal and recovery. U.S. Patent 6,171,563.
82. HO, W.S., B. WANG, T.E. NEUMULLER & J. ROLLER. 2001. Supported liquid membranes for removal and recovery of metals from waste waters and process streams. Environ. Prog. **20:** 117–121.
83. HO, W.S. & B. WANG, Inventors; Commodore Separation Technologies Inc., Assinee. 2001. Combined supported liquid membrane/stripping dispersion process for removal and recovery of metals: dialkyl monothiophosphoric acids and their use as extractants. U.S. Patent 6,291,705.
84. HO, W.S., T.K. PODDAR & T.E. NEUMULLER. 2002. Removal and recovery of copper and zinc by supported liquid membranes with strip dispersion. J. Chin. Inst. Chem. Engrs. **33:** 67–76.
85. HO, W.S., Inventor; Commodore Separation Technologies Inc., Assinee. 2002. Combined supported liquid membrane/stripping dispersion process for removal and recovery of metals. U.S. Patent 6,350,419.
86. HO, W.S., Inventor; Commodore Separation Technologies Inc., Assinee. 2002. Combined supported liquid membrane/stripping dispersion process for removal and recovery of penicillin and organic Acids. U.S. Patent 6,433,163.

Membrane Separation in Green Chemical Processing

Solvent Nanofiltration in Liquid Phase Organic Synthesis Reactions

ANDREW LIVINGSTON,[a] LUDMILA PEEVA,[a] SHEJIAO HAN,[a] DINESH NAIR,[a] SATINDER SINGH LUTHRA,[a] LLOYD S. WHITE,[b] AND LUISA M. FREITAS DOS SANTOS[c]

[a]*Department of Chemical Engineering, Imperial College, London, United Kingdom*

[b]*W.R. Grace & Co.-Conn., Columbia, Maryland, USA*

[c]*GlaxoSmithKline Pharmaceuticals, New Frontiers Science Park, Harlow, Essex, United Kingdom*

ABSTRACT: This paper describes ideas together with preliminary experimental results for applying solvent nanofiltration to liquid phase organic synthesis reactions. Membranes for organic solvent nanofiltration have only recently (during the 1990s) become available and, to date, have been applied primarily to food processing (vegetable oil processing, in particular) and refinery processes. Applications to organic synthesis, even at a laboratory feasibility level, are few. However, these membranes have great potential to improve the environmental performance of many liquid phase synthesis reactions by reducing the need for complex solvent handling operations. Examples that are shown to be feasible are solvent exchanges, where it is desired to swap a high molecular weight molecule from one solvent to another between separate stages in a complex synthesis, and recycle and reuse of homogeneous catalysts. In solvent exchanges, nanofiltration is shown to provide a fast and effective means of swapping from a high boiling point solvent to a solvent with a lower boiling point—this is a difficult operation by means of distillation. Solvent nanofiltration is shown to be able to separate two distinct types of homogeneous catalysts, phase transfer catalysts and organometallic catalysts, from their respective reaction products. In both cases the application of organic solvent nanofiltration allows several reuses of the same catalyst. Catalyst stability is shown to be an essential requirement for this technique to be effective. Finally, we present a discussion of scale-up aspects including membrane flux and process economics.

KEYWORDS: organic solvent nanofiltration; solvent exchange; homogeneous catalyst recycle

---

Address for correspondence: Andrew Livingston, Department of Chemical Engineering, Imperial College, London SW7 2BY, United Kingdom. Voice: +44 20 7594 5582; fax: +44 20 7594 5604.
   a.livingston@ic.ac.uk

INTRODUCTION

Technologies are constantly being developed for the manufacture of chemical products in more efficient and cost effective ways. During the 1990s green chemistry has come to the fore as a way of developing focus on more atom efficient chemical processes.[1] Recently, however, an analysis of many reaction types used in organic synthesis was undertaken by workers at GlaxoSmithKline.[2] This showed that the atom economy was apparently not linked to mass intensity of many reaction processes. Mass intensity is defined by

$$\text{Mass intensity} = \frac{\text{Total mass reagents employed}}{\text{Mass product produced}}. \qquad (1)$$

This somewhat surprising result suggested that the crude amount of materials used in a typical fine chemical synthesis is not strongly related to the overall atom economy of the chemical reaction. These authors go on to state that: "Our data demonstrates that the largest component of mass intensity is due to reaction and work-up solvents." This suggests that better technology for *workup* (or postreaction processing and separation) in solvent intensive processes is likely to yield significant environmental and cost benefits.

Organic solvent nanofiltration (OSN) is an emerging field, and the technical potential for molecular separations in organic solvents is vast. Commercial activity, related to (1) refinery operations[3,4] and (2) edible oil refining[5,6] is leading to a new generation of solvent resistant nanofiltration (SRNF) membranes that are able to tolerate solvents spanning the polarity range from toluene to methanol, and able to separate molecules in the MW range 200–1,000 Da. These membranes are fabricated from polymers such as polyimides, polyacrylonitrile/silicone, and polyacetylene. New work is emerging that describes research into fluxes and rejections of these membranes.[7–11] This paper describes applications of OSN to organic synthesis workups.

SOLVENT EXCHANGES

Background

Most complex organic molecules used as active pharmaceutical ingredients are synthesized through routes involving a sequence of organic chemistry reactions. Solvent exchanges are used when a first stage (for example, a reaction or crystallization) requires a particular solvent, and is followed by a second stage that requires a different solvent from the first stage. A typical method to transfer the large (non-volatile) product molecule form the first solvent to the second is by "put and take" distillation, in which some of the first solvent is boiled off, replaced with a volume of the second solvent, and the mixture is boiled again to remove additional first solvent. Ideally the temperature remains below acceptable limits otherwise the molecule may suffer thermal degradation. By a sequence of such "puts" of fresh second solvent and "takes" of the first solvent, the solvent exchange proceeds. Note, however, that this technique is most effective when the first solvent has a much lower boiling point than the second solvent—if the reverse is true, then it is very much more difficult. Thus, it is much easier to exchange methanol (bp 65°C) for toluene (bp 110°C) than it is to achieve an exchange of toluene for methanol. Furthermore, if a system exhibits azeotropic behavior, this may limit the extent to which the first solvent can be removed.

OSN can be used to effect solvent exchanges at ambient temperature and regardless of azeotropic behavior or the boiling point order of the two solvents to be exchanged. The general technique is shown in FIGURE 1. Picture a system in which we wish to swap an organic solute (OS) dissolved in a high-boiling solvent (HBS) to a low-boiling solvent (LBS). We select an OSN membrane able to retain the compound and filter the mixture through the membrane until we have removed 70–90% of the HBS. We then add an aliquot of the LBS and filter again, which serves to reduce the content of HBS (assuming no preferential permeation of LBS). As additional "shots" of LBS are added and filtered, it should be possible to reach less than 1% HBS in the final mixture. This process could be carried out at room temperature with no requirement for heating and, hence, no potential problems due to thermal decomposition of products. The HBS from the first filtration should be virtually OS-free and could be used again. The HBS/LBS mixture from the intermediate filtrations could be separated by straightforward distillation for reuse—or it may be possible to use the mixture several times before this is necessary, depending on the HBS content.

Experimental

This technique was tested in the laboratory using a SEPTA ST (Osmonics, USA) dead-end pressure filtration cell (effective membrane area $14\,cm^2$). The membrane employed was a STARMEM 122 polyimide nanofiltration (NF) membrane manufactured by WR Grace (STARMEM is a trademark of W.R.Grace). The alkylammonium salts tetraoctylammonium bromide (TOABr, MW = 546 Da) and tetrabutylammonium bromide (TBABr, MW = 322 Da) were used as model organic solutes at low concentrations (0.01 M). The concentrations of these alkylammonium salts were measured using gas chromatography. All nanofiltration runs were at a pressure of 30 bar and at ambient temperature (20–23°C). The volume of liquid present at the start of each nanofiltration was 100 ml, and nanofiltration continued until 80 ml of

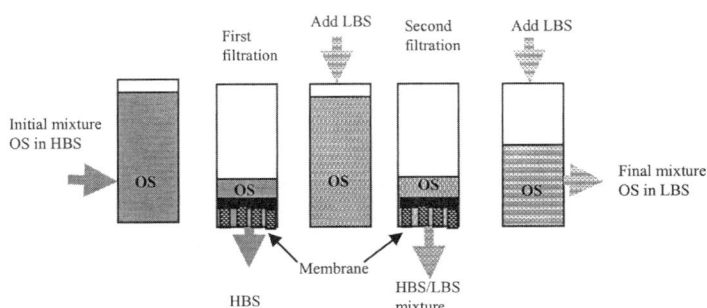

FIGURE 1. Operating principle of membrane solvent exchange: OS, organic compound; HBS, high boiling point solvent; LBS, low boiling point solvent.

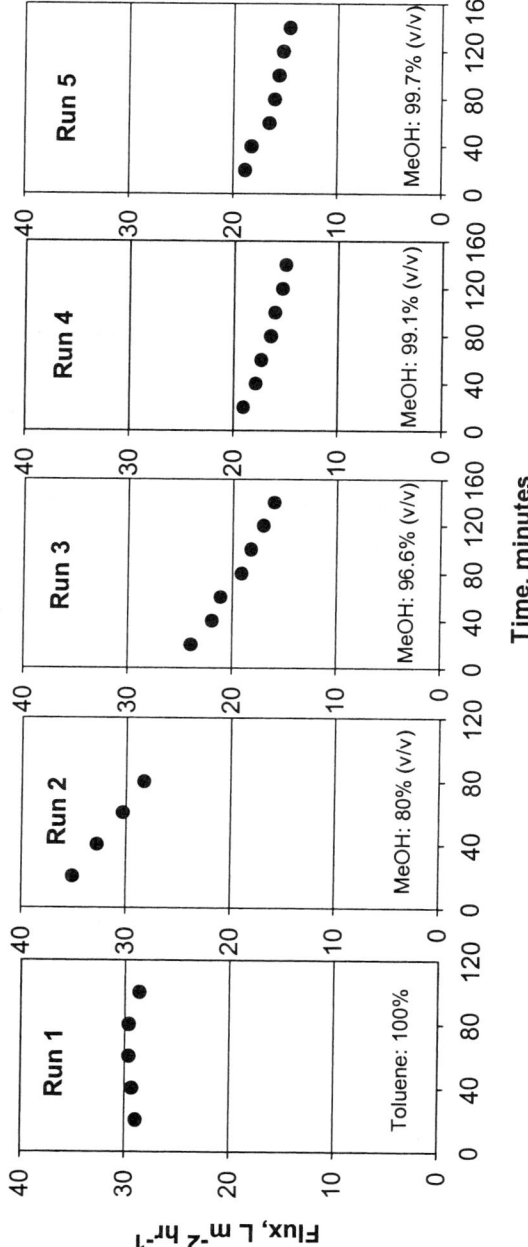

FIGURE 2. Toluene exchanged for methanol using STARMEM 122 at 30 bar and at room temperature. TOABr rejection is 100%.

liquid had been removed as permeate. Fresh second solvent was then added, and the process carried out again until the final liquid had a first solvent content below the desired level. Membranes were conditioned by immersion overnight in the solvent that was to be the first solvent in the solvent exchange to remove any storage agents present in the membrane as supplied. Multiple membrane samples were not tested for the data reported in this study, which seeks only to demonstrate the principal. Additional work is currently under way to evaluate repeatability and disc-to-disc variations.

Results and Discussion

Solvent exchange from toluene to methanol is shown in FIGURE 2. After five filtration runs, the solvent present in the cell had changed in composition from 100% toluene to 99.7% methanol. The retention of TOABr during this sequence of filtrations was determined as to exceed 99%. The flux of solvent through the membrane started at about $30 L m^{-2} h^{-1}$ for pure toluene and fell to about $15 L m^{-2} h^{-1}$ for methanol. A second solvent exchange, from methanol to ethyl acetate, is shown in FIGURE 3. Here the solute was TBABr, and the retention also exceeded 99%. The initial methanol flux was in the range $100–120 L m^{-2} h^{-1}$—considerably higher than the flux for methanol ($15 L m^{-2} h^{-1}$) obtained during the final filtration for FIGURE 2. We cannot explain this data at present; however, many experiments have shown that the order in which solvents are applied to OSN membranes can influence the flux observed. Another interesting observation was that no separation of solvents was observed; that is, the retentate and permeate concentrations of the solvents were the same at all stages.

It is important to note that the entire solvent exchange process in FIGURE 3, from 100% methanol to 99.7% ethyl acetate, took less than four hours to achieve. The liquid volume to membrane area ratio was $0.1 L/14 cm^2$ or around $70 L/m^2$ membrane area. A typical reactor containing $1 m^3$ to $10 m^3$ liquid would, therefore, require 14 to $140 m^2$ membrane area to affect the same process in less than four hours. Given the availability of modules containing $10–20 m^2$ of area, the number of modules required is modest. Finally, one can reasonably ask, is this membrane process "greener" than distillation for solvent exchange?

The energy consumption for the membrane process depends on the stage cut obtained in the membrane modules, and thus, is not easily estimated from this dead end cell data. However, since both membrane and distillation processes produce intermediate mixed solvent streams that need to go to recovery or disposal, the processes can be compared on the basis of mass intensity, Equation (**1**). The data for mass intensity of the distillation process were provided by GSK pilot plant data. It turns out that a concentration of second solvent exceeding 95% is often sufficient for chemical processes. With this as an end point, the following mass intensities obtain:

Distillation Solvent Exchange

Mass intensity, $5–10 kg$ (mixed solvents) kg^{-1} (starting solution).

Membrane Solvent Exchange

Mass intensity, $2–3 kg$ (mixed solvents) kg^{-1} (starting solution).

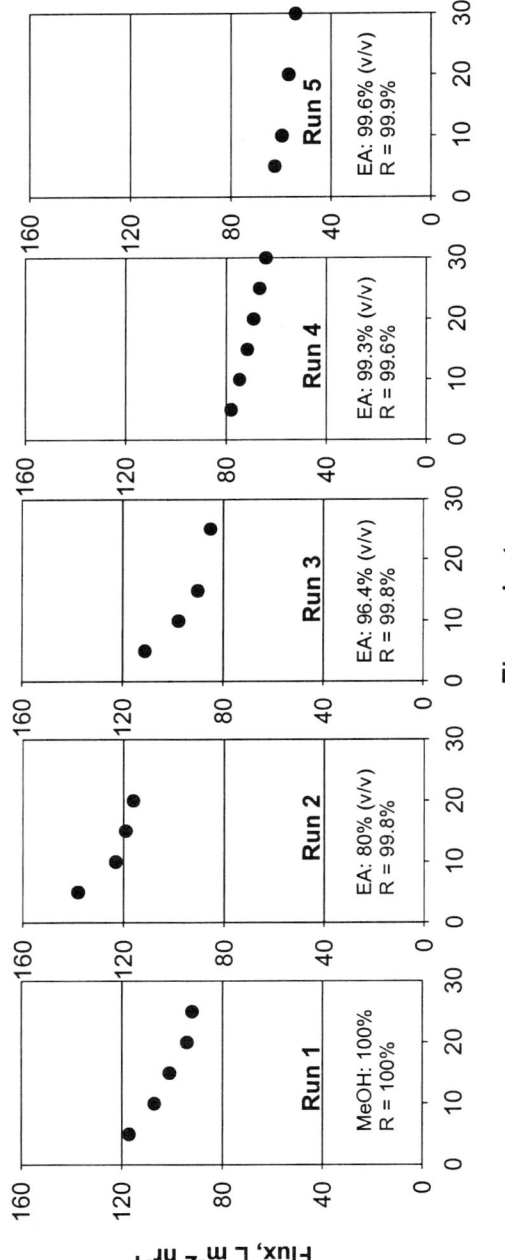

FIGURE 3. Methanol exchanged for ethyl acetate using STARMEM 122 at 30 bar and at room temperature. R denotes TBABr rejection.

Clearly the membrane process offers a considerable advantage in terms of mass intensity. In many cases, it is also considerably faster than the distillation process, even more important than mass intensity when reactor occupancy is at a premium. Hence, we might expect to see more widespread application of this technique in future.

HOMOGENEOUS CATALYST RECYCLE

Catalysis is a key technology for reducing environmental impacts of chemical processes by removing the need for stoichiometric reagents and by making desired products more directly and in purer form.[12] Homogeneous catalysis is often preferred over heterogeneous catalysis since it generally has lower mass transfer resistances. An ongoing technical issue in homogeneous catalysis is how to separate the catalyst from the product–solvent–catalyst postreaction mixture. In this work, we have applied nanofiltration membranes to homogeneous catalysts used in two distinct types of reaction.

Phase Transfer Catalysis Reactions

Background

Phase transfer catalysis, introduced by Starks,[13] is applied to reactions involving a water soluble nucleophilic reagent and an organic soluble electrophilic reagent (e.g., anions and organic substrates). In phase transfer catalysis, two immiscible phases are contacted and a phase transfer catalyst (PTC) is used to transfer a reactant from one of these phases into the other, so that reaction can occur. PTCs are often quaternary ammonium salts (*quats*), that is, tetraalkylammonium and tetraalkylphosphonium salts. A typical case, for which there are many examples,[14] is the use of a water–organic solvent system in which one reactant is water soluble and one is organic soluble, as shown in FIGURE 4. The reaction proceeds as the catalyst shuttles the nucleophilic anion into the organic phase. Typically, in the postreaction mixture, the PTC remains in the organic phase. Even PTCs that are moderately water soluble will partition into the organic phase postreaction due to the high ionic strength in the aqueous phase (FIG. 4).

Several books and reviews of phase-transfer catalysis have appeared in the recent literature,[14,15] in which the advantages of phase transfer catalysis for industrial-scale processing have been put forward. Many pharmaceuticals, agricultural chemicals, fine chemicals, and other specialities are produced via this technique. The advantages include increased productivity and quality, improved environmental performance, avoidance of need to use polar aprotic solvents (DMSO, DMF, DMAc), enhanced safety, and reduction of other manufacturing costs. However, one of the major technical problems inhibiting the use of phase transfer catalysis in industrial applications is the need to separate the product and the PTC.[16] The most commonly used methods for separation of products and PTCs on an industrial scale are extraction by water washing (when the catalyst is water soluble) and distillation (when the catalyst is lipophilic). However, extraction methods generate further toxic (catalyst containing) wastes and, in distillation methods, the catalysts may cause problems due to degradation.

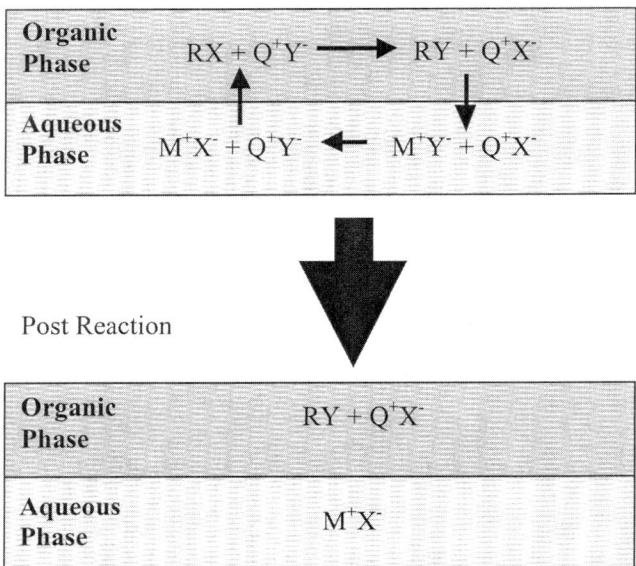

FIGURE 4. Phase transfer catalysis principle: RX, reactant; Q, phase transfer catalyst; Y, nucleophilic anion; M, counter cation in aqueous phase. In postreaction mixtures, lipophilic quat partitions into the organic phase.

We have, therefore, applied OSN to separate and recycle PTC, following the general layout shown in FIGURE 5. A biphasic reaction is carried out, and the organic and aqueous phases are then separated. The organic phase, containing the reaction product and the PTC, is subjected to OSN. The PTC remains in the retentate phase and is recycled to the next reaction. More detailed accounts of this technique can be found elsewhere.[17,18]

Experimental

The model reaction used to demonstrate the application of PTC is the conversion of bromoheptane to iodoheptane in the organic solvent toluene, via KI salt in the aqueous phase and tetraoctylammonium bromide (TOABr) as a phase-transfer catalyst, as shown in FIGURE 6. This reaction is a classic example of a nucleophilic, aliphatic substitution reaction. Toluene is a common solvent in industry and a typical solvent used in phase-transfer catalysis. At the conclusion of the reaction, the lipophilic TOABr and the iodoheptane product both partition entirely into the organic phase.

The reaction was carried out in a glass vessel of 100 mL with a 40 mL aqueous phase (2 M KI) and 40 mL organic phase (0.5 M bromoheptane + 0.05 M TOABr in toluene). The temperature was kept at $50 \pm 1°C$ using an oil bath and stirring speed

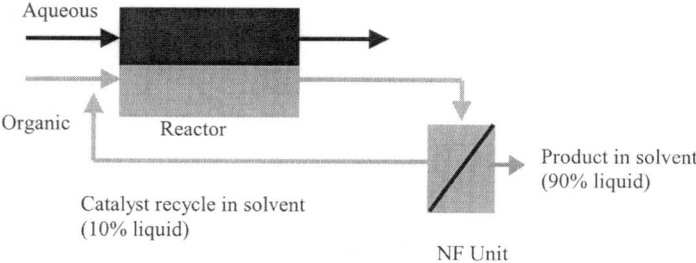

FIGURE 5. Schematic of phase-transfer catalytic reaction and catalyst recycle by OSN.

was 400 rpm. After the reaction was completed, the organic phase (40 mL) was transferred into a SEPA ST filtration cell (effective membrane area 14 cm^2) and the cell was pressurized to 30 bar at room temperature. The SEPA cell was stirred at 300 rpm and the permeate was collected. The membrane employed was a STARMEM 122 polyimide NF membrane manufactured by WR Grace. The same membrane disc was used for all three consecutive filtrations. Filtration continued until 36 mL of the permeate had been collected. Concentrations of reactant, product, and PTC were determined using GC analysis.

Results and Discussion

Profiles of reactant and product concentration are shown in FIGURE 6. Reactant concentration decreases and product concentration rises for each of reactions 1, 2, and 3. At the end of reaction 1, the postreaction mixture was filtered to remove 36 mL of permeate, leaving 4 mL of retentate in the cell. This was washed out using 36 mL of fresh organic phase reaction mixture which was then added to the reactor together with a fresh aliquot of aqueous phase containing KI. No fresh catalyst was added

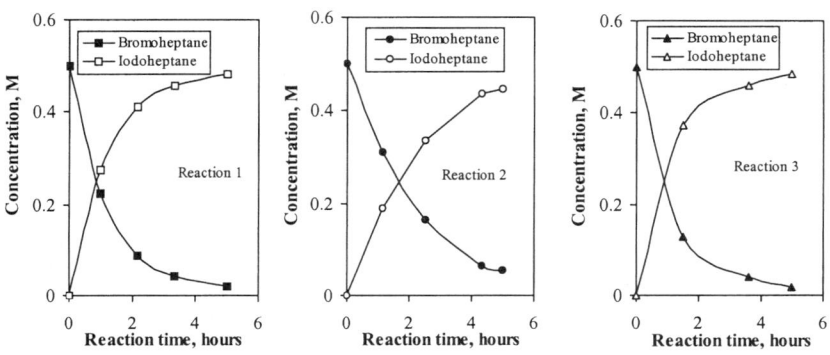

FIGURE 6. Evolution of bromoheptane and iodoheptane concentration over time for subsequent reactions in the presence of 0.05 M TOABr at 50°C.

after the initial addition; in a control reaction to which no catalyst was added, there was no reaction. Thus, the repeated reactions can be attributed to the successful retention of the catalyst by the NF membrane.

Data for the filtrations undertaken between reaction runs are shown in FIGURE 7. Typically 75–100 minutes were required to remove the 36 mL of permeate. At the end of each filtration, and after the retentate containing PTC had been washed out of the cell, toluene was added and the flux of the pure toluene was measured at 30 bar. This was between 60 and $80 \text{L m}^{-2} \text{h}^{-1}$ for both the initial and final filtrations, suggesting that there is no change in membrane properties over the cycle of three filtrations. Within each filtration, however, the flux falls from values around $40 \text{L m}^{-2} \text{h}^{-1}$ to values between $12–15 \text{L m}^{-2} \text{h}^{-1}$. By the end of the nanofiltration step, a film of organic material (assumed to be TOABr) was observed to precipitate out of the solution and appeared to attach to the membrane surface. The solubility of TOABr in toluene is 380g L^{-1}, and the starting concentration in the reaction mixture is 27g L^{-1}, so it is expected that after 36 mL of organic phase has been removed a maximum concentration of 270g L^{-1} could result. This should be below the solubility limit of TOABr in toluene. However, it is possible that effects of the counter ion and the reactant/product in the system lower the TOABr solubility, causing the catalyst to come out of the solution at the membrane surface. It is also possible that concentration polarization and osmotic pressure could exert a negative effect on flux as the concentration of TOABr increases.

Overall, the retention of the PTC was excellent, and the reuse was shown to be feasible. This suggests that OSN may have an important role to play in phase transfer catalyzed reactions. In addition, the mass intensity of the membrane process as used in this paper was compared to conventional techniques for PTC removal:

PTC removal by washing out quat using 0.2 M citric acid
  Mass intensity, 45 kg (reagents) kg^{-1} (product).

PTC removal by OSN
  Mass intensity, 22 kg (reagents) kg^{-1} (starting solution).

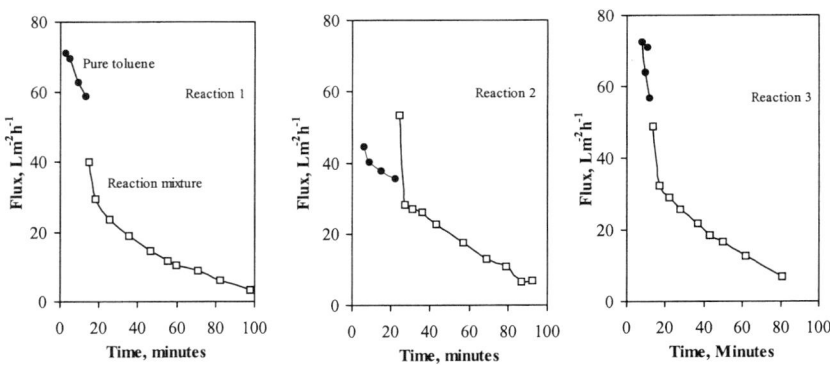

FIGURE 7. Permeate fluxes of postreaction mixture as a function of time at 30 bar for STARMEM 122.

It is important to point out that the key separation here is really between the PTC and product, not between PTC and solvent. Thus, it is necessary to have a product molecule that is small enough so that it is able to pass through the NF membrane.

Homogeneous Organometallic Catalyzed Reactions

Background

Many organic reactions (for example, C–C couplings and hydrogenations) are catalyzed by organometallic complexes of transition metals, such as Pd, Ru, and Rh. These metals are relatively expensive, and, in many cases, are attached to organic ligands that are also valuable. This is particularly true when the organic ligands are chirally-directing, such as DUPHOS or BINAP ligands. At the end of a reaction, the organometallic catalyst (OMC) must be removed. Generally this is achieved during postreaction workup, where the OMC is broken up by addition of solvents and/or acids, and the fragments removed from the system by successive extractions, and by passing the postreaction mixture through a bed of activated carbon or silica. In some cases it may be possible to recover the metal content of the catalyst by sending the beds of silica or carbon to catalyst suppliers for regeneration and metal recovery. However, the ligand value is lost totally.

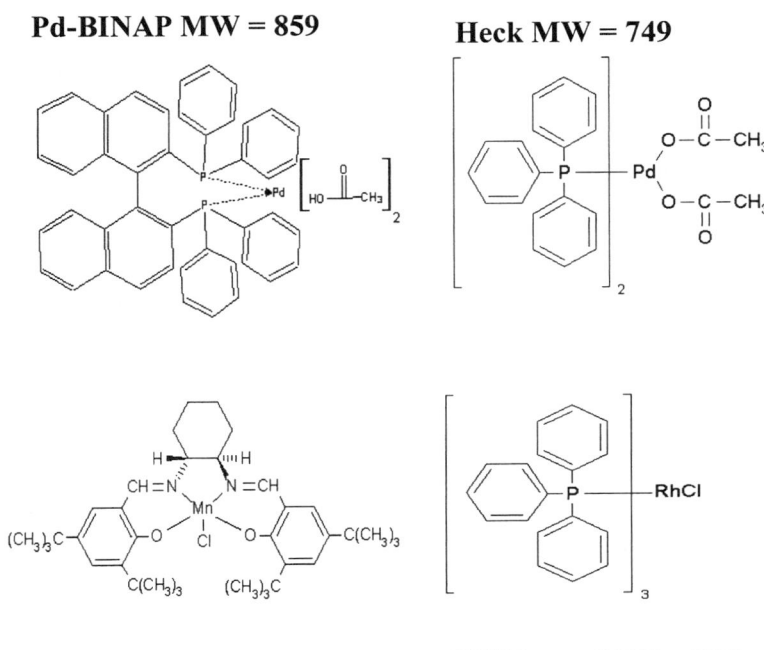

FIGURE 8. Typical organometallic catalysts.

The removal of residual catalysts is particularly important when the molecules being synthesized are for pharmaceutical use, since residual heavy metals are tightly regulated. FIGURE 8 shows some typical OMCs together with their molecular weights. Clearly, it should be possible to separate these from post reaction mixtures using OSN when the product molecules are sufficiently small that they are able to pass through the membrane. Initial attempts in this field[19] focussed on hydrogenation reactions. More recently, further work on hydrogenations[20] and Heck C–C bond coupling reactions[21] has been reported, together with work on retention of a range of OMCs.[22]

The model reaction chosen to demonstrate the recovery and reuse of OMCs is a C–C Heck coupling reaction.[23] The Heck reaction has been the subject of much interest because it is known to be the only single step method for the arylation of olefins. In addition to the problem of difficult postreaction separation (a feature of many homogeneously catalyzed reactions), a further drawback of Heck catalysis is catalyst deactivation due to the inherent instability of the catalyst at certain stages of the catalytic cycle. Consequently, the lifetime of the active catalyst is finite and low. The conventional approach to increasing the productivity of a given catalyst system in Heck chemistry has thus far mostly been by adapting the catalyst ligands, or the use of additives. The means by which the additives impart these advantageous effects remains, in most cases, unclear. Beneficial effects due to temperature and pressure have also been reported. Alternatively, Heck catalyst productivities have been increased by reusing the catalyst after separating it from the products. Liquid–liquid and solid–liquid systems are the most commonly reported. Success was limited due to the trade-off between catalyst efficiency (limited interaction with the substrate) and separation capability (catalyst transfer to reagent phase). Some individual successes have been reported in pseudohomogeneous systems utilizing *in situ* phase separation of catalysts and products, but it is not yet clear whether they will have generic applicability. In near-critical fluids, vaporization of volatile organic reactants reduces conversion, and side reactions at high temperature reduce selectivity and cause product decomposition. This can only be avoided by minimizing product exposure to this effect by accepting lower conversions (20–50%). A major problem for Heck reactions in ionic liquids is the accumulation of inorganic by-products in the reaction medium, limiting its recyclable lifetime. An excellent review of all these techniques is provided in.[23]

Experimental

The model reaction chosen for this work was the coupling of iodobenzene to styrene using $Pd(OAc)_2(P(Phe)_3)_2$ as a catalyst, shown in FIGURE 9. Reactions were performed in a 27 mL glass reactor. $Pd(OAc)_2(P(Phe)_3)_2$, $P(o\text{-tol})_3$, iodobenzene, styrene, triethylamine, and water detailed below were added to the reactor and the volume was made up to 27 mL with the desired solvent. The reactor was placed in a heating jacket controlled at $60 \pm 1.0°C$. Sampling was carried out until a 95–100% conversion of iodobenzene had been reached. The reactor was cooled to 20°C and the postreaction mixture was transferred from the reactor to the nanofiltration cell. 85% of the fed volume was permeated across the polyimide solvent resistant nanofiltration (SRNF) membrane (STARMEM 122, W.R. Grace). The retentate was recycled back to the reactor from which any precipitated salt was removed. The reactor

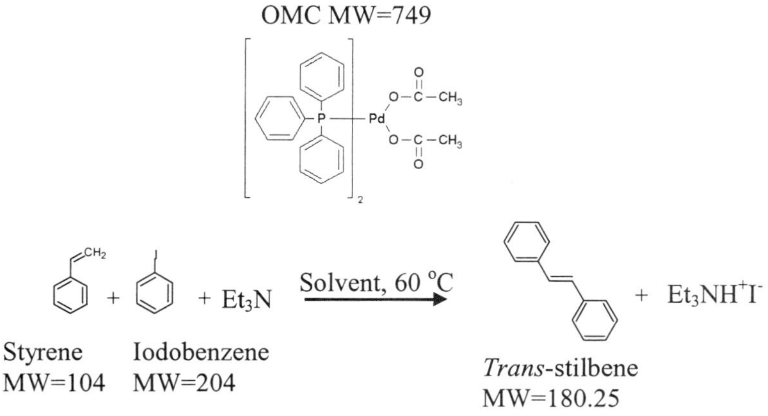

FIGURE 9. The coupling of iodobenzene to styrene using $Pd(OAc)_2(P(Phe)_3)_2$ as a catalyst.

was topped up again with fresh reactants, base and solvent. The reaction was then reinitiated. The volumes and weights used in the two different solvent systems studied are outlined below.

Reactant Concentrations: iodobenzene (2.7 mL, 24.22×10^{-3} mol, 1.5 eq); styrene (2.32 mL, 20.19×10^{-3} mol, 1.25 eq); triethylamine (2.82 mL, 20.19×10^{-3} mol, 1.25 eq); $Pd(OAc)_2(P(Phe)_3)_2$ (0.0485 g, 0.0648×10^{-3} mol, 0.4 mol%); P(o-tol)$_3$ (0.079 g, 0.194×10^{-3} mol, 1.6 mol%).

Solvent Systems: THF, tetrahydrofuran (18.08 mL) and H_2O (1.08 mL, 4 vol%). EA, 50:50 mixture of ethyl acetate and acetone (17.68 mL) and H_2O (1.485 mL, 5.5 vol%). It was found that in the THF system the membrane lost separating power after a single filtration, and so in each filtration a new membrane disc was used. However, in the EA solvent system, the membrane was stable and the same membrane disc was used for all filtrations.

The cumulative turnover number (TON) was calculated as the total moles of product synthesized over repeated reaction cycles per mole of catalyst initially added in the first run.

Results and Discussion

Data from consecutive reactions undertaken using the first solvent system THF are shown in FIGURE 10. The abrupt discontinuity of the iodobenzene conversion line at the end of reaction 1 represents a filtration, addition of fresh reagents, and the start of reaction 2 at zero iodobenzene conversion. The first batch reached near complete conversion in about three hours. In the subsequent filtration, a Pd rejection of only 90% was achieved, with a 2% rejection of *trans*-stilbene. Selectivity of the membrane towards rejecting catalyst had become noticeably worse (note that new membrane discs were used in each filtration for this system). Similar poor Pd rejections were observed after subsequent reactions, with a substantial reaction rate decline occurring—the reaction rate had fallen to below 20% of the initial value by the

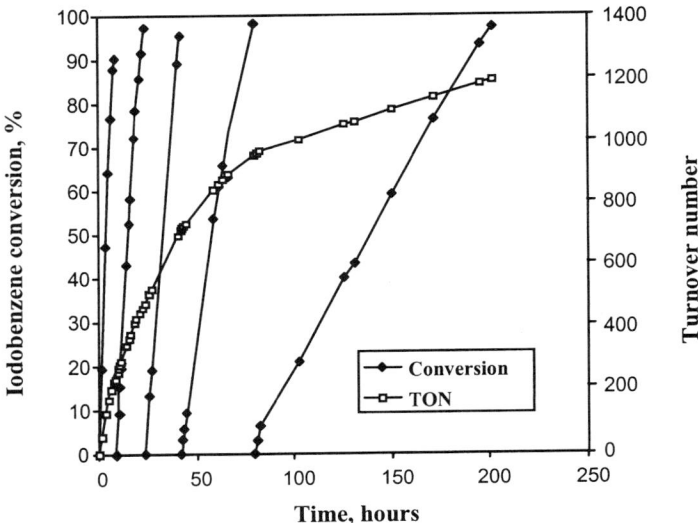

FIGURE 10. Iodobenzene conversion and cumulative turnover number versus time for the THF solvent system.

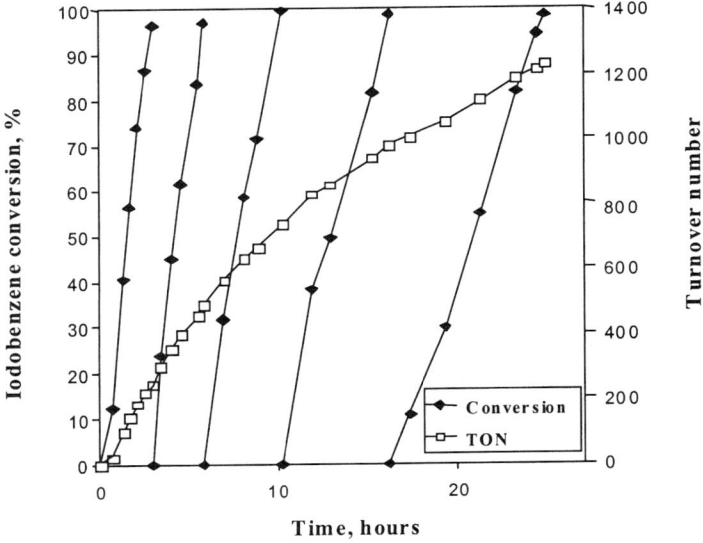

FIGURE 11. Iodobenzene conversion and cumulative turnover number versus time for the EA solvent system.

fourth catalyst recycle (which took 26 h). The sequence was arbitrarily stopped there. Nonetheless, with catalyst loadings of 0.4 mol% (based on styrene), a cumulative TON approaching 1,200 was achieved during five consecutive reactions using membrane separation to enable recycle of the same catalyst between reactions.

To overcome the problem of membrane instability in the THF system, the EA solvent system was applied. The EA solvent mixture was found empirically to provide good solubility for the reactants and catalyst. The data from sequential reactions in this solvent system are shown in FIGURE 11. As noted under EXPERIMENTAL, the membrane was more stable in the EA system and, hence, repeated reactions were performed with the same membrane disc. However, it is also apparent that the reaction rate does slow after three or four reactions, and eventually appears to decline dramatically in rate during reaction 5. A similar effect is noted in FIGURE 10.

The fall in reaction rate is most likely due to the inherent instability of Pd in the Heck catalytic cycle. Although the exact catalytic mechanism is unclear, it is assumed that smaller Pd species are forming during the cycle, including Pd(0) entities that are both small enough to permeate the membrane and unstable enough to aggregate and precipitate as nanoclusters of Pd black.[23] Because active metal centres are lost from the system by either route, the reduced activity will be reflected in a lower reaction rate. The observed solvent flux in both systems is quite different, but it has been demonstrated that this is a property of the solvent–membrane interactions.[7] Despite the loss of reaction rate, in both systems, OSN allows a product stream with lower metal content. This should reduce the downstream processing costs associated with removing the final remnants of catalyst. The same retained catalyst has also been reused multiple times, with the potential for significant cost savings, particularly for the more expensive chiral OMCs.

ENGINEERING AND SCALEUP CONSIDERATIONS

Crossflow OSN

The dead-end cell used for the work with solvent exchanges and homogeneous catalyst recycle is useful for scooping and proof of concept work. It is also interesting to consider how OSN processes might be scaled up to pilot and then fullscale operation. In any such operation, it is likely that a crossflow system, either as plate and frame or spiral wound modules, would be used. We have begun to collect data for crossflow nanofiltration in a crossflow cell with radial flow pattern. The membrane cell uses a membrane disc of 10 cm effective diameter with a channel height of 5 mm above the membrane. The area is 78.5 cm^2. The system studied was toluene with various concentrations of TOABr. The data shown in FIGURES 12 and 13 were collected using membrane discs that had been conditioned through at least one month of operation in the crossflow cell with toluene as a solvent. FIGURE 12 shows data for flux as a function of flowrate in the cross flow cell for a STARMEM 122 membrane. It is clear that at a low concentration of TOABr (0.005 M or 2.7 g L^{-1}) there is no effect of cell flowrate on membrane flux. However, at 0.35 M (191 g L^{-1}) the effect is significant, suggesting that concentration polarization, osmotic pressure, gel layer formation, or another boundary layer related phenomenon may be important.

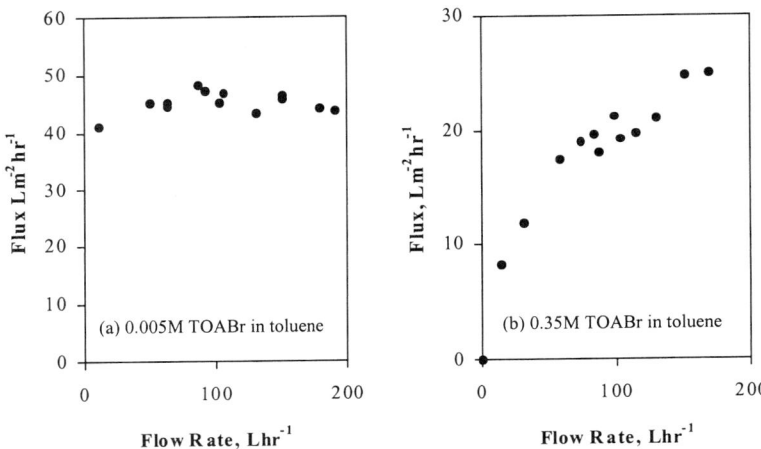

FIGURE 12. Toluene flux as a function of crossflow cell flowrate for STARMEM 122 in the crossflow cell.

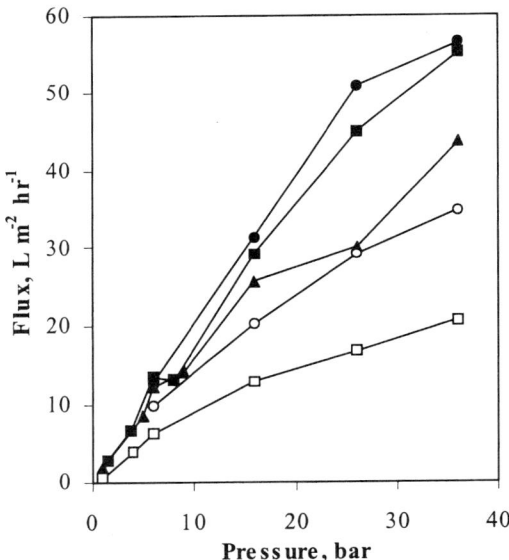

FIGURE 13. Toluene flux versus applied pressure at various TOABr concentrations in the crossflow cell. The cross flow cell flowrate was $60\,L\,h^{-1}$: ●, pure toluene; 0.005 M (■), 0.05 M (▲), 0.1 M (○), and 0.35 M (□) TOAbr.

FIGURE 13 shows toluene flux as a function of applied pressure for different TOABr concentrations. Experiments were conducted with TOABr concentration increasing in order from 0 to 0.35 M (obtained by sequential addition of TOABr into toluene solution). Applied pressures were in random order for each concentration. For each point, the data were collected after the flux reached steady state, and the flux measurement interval time was 30 minutes. One might expect that the lines for each concentration would cut the x-axis at positive applied pressures were osmotic pressure important. A simple calculation using the assumption that the Van't Hoff equation holds and rejection is 100% gives an osmotic pressure (based on bulk solution concentration of 0.35 M) of 8–9 bar. However, the data for flux versus applied pressure show a series of lines, and for each concentration, which tend toward the origin, suggesting that in fact osmotic pressure is low. This is an aspect of OSN where further work is being undertaken at present. It is also of interest to compare the toluene flux at 30 bar applied pressure for 0.35 M TOABr from FIGURE 13 to the data for batch filtrations of TOABr in FIGURE 7. The dead-end cell data ranges in concentration from 0.05 to 0.5 M, and thus the fluxes from the latter stages of the dead-end cell batch can be compared to the crossflow cell data at 0.35 M. Unsurprisingly, the crossflow fluxes are considerably higher than the dead-end cell fluxes, suggesting that boundary layer phenomena at the membrane surface exert an important influence on overall solvent flux.

At a TOABr concentration of 0.35 M, the rejection of TOABr was found to increase from 96% to 99% as applied pressure increased from 2 to 20 bar. Above 20 bar, rejection remained exceeded 99%. This increase in rejection with increasing pressure agrees with results in the literature,[9,22] which we assume to be caused by membrane compaction as pressure increases.

Process ScaleUp Configuration and Economics

FIGURE 14 shows a schematic diagram of how OSN might be linked to a typical chemical reaction system. It is likely that modules would be spiral wound and that a back pressure valve and suitable pump could be used to generate the necessary pressure. The membrane equipment would probably be mounted on a small portable

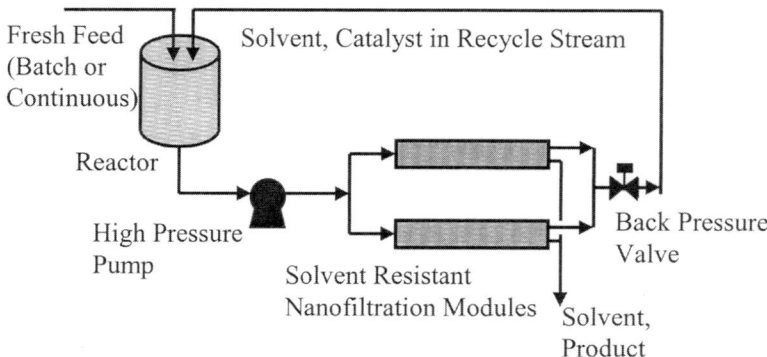

FIGURE 14. Configuration for a full-scale application.

floor skid. A setup, such as that shown in FIGURE 14, could be readily fitted to standard batch chemical reactors.

CONCLUSIONS

In conclusion, it has been demonstrated that the OSN can be applied to a range of organic synthesis operations and that it can offer both reduced environmental impacts and attractive process economics. In solvent exchange processes, it can be used to make possible an exchange of high boiling and low boiling solvents, and can make the exchange process more rapid than conventional distillation. OSN can be generically applied to homogeneous catalyst recycle, provided that there is sufficient membrane selectivity between catalyst and product. The currently available range of SRNF membranes is small but growing, and already they are versatile enough to be used across a broad range of solvents. The key to successful application of this technology to OMC systems remains catalyst stability, which should enable the concept to be applied to a wider range of organic syntheses. Overall we expect the application of OSN to organic synthesis to become an important unit operation in fine chemical and pharmaceutical manufacture.

REFERENCES

1. ANASTAS, P.T. & J.C. WARNER. 1998. Green Chemistry. Theory and Practice. Oxford University Press, Oxford.
2. CURZONS, A.D., J.C. CONSTABLE, D.N. MORTIMER & V.L. CUNNINGHAM. 2001. So you think your process is green, how do you know?—using principles of sustainability to determine what is green—a corporate perspective. Green Chem. **3**(1): 1–6.
3. WHITE, L.S. & A.R. NITSCH. 2000. Solvent recovery from lube oil filtrates with a polyimide membrane. J. Membr. Sci. **179**: 267–274.
4. KULKARNI, S.S., E.W. FUNK & N.N. LI. 1986. Hydrocarbon separations with polymeric membranes. AIChE Sym. Ser. **250**: 78–84.
5. RAMAN, L.P., M. CHERYAN & N. RAJAGOPALAN. 1996. Deacidification of soybean oil by membrane technology. JAOSC **73**: 219–224.
6. ZWIJENBURG, H.J., A.M. KROSSE, K. EBERT, et al. 1999. Acetone-stable nanofiltration membranes in deacidifying vegetable oil. JAOSC **76**: 83–87.
7. MACHADO, D.R., D. HASSON & R. SEMIAT. 1999. Effect of solvent properties on permeate flow through nanofiltration membranes. Part I: investigation of parameters affecting solvent flux. J. Membr. Sci. **163**: 93–102.
8. MACHADO, D.R., D. HASSON & R. SEMIAT. 2000. Effect of solvent properties on permeate flow through nanofiltration membranes. Part II: transport model. J. Membr. Sci. **166**: 63–69.
9. WHU, J.A., B.C. BALTZIS & K.K. SIRKAR. 2000. Nanofiltration studies of larger organic microsolutes in methanol solutions. J. Membr. Sci. **170**: 159–172.
10. BHANUSHALI, D., S. KLOOS, C. KURTH & D. BATTACHARYYA. 2001. Performance of solvent-resistant membranes for non-aqueous systems: solvent permeation results and modeling. J. Membr. Sci. **189**: 1–21.
11. YANG, X.J., A.G. LIVINGSTON & L.M. FREITAS DOS SANTOS. 2001. Experimental observations of nanofiltration with organic solvents. J. Membr. Sci. **190**: 45–55.
12. SHELDON, R.A. 1997. Catalysis—the key to waste minimisation. J. Chem. Technol. Biotechnol. **68**: 381–388.

13. STARKS, C.M. 1971. Phase-transfer catalysis. I. Heterogeneous reactions involving anion transfer by quaternary ammonium or phosphonium salts. J. Am. Chem. Soc. **93:** 195–199.
14. STARKS, C., C. LIOTTA & M. HALPERN. 1994. Phase-Transfer Catalysis: Fundamentals, Applications and Industrial Perspectives. Chapman & Hall, New York.
15. SASSON, Y. & R. NEUMANN. 1997. Handbook of Phase-Transfer Catalysis. Blackie Academic and Professional, London.
16. NAIK, S.D. & L.K. DORISWARMY. 1998. Phase-transfer catalysis: chemistry and engineering. AIChE J. **44:** 612–646.
17. LUTHRA, S.S., X. YANG, L.M. FREITAS DOS SANTOS, *et al.* 2001. Phase transfer catalyst separation and re-use by solvent resistant nanofiltration membranes. Chem. Commun. **16:** 1468–1469.
18. LUTHRA, S.S., X. YANG, L.M. FREITAS DOS SANTOS, *et al.* 2001. Homogeneous phase transfer catalyst recovery and re-use using solvent resistant nanofiltration. J. Membr. Sci. **201**(1–2): 65–75.
19. GOSSER, L.W., W.H. KNOTH & G.W. PARSHALL. 1977. Reverse osmosis in homogeneous catalysis. J. Molec. Catal. **2:** 253–263.
20. DE SMET, K., S. AVERTS, E. CUELEMANS, *et al.* 2001. Nanofiltration-coupled catalysis to combine the advantages of homogeneous and heterogeneous catalysis. Chem. Comm. **7:** 597–598.
21. NAIR, D., J.T. SCARPELLO, L.S. WHITE, *et al.* 2001. Semi-continuous nanofiltration-coupled Heck reactions as a new approach to improve productivity of homogeneous catalysts. Tet. Lett. **42:** 8219–8222.
22. SCARPELLO, J.T., D. NAIR, L.M. FREITAS DOS SANTOS, *et al.* 2002. The separation of organometallic catalysts using solvent resistant nanofiltration, J. Membr. Sci. **203**(1–2): 71–85.
23. BELETSKAYA, I.P. & A.V. CHEPRAKOV. 2000. The Heck reaction as a sharpening stone of palladium catalysis. Chem. Rev. **100**(8): 3009–3066.

Computer-Aided Simulation and Design of Nanofiltration Processes

MOHAN NORONHA,[a] VALKO MAVROV,[b] AND HORST CHMIEL[a,b]

[a]*Institute for Environmentally Compatible Process Technology, Saarbrücken, Germany*

[b]*Department of Process Technology, Saarland University, Saarbrücken, Germany*

ABSTRACT: The modelling of membrane filtration processes is often performed by applying black-box models or short-cut methods, because of the complexity of the molecular interactions on and inside the membrane. The assumptions made for short-cut methods can be applied with accuracy to reverse osmosis processes, whereas the simulation of nanofiltration can lead to unreliable results that sometimes deviate from real conditions to a great extent. A steady-state process simulation, NF-PROJECT, based on input information from membrane characterization, was developed (isothermal operation). The individual separation characteristics of each membrane element are calculated in an iterative sequence, illustrating the successive reduction in permeability and rejection between the elements arranged inside the pressure vessel. The simulation provides information on the increasing feed concentration and osmotic pressure, the hydraulic pressure loss, the deterioration of the flow conditions in the vessel, and the joint performance of the membrane elements to be analyzed. Taking an example from a practical application, a two-stage nanofiltration pilot plant was simulated, the results of which are presented in this article. Examples of optimization potentials are illustrated for the target criteria of economic efficiency (specific energy costs), permeate quality, and flow.

KEYWORDS: nanofiltration; process simulation; optimization; two-stage process

NOMENCLATURE:

A_{mem}	active membrane area, m^2
b_W	van`t Hoff coefficient defined by Equation (**4**), bar cm/mS
c	concentration measured as conductivity, mS/cm
d_h	hydraulic diameter of the feedside channel, m
D_i	diffusion coefficient, m^2/sec
J_P	permeate flow, L/m^{2}h
J_W	water flow, L/m^{2}h
k	coefficient for mass transfer, L/m^{2}h
K_W	membrane constant for water (pure water permeability), L/m^{2}h bar
L_P	effective water permeability, L/m^{2}h bar
L_S	effective salt permeability, L/m^{2}h
Q	volume flow, m^3/h
R	retention, —
u	average feedside velocity, —

Address for correspondence: Horst Chmiel, Institute for Environmentally Compatible Process Technology, Im Stadtwald 47, D-66123 Saarbrücken, Germany. Voice: 0049 681 9345 340; fax: 0049 681 9345 380.
  m.noronha@rz.uni-sb.de

Greek Symbols
α   coefficient for salt permeability, L/m²h
β   coefficient for dependence of salt permeability, —
Δp hydraulic differential pressure, bar
$\Delta \pi$ osmotic differential pressure, bar
φ   recovery, —
η   pump efficiency, —
ν   kinematic viscosity, kg/(m sec)
σ   reflection coefficient, kg/(m sec)

Indices
sys   system input
B   bulk feed
F   feed flow
P   permeate flow
R   retentate flow
W   membrane surface (wall)

INTRODUCTION

Numerous analytical methods that can describe mass transfer on a membrane are available for membrane separation processes, in particular, reverse osmosis and nanofiltration. However, these methods are confined to separation conditions in a small membrane area. In industrial processes, some of which are multistaged, a number of technical membrane elements are combined to achieve the degree of separation required via their joint operation. In process simulation with volume refluxes, calculation of the individual membrane elements with accurate models involves considerable numerical input, resulting in lengthy calculation times. Therefore, in process simulation, the models are supplemented by simplified assumptions for membrane mass transfer as well as for hydraulic flow conditions in the membrane element. It is assumed that membrane rejection is constant, concentration polarization is negligible, and pressure loss or membrane recovery are also constant. In the case of reverse osmosis, technical processes can be simulated approximately by a membrane element analysis that applies these short-cut methods. However, as far as nanofiltration is concerned, the simplified assumptions from the short-cut methods lead to approximation results that deviate considerably from real conditions, thus ruling out their use for plant design.

This paper deals with a practice-related, stationary process simulation for industrial two-stage nanofiltration processes (feed and bleed operation), termed NF-PROJECT and based on a numerical calculation of the individual membrane elements (8-inch spiral wound elements). In a sequential element-by-element analysis the volume flow, concentration, and hydraulic pressure were calculated individually at the inlet and outlet of each membrane element. From these two values average values were then determined for the membrane parameters permeability L_P, transmembrane flux J_W, membrane rejection R based on feed-side concentration c_W on the membrane surface as well as in the bulk feed c_B, permeate concentration c_P, and the hydraulic and osmotic differential pressure Δp and $\Delta \pi$. Furthermore, the retentate/permeate ratio as well as the membrane element recovery were calculated.

Membrane element characterization, which can be performed on a laboratory scale, is required for this simulation. Membrane-specific input parameters can be determined for simulation from characterization via regression methods and then be used to model and integrate the separation behavior of nanofiltration membranes efficiently into the process simulation.[1,2]

The performance of a process at industrial level is simulated by the joint operation of all membrane elements within a certain defined configuration. Examples of the simulation results from a two-stage pilot plant are presented. The major potential for improvement of multistaged membrane processes does not lie in the individual optimization of the filtration stages but rather in their interaction within the entire process. Process simulation can make a substantial contribution to detecting these potentials. For this reason, optimization potentials for the target criteria of economic efficiency (specific energy costs), permeate quality and flow are illustrated.

METHODOLOGY FOR CHARACTERIZATION OF NANOFILTRATION MEMBRANE ELEMENTS AND VERIFICATION OF MODELLING

Modelling Equations for Nanofiltration Elements

Mass transfer through the active layer of the nanofiltration membrane was described by semiempirical transfer approaches based on the solution–diffusion model and the Spiegler–Kedem model.[3] Empirical model equations for characterizing nanofiltration membranes according to Schirg and Widmer were also included.[4] The average of the feedside inlet and outlet values of the membrane elements were used in the model equations to calculate the membrane element separation performance. The model for characterization involves the following equations:

Water Flow

$$J_W = K_W(\Delta p - \Delta \pi). \tag{1}$$

Permeate Flow

$$J_P = J_W + J_S \approx J_W. \tag{2}$$

Concentration Polarization (Feed)

$$J_W = k \cdot \ln\left(\frac{c_W - c_P}{c_B - c_P}\right). \tag{3}$$

Concentration in the Bulk Feed

$$c_B = \frac{1}{2}(c_F + c_R). \tag{4}$$

Osmotic Pressure

$$\Delta \pi = b_W(c_W - c_P). \tag{5}$$

Intrinsic Rejection[3]

$$R = 1 - \frac{c_P}{c_W} = 1 - \frac{1-\sigma}{1 - \sigma \exp\left(\frac{(\sigma-1)J_W}{L_S}\right)}. \quad (6)$$

Salt Permeability[4]

$$L_S = \alpha c_W^\beta. \quad (7)$$

Pressure Loss—8 inch-element[3]

$$\Delta p_{\text{loss}} = 0.008568 \left(\frac{\dot{Q}_F + \dot{Q}_R}{2}\right)^{1.7}. \quad (8)$$

This system of equations can be solved numerically, together with simple mass balances. To model the separation behavior of nanofiltration membranes or elements according to Equations (1) to (8), only six free model parameters need to be determined: pure water permeability K_W, mass transfer coefficient k, van't Hoff coefficient b_W, reflection coefficient σ, and the parameters α and β to define salt permeability as a function of concentration. Pure water permeability K_W was determined over a period of at least 12 hours at various hydraulic pressures Δp of 5, 7, 9, and 11 bar and a temperature of 293 K. Prefiltered, fully desalinated water (ultrapure water, conductivity 0.055 µS/cm, $R = 18.2\,\text{m}\Omega\cdot\text{cm}$) was used in the measurements. The reflection coefficient σ could be adjusted by conducting experiments at low feed-side concentration c_W and high transmembrane flows J_W and was equivalent to the maximum intrinsic rejection value, measurable according to Equation (6). Equation (8) is only required to model separation characteristics and is not important for characterization.

By incorporating Equations (3) and (5) into (1), membrane permeabilities L_P, which were measured at constant temperature in a filtration experiment (e.g., with a 2.5-inch × 40-inch spiral wound element), were calculated as follows:

$$L_P = \frac{K_W}{\Delta p}\left\{\Delta p - b_W(c_B - c_P)\exp\left(\frac{J_W}{k}\right)\right\}. \quad (9)$$

The free parameters b_W and k were adjusted on the basis of experiments with model solutions, in which the set hydraulic differential pressure Δp, the water flow J_W (equivalent to the permeate flow J_P), and the average concentration difference between the feed solution and the permeate were measured.

Equation (9) can be modified so that exponential regression can be performed using the experimental results, thus allowing the parameters b_W and k to be determined,

$$\frac{\Delta p\left(1 - \frac{L_P}{K_W}\right)}{c_B - c_P} = f(J_W) = b_W \exp\left(\frac{J_W}{k}\right). \quad (10)$$

Subsequently, the concentrations on the membrane surface c_W, which allow rejection R to be determined according to Equation (6), can be calculated. A potential regression, which enables the parameters α and β to be adjusted,[1] can be determined by expressing the dependence of salt permeability L_S on concentration c_W.

Experimental Results and Discussion

The six parameters can be determined quite simply and allow separation characteristics to be modelled at various hydraulic pressures and feed concentrations. FIGURE 1 shows the test plant that was used to characterize the spiral wound 2.5-inch × 40-inch membrane elements. Filtration behavior was determined in experiments at constant pressure and temperature (293 K) and by varying the feed concentration using a model solution (NaCl). A steady-state was reached in each case after varying the concentration.

NaCl concentrations were recorded by means of the electrical conductivity. A 2.5-inch × 40-inch spiral wound membrane element (Filmtec NF70, $2.1\,m^2$ membrane area) was used in the experiments at hydraulic pressures $\Delta p = 5, 7, 9$, and 11 bar. Results for the determination of parameters b_W and k (cf., Equation **10**) are shown in FIGURE 2, illustrating that only the results obtained at hydraulic pressures Δp of 9 and 11 bar allow exponential regression. Therefore, the parameters b_W and k cannot be determined at low pressures.

In all the tests, a considerable reduction in retention occurred for high concentrations (low permeate fluxes), a phenomenon typical for nanofiltration membranes. This can be attributed to the membrane fixed charge being shielded by counter ions (Na^+), which causes a reduction in rejection of coions (Cl^-) and, thus, in NaCl. Therefore, the experimental results at high concentrations (or at low water fluxes) were not included in the exponential regression, because equations for charge effects on rejection were not incorporated into the model.

During the filtration of the NaCl solution, a drop in permeate flux occurred, caused solely by the increase in osmotic pressure (as a result of higher feed concentrations). If the mass transfer coefficient k is determined for all the tests, then osmotic

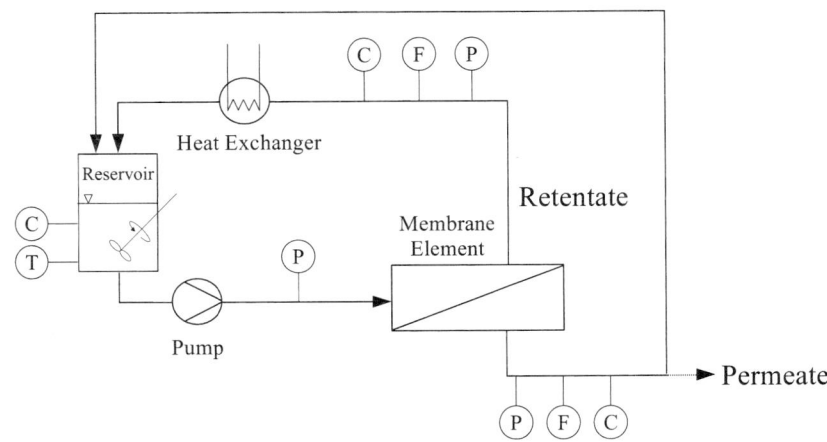

FIGURE 1. Schematic diagram of the test unit: Ⓒ, electrical conductivity; Ⓟ, pressure gauge; Ⓣ, temperature; Ⓕ, flow meter.

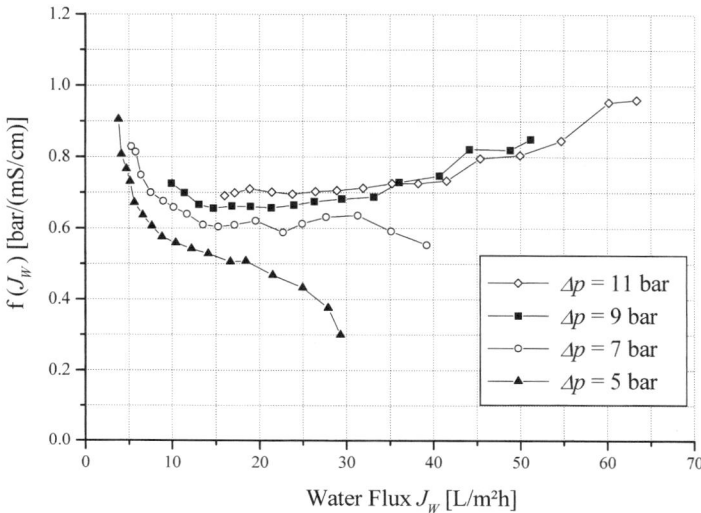

FIGURE 2. Results of characterization tests on a membrane element to determine the parameters b_W and k via exponential regression ($\Delta p = 5, 7, 9,$ and 11 bar).

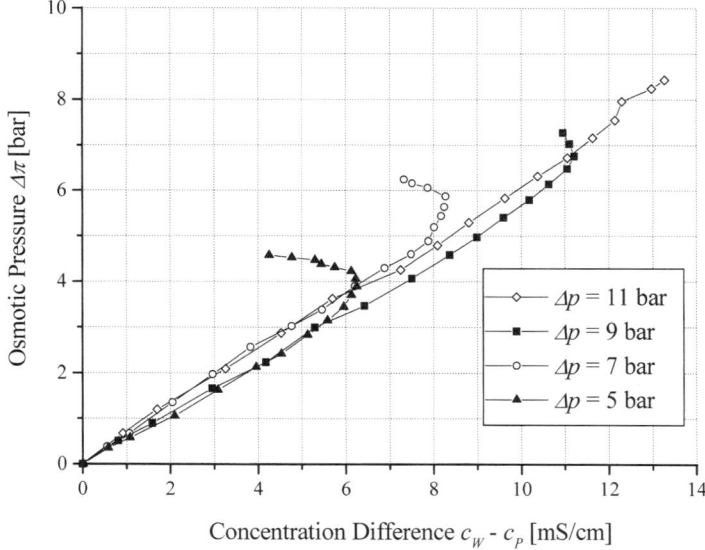

FIGURE 3. Osmotic pressure during the characterization test as a function of concentration difference $c_W - c_P$. Concentration c_W was calculated from the parameter k, determined separately for each pressure value ($k_{5\,\text{bar}}$ and $k_{7\,\text{bar}} = \infty$).

pressure can be depicted (see FIGURE 3) as a function of the concentration difference $c_W - c_P$ in order to substantiate van`t Hoff's law. Concentration polarization ($k = \infty$) was neglected for pressures Δp of 5 and 7 bar.

FIGURE 3 basically illustrates that, at low concentrations, osmotic pressure is a linear function of the concentration difference across the membrane, in accord with van`t Hoff's law. As feed concentration and osmotic pressure increase, a maximum in transmembrane concentration difference is reached across the nanofiltration membrane. Concentration difference decreases whereas osmotic pressure continues to increase. This behavior occurs because of the membrane charge, which influences the rejection property for coions. With increasing concentration the membrane fixed charge is shielded by counter-ions resulting in a lower rejection. The rapidly decreasing rejection decreases the transmembrane concentration difference $c_W - c_P$. This non-ideal behavior of the nanofiltration membrane is not included in the modelling process. At higher filtration pressures, this separation behavior occurs at higher feed concentrations.

FIGURE 4 shows salt permeability L_S as a function of concentration c_W. It can be seen that there are considerable differences for the results at low pressures (5 and 7 bar) whereas the results at high pressures harmonize well, confirming that characterization of the membrane element should be conducted at higher filtration pressures. TABLE 1 shows the parameters determined independently at $\Delta p = 9$ and 11 bar hydraulic pressure. The results of both tests showed good correspondence between the parameters.

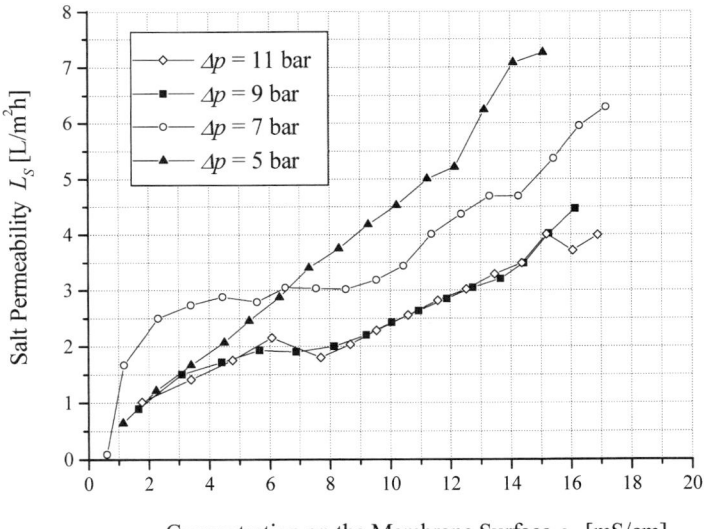

FIGURE 4. Results from the characterization test to determine the parameters α and β via potential regression.

TABLE 1. Summary of the characterization results at $\Delta p = 9\,\text{bar}$ and $\Delta p = 11\,\text{bar}$

Hydraulic Pressure, Δp:	9 bar	11 bar
K_W [L/m²h bar]	6.1	6.1
b_W [bar/(mS/cm)]	0.5841	0.6089
k [L/m²h]	166.204	169.953
σ	0.972	0.97
α	0.6499	0.6562
β	0.6136	0.6063

The characterization process was specially adjusted to a model solution so that the separation behavior of a nanofiltration membrane element could be described more appropriately, from a qualitative viewpoint, than would have been the case by using the generally valid parameter details contained in conventional short-cut methods. Moreover, little numerical input is required to solve the model equations that can be used efficiently for process simulation. The model equations can be verified using the characterization results. In this case, the simulated separation behavior of the membrane element was compared to the experimental results at hydraulic pressures $\Delta p = 5, 7, 9,$ and 11 bar while maintaining the flow conditions at the outlet (see FIGURES 5 and 6). The results from the test at $\Delta p = 11\,\text{bar}$ were used for modelling.

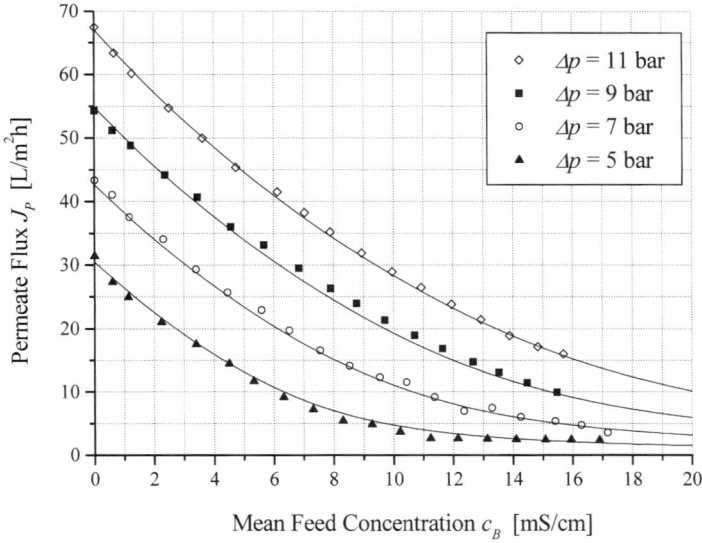

FIGURE 5. Permeate flux as a function of feed concentration c_B at various hydraulic pressure values: comparison of simulation and experimental results (simulation involved only the parameters from the test at $\Delta p = 11\,\text{bar}$).

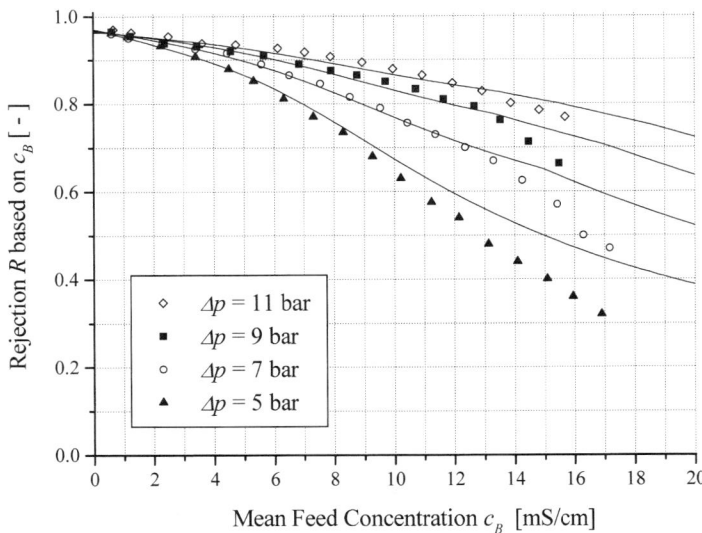

FIGURE 6. Rejection as a function of feed concentration c_B at various hydraulic pressure values: comparison of simulation and experimental results (simulation involved only the parameters from the test at $\Delta p = 11$ bar).

FIGURE 5 shows that the permeate flux J_P of the characterized membrane element in the experimental measuring range was modelled with considerable accuracy. The permeate fluxes at $\Delta p = 5$ and 7 bar could also be simulated using the parameters from the filtration experiment at 11 bar. During the modelling of membrane rejection, deviations, caused by the charge shielding effect of the fixed ions in the membrane, occurred at higher feed concentrations (FIG. 6). Nevertheless the qualitative results of the simulation of membrane rejection were satisfactory.

SIMULATION RESULTS

Simulation of Operating Conditions in a Pressure Vessel

In industrial membrane processes, several eight-inch spiral wound elements are operated in series in a pressure vessels in order to obtain the permeate recovery required (see FIGURE 7). The maximum number of elements is governed by the feed-side concentration, pressure drop and the decrease in flow through the pressure vessel. It is absolutely crucial for the design of a membrane plant, or for correct operational procedure, to check the filtration conditions in the final membrane element.[5–7] A sequential membrane element analysis for nanofiltration processes can be conducted using the model equations, to examine the operating conditions of the individual membrane elements. Since inlet feed volume flows for 8-inch membrane elements with standard feed spacer can fluctuate between 15.6 m³/h ($Re = 310$) and

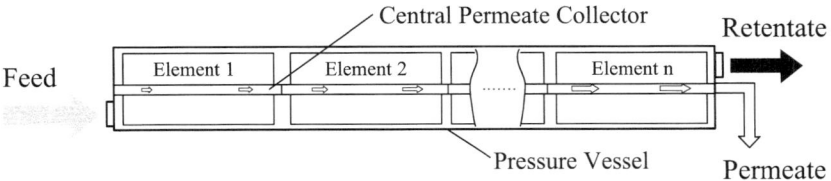

FIGURE 7. Spiral-wound elements operating in series in a pressure vessel.

TABLE 2. Analysis of operation conditions of membrane elements for a pressure vessel containing six nanofiltration spiral wound elements

Model solution H_2O / NaCl	No. of elements 6		Area per element A_{mem} 37.1 m²			
Element Number	1	2	3	4	5	6
Volume flow at inlet [m³/h]	**14.50**	13.27	12.20	11.27	10.48	9.80
Inlet concentration c_F [mS/cm]	**4.00**	4.35	4.70	5.05	5.40	5.74
Inlet pressure (hydr.) Δp_F [bar]	**10.00**	9.25	8.60	8.04	7.54	7.10
Mean hydr. pressure Δp [bar]	9.62	8.93	8.32	7.79	7.32	6.91
Pressure drop $\Delta p_{verlust}$ [bar]	0.75	0.65	0.56	0.50	0.44	0.39
Osmotic pressure $\Delta \pi$ [bar]	3.35	3.49	3.63	3.76	3.87	3.96
Feed concentration c_B [mS/cm]	4.17	4.52	4.88	5.23	5.57	5.90
Membrane surface conc. c_W [mS/cm]	5.26	5.52	5.78	6.05	6.30	6.54
Rejection R (based on c_B) [%]	93.63	93.00	92.33	91.51	90.50	89.29
Rejection R (based on c_W) [%]	94.87	94.26	93.53	92.66	91.60	90.34
Permeate volume flow $\dot{Q}_P$ [m³/h]	1.23	1.07	0.92	0.79	0.68	0.58
Permeate flux J_P [L/m²h]	33.28	28.81	24.87	21.38	18.30	15.59
Permeability L_P [L/m²h bar]	3.46	3.23	2.99	2.74	2.50	2.26
Permeate concentration c_P [mS/cm]	0.27	0.32	0.37	0.44	0.53	0.63
Volume flow at outlet $\dot{Q}_R$ [m³/h]	13.27	12.20	11.27	10.48	9.80	**9.22**
Outlet concentration c_R [mS/cm]	4.35	4.70	5.05	5.40	5.74	**6.06**
Outlet pressure (hydr.) Δp_R [bar]	9.25	8.60	8.04	7.54	7.10	**6.71**
Volume flow ratio $\dot{Q}_R/\dot{Q}_P$	10.74	11.41	12.22	13.21	14.44	15.95
Element recovery $\dot{Q}_P/\dot{Q}_F$ [%]	8.51	8.06	7.57	7.04	6.48	5.90
Total permeate from pressure vessel						
Permeate volume flow $\dot{Q}_P$ [m³/h]	5.28	Recovery				43.54
Permeate conc. c_P [mS/cm]	0.44	Rejection $\bar{R}$ (based on c_B) [%]				91.25
Permeability $\bar{L}_P$ [L/m²h bar]	2.84	Rejection $\bar{R}$ (based on c_F) [%]				89.07

Conditions at the inlet: 14.5 m³/h, c_F = 4.0 mS/cm, Δp_F = 10.0 bar.

approximately 6.0 m³/h ($Re = 119$), the feed side mass transfer coefficient k between the laminar layer at the membrane surface and the bulk feed stream can deviate by ±50%. The mean mass transfer coefficient k can be calculated for each single membrane element, using an equation given by Schock and Miquel; constant diffusion coefficient D_i and kinematic viscosity ν, with the given hydraulic diameter d_h of the feed side channel (spacer thickness), and the average velocity u,

$$k = 0.065 D^{0.75} d_h^{-0.125} \nu^{-0.625} u^{0.875}. \tag{11}$$

TABLE 2 shows examples of the results of an analysis of membrane elements for the following operating conditions at the inlet of the pressure vessel: $c_F = 4.0$ mS/cm, hydraulic pressure $\Delta p = 10.0$ bar, and volume flow $\dot{Q}_F = 14.5$ m³/h. Six spiral wound membrane elements were arranged in series in the pressure vessel. The results in TABLE 2 show that the continuous extraction of permeate throughout the length of the pressure vessel affected the operating conditions of the nanofiltration elements. There was a continuous decrease in permeate volume flow due to the drop in hydraulic pressure and to the increase in osmotic pressure. The increase in feed-side concentration and the drop in permeate flow resulted in a decrease in rejection and an increase in permeate concentration.

The analysis of the membrane elements was instrumental to the simulation of various inlet conditions for the pressure vessel, which enabled the operating conditions of the individual elements to be examined.

Simulation of a Nanofiltration Process Using NF-PROJECT

Detailed analysis of membrane elements for pressure vessels containing spiral wound elements is also conducive to the calculation and simulation of nanofiltration processes at industrial level. On the basis of this analysis, a simulation program entitled NF-PROJECT, that is instrumental to the design and simulation of nanofiltration processes, was developed for a two-stage nanofiltration process (feed and bleed operation, permeate-staged). All important process parameters were calculated on the basis of the input of the relevant information (characterization, configuration, and

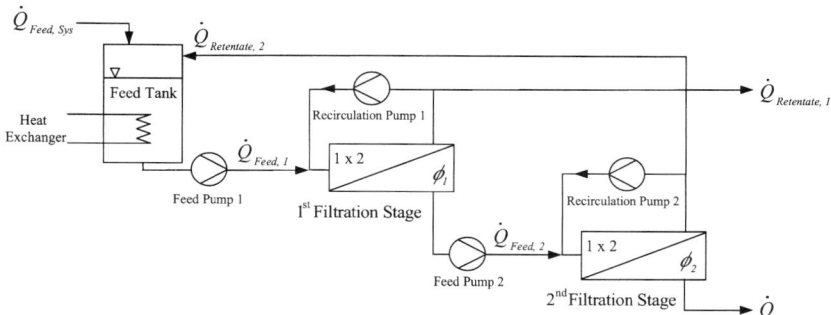

FIGURE 8. Schematic flow diagram of the operating pilot plant. First filtration stage, 2×8-inch nanofiltration spiral wound elements; second filtration stage, 2×8-inch nanofiltration spiral wound elements. $\phi_1 = \dot{Q}_{\text{feed, 2}}/\dot{Q}_{\text{feed, 1}}$, $\phi_2 = \dot{Q}_{\text{permeate, 2}}/\dot{Q}_{\text{feed, 2}}$.

TABLE 3. Configuration and constant process parameters for the pilot plant

Configuration/Process Parameter	Details
Number of pressure vessels	1 for each filtration stage
Number of spiral wound elements	2 × 8-inch nanofiltration elements for each pressure vessel
Active membrane area	37.1 m² per spiral wound element
Recirculation volume flow	10 m³/h (measured on the recirculation pump)

process parameters). In addition to process simulation, the costs, incurred by the electric power for the pumps (feed pumps and circulation pumps), can be calculated. A pilot-scale two-stage nanofiltration process, which is currently in operation within the framework of a research project for the production of water of drinking quality from low-contaminated process water in the food industry, was selected as an example.[8] A schematic representation of the two-stage membrane plant is shown in FIGURE 8.

Hydraulic pressure in the second filtration stage depends on the permeate volume flow of the first stage. To generate the necessary flow rate through the spiral wound elements, as well as high sub-recoveries, most of the retentate from the filtration stages is recirculated via recirculation pumps; only a small quantity is extracted from the filtration stages. The retentate volume flow from the second stage is fed back to the feed tank.

Due to internal recirculation, the simulation exercise of a process state can only be solved iteratively. To shorten calculation times, the iteration loops are severed once sufficient accuracy has been attained. TABLE 3 contains the configuration data of the pilot plant needed as input material for process simulation. In addition to the information contained in TABLE 3, the simulation program NF-PROJECT requires the characterization parameters of the nanofiltration membrane elements (TABLE 1) and further input information, such as operating pressure in the first filtration stage Δp_1, the partial recoveries of both filtration stages ϕ_1 and ϕ_2, the feed concentration $c_{feed,sys}$, and the overall efficiency η of the pumps.

The process simulation for the pilot plant was conducted for examples under $\Delta p_1 = 10$ bar, $f_1 = 75\%$, and $f_2 = 85\%$ (equivalent to a total system recovery of $f_{sys} = 71.8\%$), $c_{feed,sys} = 3.5$ mS/cm, and $\eta = 40\%$ for all the pumps. Some of the results are shown in the flow diagram of FIGURE 9. The exact operating data for the individual membrane elements in both filtration stages are contained in TABLE 4.

OPTIMIZATION POTENTIAL FOR VARIOUS TARGET CRITERIA

By means of NF-PROJECT, a two-stage nanofiltration process (cf., FIG. 8) can be optimized so that target criteria, such as economic efficiency (specific energy costs), permeate quality, or flow, can be improved or coordinated more appropriately. In a process analysis study, the partial recoveries ϕ_1 and ϕ_2 for both filtration stages were varied at constant system yield ϕ_{sys}. Based on practical experience, maximum partial recoveries $\phi_{1,max}$ and $\phi_{2,max}$ were set at 0.9 (90%) for both filtration stages. The

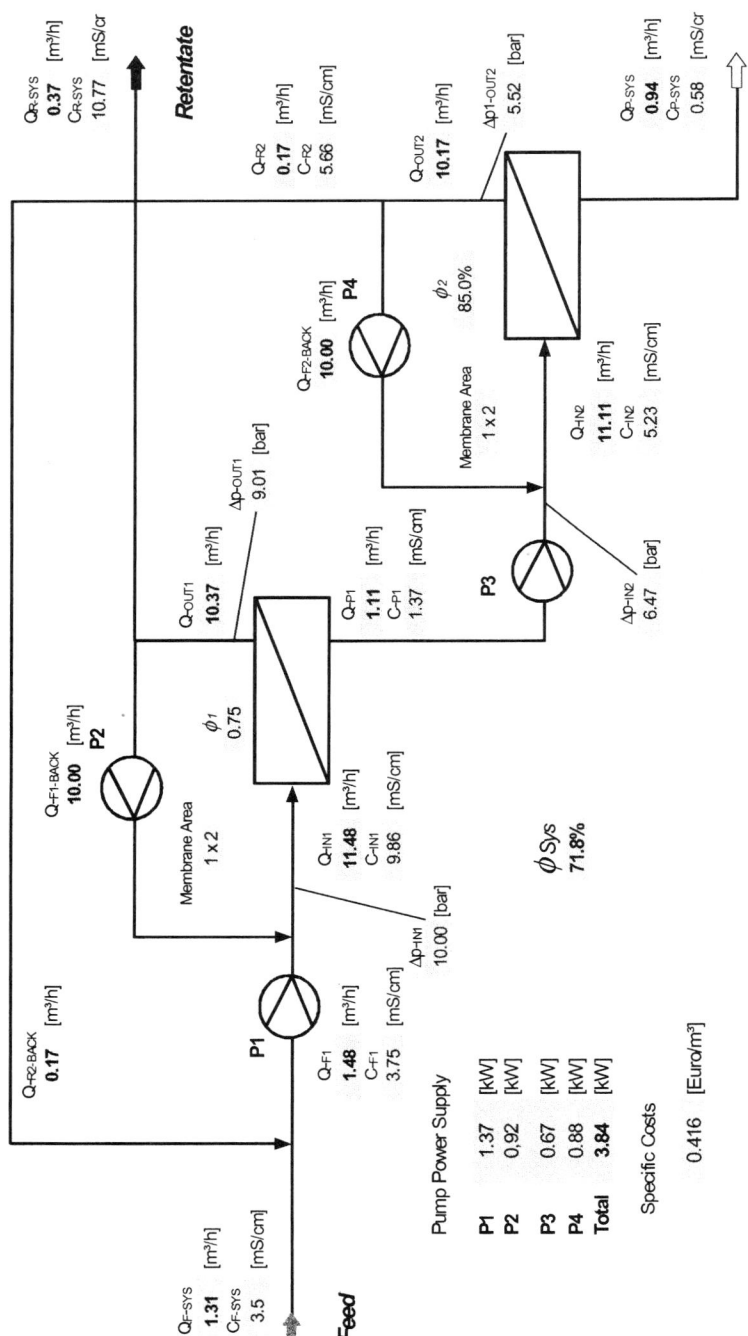

FIGURE 9. Flow sheet representation of the results of process simulation.

combinations of the partial recoveries ϕ_1 and ϕ_2 that are possible for various system recoveries (ϕ_{sys} = 50, 60, and 70%) are shown in FIGURE 10.

FIGURE 11 illustrates the effects of varying partial recoveries on potential target criteria such as specific energy costs (pure pump energy), permeate quality and flow. The results were obtained at a system yield of ϕ_{sys} = 70%, filtration pressure (hydraulic feed inlet pressure) of Δp = 10 bar in the first filtration stage, and a feed concentration of 4.0 mS/cm. The results show that, as far as specific energy costs and flow are concerned, an optimum operating point can be detected if the partial yield in the first

TABLE 4. Results of process simulation for the pilot plant

	First Filtration Stage		Second Filtration Stage	
Volume flow in stage $\dot{Q}_{feed}$ [m³/h]	**1.54**		**1.11**	
Feed concentration c_{feed} [mS/cm]	**3.76**		**1.37**	
Element Number	1	2	1	2
Volume flow at inlet $\dot{Q}_F$ [m³/h]	11.48	10.88	11.11	10.59
Inlet concentration c_F [mS/cm]	9.86	10.35	5.23	5.46
Inlet pressure (hydr.) Δp_F [bar]	10	9.48	6.47	5.97
Mean hydr. pressure Δp [bar]	9.74	9.24	6.22	5.74
Pressure drop $\Delta p_{verlust}$ [bar]	0.52	0.48	0.49	0.46
Osmotic pressure $\Delta \pi$ [bar]	6.69	6.71	3.58	3.59
Feed concentration c_B [mS/cm]	10.11	10.56	5.35	5.56
Membrane surface conc. c_W [mS/cm]	11.24	11.51	5.87	5.99
Rejection R (based on c_B) [%]	87.49	85.75	90.06	88.51
Rejection R (based on c_W) [%]	88.75	86.93	90.95	89.33
Permeate volume flow $\dot{Q}_P$ [m³/h]	0.61	0.50	0.52	0.42
Permeate flow J_P [L/m²h]	16.37	13.59	13.98	11.40
Permeability L_P [L/m²h bar]	1.68	1.47	2.25	1.98
Permeate concentration c_P [mS/cm]	1.26	1.51	0.53	0.64
Volume flow at outlet $\dot{Q}_R$ [m³/h]	10.88	10.37	10.59	10.17
Outlet concentration c_R [mS/cm]	10.35	10.77	5.46	5.56
Outlet pressure (hydr.) Δp_R [bar]	9.48	9.01	5.97	5.52
Volume flow ratio $\dot{Q}_R/\dot{Q}_P$	17.9	20.6	20.04	24.1
Element recovery $\dot{Q}_P/\dot{Q}_F$ [%]	5.3	4.6	4.7	4.0
Permeate volume flow [m³/h]	**1.11**		**0.94**	
Permeate concentration [mS/cm]	**1.37**		**0.58**	
Retentate volume flow [m³/h]	**0.37**		**0.17**	
Retentate concentration [mS/cm]	**10.77**		**5.66**	

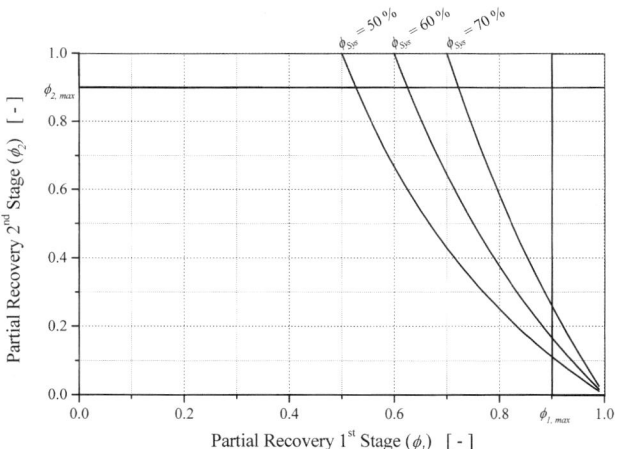

FIGURE 10. Possible combinations for the partial recoveries ϕ_1 and ϕ_2 in both filtration stages, while maintaining system recovery, ϕ_{sys}.

filtration stage can be kept at a low value. The partial yield in the second stage must be increased accordingly in order to maintain system yield. Since the maximum partial yield in the second stage was set at 0.9, the optimum operating point ($\phi_{1,opt}$, $\phi_{2,opt}$) was detected at (0.723, 0.9). However, it should be noted that a relatively poor permeate quality was obtained for this operating point. Permeate quality could be improved by reducing the partial yield in the second filtration stage (and increasing the partial yield in the first filtration stage); the flow and thus the specific energy costs decreased simultaneously. Concerning the target criteria, permeate quality, the optimum operating point ($\phi_{1,opt}$, $\phi_{2,opt}$) was obtained by increasing the partial yield in the first filtration stage as far as possible.

CONCLUSIONS

The simulation program NF-PROJECT is conducive to designing, simulating, and optimizing two-stage nanofiltration plants (feed and bleed operation, steady-state, isothermal operation). Starting from membrane screening and selection, followed by laboratory-scale membrane element characterization, six membrane parameters were determined that allowed the separation characteristics of eight-inch nanofiltration membrane elements to be modelled. Subsequently, the structure of the process could be specified including the number of pressure vessels operated in parallel and the membrane elements (eight-inch spiral wound elements) arranged in series in each pressure vessel for one or two filtration stages (a maximum of six elements in series).

Thus, a nanofiltration process for a certain separation problem can be simulated. NF-PROJECT is instrumental to plant design and process analysis in accord with

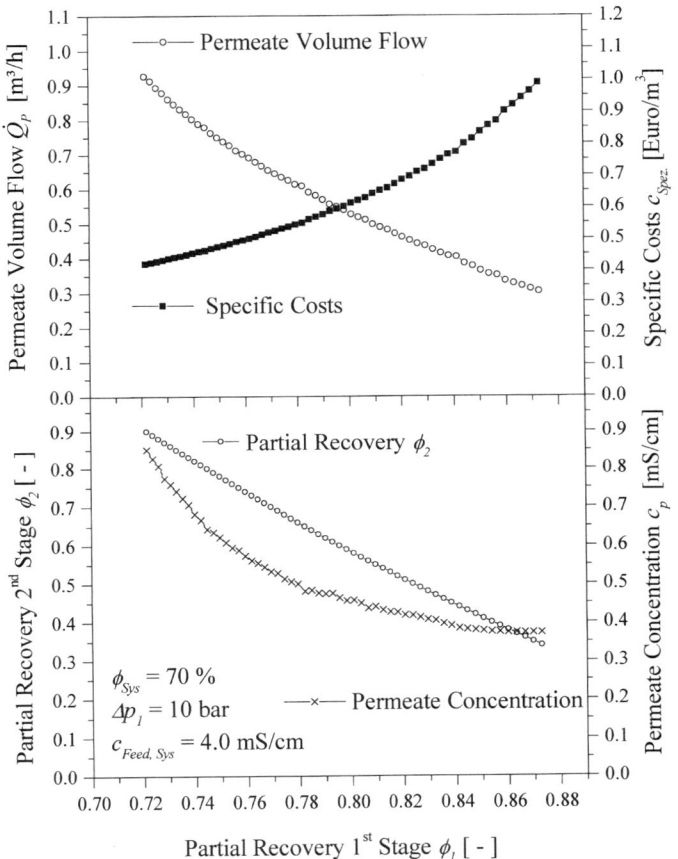

FIGURE 11. Optimization of a two-stage NF process based on various target criteria (economic efficiency, permeate quality, and volume flow) by varying the partial recoveries.

various key parameters such as maximum permissible permeate or minimum retentate concentration required. In addition, the operation of existing plant can be optimized since the flexibility of the process model enables an analysis under various aspects to be conducted (e.g., retention, permeability, differential pressure, osmotic pressure, volume flow and concentration, and power consumption).

REFERENCES

1. NORONHA, M., V. MAVROV & H. CHMIEL. 2002. Simulation model for optimisation of two-stage membrane filtration plants—minimising the specific costs of power consumption. J. Membrane Sci. **202:** 217–232.
2. NORONHA, M., V. MAVROV & H. CHMIEL. 2002. Rechnergestützte Simulation und Auslegung von Nanofiltrationsprozessen. Chemie Ingenieur Technik. **74**(1–2): 70–75.

3. KEDEM, O. & K. SPIEGLER. 1966. Thermodynamics of hyperfiltration (reverse osmosis): criteria for efficient membranes. Desalination **1:** 311–326.
4. SCHIRG, P. & F. WIDMER. 1992. Characterisation of nanofiltration membranes for the separation of aqueous dye salt solutions. Desalination **89:** 89–107.
5. WILBERT, M.C., *et al.* 19998. The Desalting and Water Treatment Membrane Manual. Water Treatment Engineering and Research, Denver, Colorado.
6. RAUTENBACH, R. 1997. Membranverfahren—Grundlagen der Modul- und Anlagenauslegung. Springer-Verlag.
7. RAUTENBACH, R. & R. ALBRECHT. 1989. Membrane Processes. John Wiley & Sons.
8. MAVROV, V., H. CHMIEL & E. BÉLIÈRES. 2001. Spent process water desalination and organic removal by membranes for water reuse in the food industry. Desalination **138:** 65–74.

Advances in Solvent-Resistant Nanofiltration Membranes

Experimental Observations and Applications

D. BHANUSHALI AND D. BHATTACHARYYA

Department of Chemical and Materials Engineering, University of Kentucky, Lexington, Kentucky, USA

ABSTRACT: Nanofiltration (NF) and reverse osmosis (RO) are well-established membrane technologies for applications involving aqueous streams. The principles of NF transport (diffusion, convection, and Donnan exclusion) are effectively used to develop novel membrane materials and applications in aqueous medium. Use of NF in a non-aqueous medium holds strong potential for the food, refining, and pharmaceutical industries because of the low energy costs involved with such membrane processes. Further understanding and development of solvent-resistant NF membranes provides opportunities for various hybrid processing ranging from reactor–membrane to distillation–membrane combinations. This paper provides a comprehensive overview of literature results and our own work in the area of non-aqueous systems. For solvent-based systems, potential membrane swelling and solvent–solute coupling needs to be considered for membrane design and transport theories. A simplified transport theory for pure solvents has been developed using solvent (molar volume, viscosity) and membrane properties (membrane surface energy). This model and has been verified with literature data for both hydrophilic and hydrophobic membranes. Membrane characterization and preconditioning aspects need to be given serious consideration for evaluating membrane performance. In addition to permeability and separation results, some novel applications of NF in non-aqueous solvents are included in this paper.

KEYWORDS: solvent-resistant membrane; nanofiltration membrane

INTRODUCTION

Traditional nanofiltration (NF) and reverse osmosis (RO) processes have found many applications in fields ranging from water and waste-water treatment to pharmaceutical to hazardous waste treatment. Because of the development of thin-film composite membranes, high flux and high separation capabilities have been achieved. NF has recently gained popularity as a membrane process for aqueous systems because of the versatile nature of its transport mechanisms (diffusion, convection, and Donnan-exclusion). These membranes have higher surface charge density and slightly open membrane morphology in comparison with dense RO membranes. As

---

Address for correspondence: D. Bhattacharyya, Department of Chemical and Materials Engineering, University of Kentucky, Lexington, KY 40506-0046, USA. Voice: 859-257-2794; fax: 859-323-1929.

db@engr.uky.edu

a result, the separation efficiency of such membranes is lower (for monovalent ions like Cl$^-$) than for RO membranes, but they can be operated at lower pressure to achieve similar fluxes. This unique nature of NF has been used in several areas such as: textile effluents, dye salts, and pigments, water softening/purification, lactic acid removal from fermentation broth and whey, and the pulp and paper industry.[1] Most current NF membranes have negatively charged groups (such as carboxylic and sulfonic) and thus provide considerable selectivity based on both size and charge. The direct extension of these membranes to non-aqueous solvents (particularly non-polar solvents) is not possible because of low flux and solvent stability problems. For a typical NF membrane the water permeability is on the order of 2×10^{-4} cm^3/cm^2 sec·bar, but unfortunately, for the same membrane (assuming solvent-resistance), the permeability to solvents such as hexane is reduced to 1/10 to 1/100 of the water value. In addition to solvent-resistant membranes, the role of backing materials and module element considerations is critical to applications in non-aqueous media.

Nanofiltration processes are finding wider applicability in non-aqueous media, although fullscale applications are rather limited because of the limited number of available commercial solvent-resistant membranes. Intuitively one would expect that the principles of conventional membrane process as would be applicable to complex non-aqueous systems, to recover and recycle solvents and compounds of interest. With the need for energy conservation and strict environmental regulations, several industries are looking at hybrid processes in which part of the stream can be treated by less energy-intensive membrane processes. Potential applications can be cited in the food and pharmaceutical industries, where the cost of downstream processing accounts for the major cost of the end product. Membrane processes are unique because of their low energy consumption, better product quality, and lower emissions to the environment; thus, can provide lucrative alternatives to traditional unit operations.

With the above comments in mind, this paper provides an overview of the recent advances and applications of NF in the area of non-aqueous systems and of our work with respect to quantifying solvent flux and solute retention behavior. In particular, the role of solvent polarity in permeate flux, transport mechanisms, importance of solvent in solute separations, and the importance of membrane preconditioning are discussed. Suitable comparisons with the aqueous-system literature are also made.

BACKGROUND AND RELEVANT LITERATURE DATA INCLUDING OUR RESULTS

Sourirajan expressed the scope for an extension of RO/NF principles in the literature as early as 1964, in his article in *Nature*.[2] The study primarily demonstrated the importance of polymer–solvent interactions and how they can be used to alter the separation behavior of mixed organic solvent feeds. The primary limitation of this extension is the solvent resistance of commercial polymeric membrane materials. Schmidt *et al.*[3] and Koseoglu *et al.*[4] summarized their findings on the solvent resistance of commercial polymers. For example, they reported a polyvinylidene fluoride (PVDF)-based membrane to be unstable in hexane media. The reason for the instability can be attributed, in certain cases, to the membrane backing material. Several research groups have reported observations in non-aqueous media and their findings are summarized in the following section, together with research work conducted in our laboratory at the University of Kentucky.

Pure and Mixed Solvent Transport

Historically, Sourirajan et al.[2] demonstrated the possibility of using RO/NF for non-aqueous media by studying separation characteristics of cellulose acetate membranes for three binary mixtures involving ethanol, n-heptane, and p-xylene. The experimental study clearly illustrated the phenomenon of preferential transport. For example, for the ethanol–n-heptane and ethanol–p-xylene systems, ethanol preferentially transported through the membrane. The direction of separation was reversed when a hydrophobic layer was coated on a cellulose acetate layer. These results enabled later researchers to understand the importance of polymer–solvent interactions. Paul et al.,[5–7] on the other hand, studied pure solvent transport through lightly crosslinked natural rubber membranes, as early as 1970. They observed reasonable amounts of swelling, corroborated by the non-linear behavior of flux versus the applied pressure. The saturation flux values and the corresponding pressures depended on the type of solvent used. The authors developed a solution–diffusion model for swollen membranes and used polymer volume fractions as the driving force, instead of applied pressure, to explain the non-linear behavior.[8] This behavior clearly illustrated the importance of physical properties of the solvent and the membrane for development of any unified transport theory for dense membranes. Aminabhavi et al.[9–11] also studied the sorption and diffusion of organic solvents through rubbery films, the results of which can be used to understand NF behavior. They studied the importance of parameters, such as crosslinking density and the effects of temperature on the sorption and diffusion of organic solvents. They reiterated the importance of hydrogen-bonding capacity of polymers containing hydrophilic groups, which enhance the solvent-uptake capacity.

Machado et al.[12] systematically studied the permeation properties of several organic solvents (polar and non-polar) by using hydrophobic silicone-based commercial membranes (Koch MPF 50). Their results indicate that the flux of non-polar solvents (pentane to octane) is considerably higher than that of the polar solvents (methanol and ethanol). FIGURE 1 shows solvent permeability (normalized with respect to ethanol) data, as reported for MPF 50[12] and our work with a recently developed Osmonics silicone-based Membrane D. The permeability, normalized with ethanol, was used to eliminate variability between membranes and for ease of comparison. As expected, the pentane flux is much greater that the water flux. The trend of our data followed that of the MPF-50 data. Machado et al. proposed a simple resistance-in-series model to explain the permeation characteristics of pure solvents.[13] They partitioned the membrane into three different transport zones, the NF surface layer, the UF sublayer, and the porous support. The final equation developed can be represented as follows:[13]

$$J_i = \frac{\Delta P}{k_1[(\gamma_C - \gamma_{1\nu}) + f_1\mu] + f_2\mu}, \quad (1)$$

where J is the solvent flux, f_1 and f_2 are solvent independent parameters respectively characterizing the NF and the UF sublayers, k_1 is a solvent parameter, γ_c is the critical

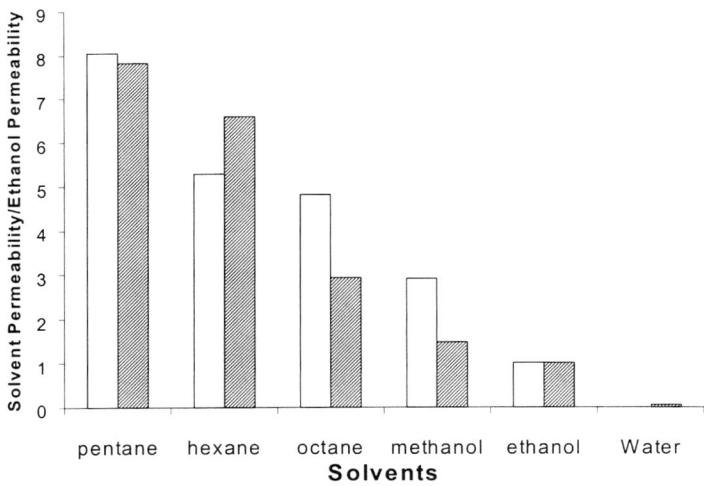

FIGURE 1. Permeabilities of polar and non-polar solvents through hydrophobic Koch MPF-50 membranes ▨ (data for MPF-50 taken from Ref. 12) and hydrophobic silicone-based (Osmonics) Membrane D ☐.

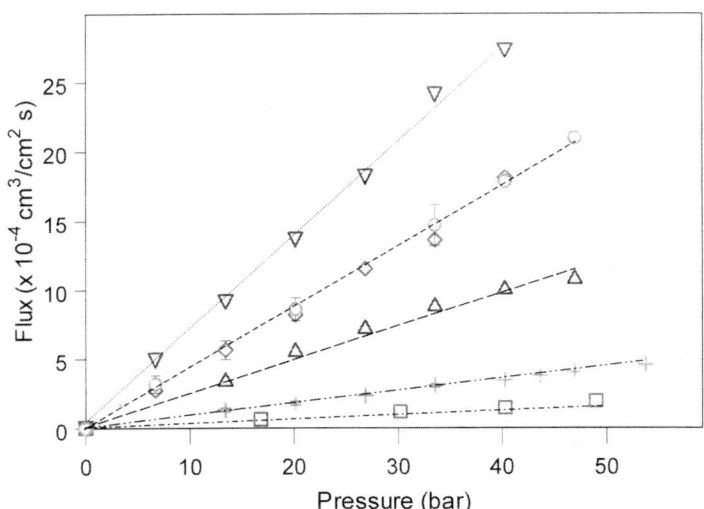

FIGURE 2. Effect of pressure on solvent flux for various solvents through a hydrophobic silicone based nanofiltration membrane (Membrane D):[14] + ethanol, △ methanol, ☐ IPA, ◇ octane, ▽ pentane, ○ hexane.

surface tension of the membrane material, and γ_{1v} is the surface tension of the solvent. The parameters f_1 and f_2 were determined from experimental data, as well as from surface tension measurements.

The observed variations in solvent flux with pressure for our experimental data for Membrane D are shown in FIGURE 2. It can be clearly seen that the solvent flux variation with pressure is linear, unlike that reported by Paul et al.[5–7] for lightly crosslinked natural rubber membranes. A comparison of solvent transport through hydrophilic and hydrophobic membranes can be made and clearly illustrates the importance of polymer–solvent interactions. For example, the solvent permeability data for a brackish water polyamide-based RO membrane (DS-11 AG, provided by Osmonics, Inc.) are shown in FIGURE 3.[14] It can be clearly seen that the trend of solvent permeability is nearly reversed (when compared to the hydrophobic membrane data shown in FIG. 1). The size dependence within a given homologous series is always prevalent.

To obtain a unified correlation of the solvent permeability for hydrophobic solvent-resistant polymeric membranes we used solvent properties (such as viscosity and molar volume) and membrane properties (such as surface energy) to describe the flux behavior. When extending the principles from water to organic solvents, affinity of the solvent for the membrane material becomes an extremely important issue. The final equation developed[14] is

$$J \propto A \propto \left(\frac{V_m}{\mu}\right)\left(\frac{1}{\phi^n \times \gamma_{sv}}\right), \tag{2}$$

where A is the solvent permeability, V_m and μ are, respectively, the molar volume and viscosity of the solvent, γ_{sv} is the membrane surface energy, and ϕ is the sorption value of the solvent for the membrane material. For example, with Membrane D

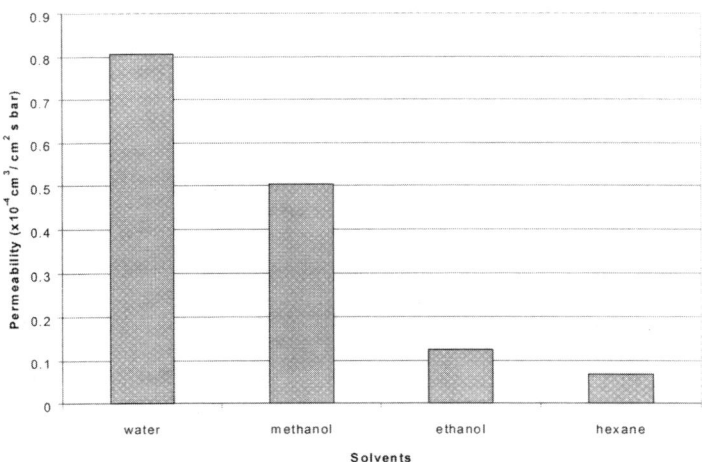

FIGURE 3. Permeabilities of various solvents through a hydrophilic reverse osmosis brackish water membrane (DS 11 AG).[14]

and hexane as a solvent the values used (at 25°C) in Equation (2) were: V_m = 131.6 cc/mole, μ = 0.32 cP for hexane, ϕ = 0.7, and γ_{sv} = 15 dyne/cm. FIGURE 4 shows the plot obtained using the above parameters.[14] A value of n = 0.25 gave the best fit to the data. As has been mentioned, a normalized permeability was used to ensure accurate comparison among various membranes. The data used to obtain the correlation includes our experimental data with Membrane D and DS 11-AG membranes, and data reported in Reference 12 for MPF-50 membranes (room temperature results together with acetone permeation data at various temperatures), and cellulose acetate data. The correlation includes 12 different organic solvents (polar and nonpolar) and four different membrane materials (hydrophilic and hydrophobic) and has a correlation coefficient of 0.78.

On the other hand, extension of pure solvent results to mixed solvents poses interesting challenges. For example, Machado et al.[12] conducted experiments with mixed solvents using the hydrophobic MPF-50 membrane. Experiments were performed first with acetone followed by the desired solvent or solvent mixture. Each set of experiments was carried out with a different membrane coupon following similar conditioning techniques to eliminate the effects of membrane history. FIGURE 5 shows the flux behavior for several polar and non-polar solvents with acetone. It can be seen that as the solvents become more non-polar, the resultant flux behavior becomes non-linear. In certain cases the flux of the mixture was lower than that of the pure solvents (minimum for the acetone–paraffin cases). The non-linear behavior

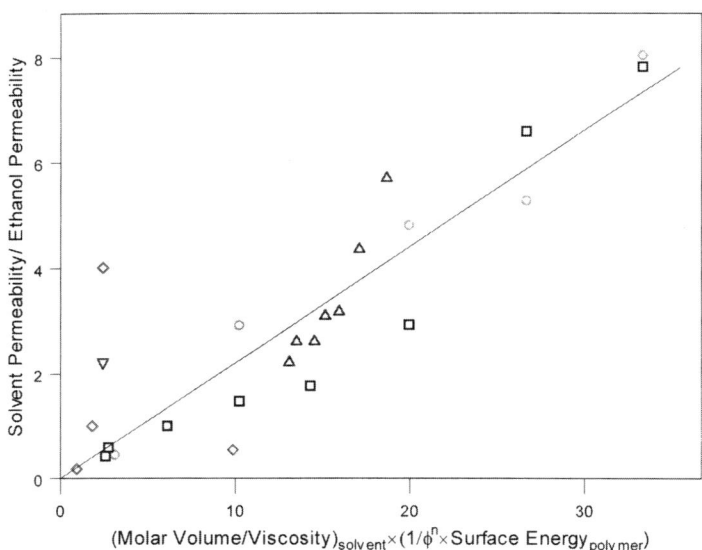

FIGURE 4. Correlation of solvent normalized flux (using ethanol flux) of polar and non-polar solvents for several hydrophilic and hydrophobic membranes:[14] ○, Membrane D (our experimental data); □, MPF 50 room temperature data (taken from Machado et al., 1999); △, MPF 50 acetone temperature data (taken from Machado et al., 1999); ◊, DS 11 AG (our experimental data); ▽, cellulose acetate data (taken from Sourirajan et al., 1985).

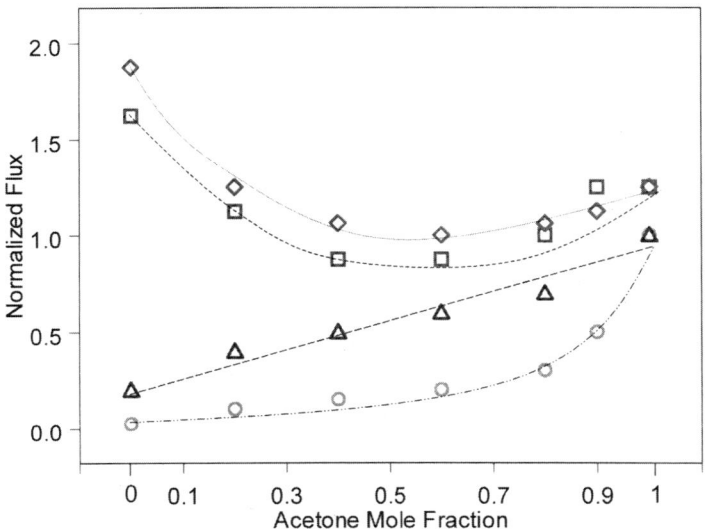

FIGURE 5. Mixture flux (normalized with respect to acetone flux) versus acetone mole fraction for a Koch MPF-50 membrane (data taken from Machado et al.[12]): ○ pentanol–acetone, △ ethanol–acetone, □ hexane–acetone, ◇ pentane–acetone.

of mixed solvent flux values can result from viscosity dependence and solvent mixture non-idealities and the preferential wettability of one solvent. For the acetone–paraffin cases, the acetone flux is slightly higher than that for the acetone–alcohol cases. This is an anomaly in the reported data that may have occurred due to the erroneous use of a single acetone flux value for all the cases.

Solute Transport

Some of the earliest work on solute transport in non-aqueous media dates back to the mid1970s with the work of Paul et al.,[15] who studied the diffusion characteristics of Sudan IV (384 MW organic dye) in various solvents through uncrosslinked natural rubber membranes. As expected, the role of solvent was quite profound. For example, the diffusivity of Sudan IV with hexane as solvent was 200 times higher than that observed for ethanol and, subsequently, the permeability in hexane is 76 times that in ethanol. They explained the results using the solution–diffusion model, taking into account the role of membrane swelling. On the other hand, Farnand et al.[16] studied the transport behavior of polar organic solutes (dimethyl aniline, acetonitrile, etc.) in a methanol medium and used the surface force–pore flow model to explain the results. The pore-flow model uses the potential function as a measure of the interaction between the solute and the membrane material. The calculations show that the constant in the potential function depends on the type of solvent implying various attractive–repulsive forces with the membrane material.

The separation behavior[17–19] of various dyes ranging from charge differences to size have been reported for various hydrophilic (MPF 44, Osmonics Desal-5, and Toray UTC-20) and hydrophobic (MPF 50 and MPF 60) commercial membranes. The specific solutes used were included positively charged Safranine O (350 MW), Brilliant Blue dye (876 MW), negatively charged Orange II dye (350 MW), and Solvent Blue 35 (350 MW). Various solvents were studied including water, methanol, ethyl acetate, and toluene. The separation behavior is also strongly dependent on the membrane type. For example, the rejection of 0.01 wt% Safranine O in methanol was 68% and 6% for the MPF-44 and MPF-50 membranes, respectively. Lower fluxes of methanol were reported for the hydrophobic membranes (MPF-50 and MPF-60) in comparison with the hydrophilic MPF-44 membrane, which is similar to our[14] observations and those of Reference 12.

Subramaniam et al.[20] studied the permeation behavior of several systems involving triglycerides and oleic acid through NTGS-2200 membrane (Nitto-Denko). The authors concluded that preferential sorption and diffusion are important considerations for possible use of membranes in the area of edible-oil processing. Koops et al.[21] arrived at similar inferences when they studied the separation behavior of laboratory-made cellulose acetate membranes. They performed permeation experiments with several solutes in ethanol and n-hexane medium and concluded that solute–membrane interactions are important for understanding transport.

The extension of solvent behavior to inorganic NF membranes was reported by Tsuru et al.[22] They used silica–zirconia membranes (pore size 1–4nm) to study the rejection characteristics of PEG molecules in methanol and ethanol media at high temperature (50°C). The critical role of solvent was further demonstrated by studying PEG600 rejection values in water and methanol solvents. The rejection in methanol was reported to be 74% in contrast to negligible rejection with water as the solvent medium.

Our experimental observations[23] primarily consist of separation of organic dyes (Sudan IV 384 MW) and triglycerides in organic solvents (polar and non-polar). The rejection of triglycerides in n-hexane at 30 and 45°C was studied, since triglycerides can be obtained in various molecular weights (554MW to 875MW) and thus the role of size can be studied effectively. FIGURE 6 shows the rejection of two triglycerides in n-hexane by silicone-based Membrane D (Osmonics, Inc.) as a function of pressure and temperature. It can be seen that temperature does not exert a strong influence on the separation. The pressure dependence is similar to that observed in aqueous systems. As expected, a distinct dependence of molecular weight on triglyceride separation was observed. The observed rejection of Sudan IV (384MW organic dye) was about 25% in n-hexane and negligible in methanol at about 15bar by silicone-based Membrane D (provided by Osmonics, Inc.). However, with a hydrophilic aromatic polyamide based NF membrane (YK, provided by Osmonics, Inc.), the rejection was 43% in n-hexane and 86% in methanol at 15bar. FIGURE 7 illustrates the comparison of the results for Membrane D and YK with respect to separation behavior of Sudan IV in hexane and methanol. This reiterates strongly the dependence of solvent and membrane type on solute separation. Appropriate modification of Membrane D (Osmonics developmental PS-18 membrane) improves the solute separation and solvent flux. This membrane gave a hexane permeability that was 1.4 times higher than that of Membrane D. Furthermore, the separation of Sudan IV in

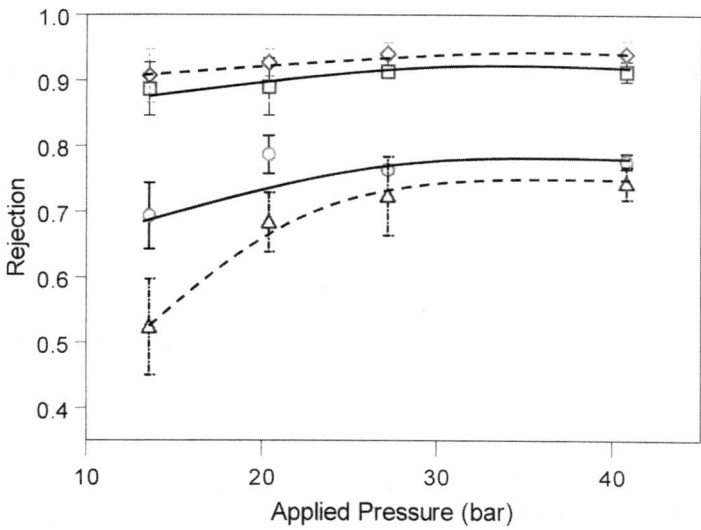

FIGURE 6. Effect of temperature and pressure on the rejection characteristics of Tripalmitin (807 MW) and Tricaprin (554 MW) in n-hexane by Membrane D:[23] □ Tripalmitin (807 MW triglyceride $T = 30°C$); ◇ Tripalmitin (807 MW triglyceride $T = 45°C$); ○ Tricaprin (554 MW triglyceride $T = 30°C$); △ Tricaprin (554 MW triglyceride $T = 45°C$).

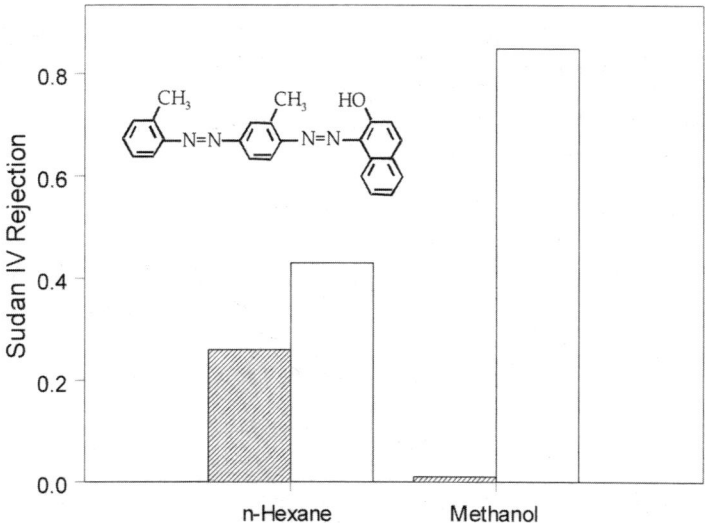

FIGURE 7. Comparison of the asymptotic rejection of Sudan IV (384 MW organic dye) in hexane and methanol medium through siloxane-based Membrane D (▨) and polyamide-based YK membrane (☐).

TABLE 1. Diffusive and convective flux contributions for our experimental data with Membrane D[a]

P (bar)	Solute Flux (mol/m^2 sec)	Convective Flux (%)
Sudan IV–hexane[b]		
3.35	5.82×10^{-8}	86.33
16.75	2.23×10^{-7}	96.43
C10 Triglyceride–hexane[c]		
13.4	7.38×10^{-6}	21.26
40.2	1.12×10^{-5}	30.71

[a]Data taken from Reference 23.
[b]Diffusive flux calculated from membrane permeation data.
[c]Diffusive Flux measured from diffusion cell apparatus.

hexane was 36% for the PS-18 membrane versus 26% for Membrane D at about 13 bar. In addition, because of potential swelling, solute–solvent coupling aspects cannot be neglected for an overall understanding of the separation behavior.

On a theoretical level, these solute–solvent coupling aspects can be evaluated by using traditional transport theories, such as the Spiegler–Kedem model and the pore-flow model. The Spiegler–Kedem model,[24–26] for example, interprets the solute flux as a combination of convection and diffusion given by

$$J_{\text{solute}} = \bar{P}\frac{dc}{dz} + J_v C(1-\sigma), \qquad (3)$$

where J_{solute} is the solute flux, $\bar{P}$ is the local solute permeability coefficient, C is the average concentration of the solute in the membrane, J_v is the total volume flux, and σ is the reflection coefficient. The relative contributions of convection and diffusion can be used to obtain greater insight into the transport mechanism.

Independent diffusion measurements were carried out in our laboratory to obtain the diffusive flux for the Sudan IV–hexane–Membrane D system. TABLE 1 shows the contributions of convection to the total flux for our experimental data. It can be clearly seen that as the solute molecular weight increases (Sudan 384 MW to tricaprin, C10 triglyceride, 554 MW), the convective contributions diminish (98% vs. 25%). The separation results for Sudan IV and tricaprin were 25% and 75%, respectively, through silicone-based Membrane D. Detailed experimental results and modeling have recently been reported elsewhere.[23]

LIMITATIONS

From the aforementioned literature data, certain key limitations on using NF for organic media can be readily identified: these include solvent-resistance, membrane–solvent–solute interactions, membrane preconditioning, membrane characterization, and other limitations. Specific membrane characterization tools need to be developed to assess membrane performance in non-aqueous media.

Preconditioning protocols need to be established for optimal membrane performance. The variations of membrane performance can be dramatic depending on the preconditioning solvents and protocols used. Thus, membrane characterization and preconditioning are discussed in a more detail here by suitable comparison with aqueous systems.

Membrane Characterization

For aqueous systems with RO/NF type membranes, pure water flux and rejection of sodium chloride/sodium sulfate are commonly used to characterize membrane performance and stability. Typical plots used to judge the performance of RO membranes consist of rejection of sodium chloride (NaCl) versus the permeability of the membrane. However, for NF, divalent ion (e.g., sodium sulfate, Na_2SO_4) rejections in excess of 90% are typically observed. A plot of this type is shown in FIGURE 8 (data taken from Refs. 1 and 27–29). Typical rejections of sodium sulfate (not exceeding 2,000 mg/L) are 90 to 99.9% (as indicated by the two dotted lines), whereas the sodium chloride rejection for these membranes is between 70 and 90%. The plot includes several commercially available membranes manufactured by Osmonics/Desal (HL, DS 5, DK, DL), Nitto Denko/Hydranautics (NTR-series), and FilmTec (NF 40, NF 45). For example, the NTR-series (Nitto-Denko) give lower rejections of divalent ions as the flux increases. A plot such as this can be used to judge the performance of NF membranes for aqueous systems. For NF membranes,

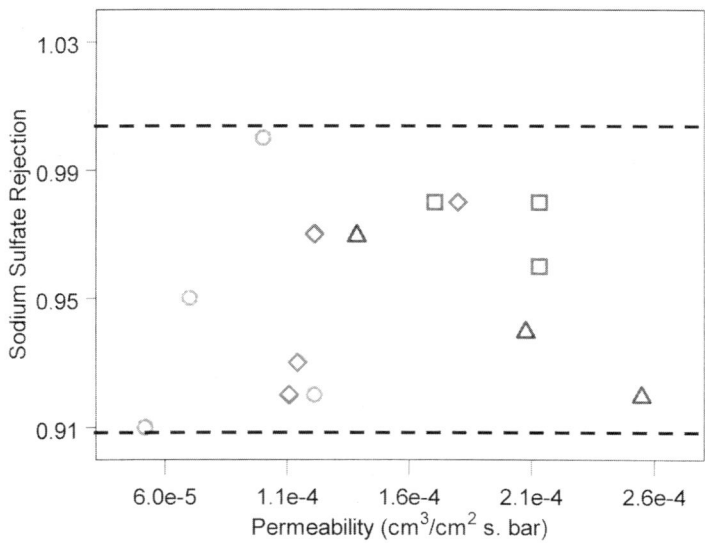

FIGURE 8. Permeability versus Sulfate rejection for commercial NF membranes: ○, FilmTec (NF-40, NF-45, NF-70, BW30); □, Osmonics (Desal HL, Desal DK, Desal DL); ▲, Nitto-Denko/Hydranautics (NTR-7410, NTR-7450, NTR-729HF); ◇, other membrane companies (HR-95, ATFRO, ATFRO-HR, ACM4).

molecular weight cutoff is also a common measure of the separation ability of the membrane.

Experimental observations in non-aqueous media have shown that the rejection of a solute depends on the type of solvent and also on the type of membrane material. With such a strong dependence of separation on the membrane–solute–solvent system, it is difficult to develop standardized characterization procedures to judge membrane performance. Thus, application-dependent standardization may be required (e.g, the use of triglycerides of different molecular weights for edible oil separation from hexane). Such tests would prove useful to researchers and membrane users in the industry in selecting suitable membranes for their application needs. In aqueous systems, PEGs, sugar molecules, and dendrimers are used as a common measure of the cutoff for the membrane. With solubility limitations of organic solutes in solvents ranging from polar to nonpolar, it becomes difficult to select a single solute for transport and characterization studies.

We prepared a molecular weight cutoff (MWCO) plot (similar to aqueous systems) for Membrane D in hexane medium, as is shown in FIGURE 9. The plot contains asymptotic rejection values for Sudan IV (384 MW organic dye), hexaphenyl benzene (554 MW organic) and triglycerides (554 MW to 890 MW). According to classical aqueous phase definitions, the membrane would have a MWCO of about 900 in *n*-hexane. It can be seen that in the 300 MW to 600 MW range, the solute–solvent and solute–membrane interactions are crucial. For example, the rejection of hexaphenyl benzene (534 MW) and Tricaprin (554 MW triglyceride) are dramatically different (40% vs. 75%, respectively). Furthermore, once the solvent changes, the same MWCO plot does not apply. In the absence of suitable characterization tools, membrane characterization and performance can be roughly estimated by using pure solvent flux and the separation of organic dyes. To create a characterization plot for

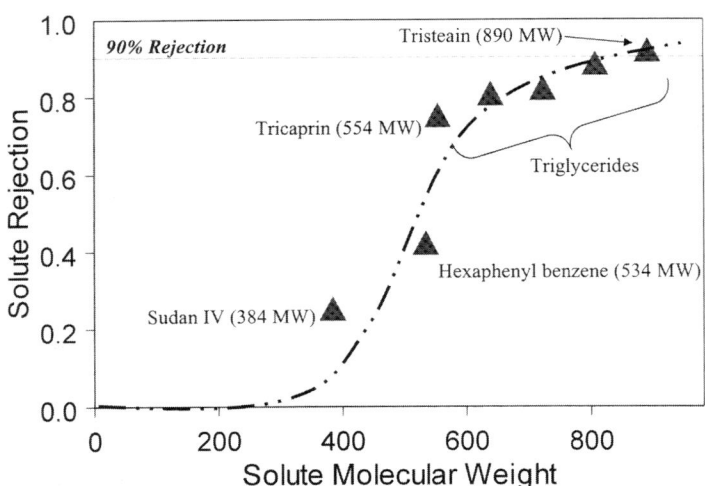

FIGURE 9. Molecular weight cutoff type plot for hydrophobic silicone-based (Osmonics) Membrane D in hexane medium.

various solvent systems one must use relative solute to solvent size information (such as, ratio of molar volumes) rather than molecular weight, which is explained further below.

Membrane Preconditioning

Membrane preconditioning is another area of significance when trying to apply NF to organic solutions. It is customary, in the case of aqueous systems, to soak the membrane in water before use to remove glycerine and other residual solvents. However, for non-aqueous systems, several membrane manufacturers prescribe conditioning techniques for optimal performance. Conditioning of the membrane becomes critical because (as reported by several investigators), the solvent–membrane interactions significantly impact the permeation characteristics of the membrane. One can easily discern that miscibility of the solvent phases is critical when organics are permeated through NF membranes. A quick illustration of this is provided by considering a membrane saturated with a methanol solution and used for hexane permeation: this would immediately form a two-phase system in the membrane pore structure causing dramatic reduction in the permeability.

The type of conditioning can also have a tremendous impact on the membrane performance. The Koch MPF-50 membrane, for example, gives varying results when exposed to different conditioning techniques. In order to show the impact of such treatment strategies on the pure solvent flux for alcohols and alkanes through the MPF-50 membrane, we compiled (see FIGURE 10) various results, including our own experimental data. The manufacturer-prescribed pretreatment strategy[30] suggested that the membrane (originally supplied in ethanol–water medium) be soaked in

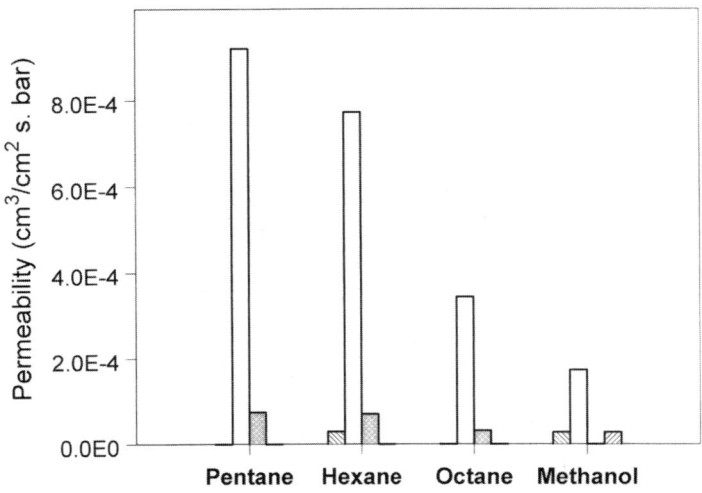

FIGURE 10. Effect of MPF-50 membrane preconditioning techniques on the solvent permeabilities: ▨, Koch literature; ▫, Machado *et al.*, 1999; ▪, our experimental data; ▨, Whu *et al.*, 2000.

methanol followed by soaking the membrane in methylene chloride. The permeability of methanol, for example, is 0.3×10^{-4} cm^3/cm^2 sec bar, as reported in the Koch literature. Our experiments with the MPF-50 membrane followed the same conditioning procedure as that prescribed by the Koch membrane literature.[30] Machado et al.[12] immersed MPF-50 membranes in distilled water followed by soaking in ethanol solution overnight and acetone was the first solvent used for permeation studies. The observed flux of methanol was about one order of magnitude higher than that reported by Koch Membrane Systems. Whu et al.[17,18] rinsed the membrane (originally received in 0.1% sodium metabisulfite and 10% glycerol solution) thoroughly by immersion in deionized water overnight, followed with membrane activation by flushing with ethanol at 440 psi for 60 to 80 min. The treated membranes were then soaked in ethanol at room temperature. The permeability of methanol observed was very similar to that reported by Koch Membrane Systems, however, the flux reached steady state after several hours of solvent permeation.

These observations suggest that suitable conditioning protocols need to be established for efficient performance of the membrane and to prevent two-phase formation due to possible immiscibility between residual unreacted solvents with the solvents of interest. For example, our experimental protocols for Membrane D conditioning was: soaking in deionized water, followed by soaking in isopropanol, followed by soaking in 50 vol% isopropanol–hexane mixture, followed by soaking in 100% hexane. Each of the soaking periods was performed for 30 minutes. When more than one solvent is used for transport studies, suitable solvent-exchange procedures need to be employed to ensure proper miscibility and to prevent two-phase formation. Tsuru et al.[22] observed the importance of membrane history on permeation character for their silica–zirconia membranes. They reported that, after solvent exchange, it requires several hours for the membrane to attain a steady performance. Whu et al.[17,18] also reported significant delay in reaching steady performance for the MPF-series of membranes in methanol and water medium. To eliminate such variability in membrane performance, several researchers have adopted a technique of using normalized flux values for comparison of flux values obtained with various membrane coupons.

COMPARISON OF NON-AQUEOUS AND AQUEOUS SYSTEMS

Summarizing the foregoing solute and solvent transport studies: for non-aqueous systems, membrane–solute, membrane–solvent, and solute–solvent interactions cannot be neglected. For organic solvents, due to the presence of hydrophilic and hydrophobic segments, the importance of interactions becomes apparent. For example, alcohols are polar compounds but also have hydrophobic segments and the length of these segments varies within a particular homologous series. These varying lengths of hydrophobic segments lead to several different levels of interaction with various polymeric materials.

To compile conventional aqueous NF molecular weight cutoff data with solvent-based systems, the size ratios of the permeating species must be considered. Thus, we compiled our own experimental data for both solvent and aqueous systems in a single plot (see FIGURE 11). The figure contains data for water permeating hydrophilic

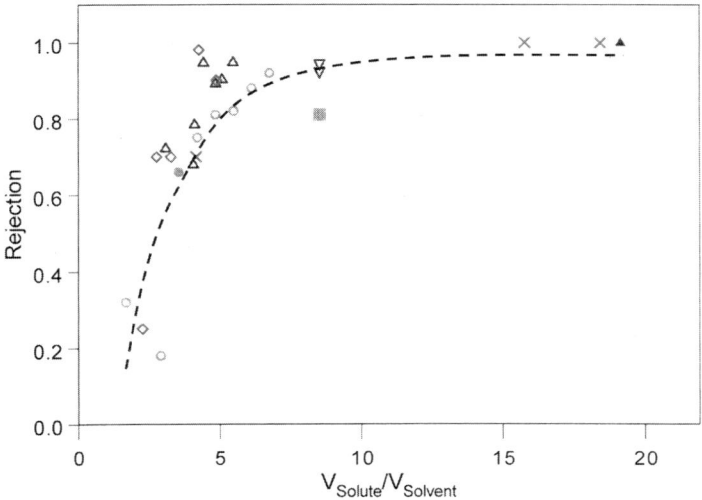

FIGURE 11. Comparison of rejection behavior for aqueous and non-aqueous systems as a function of the ratio of molar volumes of minor and major component:[23] ○, Membrane D (organic, our experimental data); ▲, polyacrylic acid RO membrane[32] (aqueous); ◇, FT 30 membrane[31] (aqueous); ×, MPT 30 membrane[33] (aqueous); ▽, MPF 60[19] (350 MW charged dyes in methanol); ●, MPF 60[19] (350 MW uncharged dye in ethyl acetate); ▲, MPF 44 hydrophilic[19] (aqueous, 350 MW dye); ■, MPF 60[19] (350 MW uncharged dye in methanol).

membranes FT-30,[31] a crosslinked polyacrylic acid based RO membrane,[32] a pH-stable membrane MPT 30,[33] and a MPF 44 hydrophilic membrane.[19] The figure also contains data for non-aqueous systems for Membrane D[23] and a MPF 60 membrane.[19] The data for Membrane D consists of solute permeation with n-hexane and Sudan IV and various triglycerides. All the membranes discussed have high fluxes for the major species in the feed and are, thus, comparable. The correlation strongly reflects the similar nature of separation for aqueous and non-aqueous systems; however, certain discrepancies can be observed in the region between 70 and 95% rejection interval. This discrepancy is probably due to the fact that the correlation does not consider the solute–membrane interactions, solute conformation, and other factors that may exert a significant impact on rejection, as has been observed in the literature.

APPLICATIONS

Solvent-resistant membranes have a strong potential for a variety of applications ranging from pharmaceutical to chemical to food industries. Some selected examples for the need of solvent-resistant membranes for material recovery and recycle of solvents are shown in FIGURE 12. An immediate application that can be identified is the separation of vegetable oil/soybean oil from hexane.[4,34,35] With increasing energy costs, industries are looking into alternate processes for recovery of valuable chemicals. Pharmaceutical industries, for example, spend most of their research efforts in downstream processing to recover valuable drugs. Membranes and membrane processes offer an excellent avenue in downstream processing with their versatility and low energy costs.

Solvent lube oil dewaxing[36,37] processes are commonly used in refinery operations. Current recovery processes employ dissolution in solvent blends followed by precipitation of the wax by cooling. ExxonMobil in conjunction with W.R. Grace Company have developed a membrane-based process in which the membrane reduces the residence time for the solvent mixtures which allows faster processing of the solvent.[36,37] The membrane was typically operated at $-10°C$ (14°F) and a pressure of 600 psi (41 bar) with 20 wt% lube oil in the feed. The cold solvent was directly recovered from the permeate and recycled back into the process. The membrane used in this process was a developmental asymmetric polyimide-based membrane that can withstand the corrosive solvent blend (methyl-ethyl-ketone and toluene) used in the application. The authors reported better than expected flux and rejection behavior. The initial data gives better than 95% rejection of lube oil and better than 99% pure solvent under the operating conditions stated above. The process has been scaled-up from a laboratory scale to a commercial scale by ExxonMobil in Beaumont, Texas, and is called the MAX-DEWAX process. The authors also report an increase in the base oil production by 25 vol% and that energy consumption per unit volume of product was reduced by about 20%.

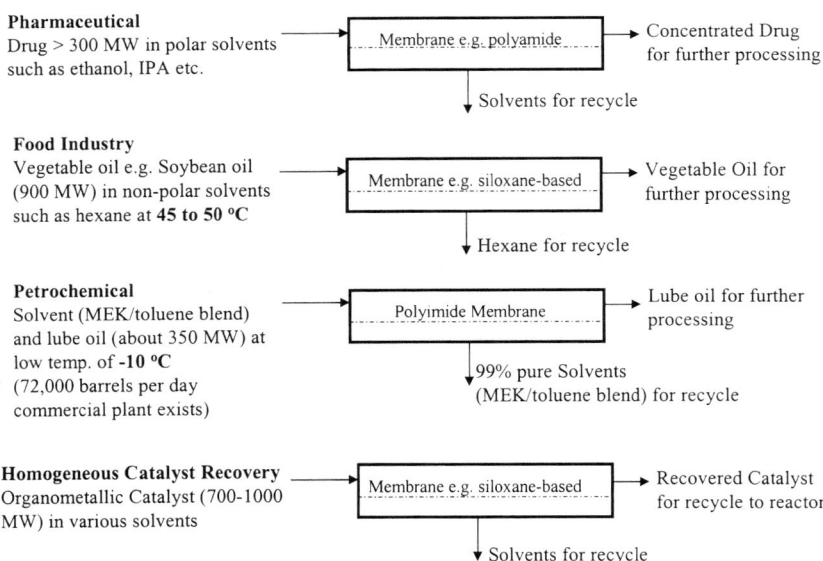

FIGURE 12. Selected examples of solvent-resistant membranes for material recovery and solvent recycle.

Homogeneous catalyst recovery is another important area in chemical/pharmaceutical industries. Use of organometallic catalysts is extremely common in such industries and their recovery is essential because of their high costs. Membrane processes like NF can be used in such applications to recover the catalyst, which can then be reused without sacrificing activity and product quality. This would also enable wide use of homogeneous catalysis (such as Ru or Pd containing organometallic catalysts) to eliminate mass-transfer limitations for the reaction rate. Several researchers[38–43] have ventured into this area to recover these valuable organometallic catalysts.

Nair et al.,[38] for example, studied the reaction of styrene and iodobenzene to form *trans*-stilbene with a palladium-based Heck catalyst. They used developmental polyimide membranes (SR NF, W.R. Grace and Co.) and studied the reaction in three different organic solvents (ethyl acetate, MTBE, and THF) to match catalyst and membrane properties. The authors also reported that these developmental membranes were preferred to the MPF-series of membranes because of better flux and separation capabilities. Luthra et al.[39] used a similar approach, but for a phase-transfer catalyst (PTC) system involving the reaction of bromoheptane and potassium iodide to form iodoheptane. In this reaction, tetraoctylammonium bromide (TOABr) was used as the PTC and toluene was used as the organic solvent. After the reaction, the iodoheptane and the bromoheptane partition into the organic phase and this organic phase is the feed to the NF membrane (142A, polyimide-based 220Da cutoff, W.R. Grace and Co.), which concentrates the catalyst in the retentate.

ACKNOWLEDGMENTS

We would like to thank the NIST-ATP (Cooperative Agreement Number 70NANB8H4028) and funding to the University of Kentucky from Osmonics, Cargill and GlaxoSmithKline for partial support of this work. We would also like to thank Dr. Steven D. Kloos from Osmonics, Inc. for providing valuable technical support throughout this project.

REFERENCES

1. NYSTROM, M., L. KAIPIA & S. LUQUE. 1995. Fouling and retention of nanofiltration membranes. J. Membr. Sci. **98**: 249–262.
2. SOURIRAJAN, S. 1964. Separation of hydrocarbon liquids by flow under pressure through porous membranes. Nature **203**: 1348–1349.
3. SCHMIDT, M., S. MIRZA, R. SCHUNERT, et al. 1998. Nanofiltration membranes for separation problems in organic solutions. Chemie Ingenieur Technik. **71**: 199.
4. KOSEOGLU, S.S., J.T. LAWHON & E.W. LUSAS. 1990. Membrane processing of crude vegetable oils: pilot plant scale removal of solvent from oil miscellas. J. Am. Oil Chem. Soc. **67**: 315–322.
5. PAUL, D.R. & O.M. EBRA-LIMA. 1970. The mechanism of liquid transport through highly swollen polymeric membranes. J. Appl. Polym. Sci. **15**: 2199–2210.
6. PAUL, D.R. & J.D. PACIOTTI. 1975. Driving force for hydraulic and pervaporative transport in homogeneous membranes. J. Polym. Sci. **13**: 1201–1214.
7. PAUL, D.R., J.D. PACIOTTI & O.M. EBRA-LIMA. 1975. Hydraulic permeation of liquids through swollen polymeric networks. J. Appl. Polym. Sci. **19**: 1837–1845.

8. PAUL, D.R. 1976. The solution–diffusion model for swollen membranes. Sep. Purif. Methods **5**(1): 33–50.
9. AMINABHAVI, T.M., S.B. HAROGOPPAD & R.S. KHINNAVAR. 1991. Rubber–solvent interactions. J. Macromol. Sci.–Rev. *In* Macromol. Chem. Phys. **C31**(4): 433–498.
10. AMINABHAVI, T.M. & R.S. KHINNAVAR. 1993. Diffusion and sorption of organic liquids through polymer membranes: 10. polyurethane, nitrile-butadiene rubber and epichlorohydrin versus aliphatic alcohols (C1–C5). Polymer **34**(5): 1006–1018.
11. AMINABHAVI, T.M. & H.T.S. PHAYDE. 1995. Sorption, desorption, diffusion and permeation of aliphatic alkanes into santoprene thermoplastic rubber. J. Appl. Polym. Sci. **55**: 17–37.
12. MACHADO, D.R., D. HASSON & R. SEMIAT. 1999. Effect of solvent properties on permeate flow through nanofiltration membranes. Part 1: investigation of parameters affecting solvent flux. J. Membr. Sci. **163**: 93–102.
13. MACHADO, D.R., D. HASSON & R. SEMIAT. 1999. Effect of solvent properties on permeate flow through nanofiltration membranes. Part II. transport model. J. Membr. Sci. **166**: 63–69.
14. BHANUSHALI, D., D. BHATTACHARYYA, S.D. KLOOS & C.J. KURTH. 2001. Performance of solvent-resistant membranes for non-aqueous systems: solvent permeation results and modeling. J. Membr. Sci. **189**: 1–21.
15. PAUL, D.R., M. GARCIN & W.E. GARMON. 1976. Solute diffusion through swollen polymer membranes. J. Appl. Polym. Sci. **20**: 609–625.
16. FARNAND, B.A., F.D.F. TALBOT, T. MATSUURA & S. SOURIRAJAN. 1983. Reverse osmosis separations of some organic and inorganic solutes in methanol solutions using cellulose acetate membranes. Ind. Eng. Chem. Process Design Develop. **22**: 179–187.
17. WHU, J.A., B.C. BALTZIS & K.K. SIRKAR. 1999. Modeling of nanofiltration—assisted organic synthesis. J. Membr. Sci. **163**: 319–331.
18. WHU, J.A., B.C. BALTZIS & K.K. SIRKAR. 1999. Nanofiltration studies of larger organic microsolutes in methanol solutions. J. Membr. Sci. **170**: 159–172.
19. YANG, X.J., A.G. LIVINGSTON & L. FREITAS DOS SANTOS. 2001. Experimental observations of nanofiltration with organic solvents. J. Membr. Sci. **190**: 45–55.
20. SUBRAMANIAM, R., K.S.M.S. RAGHAVRAO, H. NABETANI, *et al.* 2001. Differential permeation of oil constituents in nonporous denser polymeric membranes. J. Membr. Sci. **187**: 57–69.
21. KOOPS, G.H., S. YAMADA & S.-I. NAKAO. 2001. Separation of Linear hydrocarbons and carboxylic acids from ethanol and hexane solutions by reverse osmosis. J. Membr. Sci. **189**: 241–254.
22. TSURU, T., T. SUDOH, T. YOSHIOKA. & M. ASAEDA. 2001. Nanofiltration in non-aqueous solutions by porous silica-zirconia membranes. J. Membr. Sci. **185**: 253–261.
23. BHANUSHALI, D., D. BHATTACHARYYA & S.D. KLOOS. 2002. Solute transport in solvent-resistant nanofiltration membranes for non-aqueous systems: experimental results and the role of solute-solvent coupling. J. Membr. Sci. **208**(1–2): 343–359.
24. JAGUR-GRODZINSKI, J. & O.KEDEM. 1966. Transport coefficients and salt rejection in hyperfiltration membranes. Desalination **1**: 327.
25. BURGHOFF, H.G., K.L. LEE & W. PUSCH. 1980. Characterization of transport across cellulose acetate membrane in the presence of strong solute–membrane interactions. J. Appl. Polym. Sci. **25**: 323–347.
26. GILRON, J., N. GARA & O. KEDEM. 2001. Experimental analysis of negative salt rejection in nanofiltration membranes. J. Membr. Sci. **185**: 223–236.
27. LEE, S. & R.M. LUEPTOW. 2001. Membrane rejection of nitrogen compounds. Environ. Sci. Technol. **35**: 3008–3018.
28. LEE, S. & R.M. LUEPTOW. 2001. Reverse osmosis filtration for space mission wastewater: membrane properties and operating conditions. J. Membr. Sci. **182**: 77–90.
29. XU, Y. & R.E. LEBRUN. 1999. Comparison of nanofiltration properties of two membranes using electrolyte and non-electrolyte solutes. Desalination **122**: 95–106.
30. KOCH MEMBRANE SYSTEMS. 1999. Literature provided with the MPF-50 membrane.
31. SIRKAR, K.S. & W.S. HO. 1993. Membrane Handbook. Chapman & Hall.

32. HUANG, J., Q. GUO, H. OHYA & J. FANG. 1998. The characteristics of cross-linked PAA composite membrane for separation of aqueous organic solutions by reverse osmosis. J. Membr. Sci. **144:** 1–11.
33. PERRY, M. & C. LINDER. 1989. Intermediate reverse osmosis ultrafiltration (RO UF) membranes for concentration and desalting of low molecular weight organic solutes. Desalination **71:** 233–245.
34. RAMAN, L.P., M. CHERYAN & N. RAJAGOPALAN. 1996. Deacidification of soybean oil by membrane technology. J. Am. Oil Chem. Soc. **73:** 219–224.
35. RAMAN, L.P., M. CHERYAN & N. RAJAGOPALAN. 1996. Solvent recovery and partial deacidification of vegetable oils by membrane technology. Lipid **98:** 10.
36. WHITE, S.L. & A.R. NITSCH. 2000. Solvent recovery from lube oil filtrates with a polyimide membrane. J. Membr. Sci. **179:** 267–274.
37. GOULD, R.M., S.L. WHITE & C.R. WILDEMUTH. 2001. Membrane separation in solvent lube dewaxing. Environ. Prog. **20**(1): 12–16.
38. NAIR, D., J.T. SCARPELLO, L.S. WHITE, et al. 2001. Semi-continuous nanofiltration-coupled Heck Reactions as a new approach to improve productivity of homogeneous catalysts. Tetrahed. Lett. **42:** 8219–8222.
39. LUTHRA, S.S., X. YANGA, L.M. FREITAS DOS SANTOS, et al. 2001. Phase-transfer catalyst separation and reuse by solvent resistant nanofiltration membranes. Chem. Commun. **16:** 1468–1469.
40. BRINKMANN, N., D. GIEBEL, G. LOHMER, et al. 1999. Allylic substitution with dendritic palladium catalysts in a continuously operating membrane reactor. J. Catal. **183:** 163–168.
41. GIFFELS, G., J. BELICZEY, M. FELDER & U. KRAGL. 1998. Polymer enlarged oxazaborolidines in a membrane reactor: enhancing effectivity by retention of the homogenous catalyst. Tetrahed. Asymmetry **9:** 691–696.
42. SCARPELLO, J.T., D. NAIR, L. FREITAS DOS SANTOS, et al. 2002. The separation of homogeneous organomettalic catalysts using solvent-resistant nanofiltration membranes. J. Membr. Sci. **203:** 71–85.
43. SMET, K.D., S. AERTS, E. CEULEMANS, et al. 2001. Nanofiltration-coupled catalysis to combine the advantages of homogeneous and heterogeneous catalysis. Chem. Commun. **7:** 597–598.

Characteristics and Application of Ceramic Nanofiltration Membranes

RALPH WEBER,[a,b] HORST CHMIEL,[a] AND VALKO MAVROV[b]

[a]*Institute for Environmentally Compatible Process Technology, Saarbrücken Im Stadtwald, Saarbrücken, Germany*

[b]*Department of Process Technology, Saarland University, Saarbrücken, Germany*

> ABSTRACT: In this paper we report on the characteristic and filtration behavior of a newly developed ceramic nanofiltration membrane and compare it with other commercial ceramic nanofiltration membranes currently available. It is shown that it is possible to produce a ceramic membrane with separation properties in the nanofiltration range and with permeability rates that are clearly superior to those of polymer nanofiltration membranes. The ceramic membrane was used in tests involving the treatment of textile wastewater, alkaline solutions from bottle washing machines, and pickling bath solutions.
>
> KEYWORDS: ceramic NF membranes; polymer NF membranes; wastewater treatment

INTRODUCTION

During the past decade, nanofiltration has grown in significance as a pressure-driven membrane separation process, becoming more important in wastewater treatment. For this purpose, a series of attractively priced polymer membranes is available. However, if these membranes are to be applied successfully, the limitations for chemical, thermal, and mechanical stability in membrane operation or cleaning should not be exceeded by the medium to be treated. Until now, it has been mainly these limitations that have prevented nanofiltration from gaining greater significance in the field of production-integrated treatment of process streams. Membrane manufacturers have been producing polymer membranes with greater chemical and thermal resistance for some years now. Some of these membranes have been used successfully to treat acids,[1] bases,[2,3] and solvents. However, their long-term resistance often prove inadequate for separation problems on an industrial level that requires high thermal and chemical stability. These membranes are considerably more expensive than standard nanofiltration membranes; nevertheless, ceramic membranes have the combined advantage of high chemical, mechanic and thermal resistance. For this reason, many companies and research institutes are working on the development of ceramic membranes. Some manufacturers can now supply industrial-scale ceramic membranes that, with a cutoff of 1,000 g/mol, can be used

---

Address for correspondence: Ralph Weber, Institute for Environmentally Compatible Process Technology, D-66123 Saarbrücken Im Stadtwald 47, D-66123 Saarbrücken, Germany. Voice: 0049 681 9345 317; fax: 0049 681 9345 380.
  h.chmiel@rz.uni-sb.de

as nanofiltration membranes. In this paper, the characterization and application behavior of a new ceramic NF membrane (developed by HITK[4]) are discussed and compared with commercial ceramic NF membranes.

EXPERIMENTAL

Membranes Studied

The tests described were conducted with the ceramic membranes defined in TABLE 1. These membranes are single or multi-channel membranes with an active layer made of TiO_2 or ZrO_2. The membrane k-NF-new was available in the form of a single-channel, 19-channel, or flat-sheet membrane but no distinction is made between these in the study because similar cutoff values were determined for all three types. The commercial membranes k-NF-3 and k-NF 5 are of similar composition, but their geometry is different. They will be dealt with separately since their retention and permeability are very different.

Zeta Potential Measurements

Information on the extent and polarity of the zeta potential allows conclusions to be made about the charge on the membrane surface. Various measurement methods are available to determine this potential, however only the measurement of the streaming potential is suitable for characterizing the flat surfaces of membranes. All

TABLE 1. Ceramic membranes under study

Membrane	Material		Membrane Geometry	
	Active Layer	Substructure	Module Type	External Diameter
k-NF-new	TiO_2	TiO_2, Al_2O_3	single channel tubular membrane	10 mm
	TiO_2	TiO_2, Al_2O_3	19 channel tubular membrane	25 mm
	TiO_2	TiO_2, Al_2O_3	flat sheet membrane	90 mm
k-NF 1	TiO_2	no details	single channel tubular membrane	10 mm
k-NF 2	ZrO_2	no details	19 channel tubular membrane	25 mm
k-NF 3	ZrO_2	TiO_2, ZrO_2, Al_2O_3	3 channel tubular membrane	10 mm
k-NF 4	ZrO_2	no details	19 channel tubular membrane	20 mm
k-NF 5	ZrO_2	TiO_2, ZrO_2, Al_2O_3	23 channel tubular membrane	25 mm

of the other methods are mainly used to characterize powder and fibers. For this reason, the zeta potentials of ceramic nanofiltration membranes were determined by the streaming potential method. The significance of this measurement method for membrane characterization has already been verified in numerous publications.[5–7]

In evaluating the results, it should be taken into account that ultrafiltration and microfiltration membranes are usually characterized by transmembrane determination of the streaming potential that is induced when an electrolyte passes through the membrane pores. In contrast, in this study, the streaming potential was measured when the electrolyte was pumped tangentially across the surface. This determination method can be recommended for characterizing nanofiltration and reverse osmosis membranes for several reasons.

1. Only the upper layer (active layer) is characterized.
2. The streaming potential can be measured directly by using Ag/AgCl electrodes, because no difference in the concentration of the electrodes is caused by separation across the membrane.

The measurements were conducted using an electrokinetic analyzer (EKA) made by Anton Paar. To characterize the ceramic membranes, a cell allowing flat sheet membranes of 90 mm diameter to be characterized was designed. However, tubular membranes cannot be characterized by this apparatus. Therefore, only measurements for the membranes k-NF-new, as well as the structurally identical membranes k-NF-3 and k-NF-5 that were also available as flat sheet membranes, could be conducted with this measuring device.

Zeta potentials were determined using NaCl, KCl, and Na_2SO_4 electrolytic solutions in a concentration range between 0.001 and 0.1 mol/L and a pH range between 2.5 and 11. During the measurements, the temperature of the electrolytic solutions was constant at 293 K.

Prior to each measurement, the pH value was adjusted and the cell was rinsed for approximately 90 min. The pH value was adjusted using HCl (NaCl or KCl solutions), H_2SO_4 (Na_2SO_4), NaOH (Na_2SO_4, NaCl), and KOH (KCl). Subsequently eight measurements were taken and an average value was calculated.

Membrane Test Unit

A membrane test unit, the flow diagram of which is shown in FIGURE 1, was used for the experiments on filtration properties of the ceramic nanofiltration membranes. In this unit, the solution to be filtered was extracted from the storage tank (25 L) by a piston diaphragm pump (made by Verder, 2,000 L/h). A pulsation damper, which evens out the pressure peaks occurring behind the pump, was fitted on the pressure side.

Part of the volume flow could be rerouted directly to the storage tank via a bypass valve (V5). The volume flows and the pressures in both cells could be set separately via pump frequency and the valves fitted on the feed (V1, V3) and retentate side (V2, V4). A frequency converter controlled the pump so that feed pressure remained constant. Volume flows through both cells were measured by float-type flowmeters.

Transmembrane flow was determined gravimetrically by cyclically switching the electrovalves MV1 and MV2. Measurement values for pressures, temperature, and balances were recorded by computer via a data acquisition system. Conductivity and

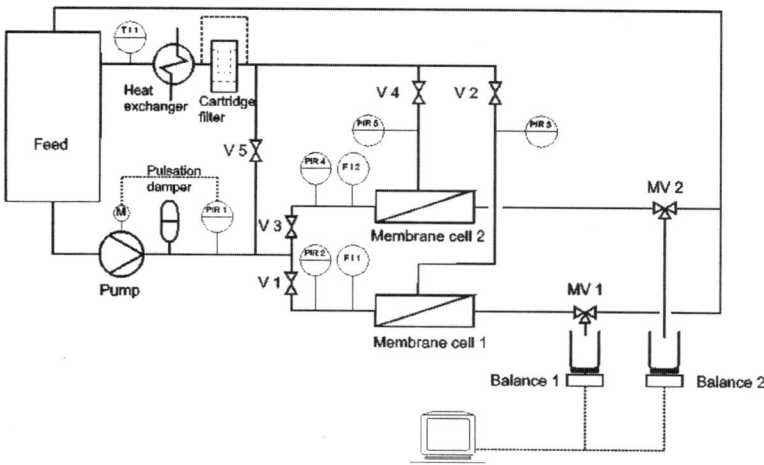

FIGURE 1. Flow diagram of the membrane test unit used in the tests.

pH values were recorded manually. Various cells and modules for flat sheet membranes and tubular membranes were available for the tests.

Pure Water Permeability Measurements

Pure water permeability was determined over a period of at least 12 hours at a pressure of 6 bar and a temperature of 293 K. Prefiltered, fully desalinated water (ultrapure water, conductivity 0.055 µS/cm, $R = 18.2$ mΩcm) was used in the measurements.

Determination of Molecular Weight Cutoff Using Polyethylene Glycol

Determination of molecular weight cutoff was conducted in accord with the French AFNOR standard NF X 45-103.[8] This process allows the cutoff to be determined in a molecular range between 100 and 3,000 g/mol and can be applied to all nanofiltration membranes, provided there is no interaction (e.g., adsorption) with the test substance. However, this type of adsorption effect can be ruled out for the membranes under study. The compositions of the feed and permeate solutions were determined by gel permeation chromatography (GPC).

The filtration experiments were performed in accord with the AFNOR standard at the lowest pressure possible, selected so that a linear correlation existed between pressure and permeate flux. A pressure of 2–4 bar proved to be viable for ceramic nanofiltration membranes. However, the tests in this study deviated essentially from the AFNOR standard in that:

1. The feed mixture was composed of 6–10 fractions PEG rather than three fractions (AFNOR), which resulted in a more homogeneous molecular weight distribution.

2. To prevent coating formation and concentration polarization, a distinctly lower total PEG concentration was chosen than specified in the AFNOR standard. The feed mixture contained ethylene glycol fractions with a total concentration of 0.5 wt%, whereas in the AFNOR standard, a total concentration of 1.5 wt% is suggested for characterization processes below 10,000 g/mol.

All permeate samples were taken thirty minutes after testing started, directly at the permeate outlet. At the same time, the respective feed samples were taken from the storage tank.

Determination of Salt Retention

Salt retention was determined using various electrolytic solutions (NaCl, KCl, Na_2SO_4, $CaCl_2$, and $NaNO_3$) at various pH values in a pressure range 4–15 bar. In each of the tests, the pH was adjusted using 0.1 molar acids and bases of the respective electrolytes (NaOH, KOH, $Ca(OH)_2$, HCl, H_2SO_4, and HNO_3).

EXPERIMENTAL RESULTS

Characterization Results of the Ceramic NF Membranes

Separation behavior (permeability, retention of organics, and salt) and the charge properties (zeta potential measurements) of five commercial ceramic membranes (k-NF-1 through k-NF-5) as well as a newly developed ceramic nanofiltration membrane made of TiO_2 (k-NF-new)[4] were characterized. The morphology of these membranes (REM, AFM, and N_2-adsoption/desorption) is dealt with elsewhere.[9]

Membrane Permeability

The results of pure water permeability (10–24 L/m²h bar) (see FIGURE 2) showed that the permeability rates of all the ceramic membranes under study, with the exception of k-NF-2, were considerably higher than the rates of currently available polymer membranes, whose pure water permeability is generally in the range of about 1–8 L/m²h bar.

Membrane Molecular Weight Cutoff

The tests to determine molecular weight cutoff (MWCO), which were performed by retention measurements using a mixture of polyethylene glycols (PEG),[4,9,10] showed that the newly developed membrane k-NF-new, with a 90% MWCO of roughly 450 g/mol for the retention of organics, could clearly be defined as a nanofiltration membrane (see FIGURE 3).

In contrast with this new membrane, the membrane retention determined for all the other commercial ceramic membranes (FIG. 3) was distinctly poorer, although membrane permeability was similar (k-NF-1, k-NF-3) or even lower (k-NF-2) (FIG. 2). The cutoffs for these membranes, determined as exceeding 2,500 g/mol, clearly deviate from the specifications of the manufacturer (MWCO 1,000 g/mol) and lie in the range between ultrafiltration and nanofiltration.

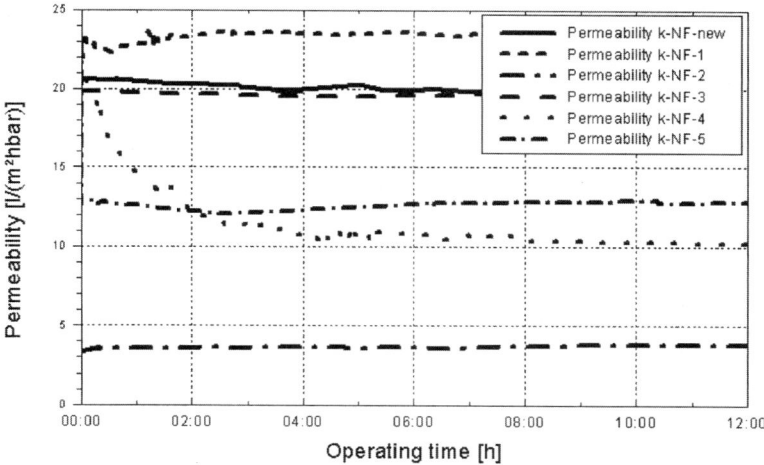

FIGURE 2. Pure water permeabilities of the ceramic membranes under study; $T = 293$ K, $p = 6$ bar.

Membrane Retention

As far as ceramic membranes are concerned, salt retention is controlled by the charge on the membrane. This charge was determined by measuring the zeta potential on the ceramic nanofiltration membranes at various pH values for NaCl, KCl, and Na_2SO_4 electrolytes (see FIGURE 4).

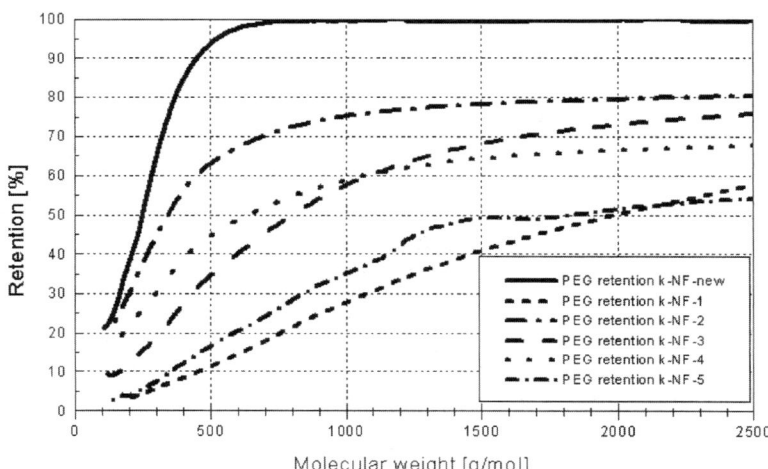

FIGURE 3. PEG retention of the ceramic membranes under study.

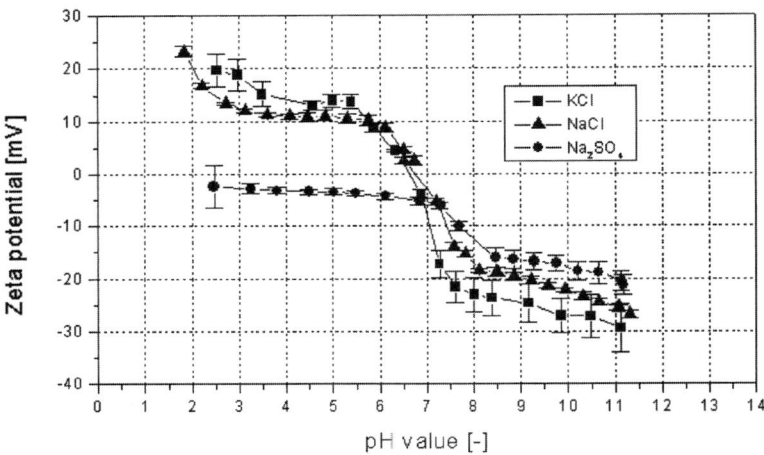

FIGURE 4. Zeta potential of k-NF-new for NaCl, KCl, and Na_2SO_4 ($c = 10^{-3}$ mol/L).

Similar trends for zeta potentials with an isoelectric point (i.e.p) at a pH value between 6.5 and 7 (FIG. 4)[9,10] were recorded for NaCl and KCl. In contrast, the values for zeta potential were negative throughout the entire pH range for a Na_2SO_4 solution, in which case values at pH > 7 clearly continued to drop even further below the values in the acid range so that no isoelectric point could be determined.

The trend of this electric charge was reflected in the salt retention values determined. For NaCl solutions, only negligible salt retention values were recorded for the ceramic nanofiltration membrane k-NF-new near the isoelectric point (pH ≈ 6.5),

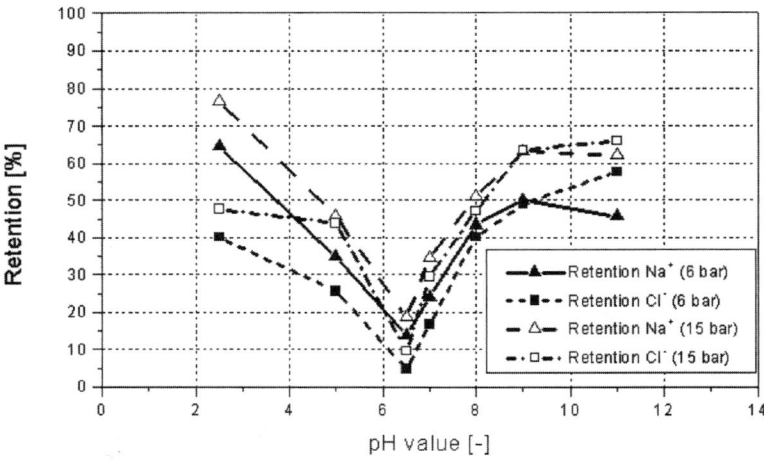

FIGURE 5. NaCl retention for k-NF-new at various pH values; c(NaCl) = 0.01 mol/L.

whereas retention in higher charge ranges, that is, at high and low pH values, increased greatly (see FIGURE 5). The relation between membrane charge and salt retention is described in more quantitative detail in Reference 9.

In contrast, a constant increase in Na^+ and SO_4^{2-} retention was observed as pH value increased from 2.5 to 11 in tests using a Na_2SO_4 solution (see FIGURE 6). This correlated to the zeta potential trend for the membrane k-NF-new in a Na_2SO_4 solution for which negative values were recorded throughout the entire pH range. As pH values increased, these values decreased from near zero at pH 2.5 to $-30\,mV$ at pH 11. The retention value (greater than 90%) determined at pH 11 is comparable to the usual sulfate retention obtained with polymer nanofiltration membranes throughout the entire pH range.

FIGURE 5 clearly illustrates that electrolyte retention, particularly for NaCl solutions, increases as the pressure rises from 6 to 15 bar. As well as dependence on pH, membrane retention was also strongly influenced by electrolyte concentrations in all the electrolyte solutions under study. As concentration was increased in the NaCl solution from 0.01 to 0.1 mol/L, membrane retention decreased noticeably (see FIGURE 7).[9] The qualitative trend in the retention curve was not affected by an increase in concentration. The reason for this effect, which also occurred with other salts (Na_2SO_4)[9] is that the degree of salt retention is determined by the degree of the relative membrane charge (membrane charge/feed concentration). This decreases as feed concentration is increased.

Membrane Filtration of Real Media

Because ceramic nanofiltration membranes are generally more expensive than standard commercial polymer membranes, their use should focus on fields of application that demand greater thermal or chemical resistance. As already shown, the

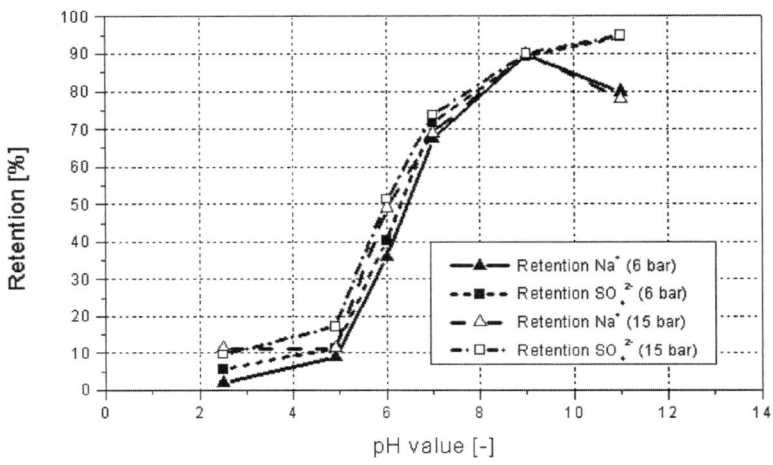

FIGURE 6. Na_2SO_4 retention for k-NF-new at various pH values; $c(Na_2SO_4) = 0.01\,mol/L$.

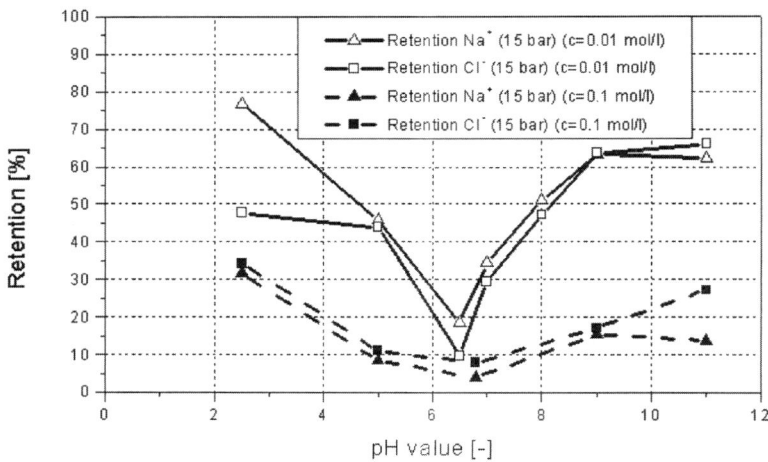

FIGURE 7. NaCl retention for K-NF-new at various concentrations and pH values; $c(NaCl) = 0.01$ or 0.1 mol/L.

newly developed ceramic membrane (k-NF-new) can reject organic molecules with a molecular weight of roughly 450 g/mol and larger. Furthermore, it was shown that it is only feasible to use this membrane at moderate electrolyte concentrations for the purpose of salt retention. In the following, the use of new ceramic nanofiltration membranes is investigated for several fields of application, focussing on the decolorization of textile wastewater and the treatment of hot alkaline solutions from bottle washing machines. The treatment of solutions from pickling baths in the metalworking industry highlights the extent to which ceramic NF membranes can be used.

Treatment of Textile Wastewater Containing Dyes

Despite the numerous measures implemented in textile companies to promote production-integrated environmental protection, large quantities of wastewater are still generated. When surveying the wastewater situation for the textile finishing industry, the wastewater color adopts particular significance because the colorants used in dyeing processes are generally difficult to degrade biologically[11] and, thus, are often not completely degraded in the downstream wastewater treatment plant. The color of the wastewater is defined, according to the standard DIN EN ISO 7887, by the official color index at wavelengths 436, 525, or 620 nm which is identical to the more frequently used spectral absorption coefficients (SAC $[m^{-1}]$) at these same wavelengths. The typical composition of textile wastewater containing dye is shown in TABLE 2.

Only nanofiltration or reverse osmosis membranes are suitable for the recovery of dyes, due to the fact that reactive and direct dyes used in textile dyeing are present in dissolved form with molecular weights of approximately 400–1,300 g/mol. Although some publications have reported on the success of polymer nanofiltration membranes used to treat wastewater containing dyes from the textile industry,[12–14] temperatures of up to 90°C, strong fluctuations in pH values of about 3–12.5, as well

TABLE 2. Typical composition of textile wastewater containing dye

Temperature (°C)	pH	Electrical Conductivity (mS/cm)	COD (mgO$_2$)	Spectral Adsorption Coefficient (m^{-1})		
				SAC$_{436}$	SAC$_{525}$	SAC$_{620}$
37–62	4.7–11.9	0.4–8.4	110–6,100	0–240	0–323	0–318.9

as oxidants (H$_2$O$_2$) in certain partial streams, frequently hinder the successful use of such polymer membranes. Furthermore, despite implementing pretreatment measures, there is still a high concentration of suspended or colloidal particles, which often causes fouling problems[13,15] in the spiral-wound modules normally used for polymer membranes. The membranes k-NF-new, k-NF-2, and k-NF-5 (TABLE 1) were used in tests on the treatment of wastewater contaminated with dyes from the textile industry. These tests confirmed the results obtained for the retention of ceramic nanofiltration membranes using PEG. From among all the ceramic membranes tested, the newly developed membrane k-NF-new (cutoff approximately 450 g/mol) achieved the highest degree of retention for dyes and COD (see TABLE 3).

Retention was poorest for the membrane k-NF-5, which also showed the lowest membrane retention value during characterization with PEG. As far as its retention properties are concerned, the ceramic membrane k-NF-2 ranks between the two membranes. From an optical viewpoint, the membrane k-NF-new produced an almost uncolored permeate, whereas the permeates for all the other ceramic membranes clearly contained residual dye. When pressure was increased from 6 to 15 bar, no significant influence of transmembrane pressure on organic retention could be detected for any of the ceramic membranes tested.

The permeability rates of all the membranes depended strongly on wastewater composition and yield. In contrast to the results obtained for the determination of pure water permeability, a higher permeability was recorded for the membrane k-NF-5 at 6 bar than for the membrane k-NF-new (see FIGURE 8) in wastewater treatment. With the exception of k-NF-new, the permeability of all membranes depended strongly on transmembrane pressure. An increase in pressure from 6 to 15 bar at identical feed flow rates of 1,500 L/h caused a reduction in permeability rates of

TABLE 3. COD or SAC retention (percent) in the treatment of textile wastewater containing dyes

Membrane	COD	SAC$_{436}$	SAC$_{525}$	SAC$_{620}$
k-NF-new	95–97	95.5–99.9	98.5–99.9	99–99.9
k-NF-2	86–92	94–98.5	96–99	91–98
k-NF-5	79–90	89–92	91–93.5	83.5–96
NF-1	95–98	99–99.9	99–99.9	92.5–99.5
NF-2	68–89	80–96	81–99	83–96

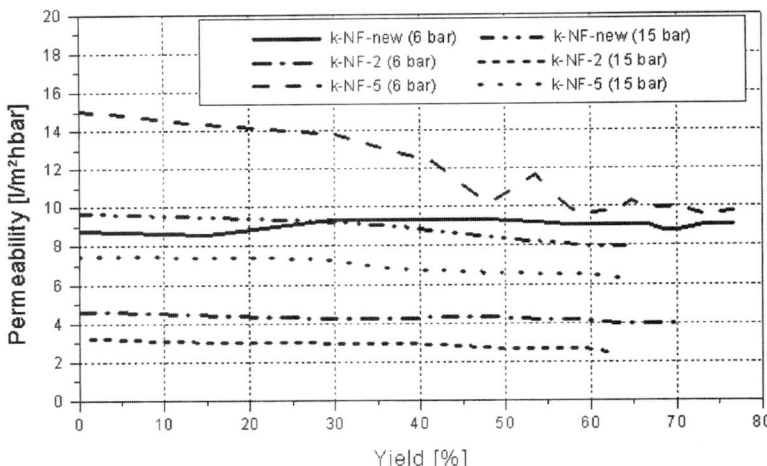

FIGURE 8. Membrane permeabilities for k-NF-new, k-NF 2, and k-NF 5 during the treatment of textile wastewater; θ = 40°C.

20–50% for these membranes (FIG. 8). At a transmembrane pressure of 15 bar, this drop in membrane permeability again resulted in the same relation between membrane permeability rates as was the case when determining pure water permeability:

$$L(\text{k-NF-new}) > L(\text{k-NF-5}) > L(\text{k-NF-2}).$$

For the treatment of textile wastewater, upscaling to pilot scale was conducted using the membrane k-NF-new, the results of which were published by Voigt.[16] Judging from the values recorded for retention and permeability, it can be stated that the membrane k-NF-new is a good alternative to polymer membranes and that the residence time assumed for the actual case of application in question, as well as the operating costs based on this, are the main factors to be considered when deciding on the use of this membrane. However, if the commercial membranes k-NF-2 and k-NF-5 are to be used instead of polymer membranes, considerably reduced membrane retention rates need to be taken into account.

Treatment of Alkaline Solutions from Bottle Washing Machines

According to known facts, returnable bottles in well organized systems are superior to non-returnable bottles from an ecological viewpoint. However, one requirement for bottle reuse in the beverage industry stipulates that these bottles have to be cleaned effectively before refilling.

According to the state of the art, the bottles are firstly emptied of residual substances prior to entering a soaking stage, during which the bottles are precleaned. Subsequently, the bottles are routed to a multistage alkaline cleaning process using a hot caustic soda solution, followed by downstream cascade-like warm water rinsing to remove any remaining impurities. Finally, the bottles enter the filling plant. The impurities, having entered the washing solution via multistage transfer, lead to an increase in water consumption in the downstream water rinsing stage. In addition,

TABLE 4. Composition of the alkaline solution from bottle washing machines

pH	Electrical Conductivity (mS/cm)	Temperature (°C)	Ca (mg/L)	Al (mg/L)	COD (mg/L)
13–13.5	60–80	65–80°C	100–200	100–200	4,000–9,000

more cleaning additives are used as the concentration of impurities in the alkaline solution increases.[17] The fact that the material and energy streams in the alkaline cleaning and warm water rinsing stages are closely linked, provides good reason to close material streams in bottle washing machines by treating washing water and alkaline cleaning solution simultaneously.[2]

For the treatment of washing water by membrane separation processes, a variety of polymer nanofiltration and reverse osmosis membranes can be used because of the moderate chemical and thermal conditions of this partial stream ($T < 50°C$, pH < 10). In this case, the use of ceramic nanofiltration membranes cannot be justified for economic reasons due to the high specific costs. However, only a few polymer nanofiltration membranes are available for the treatment of alkaline cleaning solutions because this process requires membranes with high chemical and thermal resistance. Alternatives to polymer membranes have to be sought because prices are high and operational problems involving residence time and membrane fouling often occur in industrial applications.[2,17] In view of this, the treatment of alkaline solutions from bottle washing machines in a brewery was studied using ceramic nanofiltration membranes (k-NF-new, k-NF-2, and k-NF-5). TABLE 4 gives an example of the composition of this alkaline cleaning solution.[10,17] The tests were conducted in batch

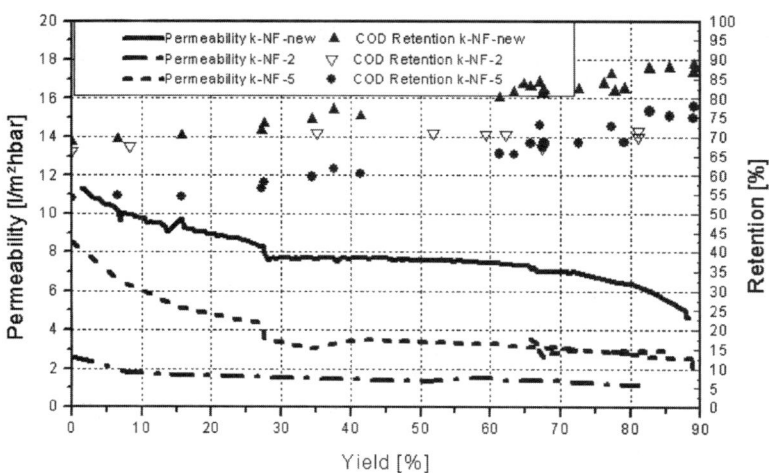

FIGURE 9. Membrane permeabilities and COD retention during treatment of alkaline solution from bottle washing machines using k-NF-new, k-NF 2, and k-NF 5 membranes; $\theta = 60°C$, $p = 6\,\text{bar}$.

TABLE 5. Separation efficiency for the treatment of alkaline solution from bottle washing machines

Membrane Type	Membrane Permeability (L/m^2 h bar)	COD Retention (%)	Al Retention (%)
k-NF-new	5–20	70–90	<20
k-NF-2	1–4	65–75	<20
k-NF-5	2–9	55–75	<10

operation at a temperature of 65°C and at turbulent flow (1–2.5 m/sec). Pretreatment was performed using cartridge filters (10–50 μm). The results of the tests (see FIGURE 9) confirmed the separation properties of the ceramic membranes determined for membrane characterization.

Compared to k-NF-2, k-NF-5, and k-NF-new (see TABLE 5), the highest membrane permeability rates could be recorded for k-NF-new as well as the highest COD retention values for organics that were in the range of a polymer membrane tested for purposes of comparison.[9] The use of the ceramic nanofiltration membrane k-NF-new for industrial applications seems promising if permeability, which was four to ten times higher than the polymer membrane, is also taken into account. However, for the electrolyte concentrations occurring in this case, all the ceramic nanofiltration membranes displayed distinctly lower retention of ions (Al) than the polymer membrane, thus confirming the mass transfer properties for salts as defined earlier.

Treatment of Pickling Bath Solutions from the Metal-Working Industry

Metal work pieces have to be cleaned several times in the course of production because, during manufacture or storage, they come into contact with numerous substances that partly remain on the surface as impurities.[18] These impurities contain oxidic residues, corrosion products, as well as organic and inorganic residues. In the cleaning process, work pieces usually pass through a degreasing bath to remove oily impurities. Following rinsing, the work pieces are routed to a pickling bath and then to an additional rinsing process.

Pickling agents generally consist of strong acid solutions or acid mixtures[18] in which all oxidic layers are dissolved chemically or separated. In addition, these acids contain organic pickling inhibitors and surfactants. During these operations, the acids are used up by chemical or electrolyte decomposition involving the work pieces undergoing treatment. Because of this, the acid concentration in the pickling bath is checked regularly and increased if necessary, until the metallic salt concentration is so high that further use is no longer possible because of precipitation. In particular, hydrochloric acid, sulfuric acid, phosphoric acid, hydrofluoric acid, as well as mixtures of these acids are of technical significance for the pickling of ferrous metals. The objective of a recycling process for pickling baths is the recovery of the pickling acids. To this end, metal ions and impurities must be removed from the acids.

Fe concentrations exceeding 200 g/L can occur in HCl pickling agents because iron chloride is highly soluble in hydrochloric acid, although this has practically no adverse effect on the rate of pickling, even at high concentrations. TABLE 6 shows the typical composition of a HCl solution with high Fe and Zn concentrations in a pickling bath in a metal-working plant.

TABLE 6. Composition of a pickling bath HCl solution

Acid Content (mol/L)	Electrical Conductivity (mS/cm)	TOC (mg/L)	Cl (g/L)	PO_4 (mg/L)	Na (mg/L)	Fe (g/L)	Zn (g/L)
3.86	445	862	231	2,711	32	78	29

Tests on salt retention were performed using this pickling bath solution, the results of which are shown in FIGURE 10. A relatively stable permeability rate of 10^{-6} L/m²h bar could be achieved with the membrane k-NF-new, but retention of Fe and Zn at 4–12% proved to be too low for commercial application in the treatment of pickling bath solutions.

The results of these tests confirmed that salt retention decreases strongly as electrolyte concentration increases, as illustrated earlier. In view of this, it would not be a viable option to use the currently available nanofiltration membranes for the purpose of salt retention at the electrolyte concentrations present in HCl pickling baths.

CONCLUSIONS

These tests proved that it was possible to develop the ceramic membrane k-NF-new, which can be classified as a nanofiltration membrane because of its retention properties for organics. In contrast, the comparative tests on all other commercial ceramic membranes showed that these membranes fall into an intermediate area between ultrafiltration and nanofiltration, clearly falling short of the 1,000 g/mol

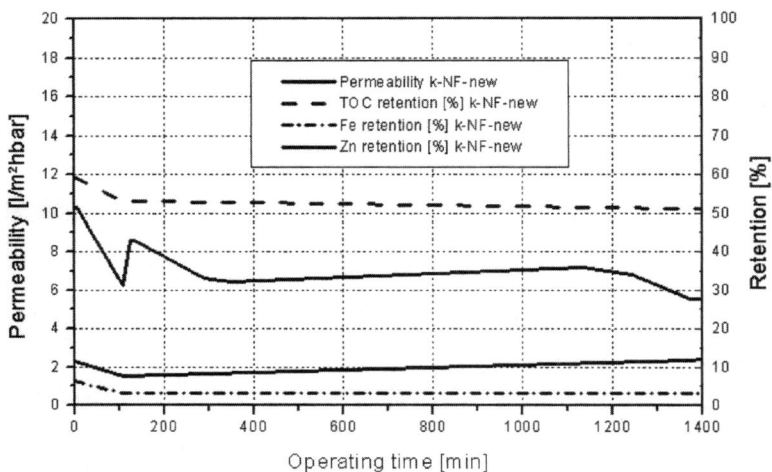

FIGURE 10. Membrane permeabilities and retention during treatment of a HCl solution in a pickling bath using k-NF new ($\theta = 25°C$, p = 6 bar).

cutoff specifications of the manufacturers. The salt retention of all ceramic NF membranes is controlled by the membrane charge and depends, to a great extent, on the type of salt, salt concentration, as well as the pH value of the solution.

The results of the tests conducted on the treatment of real media confirm the results of membrane characterization. High permeability rates, good retention of organics, as well as a low fouling tendency were confirmed particularly in the case of the newly developed ceramic membrane k-NF-new. These advantages over commercial polymer and ceramic membranes provide a good argument in favour of the industrial application of the newly developed ceramic membrane, particularly for the treatment of textile wastewater containing dyes, as well as the treatment of alkaline solutions from bottle washing machines. However, in contrast, salt retention decreases strongly as electrolyte concentrations rise during the treatment of pickling bath solutions so that the use of ceramic nanofiltration membranes in this field is not a viable option.

REFERENCES

1. KYBURZ, M. 1995. Säureaufbereitung mit Umkehrosmose und Nanofiltration. Proceedings of the 5th Aachener Membrankolloquium, Aachen, Germany. 1–25.
2. RÖGENER, F. 2002. Untersuchungen zur Separation von Verunreinigungen aus Waschlauge und Nachspülwasser von Flaschenwaschmaschinen Saarbrücken. Thesis.
3. TRÄGARDH, G. & D. JOHANSSON. 1998. Purification of alkaline cleaning solutions from the dairy industry using membrane separation technology. Desalination **119:** 21–29.
4. PUHLFÜRß, P., I. VOIGT, R. WEBER & M. MORBÉ. 2000. Microporous TiO_2 membranes with a cut off < 500 Da. J. Membr. Sci. **174:** 123–133.
5. KIM, K.J., A.G. FANE, M. NYSTROM, *et al.* 1996. Evaluation of electroosmosis and streaming potential for measurement of electric charges of polymeric membranes. J. Membr. Sci. **116:** 149–159.
6. RICQ, L. & J. PAGETTI. 1999. Inorganic selectivity to ions in relation with streaming potential, J. Membr. Sci. **155:** 9–18.
7. SZYMCZYK, A., P. FIEVET, M. MULLET, *et al.* 1998. Comparison of two electrokinetic methods—electroosmosis and streaming potential to determine the zeta potential of plane ceramic membrane. J. Membr. Sci. **143:** 189–195.
8. AFNOR. 1997. NF X 45-103 Membranes Poreuses. Beuth-Verlag, Berlin.
9. WEBER, R. 2001. Charakterisierung, Stofftransport und Einsatz keramischer Nanofiltrationsmembranen. Thesis.
10. Abschlussbericht zum Forschungsvorhaben, Entwicklung keramischer Nanofiltrationsmembranen für den produktionsintegrierten Umweltschutz am Beispiel farbstoffbelasteter Abwässer der Textilindustrie, Förderprogramm: Produktionsintegrierter Umweltschutz (PIUS) Förderkennzeichen: 01 RV 9641/9.
11. BRAUER, H. 1996. Behandlung von Abwässern, Band 4. Springer (Additiver Umweltschutz), Berlin, Heidelberg, New York.
12. DHALE, A.D. & V.V. MAHAJANI. 2000. Studies in treatment of disperse dye waste: membrane wet oxidation process. Waste Manage. **20:** 85–92.
13. SCHÄFER, T., R. GROSS, J. JANITZA & J. TRAUTER. 1999. Nanofiltration von Färbereiabwasser, Filtrieren und Separieren, 13, **1:** 9–16.
14. SÓJKA-LEDAKOWICZ, J., T. KOPROWSKI, W. MACHNOWSKI & H.H. KNUDSEN. 1998. Membrane filtration of textile dyehouse wastewater for technological water reuse. Desalination **119:** 1–10.
15. FRITSCH, J. 1993. Untersuchungen der Aufbereitung von Abwässern aus Textilbleichereien mittels Nanofiltration. Proceedings of the 5th Aachener Membrankolloquium, Aachen, Germany. 53.

16. VOIGT, I., M. STAHN & S. WOEHNER. 2000. Produktionsintegrierte Reinigung aggressiver farbstoffhaltiger Abwässer mit keramischer Nanofiltration. Chemie Ingenieur Technik. **72:** 1127–1128.
17. Abschlussbericht zum Forschungsvorhaben, Produktionsintegrierter Umweltschutz in der Lebensmittelindustrie am Beispiel der Brauindustrie, Förderprogramm: Produktionsintegrierter Umweltschutz (PIUS) Förderkennzeichen: 01 ZF 9501/3.
18. RITUPER, R. 1993. Beizen von Metallen Saulgau/Württ. Eugen G. Leuze Verlag (Galvanotechnik).

Nanocomposite Lithium Ion Conducting Membranes

FAUSTO CROCE AND BRUNO SCROSATI

Department of Chemistry, University "La Sapienza", Rome, Italy

ABSTRACT: This review describes the properties and characteristics of a class of membranes formed by blends of a lithium salt, LiX, where X is preferably a large soft anion, such as ClO_4 or $N(CF_3SO_2)_2$, and a high molecular weight polymer containing Li^+-coordinating group, such as polyethylene oxide (PEO) with the dispersion of selected ceramic powders, such as TiO_2, Al_2O_3, and SiO_2, at the nanoscale particle size. These nanocomposite membranes behave as lithium polymer electrolytes, that is, they exhibit a high lithium ion conductivity. Because of this property, the PEO-LiX nanocomposite electrolytes may find an important application as separators in advanced, rechargeable lithium polymer batteries.

KEYWORDS: nanocomposite; lithium ion membrane

POLYMER ELECTROLYTES AND LITHIUM POLYMER BATTERIES

Ionically conductive membranes are a class of materials that play a key role in modern energy technology. This prominent position results from the latest developments in the field, which have produced a series of new polymers with exceptional electric properties. Accordingly, ionic polymers are widely studied as electrolytes for the development of high performance fuel cells and of high energy density batteries, with particular interest in lithium batteries.

Novel design, high energy density batteries are in increasing demand. The aggressive and growing market for portable electronic products, as well as the environmental necessity for zero emission vehicles, for example, electric vehicles (EVs), has motivated research to develop electrochemical power sources characterized by high energy density, long cyclability, reliability, and safety. A recent breakthrough has been the commercialization of rechargeable lithium batteries, the so called lithium-ion batteries,[1–6] that are now produced at a rate of about a million units per month.

Most of the lithium ion battery projects are focused on the fabrication of prototypes using liquid electrolytes. An important step forward in this technology is the replacement of the liquid electrolyte with an ionic membrane in order to produce novel devices having a fully plastic configuration. This is an interesting concept since it provides the prospect of a favorable combination of the high energy and long life, typical of lithium or lithium ion cells, with the reliability and easy manufacturing, typical of polymer-based, all-plastic structures. The practical development of

Address for correspondence: Fausto Croce, Department of Chemistry, University "La Sapienza", 00185 Rome, Italy.

this concept requires the availability of polymer membranes with ionic conductivity approaching that of the conventional liquid solutions.

Ionically conducting membranes with these characteristics are generally termed *polymer electrolytes*. With emphasis on batteries, the development of polymer electrolytes has been mainly confined to lithium ion conducting systems; whereas modern fuel cell technology relies on polymer electrolytes with proton transport.

In this review we mainly focus on the former; that is, on membranes having transport properties comparable with those of the common liquid ionic solutions formed by the dissolution of lithium salts in organic aprotic solvents.[5] Classical examples of lithium polymer electrolytes are blends of a lithium salt, LiX, where X is preferably a large soft anion, such as ClO_4 or $N(CF_3SO_2)_2$, and a high molecular weight polymer containing Li^+-coordinating groups, such as polyethylene oxide (PEO).[7–10]

Conventional PEO–LiX membranes are typically prepared by casting an acetonitrile solution of the two components or by directly hot-pressing their intimate mixture. The polar oxygen atoms in the sequential oxyethylene groups of the PEO chains coordinate the Li^+ cations, thus separating them from their X^- counter anions. This favors the dissolution of the LiX salt in the PEO matrix with a solvating mechanism that is, in effect, similar to that occurring in the liquid electrolytes.[5,11–14]

The structure of the PEO-LiX complexes may be broadly considered as a sequence of polymer chains coiled around the lithium ions, whereas the anions are more loosely coordinated.[12] FIGURE 1 illustrates this structural sequence.

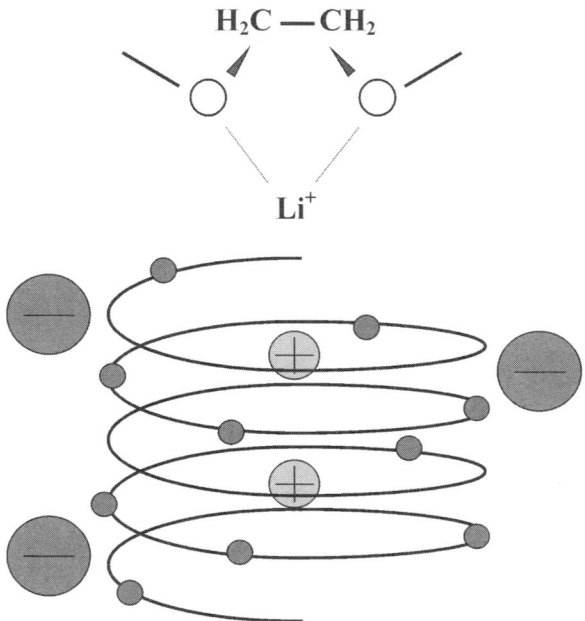

FIGURE 1. Schematic structural arrangements of the PEO chains coiling around Li^+ cations: +, Li^+; –, X^-.

This oversimplified picture progressively diverges from reality depending upon the relative concentration of the two components. As the concentration of LiX (expressed as the ratio between oxygen atoms in PEO and lithium ions in LiX) increases, the overall structure becomes more and more complicated because of ion–ion interactions, which may even lead to large ion–cluster formation.[15]

As is the case for all conductors, the conductivity of the PEO–LiX polymer electrolytes depends on the number of ionic carriers and on their mobility. The number of the Li^+ carriers increases as the LiX concentration increases, but their mobility is greatly diminished by the progressive occurrence of ion–ion interactions and association phenomena.[15]

Because of their particular structural position (FIG. 1), the Li^+ ions can be released to transport the current only on unfolding the coordinating PEO chains. In other words, this type of polymer electrolytes require local relaxation and segmental motion of the solvent (i.e., PEO) chains to allow fast Li^+ ion transport. Thus, high conductivity is restricted to the amorphous state of the PEO component, which on average occurs at temperatures above 70°C. This is represented in FIGURE 2, which illustrates the ionic-conductivity Arrhenius plot for a typical PEO-based polymer electrolyte.

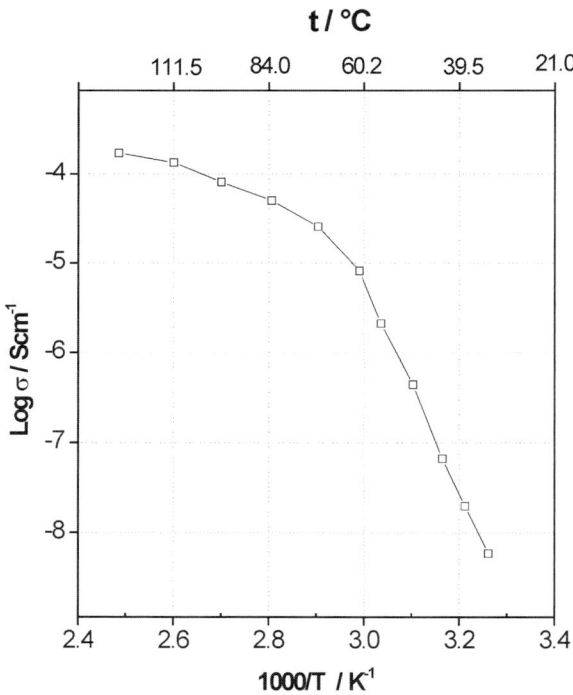

FIGURE 2. Typical conductivity Arrhenius plot for the PEO–LiClO$_4$ polymer electrolyte.

It can be clearly seen that at ambient temperature, when the system is in its crystalline state, conductivity is very low (i.e., in the 10^{-8}–10^{-7} Scm^{-1} range). However, at about 70°C, that is, at the PEO crystalline to amorphous transition,[12] the conductivity increases by several orders of magnitude to reach, at about 100°C, the order of 10^{-3} Scm^{-1}, values that are of practical interest. This implies that the use of the PEO–LiX electrolytes is restricted to batteries for which a relatively high operation temperature does not represent a major problem; for example, batteries designed for EV traction. Indeed, various research and development projects aimed at the production of PEO-based, EV lithium batteries are in progress worldwide.[13] These batteries typically use a lithium metal anode and a Li-intercalation cathode. The latter may be basically described as a compound with an *open structure*—that is, a layered (e.g., TiS$_2$, V$_2$O$_5$, or LiCoO$_2$) or a tunnel (e.g., V$_6$O$_{13}$ or LiMn$_2$O$_4$) structure that provides channels for the reversible insertion–deinsertion of lithium ions.[16–18] We identify this intercalation cathode by the generic notation $A_y B_z$. Then the electrochemical process of the battery can be illustrated by the schematic diagram in FIGURE 3.[2,19] On discharge, the Li$^+$ ions, produced at the Li metal anode, travel across the electrolyte to reach the $A_y B_z$ cathode and are inserted into its structure; the electrons travel through the external circuit to reach the cathode and modify its electronic density of states.[18] The charging process is just the opposite and, thus, the overall process may be described as follows:

$$x\text{Li} + A_y B_z \Leftrightarrow \text{Li}_x A_y B_z. \qquad (1)$$

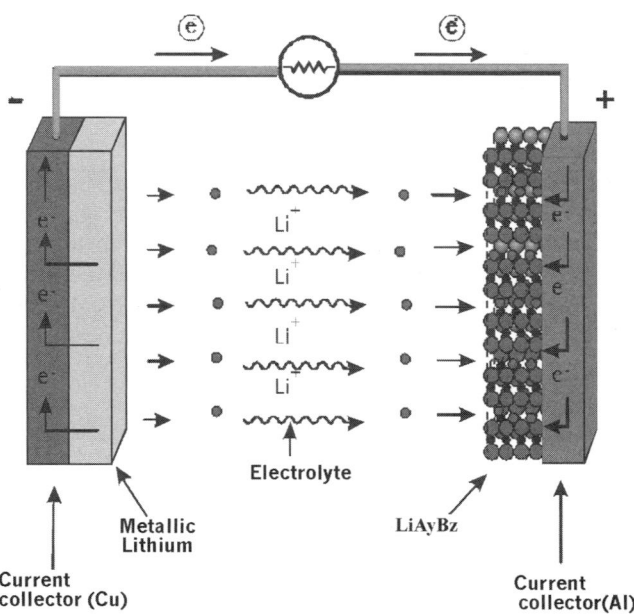

FIGURE 3. Schematic diagram of the Li–AyBz lithium battery.

The open circuit voltage of this battery is associated with the difference in the Fermi levels of the two electrodes and, if the overall process does not induce phase changes in the host A_yB_z cathode, the discharge voltage decreases on increasing lithium intercalation level (x in Equation (**1**)).

NANOCOMPOSITE POLYMER ELECTROLYTES

Although the batteries described above are of relevant interest for EV application, the relatively high operating temperature limits their overall practical output. Accordingly, many studies have been carried out with the goal of improving the low temperature conductivity of PEO–based polymer electrolytes. Various approaches have been considered in attempting to achieve this goal; for example, the use of modified PEO polymer architectures to reduce the crystallinity at room temperature. These modifications include block copolymers, crosslinked polymer networks and comb-shaped polymers having short oligo-oxyethylene chains attached to the polymer backbone.[12,20–22] Other approaches have considered the addition of plasticizers, such as organic liquids (e.g., propylene carbonate PC or ethylene carbonate EC)[23,24] or low molecular weight ethylene glycols.[25] However, in all these modified PEO electrolytes and, in particular, the plasticizer-added modifications, the gain in conductivity is adversely accompanied by a loss of the solid-state configuration and by a loss of compatibility with the lithium electrode, that is, by a loss of the most important intrinsic features of a polymer electrolyte. This is reflected in the fact that liquid plasticizer-added PEO–LiX electrolytes cannot be generally used in lithium metal batteries since they are affected by limited manufacturing and by high reactivity with the metal anode, both drawbacks resulting in serious problems in terms of battery cycle life and safety hazard. Therefore, one reaches the important conclusion that only dry (i.e., liquid plasticizer free) polymer electrolytes can ensure an efficient cyclability of the lithium metal electrode.[19,26,27]

On the other hand, this interfacial stability is adversely accompanied by a decay in room temperature conductivity, since PEO-based dry electrolytes conduct only at temperatures above 70°C. Therefore, the ideal solution in this field would be the use of *solid plasticizers*; that is, solid additives that would promote amorphicity at ambient temperature without affecting the mechanical and the interfacial properties of the electrolyte. A result that approaches this ideal condition has been obtained by dispersing selected ceramic powders, such as TiO_2, Al_2O_3, and SiO_2, at the nanoscale particle size, in the PEO–LiX matrix.[28–34]

The general concept of adding ceramic powders to PEO–LiX polymer electrolytes dates back to the early 1980s when the procedure was successfully employed to improve their mechanical properties.[35] Later, various authors demonstrated the favorable role of the dispersed ceramics in improving the interface with the lithium electrode.[36–42] In the early 1990s, Raman spectroscopy studies suggested that the structural properties of solid composite PEO–LiX electrolytes containing dispersed, low particle size ceramic filers (e.g., γ-$LiAlO_2$ fillers) could be comparable with those of liquid electrolytes formed by low molecular weight polyethylene glycol–lithium salt solutions.[43] More recently, Day *et al.*[44] demonstrated by ^{7}Li-NMR studies that the addition of nanoparticle size Al_2O_3 to concentrated PEO–LiI polymer electrolytes suppresses the formation of crystalline phases. Similarly, Wieczorek et al.[45,46] reported that a reduction to below 4μm in the particle size of added Al_2O_3 ceramics in composite PEO–NaI electrolytes is accompanied by a consistent increase in ionic conductivity. These authors also suggested that there is an active role of the surface groups of the ceramic particles in promoting local structural modifications.

However, only recently has the role of the dispersed ceramics in influencing the transport properties of the PEO–LiX polymer electrolytes been clearly demonstrated. This has been achieved by selecting inorganic fillers having appropriate chemical and morphologic properties. The concept was initially tested by developing composite polymer electrolytes having proper nanoscale dimensions, as well as suitable surface characteristics.[31]

Basically, the preparation of these new *nanocomposite* polymer electrolytes involves first the dispersion of the selected ceramic powder (e.g., TiO_2, SiO_2, or Al_2O_3) and of the LiX (e.g., $LiClO_4$ or $LiCF_3SO_3$) lithium salt in acetonitrile, followed by the addition of the PEO polymer component, and by thoroughly mixing the resulting slurry. The slurry is then cast on a plate to finally yield homogenous and mechanically stable membranes.[47]

FIGURE 4 shows the conductivity Arrhenius plot of a representative example of nanocomposite polymer electrolytes; that is, $PEO_8LiClO_4.10$ wt% Al_2O_3 sample. A plot for a ceramic-free PEO_8LiClO_4 polymer electrolyte is also included for comparison. The heating scan of the latter shows a break about 70°C, reflecting the cited transition from the PEO crystalline to the amorphous state, which is accompanied by

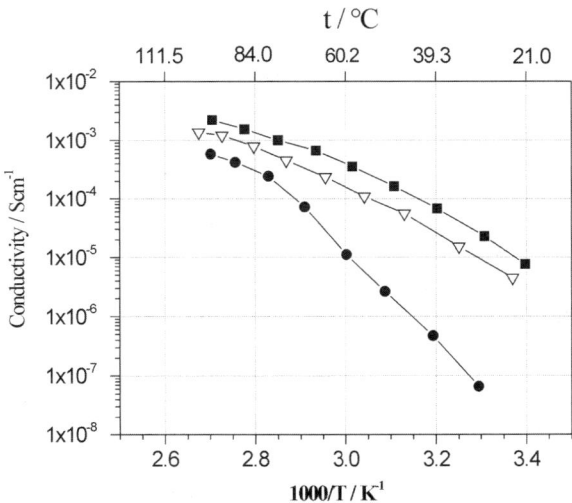

FIGURE 4. Conductivity Arrhenius plots of ceramic-free and composite PEO-based polymer electrolytes: ■, TiO_2-added; ▽, Al_2O_3-added; ●, ceramic free.

a relevant increase in ionic conductivity. When brought back to below 70°C, the common PEO_8LiClO_4 electrolyte initially remains in the amorphous state because of the slow recrystallization kinetics. However, after a short delay, this electrolyte readily recrystallizes and, consequently, its conductivity decays to low values. As prepared, the nanocomposite electrolytes have a room temperature conductivity and a first heating scan similar to those of the ceramic-free electrolyte. However, the behavior of the following cooling scan is quite different, since no break occurs at about 70°C and the conductivity remains consistently higher (between 10^{-3} and $10^{-5}\,Scm^{-1}$ versus 10^{-4} and $10^{-8}\,Scm^{-1}$ in the 80–30°C temperature range. This conductivity trend is reproduced in the following heating and cooling scans. Similar behavior is also displayed by other types of nanocomposite polymer electrolytes using various types of ceramic fillers.[48–53] The enhancement in conductivity is quite stable and does not decay with time.[48] In addition, stress–strain measurements have revealed a large enhancement of the Young's modulus and of the yield point stress when passing from ceramic-free to nanocomposite polymer electrolyte samples.[47] This demonstrates that the conductivity state of the latter is not due a polymer degradation but, rather, is accompanied by a substantial increase in the mechanical properties of the electrolyte. It is reasonable, therefore, to conclude that the favorable transport behavior is an inherent feature of the nanocomposite structure.

INTERPRETATION MODELS OF THE TRANSPORT MECHANISM IN NANOCOMPOSITE POLYMER ELECTROLYTES

Various models have been proposed to account for the effects of the ceramic fillers. For instance, one may assume that the conductivity enhancement in the nanocomposite electrolytes is due to the promotion of a large degree of amorphicity in the polymer. Accordingly, once the electrolytes are annealed at temperatures higher than the PEO crystalline to amorphous transition (i.e., above 70°C), the ceramic additive, due to its large surface area, prevents local PEO chain reorganization to result in freezing at ambient temperature with a high degree of disorder that is accompanied by a consistent enhancement of the ionic conductivity.

FIGURE 5 shows differential scanning calorimetry (DSC) traces of two electrolyte samples; that is, a ceramic-free PEO_8LiClO_4 and a PEO_8LiClO_4.10 wt% Al_2O_3 nanocomposite. In both cases, the figure shows traces referring to the first heating–cooling scan of samples as prepared and to heating–cooling scans run after a number of days of storage at room temperature.[47] Consider first the response of the ceramic-free sample. The first heating scan shows the expected crystalline to amorphous peak at 60–70°C and the immediately following cooling scan from 100°C to room temperature is peakless, since even ceramic-free electrolytes have a relatively slow recrystallization kinetics.[54] However, this electrolyte does recrystallize, as is indeed shown by the peak of the trace obtained after six days storage at room temperature. Quite different is the DSC response for the Al_2O_3-added composite electrolyte, FIGURE 5B. As expected, and in agreement with the conductivity results (FIG. 4), a peak due to the crystalline to amorphous transition is shown at about 60–70°C in the first heating scan. However, in all following heating and cooling scans, no peaks are revealed, even after prolonged storage times (i.e., exceeding two weeks). This again

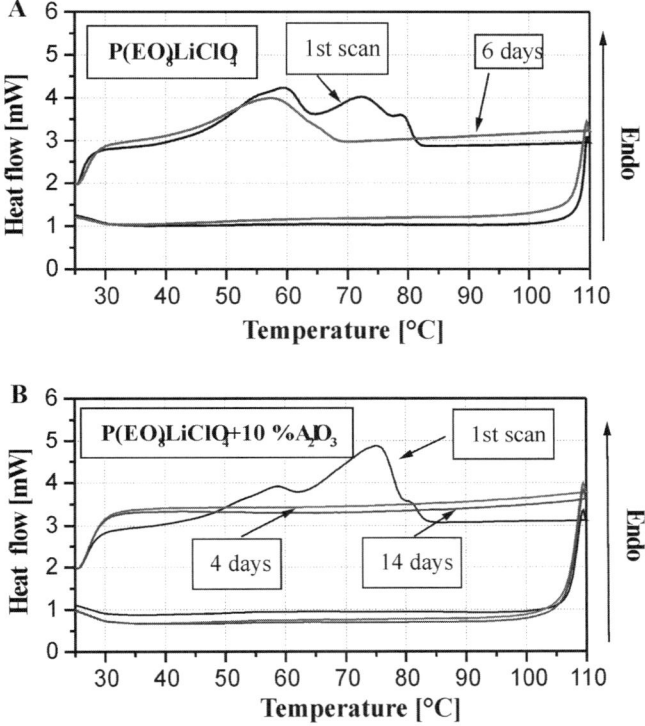

FIGURE 5. DSC traces for **(A)** ceramic-free PEO_8LiClO_4 and **(B)** of $PEO_8LiClO_4 \cdot 10$ wt% Al_2O_3 nanocomposite polymer electrolyte samples. The traces refer to the first heating–cooling scan of samples as prepared and to heating–cooling scans after a number of days of storage at room temperature. The heating–cooling rate was $10°C\,min^{-1}$.

confirms the conductivity results and, ultimately, that after being annealed at temperatures above the PEO transition, the nanocomposite polymer electrolytes retain their amorphous state even if kept at room temperature for many days. Thus, the role of the added nanoceramics is indeed that of influencing the recrystallization kinetics of PEO–LiX polymer electrolytes.

The model has been further confirmed by impedance spectroscopy analysis.[55] FIGURE 6 shows the evolution at 30°C and at progressive time intervals of the impedance spectra for a ceramic-free $P(EO)_8LiClO_4$ polymer electrolyte previously annealed at 100°C. As already noted, the recrystallization kinetics of these concentrated polymer electrolyte samples are relatively slow due to the salt plasticizing effect; however, crystallization does occur after sufficient induction time. Accordingly, evolution of the impedance spectra is characterized by intermediate states that evolve toward a final steady-state configuration represented by a large semicircle in the high-frequency-region. The latter may in fact be representative of grain boundary effects such as those expected from non-homogeneous, crystallized phases.

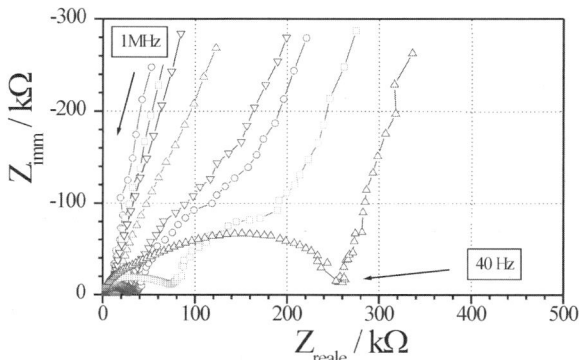

FIGURE 6. Time evolution at 30°C of the impedance spectra for an annealed ceramic-free P(EO)$_8$LiClO$_4$ polymer electrolyte: ○, 0.1-; □, 3-; ▽, 5-; △, 7-; ▽, 8-; ○, 10-; □, 11-; and △, 14-days annealing.

Quite opposite is the impedance response of the parent nanocomposite electrolytes, as shown in FIGURE 7, which illustrates the time evolution of the related spectra. In the course of the first 23 days the impedance spectra shift toward a low resistance intercept (FIG. 7A) to then evolve to the steady-state configuration (FIG. 7B), which is consistently different than the behavior of the parent ceramic-free electrolyte (compare FIG. 6). In fact, the high-frequency response of the nanocomposite samples remains almost linear for many days, only after long storage times showing a tendency to a slight curvature without ever closing on the real axis in a complete semicircle fashion. This is a convincing evidence of the role of the nano-sized ceramic filler on influencing the structural reorganization of the PEO chains, a result favoring the long term stability of the amorphous phase and, thus, the enhancement of the room temperature conductivity of the polymer electrolyte.

However, as one may observe from FIGURE 4, the conductivity enhancement in the composite polymer electrolytes occurs throughout the entire temperature range; that is, not only below but also above 70°C where the PEO is amorphous in nature. Therefore, the model must be extended by assuming that the role of the ceramic cannot be limited to the sole action of preventing crystallization of the polymer chain. One needs also to consider the occurrence of specific interactions between the surface groups of the ceramic particles and both the PEO segments and the lithium salt.

Indeed, the Lewis acid groups of the added ceramics (e.g., the –OH groups on the Al$_2$O$_3$ surface) may quite likely compete with the Lewis-acid lithium cations for the formation of complexes with the PEO chains, as well as with the anions of the added LiX salt; see FIGURE 8.

More specifically, the structural modifications occurring at the ceramic surface, because of the specific actions of the polar surface groups of the inorganic filler, may result in

1. crosslinking centers for the PEO segments and for the X$^-$ anions, lowering the PEO reorganization tendency and, hence, promoting a structure modifications of the polymer chains; the expected effect being the promotion of Li$^+$ conducting pathways at the ceramic surface; and

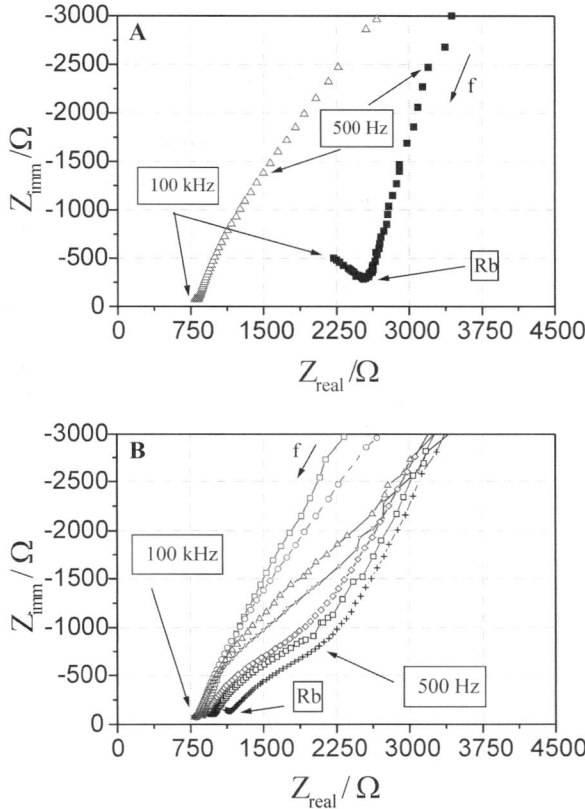

FIGURE 7. Time evolution at 30°C of the impedance spectra for an annealed, P(EO)$_{30}$LiClO$_4$ +10 wt% SiO$_2$ nanocomposite polymer electrolyte: **(A)** initial trend at 1 day (■) and at 23 days (△); and **(B)** steady state behavior at 21 (□), 23 (○), 26 (△), 33 (▽), 47 (◇), 56 (□), and 68 (+) days.

2. Lewis acid–base interaction centers with the electrolyte ionic species, lowering ionic coupling; the expected effect being the promotion of salt dissociation via a sort of "ion–ceramic complex" formation.

These two effects may favor the formation of "free" ions and, thus, they may indeed account for the observed enhancement of the conductivity of the nanocomposites throughout the entire temperature range examined. This extended, structural model has been confirmed by a series of *ad hoc* measurements. These include determination of conductivity and of lithium ion transference number of various composite electrolyte samples that differ in type and nature from the ceramic filler.[56]

FIGURE 9 shows a comparison of the Arrhenius conductivity plots for three nanocomposite samples differing in the type of the dispersed ceramic; e.g., Al$_2$O$_3$ in its acidic, basic, and neutral forms. These three forms vary in the extent of their surface group arrangements in the order: Al$_2$O$_3$ acidic ≥ Al$_2$O$_3$ neutral ≥ Al$_2$O$_3$ basic.[56]

Since all of the types of Al_2O_3 ceramics used had comparable particle size, one may safely assume that the dilution effect is similar for all three cases and, consequently, that differences in conductivity are directly related to differences in the extent of the specific ceramic surface interactions. Indeed, differences are clearly revealed in FIGURE 9 when passing from one electrolyte to another: although the conductivity of the composites based on acidic and neutral, Al_2O_3 filler is higher than that of the corresponding ceramic-free sample, no substantial differences are noticed for the Al_2O_3 basic added composite. These differences are clearly related to changes in the surface state distribution of the three ceramic types.

The structural model is further supported by the measured values of the Li^+ transference number T_+ in the four samples; that is, the Al_2O_3 acidic, basic, and neutral, composite, and the ceramic-free sample. FIGURE 10 reports the results in terms of percentage of changes of T_+ associated with each sample. The trend confirms the specific role of the surface groups of the filler by showing progressive increase in T_+ when passing from the ceramic-free to the composite electrolytes with, in sequence, basic, neutral, and acidic Al_2O_3 additions.

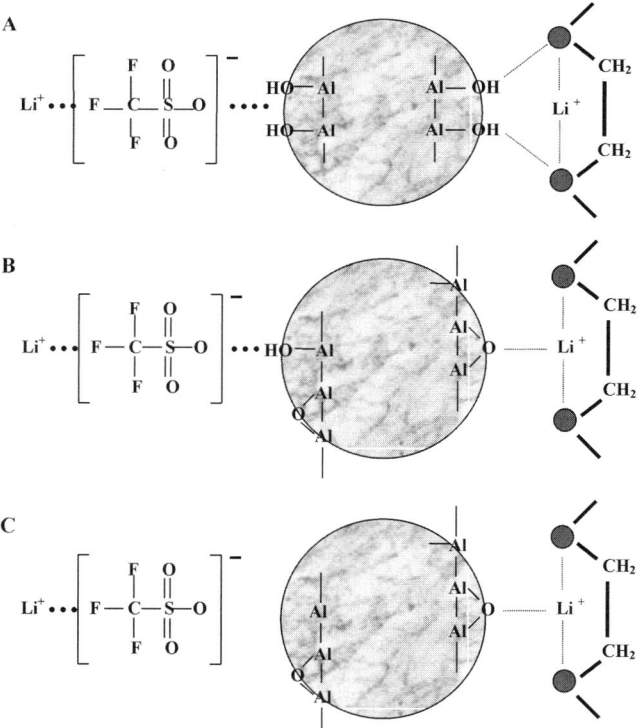

FIGURE 8. Schematic model of the surface interactions between ceramic particles and the polymer chain and with the salt anion in PEO–LiX nanocomposite electrolytes under (**A**) acidic, (**B**) neutral, and (**C**) basic conditions.

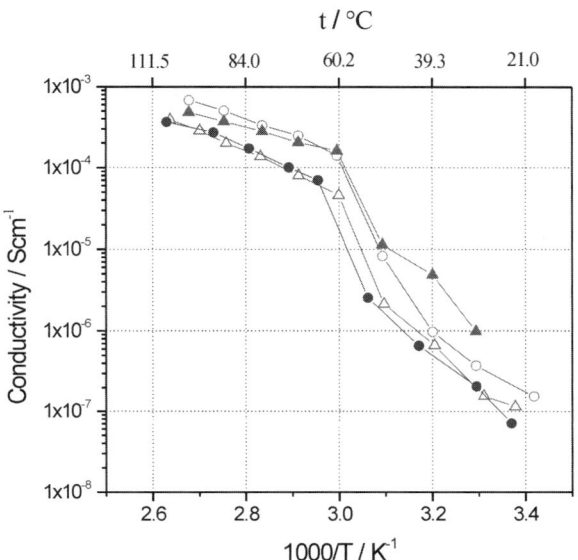

FIGURE 9. Conductivity Arrhenius plots for $P(EO)_{20}LiSO_3CF_3$ + 10wt% Al_2O_3 nanocomposite samples differing only in the type of surface states of the selected ceramic filler. The conductivity of a $P(EO)_{20}LiSO_3CF_3$ ceramic-free sample is also shown for comparison. The data were obtained from impedance measurements. ●, Al_2O_3 basic-added; ○, Al_2O_3 neutral-added; △, without filler; ▲, Al_2O_3 acidic-added.

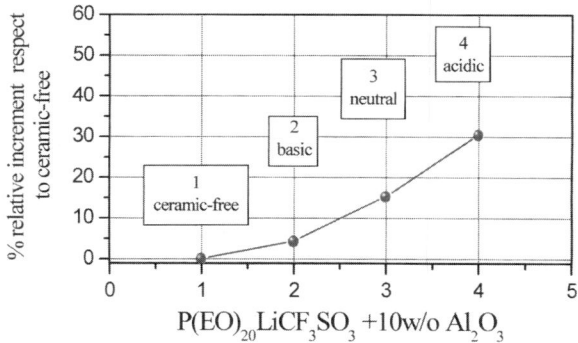

FIGURE 10. Percentage variation of the lithium ion transference number in passing from ceramic-free to nanocomposite polymer electrolytes with various types of ceramic additives.

Finally, the model is confirmed by spectroscopic analysis. Recent Raman results reported by Best et al.[57] demonstrate the specific action of nanometric TiO_2 powders in promoting ceramic–salt interactions. In addition, NMR data[58] have shown that the diffusion of Li^+ ions, and thus, the related T_+ value, in the nanocomposite electrolytes is considerably higher than that of the parent ceramic-free electrolytes.

CONCLUSION

The dispersion of selected, nano-particle size ceramics yield to the development of true solid-state PEO–LiX nanocomposite polymer electrolytes that, in the 30–80°C range, possess excellent mechanical stability (promoted by the network of the fillers into the polymer bulk) and high ionic conductivity (induced by the high surface area of the dispersed fillers). These electrolytes appear to be particularly suitable for use as separators in advanced, rechargeable lithium batteries characterized by a long cycle life and high power capabilities. Preliminary results recently obtained in our laboratory confirm this expectation. Lithium battery prototypes, based on using an Al_2O_3-added polymer electrolyte, show electrochemical performances very superior to those obtained with conventional, ceramic-free systems.[59,60]

ACKNOWLEDGMENTS

We would like to thank our students and collaborators: Giovanni Battista Appetecchi, Franco Bonino, Stefania Panero, Luigi Persi, Priscilla Reale, and Josef Hassoun for their important and dedicated research work, some results of which are reported in this paper. The financial support of U.S. Army, under contract No. DAAB07-01-C-D414, is also deeply acknowledged.

REFERENCES

1. MEGAHED, S. & B. SCROSATI. 1994. J. Power Sources **51:** 79.
2. MEGAHED, S. & B. SCROSATI. 1995. Interface **4:** 34.
3. PISTOIA, G. 1997. Lithium Batteries. Elsevier Science, London.
4. LINDEN, D. 1995. Handbook of Batteries, 2nd edit. McGraw-Hill, Inc., New York.
5. VINCENT, C.A. & B. SCROSATI. 1997. Modern Batteries. An Introduction to Electrochemical Power Sources, 2nd edit. Arnold, London.
6. WAKIHARA, M. & O. YAMAMOTO, Eds. 1998. Lithium Ion Batteries. Kodansha & Wiley-VCH, Weinheim.
7. FENTON, D.E, J.M. PARKER & P.V. WRIGHT. 1973. Polymer **14:** 589.
8. ARMAND, M.B., S.M. CHABAGNO & M. DUCLOT. 1978. Presented at the Second International Meeting on Solid Electrolytes, St. Andrews, Sept. 20–22.
9. SCROSATI, B. 1993. Mater. Sci. Engin. B-Solid State Mater. Adv. Technol. **12:** 369.
10. ARMAND, M. 1990. Adv. Mater. **2:** 278.
11. MACCALLUM, J.R. & C.A. VINCENT, Eds. 1987–1989. Polymer Electrolytes Reviews, I & II. Elsevier Science, London.
12. GRAY, F.M. 1997. Polymer Electrolytes, Royal Society of Chemistry Monographs, Cambridge.
13. NEAT, R. & B. SCROSATI. 1993. *In* Applications of Conductive Polymers, B. Scrosati, Ed.: 182. Chapman & Hall, London.

14. SCROSATI, B. 1995. Chim. Ind. Milan **77:** 285.
15. VINCENT, C.A. 1999. IEE Rev. **65**.
16. SCROSATI, B. 1994. *In* Novel Materials—Frontiers of Electrochemistry. J. Lipkowski & P.N. Ross, Eds.: 111. VCH, Weinheim.
17. MCKINNO, W.R. 1995. *In* Solid State Electrochemistry. P.G. Bruce, Ed.: 163. Cambridge University Press, Cambridge.
18. GOODENOUGH, J.B. 1971. *In* Progress in Solid State Chemistry, Vol. 5. H. Resiss, Ed.: 279. Pergamon Press, Oxford.
19. SCROSATI, B. 1997. Chim. Ind. Milan **79:** 463.
20. BOURIDAH, A., F. DALARD, D. DEROO, *et al.* 1985. Solid State Ionics **15:** 6854.
21. ABRAHAM, K.M. & M. ALAMGIR. 1991. Chem. Mater. **3:** 339.
22. MARCHESE, L., M. ANDREI, A. ROGGERO, *et al.* 1992. Electrochim. Acta **37:** 1559.
23. MUNSHI, M.Z.A. & B.B. OWENS. 1998. Solid State Ionics **26:** 41.
24. CHINTAPALLI, S. & R. FRECH. 1996. Solid State Ionics **86–88:** 341.
25. DAUTZENBERG, G., F. CROCE, S. PASSERINI & B. SCROSATI. 1996. J. Electrochem. Soc. **143:** 6.
26. CROCE, F. & B. SCROSATI. 1993. J. Power Sources **43–45:** 9.
27. FAUTEAUX, D. 1985. Solid State Ionics **17:** 133.
28. SCROSATI, B. & F. CROCE. 1993. Pol. Adv. Techn. **4:** 198.
29. CAPUANO, F., F. CROCE & B. SCROSATI. 1996. US Patent, N. 5,576,115, Nov. 19.
30. KUMAR, B. & L.G. SCANLON. 1994. J. Power Sources **52:** 261.
31. CROCE, F., G.B. APPETECCHI, L. PERSI & B. SCROSATI. 1998. Nature **394:** 456.
32. KUMAR, B. & L.G. SCANLON. 1999. Solid State Ionics **124:** 239.
33. CAPIGLIA, C., P. MUSTARELLI, E. QUARTARONE, *et al.* 1999. Solid State Ionics **118:** 73.
34. BEST, A.S., A. FERRY, D.R. MACFARLANE & M. FORSYTH. 1999. Solid State Ionics **126:** 269.
35. WESTON, J.E. & B.C.H. STEELE. 1982. Solid State Ionics **7:** 75.
36. SCROSATI, B. 1990. J. Electrochem. Soc. **136:** 2774.
37. CROCE, F. & B. SCROSATI. 1993. J. Power Sources **43:** 43.
38. BORGHINI, M.C., M. MASTRAGOSTINO, S. PASSERINI & B. SCROSATI. 1995. J. Electrochem. Soc. **142:** 2118.
39. CAPUANO, F., F. CROCE & B. SCROSATI. 1996. US Patent, N. 5,576,115. Nov. 19.
40. PELED, E., D. GOLODINITSKY, G. ARDEL & V. ESHKENAZY. 1995. Electrochim. Acta **40:** 2197.
41. KUMAR, B. & L.G. SCANLON. 1994. J. Power Sources **52:** 261.
42. QUARTARONE, E., P. MUSTARELLI & A. MAGISTRIS. 1998. Solid State Ionics **110:** 1.
43. CROCE, F., B. SCROSATI & G. MARIOTTO. 1992. Chem. Mat. **4:** 1134.
44. DAI, Y., S. GREENBAUM, D. GOLODNITSKY, *et al.* 1988. Solid State Ionics **106:** 25.
45. WIECZOREK, W., Z. FLORJANCYK & J.R. STEVENS. 1995. Electrochimica Acta **40:** 2251.
46. PRZYLUSKI, J., M. SIEKIERSKI & W. WIECZOREK. 1995. Electrochimica Acta **40:** 2101.
47. CROCE, F., L. PERSI, F. RONCI & B. SCROSATI. 1999. J. Phys. Chem. B **103:** 10632.
48. CROCE, F., L. PERSI, F. RONCI & B. SCROSATI. 2000. Solid State Ionics **135:** 47.
49. APPETECCHI, G.B., F. CROCE, L. PERSI, *et al.* 2000. Electrochim. Acta **45:** 1481.
50. MACFARLANE, D.R., P.J. NEWMAN, K.M. NAIRN & M. FORSYTH. 1998. Electrochim. Acta **43:** 1333.
51. KUMAR, B. & L.G. SCANLON. 1999. Solid State Ionics **124:** 239.
52. PLOCHARSKI, J. & W. WIECZOREK. 1998. Solid State Ionics **28–30:** 979.
53. TAMBELLI, C.C., A.C. BLOISE, A.V. ROSARIO, *et al.* 2002. Electrochim. Acta **47:** 1677.
54. CROCE, F., L. PERSI, B. SCROSATI, *et al.* 2001. Electrochim. Acta **46:** 2457.
55. SCROSATI, B., F. CROCE & L. PERSI. 2000. J. Electrochem. Soc. **147:** 1718.
56. CROCE, F., L. PERSI, B. SCROSATI, *et al.* 2001. Electrochim. Acta **46:** 2457.
57. BEST, A.S., A. FERRY, *et al.* 1999. Solid State Ionics **126:** 269.
58. CHUNG, S.H., Y. WANG, L. PERSI, *et al.* 2001. J. Power Sources **97–98:** 644.
59. CROCE, F., F. SERRAINO FIORY, L. PERSI & B. SCROSATI. 2001. Electrochem. Solid State Lett. **4:** A121.
60. PERSI, L., B. SCROSATI, E. PLICHTA & M.A. HENDRICKSON. 2002. J. Electrochem. Soc. **149:** A212.

Preparation of Nano-Structured Polymeric Proton Conducting Membranes for Use in Fuel Cells

GIULIO ALBERTI, MARIO CASCIOLA, MONICA PICA, AND GIUSI DI CESARE

Chemical Department, University of Perugia, Perugia, Italy

ABSTRACT: We briefly discuss the state of the art of polymer electrolyte membrane fuel cells and suggest that the main obstacles to the commercial development of these fuel cells are essentially the high costs and poor characteristics of present proton conducting membranes. A strategy for the preparation of improved nanocomposite membranes based on the introduction of proton conducting lamellæ in the polymeric matrix of present ionomeric membranes is then discussed. Due to their high proton conductivity (in some cases even higher than $10^{-1} S cm^{-1}$), tailor made lamellæ obtained by exfoliation of superacid metal (IV) phosphonates are particularly suitable for the preparation of these hybrid membranes. The expected positive influence of the dispersed lamellæ on important properties of proton conducting membranes, such as swelling, mechanical resistance, proton transport, and diffusion of methanol, are also discussed. The methods used to obtain good lamellar dispersions into ionomeric polymers and the preparation and main characteristics of some hybrid membranes are also briefly described. The presence of nanoparticles of metal phosphonates in the electrodic interfaces Nafion/Pt already considerably improves the electrochemical characteristics of fuel cells in the temperature range 80–130°C. The increased working temperature of the fuel cell considerably reduces CO poisoning of the platinum electrodes and allows better control of the cooling system, thus overcoming important obstacles to the development of medium temperature PEM fuel cells.

KEYWORDS: nano-structured membrane; polymeric proton conducting membrane; fuel cells; zirconium phosphonates

INTRODUCTION

The greater part of energy is today obtained by fossil fuel combustion with consequent emission of a large amount of carbon dioxide to the atmosphere. In addition to the *green-house effect*, intense vehicular traffic in large towns produces the well-known type of pollution that contains significant amounts of nitric oxide, carbon monoxide, benzene, unburned hydrocarbons, and so forth. A reduction of this pollution without a drastic reduction in car mobility is a problem of primary importance that requires a quick solution to avoid further lowering of our life quality.

Address for correspondence: Giulio Alberti, Chemical Department, University of Perugia, via Elce di Sotto 8, 06123 Perugia, Italy. Voice: +39/075/5855562; fax +39/075/5855566.
alberti@unipg.it

Among various possible solutions, the gradual substitution of vehicles by electric cars fed with fuel cells is under serious consideration. However, for large-scale commercialization of these electric cars, the price of fuel cells should be considerably reduced and, at the same time, they should possess most of the following characteristics: to be able to operate at low temperatures; to be sufficiently small, light, and easy to cool; to have long life; to possibly be fed with a non-toxic liquid of low flammability. Unfortunately, although several years of research have been already dedicated to the development of fuel cells for electric cars, the present state of the art is still unable to produce cells with the above characteristics and further research is, therefore, necessary. In particular, since proton conducting membranes play a very important role, new membranes with improved characteristics and lower cost than Nafion must be found.

In this paper, after summarizing the present state of the art on polymer electrolyte membranes for fuel cells, a brief account of *tailor-made* lamellar nanoparticles based on zirconium phosphate-phosphonates is offered. We finally discuss use of these nanoparticles for preparation of innovative nanostructured membranes, planned in our laboratory to improve the characteristics of present membranes for their use in fuel cells for mobile applications.

STATE OF THE ART ON FUEL CELLS FOR MOBILE APPLICATIONS

Many different kinds of fuel cells are presently known. For their use in electric cars, or as energy source for portable electrical apparatus (personal computers, cellular telephones, etc.), the most convenient fuel cells for weight and low working temperature are at present the so-called polymer electrolyte membrane fuel cells (PEM FC).[1–3] In these cells, electricity is produced by an electrochemical reaction that involves oxidation of the fuel at the anode with the production of protons. The protons diffuse to the cathode, through the proton conducting membrane, where react with oxygen to form water. In PEM FC the choice of the fuel is, therefore, limited to hydrogen or to hydrogenated molecules. Because of various technologic problems related to the choice of the fuel, PEM FC using hydrogen or hydrogenated molecules are discussed separately.

PEM FC Using Hydrogen as Fuel

These cells use Nafion or other Nafion-like sulfonated perfluorinated polymers that exhibit high proton conductivity at temperatures lower than 100°C ($\sigma \geq 10^{-2}\,\mathrm{S\,cm^{-1}}$) combined with high chemical stability.[3,4] Dispersed platinum, which assures a high catalytic efficiency for hydrogen and oxygen even at room temperature, is used for the electrodes. Today, many technological problems related to the hydrogen PEM FC have been solved. Typical hydrogen PEM FCs usually work at 70–90°C since, at higher temperatures, a consistent reduction in the energy delivered is observed. The phenomenon is usually attributed to a loss of hydration from the membrane with consequent drastic decrease in its proton conductivity. The above limit on the maximum temperature at which the cell can work under optimal conditions may create serious problems in the cooling of cells, especially when high performance is required. Very recently this limit has been increased to 110–130°C, thus simplifying the cooling of the stacks (see later in this paper) and, therefore, eliminating the last important technologic problem in the realization of efficient hydrogen PEM FCs. Thus, large-scale diffusion of hydrogen fuel cells is today essentially limited by the high cost of both membranes[4] and hydrogen.[5] Many different routes are presently employed to develop more economic ionomers than Nafion. Nevertheless, the solution of this problem is not simple because the ionomer, as well as being economic, must exhibit a good proton conductivity (at least $10^{-2}\,S\,cm^{-1}$) and a very high stability toward its oxidation. At present, no really good substitutes for Nafion membranes have been found, although encouraging results have been obtained with sulfonated polyetherketones[6–9] and polysulfones.[10–14]

PEM FC Using Hydrogenated Compounds as Fuels

Even in case in which the economic aspects of proton conducting membranes can be positively solved in the near future, it is reasonable to forecast that a large-scale use of cars employing hydrogen as fuel will be hindered by the difficulties in car refueling at filling stations and by the diffidence of many drivers toward the risks associated with the use of gaseous hydrogen. The expectation for commercialization of direct hydrogen PEM FC for portable electrical apparatus is even lower. The use of liquid fuels, possibly of low flammability, could considerably facilitate car refueling at common petrol stations and is expected to considerably reduce the diffidence toward these fuel risks. Thus, there is strong interest in PEM FC using hydrogenated compounds as fuels. Two kinds of PEM FC are presently under development. In the first, hereafter called indirect hydrogenated fuel PEM FC, the hydrogen necessary for feeding the PEM FC is obtained by reforming a hydrogenated liquid compound. In the second, hereafter called direct hydrogenated fuel PEM FC, the PEM FC is directly fed with pure (or with an aqueous solution of) hydrogenated fuel. We examine technologic problems of these two kinds of PEM FC.

1. Indirect Hydrogenated Fuel PEM FC: hydrogen can be obtained by reforming many hydrogenated compounds. However, for our purposes, attention will be limited here to methanol. A typical composition of the gaseous mixture obtained by reforming methanol is hydrogen 63%, carbon dioxide 35%, water 2%, and carbon monoxide 150–300 ppm.

Unfortunately, as well as an additional cost due to the reforming apparatus, the presence of a non-negligible concentration of CO introduces a new technologic problem: the poisoning of the platinum electrodes. The purification of hydrogen from carbon monoxide is possible, but costs are further increased. Furthermore, hydrogen purification *in situ* is not a good choice for mobile applications since the total weight is increased and more room is required for the separator apparatus.

The problem may be partially solved by the catalytic oxidation of CO to CO_2. Since the rate of the oxidation increases considerably with increasing temperature, an increase in the working temperature of the PEM FC is, however, desirable. In any case, about 10 ppm of CO remain in the reforming gas, even after the catalytic oxidation and, unfortunately, this small amount is sufficient to cause consistent Pt-poisoning. Taking into account that the adduct Pt–CO is thermolabile, working temperatures of the indirect methanol PEM FC at about 120–130°C could drastically reduce poisoning. The realization of cells working at temperatures higher than 100°C is discussed below.

2. Direct Methanol PEM FC: direct use of gaseous methanol is attractive since weight, size, and costs of the reformer and separator apparatus are avoided.[15] Unfortunately, as well as the poisoning of the platinum electrodes, a new technologic problem arises: the high methanol permeability through proton conducting membranes, such as Nafion.[4] Both Pt-poisoning and methanol-crossover are reduced if dilute aqueous solutions of methanol are used. The development of this kind of cell, especially for low energy electrical mobile apparatus is in progress.

From the state of the art discussed above it is clear that to make PEM FC commercially competitive requires proton conducting membranes with lower cost and/or lower methanol permeability and/or able to work well in the temperature range 90–160°C. Thus, the cost and performance of pure Nafion membranes may no longer be acceptable and new membranes need to be developed. In this context, two main directions can be followed: (1) development of membranes made with new ionomers exhibiting the required characteristics; and (2) improvement of the characteristics of known ionomers by dispersing suitable nanoparticles in their polymeric matrix (hybrid membranes).

Resulting from 10 years of experience of our laboratory with the preparation of *tailor-made* lamellar nanoparticles, we are pursuing the latter.

PROTON CONDUCTING HYBRID MEMBRANES CONTAINING TAILOR-MADE LAMELLAR PARTICLES DISPERSED IN AN IONOMERIC MATRIX

A polymeric nanocomposite (or hybrid polymer) is a polymer that contains an appreciable amount of inorganic nanoparticles homogeneously dispersed in its polymeric matrix.[16] Because of the high specific surface of the nanoparticles (see TABLE 1) the interface interaction between the polymer matrix and the nanoparticles can become very high and can therefore exert a strong influence on the properties of the polymer itself. The higher the interface interaction, the greater is the expected influence of the nanoparticles on the polymer characteristics. In this respect, the high specific surface of lamellar nanoparticles seems to be very convenient (TABLE 1). The recent industrial success with polymeric nanocomposites containing exfoliated organophilic clays[17] has clearly demonstrated that the presence of the dispersed lamellar nanoparticles may indeed greatly modify important properties of the polymers, such as inflammability, permeability to neutral or ionic species, mechanical resistance, and thermal stability. Thus, it was hoped that the presence of tailor-made lamellar nanoparticles dispersed in the matrix of known ionomers might modify the

TABLE 1. Examples of specific surface area of nanoparticles of various dimensionalities

Nanoparticle	Area (m^2) per Unit Volume of Nanoparticles (cm^3)
One-dimensional (or fibrous) particle (diameter, 10 nm)	400
Two-dimensional (lamellar) particle (layer thickness, 1 nm)	2,000
Three-dimensional spherical particle (diameter, 10 nm)	150

properties so as to overcome some of the present technologic restrictions to large-scale expansion of PEM FCs for mobile applications.

Experience accumulated during many years of research on functional layer solids obtained by using octahedra–tetrahedra building block chemistry stimulated us to investigate this possibility. Our attention has especially focused on the chemistry of lamellar metal (IV) phosphate–phosphonates (the first zirconium phosphonates were prepared in our laboratory in 1978[18]) in which a tetravalent metal ion shares six oxygens with oxoacids of tetravalent phosphorus. The chemistry of many zirconium (titanium) phosphates, phosphonates, and diphosphonates with α-, γ-, and λ-layered structures is today well known. We recall here that metal (IV) phosphonates can be seen as organic derivatives of the corresponding layered metal (IV) phosphates among which the most investigated compound is zirconium phosphate monohydrate, $Zr(O_3P\text{-}OH)_2 \cdot H_2O$, exhibiting the typical layered α-structure, schematically illustrated in FIGURE 1A. The compositions of possible metal(IV) phosphate–phosphonates can be very complicated, since the phosphonate groups can be chosen among a large variety of species and even mixtures of two or more phosphonate groups in various ratios can be used to obtain tailor made lamellæ. Thus, for the sake of brevity, when a more precise definition of the composition is not necessary, the zirconium phosphate–phosphonates are here simply indicated by the general formula $M(IV)(O_3P\text{-}G)_{2-x}(O_3P\text{-}ArX)_x$ where –G may be an inorganic group (e.g.,–OH), or organic (e.g., –CH_2OH), or an inorganic–organic group (e.g., –$CH_2\text{-}SO_3H$ or –$CF_2\text{-}PO_3H_2$);

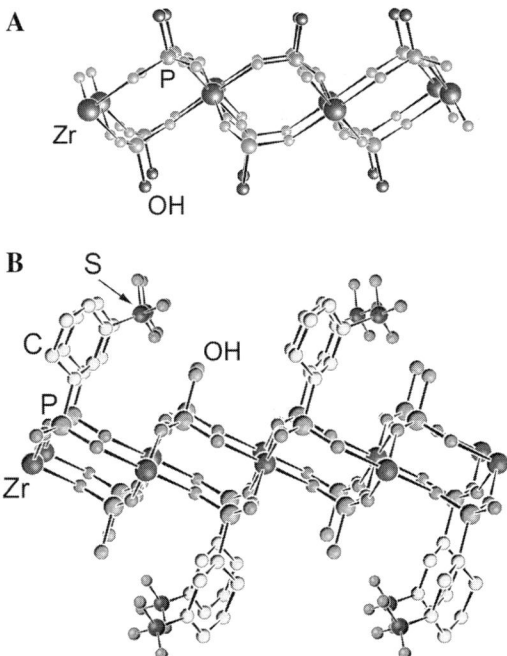

FIGURE 1. (**A**) Schematic view of a single lamella of $\alpha\text{-}Zr(O_3P\text{-}OH)_2$; (**B**) schematic view of a single lamella of $\alpha\text{-}Zr(O_3P\text{-}OH)(O_3P\text{-}C_6H_4\text{-}SO_3H)$.

TABLE 2. Composition and interlayer distance of typical zirconium phosphonates

Layered Compound	Interlayer Distance (nm)
$\alpha\text{-Zr}(O_3PH)_{0.75}(O_3PC_6H_5)_{1.25}$	1.50
$\gamma\text{-Zr}(PO_4)(O_3POH)_{0.33}(O_2POHC_6H_5)_{0.67} \cdot 2H_2O$	1.54
$\alpha\text{-Zr}[O_3P(CH_2)_5COOH]_2$	1.87
$\alpha\text{-Zr}(O_3PC_6H_4SO_3H)_{0.7}(O_3PCH_2OH)_{1.3} \cdot 5H_2O$	1.90
$\alpha\text{-Zr}(O_3PC_6H_4SO_3H)_{1.35}(O_3POH)_{0.65} \cdot 10H_2O$	2.21

Ar is an arylen group (e.g., phenylen); X is an acid group preferably chosen among $-SO_3H$, $-PO_3H_2$, and $-COOH$; x is a coefficient that can vary between 0 and about 1.5. A schematic view of a single lamella of the compound $Zr(O_3P-OH)(O_3P-C_6H_4-SO_3H)$ (i.e., where $G = -OH$; $-Ar- = -C_6H_4-$; $-X = -SO_3H$) is shown in FIGURE 1B. Some zirconium phosphonates and zirconium phosphate-phosphonates are listed in TABLE 2. More detailed information on the chemistry of these lamellar compounds can be found, for example, in References 19–21 and in references therein. Here we are essentially concerned with the effect of tailor-made lamellar zirconium phosphate–phosphonates on the mechanical stability, methanol permeability, and proton conductivity of ionomers of the state of the art. The use of tailor made lamellar phosphate–phosphonates for preparing nanopolymers with other properties is reported elsewhere.[22] For the presence of pendant organic groups covalently bonded to the backbone of the inorganic lamella, strong interaction of some zirconium phosphonates with a polymeric matrix is expected, as schematically shown in FIGURE 2. Thus, zirconium phosphonates seem to be particularly suitable for the preparation of nanopolymers with improved mechanical properties, although the research in this field is still in its infancy. The first experimental determinations are very encouraging.[22]

Knowledge of the improvement of the mechanical properties for a given ionomer is of great importance both for preparing thinner membranes, and hence, membranes with higher current intensity per unit surface, or for shifting the critical swelling of the ionomers to higher temperature, thus increasing the possible working temperature of PEM FCs.

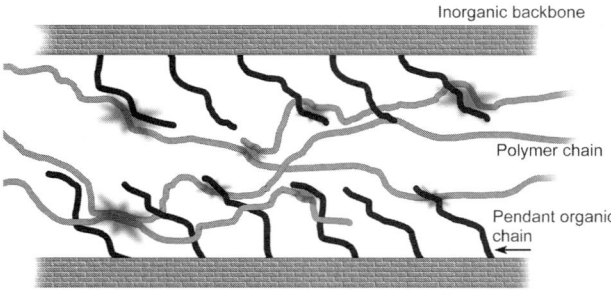

FIGURE 2. Schematic representation of the interaction between the polymer chains and the pendant groups of a generic M(IV) phosphonate.

Our present knowledge about the effects of lamellar particles on the protonic conductivity of the ionomers is still scarce. The orientation of the lamellæ in the ionomeric matrix, the presence of hydrophilic and proton acceptor groups in the dispersed lamellæ, as well as the proton conductivity of the lamellar compound, could play an important role. As is discussed below, for the permeability of methanol, an *iso*-orientation of the dispersed lamellar particles parallel to the faces of the membrane is expected to reduce proton conductivity. Concerning the effect of the proton conductivity of the lamellar particles (LC) on the ionomer protonic conductivity (IC) it is reasonable to suppose that the global effect depends on the ratio LC/IC. Thus, for a random distribution of the particle orientation, and in the absence of strong specific interactions with the ionomeric matrix, the conductivity of a given nanoionomer in comparison to that of the pure ionomer is expected to increase, to remain about the same, or to decrease when the ratio LC/IC is much greater than, equal to, or much lower than 1, respectively. Furthermore, the total effect also depends on the number of particles dispersed in the polymeric matrix. Accordingly, taking into account that the protonic conductivity of the greater part of the state of the art ionomers under investigation for use in PEM FC is in the range 10^{-3} to $10^{-1}\,\mathrm{S\,cm^{-1}}$, the choice of the lamellar compound must be limited to those exhibiting the highest proton conductivity. For this reason, we are here particularly concerned only with lamellar zirconium phosphate–phosphonates exhibiting high protonic conductivity. This property can be achieved with phosphonates containing acid groups, such as $-SO_3H$, $-PO_3H_2$, $-P-OH$, $-COOH$, or their mixtures.[23–25] As shown in TABLE 3, the proton conduction properties of some zirconium phosphate–sulfophenylenphosphonates are indeed very high (in some cases exceeding $9 \times 10^{-2}\,\mathrm{S\,cm^{-1}}$). Thus, zirconium acid phosphonates seem to be very suitable for also increasing the proton conductivity of less expensive ionomers, but unfortunately not as well as Nafion; as is the case of sulfonated PEKs. This fact has been experimentally confirmed. In TABLE 4 we report a comparison of the proton conductivity of pure s-PEK-1.3 membrane with a s-PEK/zirconium phosphonate and a s-PEK/zirconium phosphate nanoionomeric membrane. Note that, in contrast to the ionomer containing particles of phosphonate, the presence of lamellar particles of zirconium phosphate appreciably decreases the total proton conductivity. In agreement with the previous discussion, this decrease can be related to the fact that the proton conductivity of pure zirconium phosphate (of the order of $10^{-3}\,\mathrm{S\,cm^{-1}}$) is lower than that of the pure ionomer. Lamellar particles with high protonic conductivity can also be used to prepare Nafion composite membranes if the improvement due to the mechanical properties compensates for the small decrease in the proton conductivity. Finally, the filling of an ionomer with

TABLE 3. Conductivity (σ) of layered zirconium phosphates and phosphate–sulfophenylenphosphonates (100°C; 95% relative humidity)

Layered Compound	σ ($\mathrm{S\,cm^{-1}}$)
γ-Zr(PO$_4$)(O$_2$P(OH)$_2$)	4.4×10^{-4}
Zr(O$_3$POH)$_2$ (amorphous)	8.9×10^{-4}
γ-Zr(PO$_4$)(O$_2$P(OH)$_2$)$_{0.5}$(O$_2$P(OH)CH$_2$C$_6$H$_4$SO$_3$H)$_{0.5}$	3.0×10^{-2}
Zr(O$_3$POH)(O$_3$PC$_6$H$_4$SO$_3$H) (amorphous)	1.2×10^{-1}

TABLE 4. Comparison between proton conductivity of pure s-PEK 1.3 and composite membranes with zirconium phosphate and s-PEK/zirconium phosphate sulfophenylen phosphonate

Membrane	σ (S cm^{-1})
s-PEK 1.3	4.2×10^{-3}
s-PEK 1.3 + 20% α-Zr(O$_3$POH)$_2$	3.0×10^{-4}
s-PEK 1.3 + 20% Zr(O$_3$POH)(O$_3$PC$_6$H$_4$SO$_3$H)	4.2×10^{-3}

Experimental conditions: 100°C; 90% relative humidity.

lamellar particles is also expected to reduce the methanol crossover through the membrane. Because of the great importance of this problem, some considerations on the effect of dispersed lamellar particles on the methanol permeability through the nanopolymeric membranes are reported here.

Insertion of Lamellar Nanoparticles in the Polymeric Matrix of Ionomers for Decreasing Methanol Crossover

As already discussed, development of direct methanol PEM FCs greatly depends on the future availability of proton conducting membranes with low methanol permeability. The problem is difficult to solve since methanol, being a proton acceptor, exhibits a strong tendency to enter inside the polymeric matrix of the ionomers. The presence of lamellar particles dispersed in the polymeric matrix parallel to the membrane faces is expected to increase the tortuosity of the diffusion path in the perpendicular direction. The prolongation of distance covered between the two sides of the membrane is expected to increase with an increasing number of lamellar particles per volume unit of membrane (compare FIG. 3A and B) and with the dimensions of the particles (compare FIG. 3B and C). Hence, for a given amount of a lamellar compound in a polymeric matrix, the effect depends on dimensions of the initial crystals, their degree of exfoliation, and on the formula weight and layer thickness of the inserted compound. Success in the preparation of the hybrid membranes with low methanol permeability, therefore, depends on the ability to obtain homogeneous dispersions of large (e.g., several μm^2) and well-exfoliated lamellar particles. Parallel disposition of the lamellar particles is fortunately that expected from the casting of the membrane. On the other hand, it is also expected that the parallel orientation provokes a decrease in proton conductivity. Thus, to avoid an appreciable decrease in proton conductivity of the ionomer, it is highly advisable to use lamellar particles that exhibit very high protonic conductivity. In this connection, lamellar particles obtained by exfoliation of zirconium phosphate–sulfophenylenphosphonates seem to be particularly suitable for this application because of their high proton conductivity.

Exfoliation of Layered Zirconium Compounds

As mentioned already, the preparation of a nanoionomer with the desired properties not only requires specific lamellar particles that exhibit the proper characteristics, but it is also very important to achieve uniform particle dispersions, possibly well oriented, within the polymeric matrix. In this connection, we are exploring various strategies. We limit our attention here to only the exfoliation procedure.

The purpose of exfoliation is the preparation of colloidal dispersions of exfoliated lamellar particles in various solvents, with a particular preference to the solvents of the ionomers, in order to facilitate the filling of the ionomeric matrix with these particles. Exfoliation of a layered compound can be considered as an extreme case of intercalation in which interactions between the lamellæ become weakly attractive or even repulsive. Consequently, the lamellæ can be easily separated by solvent molecules to form discrete lamellæ.[26] These lamellar colloids have the same basal planes as the lamellar microcrystals from which they are obtained (e.g., from $100\,nm^2$ to several μm^2). The thickness of the lamella is always of nanometric order (typical lamellar thickness of zirconium phosphate–phosphonates is in the range 1–3 nm). The reader interested in exfoliation processes of layered compounds and in the exfoliation of γ-zirconium phosphate dihydrogenphosphate in water and in 1:1

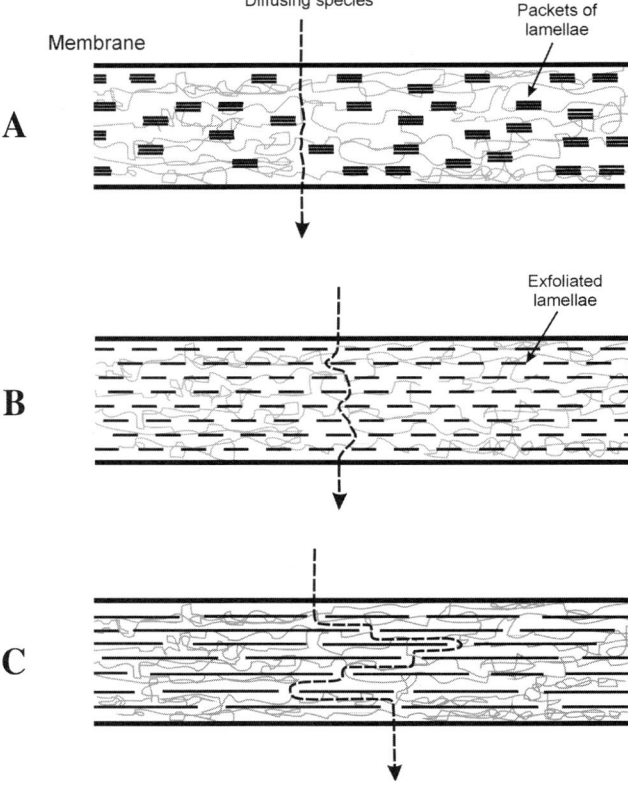

FIGURE 3. Schematic representation showing how the tortuosity encountered by diffusing species increases with the degree of exfoliation ($A \to B$) and dimension ($B \to C$) of the lamellar particles.

acetone/water could consult Reference 27 and References 28 and 29, respectively. We limit attention here to preparation of colloidal dispersions of α-zirconium *bis*(monohydrogenphosphate), and mixed α-zirconium phosphate–sulfophenylenphosphonate in water and in basic organic solvents, such as DMF and *N*-methyl pyrrolidone (NMP), commonly used for the solubilization of ionomers. In the following, examples of exfoliation are provided.

Exfoliation of α-Zirconium Phosphate in Water

α-Zirconium phosphate, unlike clays such as montmorillonite, does not exfoliate directly when dispersed in water. It is, however, sufficient for the intercalation of a short chain alkylamine, such as propylamine,[27] to provoke the exfoliation in water. The exfoliation process in water and the formation of a colloidal dispersion in DMF is schematically shown in FIGURE 4.

Preparation of Colloidal Dispersions of α-Zirconium Phosphate in DMF

A colloidal dispersion of zirconium phosphate in water, prepared as described above, is treated with 1 M HCl until the pH is less than 2 to provoke flocculation of the lamellar colloidal particles. The solid is separated from the solution and washed with dimethylformamide under vigorous stirring. A colloidal dispersion in this solvent is thereby obtained. A gelatinous precipitate, containing 4% α-zirconium phosphate, settles by centrifugation at 3,000 rpm. Washing is repeated two or three times in order to eliminate the greater part of propylamine and chloride ions.

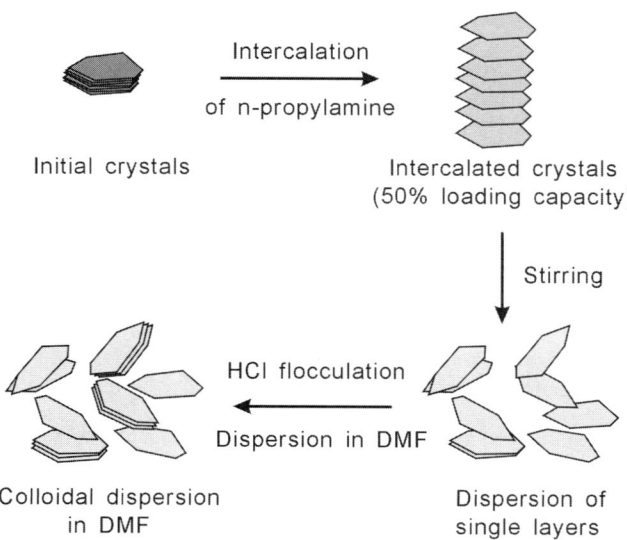

FIGURE 4. Schematic representation of the exfoliation of α-Zr(O$_3$P–OH)$_2$ microcrystals by intercalation of *n*-propylamine in aqueous solution. The flocculation of the lamellæ by acidification and their dispersion in *N,N'* dimethylformamide is also shown.

*Preparation of Colloidal Dispersions of α-Zirconium Phosphate–
Sulfophenylenphosphonate in a Basic Organic Solvent, such as Dimethylformamide*

To achieve very small lamellar particles, the compound $Zr(O_3P-OH)_{0.6}(O_3P-C_6H_4-SO_3H)_{1.4}$ was prepared in semicrystalline or amorphous form (particles smaller than 0.1 µm) as follows: 7.5 mL of 1 M H_3PO_4 and 15 mL of 1 M metasulfophenylenphosphonic acid are mixed and concentrated by heating overnight at 80°C. The dense liquid thus obtained is mixed with 50 mL of acetonitrile, and water is added until a clear solution is obtained; 13.6 mL of an aqueous solution of 0.75 M $ZrOCl_2$ is then added dropwise to the acetonitrile solution. The white precipitate formed is held under vigorous stirring for half an hour and washed twice with 2 M HCl (2 × 50 mL) and twice with acetonitrile (2 × 50 mL). A weighed amount of the slurry obtained after centrifugation at 3,000 rpm, is mixed with an equal amount of DMF and stirred overnight. The mixture is left to rest for one day to allow sedimentation of the solid. The supernatant colloidal dispersion contains 9% $Zr(O_3P-OH)_{0.6}(O_3P-C_6H_4-SO_3H)_{1.4}$, 50% DMF, and 41% acetonitrile.

Examples of Preparing Hybrid Ionomeric Membranes Using Exfoliated Lamellar Zirconium Phosphates and Phosphonates

The colloidal dispersions of lamellar nanoparticles described above can be used to prepare hybrid membranes. Attention here is limited to preparation examples of hybrid Nafion and hybrid sulfonated polyetherketones containing dispersed lamellar zirconium phosphates and/or phosphonates in their polymeric matrix.

Hybrid Membranes of Sulfonated Polyetherketone and α-$Zr(O_3P-OH)_2$

A weighted amount of sulfonated polyetherketone with ion-exchange capacity 1.3 meq/g (s-PEK 1.3, corresponding to 9 g of this anhydrous ionomer) are dissolved in 40 g of NMP under nitrogen atmosphere at 130°C. Then, 25 g of the gelatinous precipitate containing 4% of α-zirconium phosphate is dispersed into the polymer solution. The polymer/zirconium phosphate dispersion in DMF is processed to a membrane using a semiautomatic film casting processor (type, Erichsen; gap, 400 µm; proceeding velocity, 10 mm/sec). The solvent is eliminated by drying for one hour at 80°C and half-hour at 120°C. The composite membrane thus obtained (thickness 0.035 mm; 10%wt of inorganic particles) is stored in water. The process of preparation of the hybrid membrane is shown schematically in FIGURE 5.

*Hybrid Membranes of Sulfonated Polyetherketone and
Semicrystalline or Amorphous $Zr(O_3P-OH)_{0.6}(O_3P-C_6H_4-SO_3H)_{1.4}$*

A weighted amount of s-PEK 1.3 (corresponding to 1.2 g of anhydrous ionomer) is dissolved under vigorous stirring in 8 g of NMP at 130°C. Then, 3.4 g of the above colloidal dispersion of amorphous zirconium phosphate sulfophenylenphosphonate is mixed with 9.0 g of the s-PEK1.3 solution. This mixture is held under stirring for half an hour at 130°C and then cast on a glass plate. The solvent is eliminated by heating one hour at 80°C and half an hour at 120°C. The resulting membrane, containing about 20%wt inorganic material, is detached by immersing the glass plate in water. The membrane conductivity at 100°C is 1.2×10^{-3} S cm^{-1} and 1.3×10^{-2} S cm^{-1}

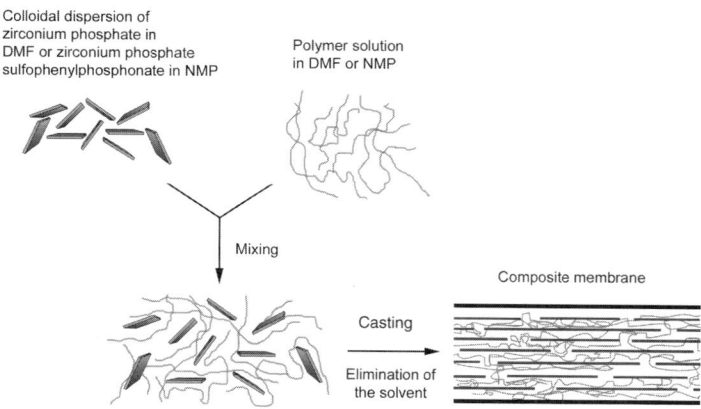

FIGURE 5. Schematic representation of the formation of hybrid ionomeric membranes containing oriented lamellæ of zirconium-phosphate–phosphonates.

at 80% and 95% relative humidity (r.h.), respectively. The proton conductivity at 100°C of a pure s-PEK 1.3 membrane obtained with an analogous procedure was found to be $1.2 \times 10^{-2} \, S \, cm^{-1}$ (95% r.h.). It can be concluded that the addition of 20%wt of inorganic lamellar particles with high proton conductivity does not decrease the proton conductivity of this ionomer. This fact could be significant if, as expected, it is also accompanied by an improvement in the mechanical properties and/or by a reduction in methanol permeability. Research is, therefore, in progress in our laboratory and will be reported elsewhere.[22]

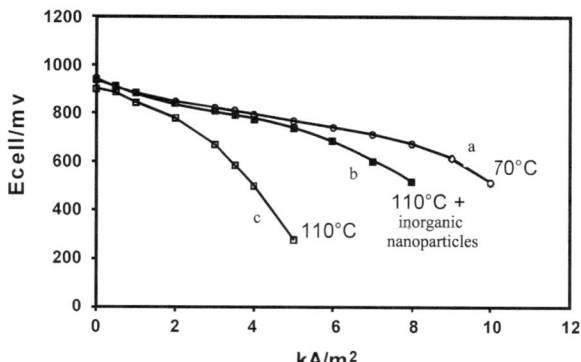

FIGURE 6. Typical polarization curves of a PEM FC using Nafion membranes at 70°C (*a*) and at 110°C (*b*); improvement obtained by inserting inorganic particles in the electrode–Nafion film interfaces (*c*).

Effect of the Insertion of Inorganic Nanoparticles in the Nafion Films Sprayed over the Electrodes on the Performance of PEM FC at Temperatures above 100°C

As previously emphasized, present fuel cells using Nafion as proton conducting membrane usually operate at 70–90°C. A typical polarization curve at 70°C for PEM FC using Nafion 117 as membrane is shown in FIGURE 6 (curve a).

Curve (c) in FIGURE 6 shows the polarization curve at 110°C. As already discussed, the drastic reduction in the delivered electrical energy is generally attributed to the fact that at temperatures greater than 70–90°C it is difficult to maintain full hydration of the Nafion membrane. Thus, determination of the proton conductivity of Nafion membranes at various relative humidity (r.h.) values in the temperature range 80–160°C is very important in order to better understand the above phenomenon. A special cell for the determination of proton conductivity of membranes was, therefore, assembled in our laboratory (see FIGURE 7). FIGURE 8 shows conductivity curves as function of temperature for r.h. values of 100, 91, 85, 75, 50, and 30%.[30] As expected, there is marked effect of relative humidity on the protonic conductivity of Nafion. It is of interest to note that when r.h. is maintained at 100%, a strong conductivity decrease was found at temperatures higher than 120–125°C (dotted curves in FIG. 8). Lowering the relative humidity caused this decrease to shift to higher temperatures. At r.h. values of 91, 85, and 75% the decrease was found for temperatures greater than 130, 145, and 155°C, respectively, whereas for r.h. not exceeding 50%, no appreciable decrease was observed, even at the maximum temperature investigated (160°C). Furthermore, it was found that the conductivity decrease was not reversible since the initial value was not recovered when the temperature was again decreased below 110°C. An examination of the membranes after the measurements showed that the decrease in conductivity was associated with a deformation of the membranes, which in some cases were partially squeezed out of the electrodes. These results suggest that the above phenomenon is related to excessive swelling of

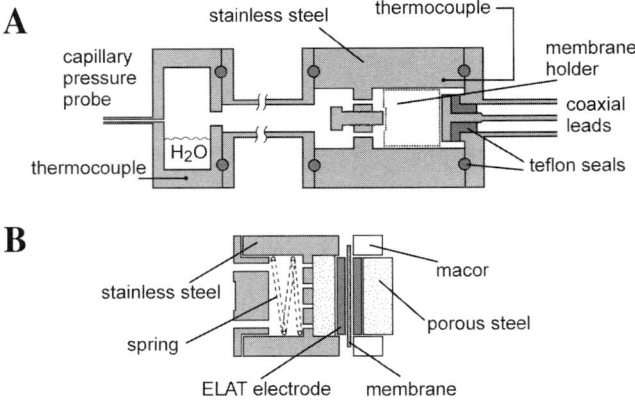

FIGURE 7. (**A**) Schematic representation of the conductivity cell used to control the relative humidity in the temperature range 70–160°C. (**B**) A detail of the membrane holder, from Reference 30.

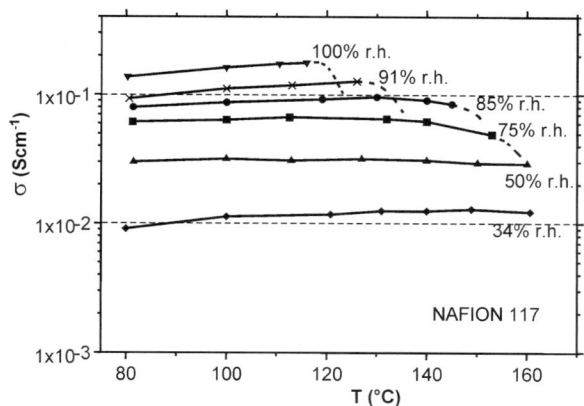

FIGURE 8. Conductivity of Nafion 117 membrane in the temperature range 80–160 °C at the indicated relative humidity values (from Ref. 30).

the ionomer. Thus, the resulting mechanical stress at the electrode/membrane interface is expected to considerably reduce the effective contact area with the electrodes and to alter the determination of the electrical conductivity. The large swelling of the Nafion at medium temperatures is in agreement with the results of Kreuer, who found irreversible swelling for temperatures greater than 130°C corresponding to the beginning of the glass transition of this polymer.[31] Since our measurements were carried out after long equilibration times, it is not surprising that the irreversible swelling at 100% r.h. was found to be 5–10°C below the value reported by Kreuer. Furthermore, it is reasonable to think that the irreversible swelling takes place at higher temperatures when the r.h. is decreased. Thus, the drastic conductivity decrease found when the membrane was heated in the conductivity apparatus was essentially attributed to failure of electrical contact caused by the large swelling mentioned above. Hence, very high r.h. values must be avoided at temperatures higher than 120–130°C.

In the light of the above results it must be concluded that, under equilibrium conditions and in the absence of dc current, Nafion retains high proton conductivity even at 160°C. Thus, why is there a dramatic decrease of the delivered energy in cells operating at temperatures greater than 70–90°C? The phenomenon is very likely related to kinetic problems associated with the passage of an intense dc current flow through the membrane under working conditions. It is known that the flow of current is accompanied by water transport from the anode to the cathode.[32] The resulting dehydration of the anodic side of the membrane can be completely or only partially compensated by an opposite flow of water produced at the cathodic compartment (backdiffusion) until the profile of the water content of the membrane reaches a steady state. In particular, the higher the electric current and the lower the membrane permeability to water, the larger the dehydration of the anodic side of the membrane under the steady state conditions, as shown schematically in FIGURE 9.

On the basis of these considerations, research in the preparation of hybrid membranes with improved characteristics of water backdiffusion are in progress in our laboratory. In any case, taking into account the kinetic nature of the PEM FC failure at temperature greater than 70–90°C, we had the intuition that the electrodic interfaces could play an important role in determining the steady state profile of the water content of the membrane. Therefore, in collaboration with De Nora and Nuvera Europe Fuel cell firms, experiments were carried out by simply replacing the interface Nafion films with hybrid Nafion films. The results showed that the presence of the inorganic or lamellar inorgano–organic nanoparticles in the electrodic interfaces indeed may considerably improve the PEM FC characteristic at temperature greater than 100°C.[33] FIGURE 6 (curves b and c) shows the polarization curves obtained at 110°C in the presence and absence of inorganic nanoparticles. The method is particularly simple since it allows the use of conventional electrodes and membranes, consequently no changes in the present PEM FCs are required. It is sufficient to add the proper nanoparticles to the solution of Nafion used for the spray over the electrodes. These results are of great importance since, as discussed, cooling of the stacks of hydrogen fed PEM FC is greatly simplified at temperatures greater than 100°C. Furthermore, CO poisoning of the platinum is expected to be reduced considerably due both to higher temperature and high hydration of the anode under working conditions.

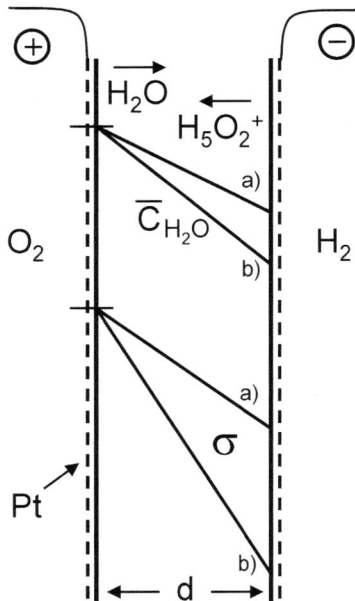

FIGURE 9. Schematic view of the water concentration profiles (C_{H_2O}) and proton conductivity profiles (σ) at increasing current density (curves *a* and *b*).

CONCLUSION

The preparation of hybrid membranes obtained by dispersing lamellar nanoparticles in the matrix of commercial proton conducting polymers seems to be a promising road for solving some of the present technologic limitations to a larger diffusion of PEM FC.

Taking into account that all problems cannot be overcome with a single type of hybrid membrane, research is in progress in our laboratory for preparing hybrid membranes specifically planned with the following characteristics: (1) To exhibit chemical stability and proton conductivity comparable to those of Nafion, but being more economic. At the moment, an intense research has been already carried out within an ENERGY European project in the field of hybrid membranes consisting of sulfonated polyetherketones/lamellar tetravalent metal phosphate–sulfophosphonates.[34–35] (2) To exhibit chemical stability and proton conductivity comparable with those of Nafion but having a considerably lower permeability to methanol (e.g., hybrid Nafion membranes). (3) To be able to operate at least up to 140–150°C, where the CO poisoning of platinum is no longer expected even with unpurified reforming gas.

Many problems remain to be solved, but the day on which many PEM FC electric cars will circulate in our towns is not so far in the future if the present intense effort of research on proton conducting membranes is continued.

ACKNOWLEDGMENTS

This work was partially supported by ENERGY European Project ERK6-CT1999-00025.

REFERENCES

1. Voss, H. & J. Huff. 1997. Portable fuel cell powder generator. J. Power Sources **65**: 155.
2. Heinzel, A., R. Nolte, K. Ledjeff-Hey & M. Zedda. 1998. Membrane fuel cell-concept and system design. Electrochim. Acta **43**: 3817.
3. Savadogo, O. 1998. Emerging membranes for electrochemical systems. Solid polymer membranes for fuel cell systems. J. New Mater. Electrochem. Syst. **1**: 47.
4. Kerres, J.A. 2001. Development of ionomer membranes for fuel cells. J. Membr. Sci. **185**: 3–27.
5. Babir, F. & T. Gomez. 1996. Efficiency and economics of proton exchange membrane fuel cells. Int. J. Hydrogen Energy **21**: 891.
6. Bailly, C., D.J. Williams, F.E. Karasz & W.J. MacKnight. 1987. The sodium salts of sulfonated poly(aryl-ether-ether-ketone) (PEEK): preparation and characterisation. Polymer **28**: 1009–1016.
7. Kreuer, K.D., Th. Dippel & J. Maier. 1995. Membrane materials for PEM fuel cells: a microstructural approach. Proc. Electrochem. Soc. **95**(23): 241.
8. Kreuer, K.D. 1997. On the development of proton conducting materials for technological applications. Solid State Ionics **97**: 1.
9. Linkous, C.A., H.R. Anderson, R.W. Kopitzke & G.L. Nelson. 1998 Development of new proton exchange membrane electrolytes for water electrolysis at higher temperatures. Int. J. Hydrogen Energy **23**: 525.

10. JONES, D.J. & J. ROZIÈRE. 2001. Recent advances in the functionalisation of polybenzimidazole and polyetherketone for fuel cell applications. J. Membr. Sci. **185:** 41–58.
11. NOSHAY, A. & L.M. ROBESON. 1976. Sulfonated polysulfone. J. Appl. Polym. Sci. **20:** 1885–1903.
12. COPLAN, M.J. & G. GÖTZ, Inventors. 1983. Heterogeneous sulfonation process for difficulty sulfonatable Poly(ether sulfone). US Patent 4,413,106. November 1.
13. NOLTE, R., K. LEDJEFF, M. BAUER & R. MULHAUPT. 1993. Partially sulfonated (polyarilene ether sulfone), a versatile proton conducting membrane for modern E-conversion technology. J. Membr. Sci. **83:** 211.
14. LUFRANO, F., I. GATTO, P. STAITI, et al. 2001. Sulfonated polysulfone ionomer for fuel cells. Solid State Ionics **145:** 45–71.
15. SURAMPUDI, S., S.R. NARAYANAN, E. VAMOS, et al. 1994. Advances in direct oxidation methanol fuel cells. J. Power Sources **47:** 377–385.
16. DAGANI, R. 1999. Putting the "nano" into composites. Chem. Eng. News **77:** 25–37.
17. ALEXANDRE, M. & P. DUBOIS. 2000. Polymer-layered silicate nanocomposites: preparation, properties and uses of a new class of materials. Mat. Sci. Eng. Rev. **28:** 1–63.
18. ALBERTI, G., U. COSTANTINO, S. ALLULLI & N. TOMASSINI. 1978. Crystalline Zr(R–PO$_3$)$_2$ and Zr(R–OPO$_3$)$_2$ compounds: a new class of materials having layered structure of the zirconium phosphate type. J. Inorg. Nucl. Chem. **40:** 1113–1117.
19. ALBERTI, G. 1996. Layered metal phosphonates and covalently pillared diphosphonates. In Two- and Three-dimensional Inorganic Networks. G. Alberti & T. Bein, Eds.: Vol. 7. Chapt. 5 of Comprehensive Supramolecular Chemistry, L. Jean-Marie, Ed. Pergamon, Elsevier Science, Oxford.
20. CLEARFIELD, A. 1996. Recent advances in metal phosphonate chemistry. Curr. Opin. Solid State Mater. Sci. **1**(2): 268–278.
21. ALBERTI, G. & M. CASCIOLA. 1997. Layered metal IV phosphonates, a large class of inorgano-organic proton conductors. Solid State Ionics **97:** 177–186.
22. ALBERTI, G., M. CASCIOLA & M. PICA. 2003. To be published.
23. ALBERTI, G., M. CASCIOLA & R. PALOMBARI. 2000. Inorgano-organic proton conducting membranes for fuel cells and sensors at medium temperatures. J. Membr. Sci. **172:** 233–239.
24. ALBERTI, G. & M. CASCIOLA. 2001. Solid state protonic conductors, present main applications and future prospects. Solid State Ionics **145:** 3–16.
25. ALBERTI, G., U. COSTANTINO, M. CASCIOLA, et al. 2001. Preparation and proton conductivity of titanium phosphate sulfophenylenphosphonate. Solid State Ionics **145:** 249–255.
26. JACOBSON, A.J. 1996. Colloidal dispersion of compounds with layer and chain structures. In Two- and Three-dimensional Inorganic Networks. G. Alberti & T. Bein, Eds.: Vol. 7. Chapt. 10 of Comprehensive Supramolecular Chemistry, L. Jean-Marie, Ed. Pergamon, Elsevier Science, Oxford.
27. ALBERTI, G., M. CASCIOLA & U. COSTANTINO. 1985. Inorganic ion-exchange pellicles obtained by delamination of α-zirconium phosphate crystals. J. Colloid Interf. Sci. **107:** 256–263.
28. ALBERTI, G., S. CAVALAGLIO, C. DIONIGI & F. MARMOTTINI. 2000. Formation of aqueous colloidal dispersions of exfoliated γ-zirconium phosphate by intercalation of short alkylamines. Langmuir **16:** 7663–7668.
29. ALBERTI, G., C. DIONIGI, E. GIONTELLA, et al. 1997. Formation of colloidal dispersions of layered γ-zirconium phosphate in water/acetone mixtures. J. Colloid Interf. Sci. **188:** 27–31.
30. ALBERTI, G., M. CASCIOLA, L. MASSINELLI & B. BAUER. 2001. Polymeric proton conducting membranes for medium temperature fuel cells (110–160°C). J. Membr. Sci. **185:** 73–81.
31. KREUER, K.D. 2001. On the development of proton conducting polymer membranes for hydrogen and methanol fuel cells. J. Membr. Sci. **185:** 29–39.
32. EIKERLING, M., YU.I. KHARKATS, A.A. KORNYSHEV & YU.M. VOLFKOVICH. 1998. Phenomenological theory of electro-osmotic effect and water management in polymer electrolyte proton-conducting membranes. J. Electrochem. Soc. **145:** 2684–2698.

33. ALBERTI G., M. CASCIOLA, E. RAMUNNI & J.R. ORNELAS, Inventors. 2000. Assieme membrana-elettrodo per cella a combustibile a membrana polimerica. Italian patent 20134. October.
34. BONNET, B., D.J. JONES, J. ROZIERE, et al. 2000. Hybrid organic-inorganic membranes for a medium temperature fuel cell. J. New Mat. Electrochem. Syst. **3:** 87–92.
35. BAUER, B., D.J. JONES, J. ROZIERE, et al. 2000. Electrochemical characterization of sulfonated polyetherketone membranes. J. New Mat. Electrochem. Syst. **3:** 93–98.

High Performance Perfluoropolymer Films and Membranes

VINCENZO ARCELLA, ALESSANDRO GHIELMI, AND GIULIO TOMMASI

Solvay Solexis, Bollate (MI), Italy

ABSTRACT: Membrane processes are receiving increasing attention in the scientific community and in industry because in many cases they offer a favorable alternative to processes that are not easy to achieve by conventional routes. In this context, membranes made with perfluorinated polymers are of particular interest because of the unique features demonstrated by these materials. Both highly hydrophobic and hydrophilic membranes have been developed from appropriate perfluoropolymers that were, in turn, obtained by copolymerizing TFE with special monomers available on an industrial scale. Highly hydrophobic membranes obtained from the glassy copolymers of TFE and 2,2,4-trifluoro-5 trifluoromethoxy-1,3 dioxole (Hyflon® AD) exhibit properties that make them particularly well suited for use in optical applications, in the field of gas separation, and in gas–liquid contactors. Conditions for preparing membranes that are adequate for use in various applications are exemplified. Hydrophylic highly conductive proton exchange membranes obtained from the copolymer of TFE and a short-side-chain (SSC) perfluorosulfonylfluoridevinylether (Hyflon Ion) find interesting application in the field of fuel cells, especially in view of the current tendency to move to high temperature operation. The advantages offered by these hydrophobic and hydrophylic perfluorinated materials for use in membrane technology are discussed. Comparison of membrane properties and performance is made with other membranes available on the market.

KEYWORDS: perfluorinated membranes; Hyflon AD; gas separation; Hyflon Ion; membrane contactors; hydrophobic membranes; ion exchange membranes; fuel cells

INTRODUCTION

Fluoropolymer materials are capturing greater and greater interest in industrial applications because of the remarkable combination of properties that they exhibit when compared to other polymeric materials. The most well-known property for which fluoropolymers are employed in high-demanding applications is their outstanding thermal and chemical resistance. However, the peculiar nature of the carbon–fluorine bond confers on these materials other unique physical properties (e.g., electrical, optical, and superficial) that can be valuably exploited in the most variegated fields.

Perfluoropolymers represent the ultimate in resistance to hostile chemical environments and high service temperature because of the high bond energy of C–F and

Address for correspondence: Vincenzo Arcella, Solvay Solexis, viale Lombardia 20, 20021 Bollate (MI), Italy.

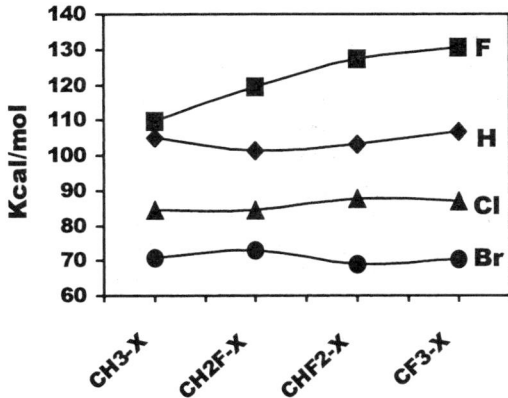

FIGURE 1. C–X bond energy in the model molecules CH_3–X, CH_2F–X, CHF_2–X, and CF_3–X, with X = H, F, Cl, and Br.

C–C bonds of fluorocarbons, equal to 485 kJ/mol and 360 kJ/mol, respectively. C–X bond energy data reported for the substituted methane model molecules CH_3–X, CH_2F–X, CHF_2–X, and CF_3–X, where X = H, F, Cl, or Br, show that the maximum C–X bond energy is attained for the CF_3X molecule when X = F[1] (see FIGURE 1). In many applications fluoropolymers offer the highest protection afforded by any polymer available today, to a huge variety of chemicals, such as acids and alkalis, fuels and oils, low molecular weight esters, ethers and ketones, aliphatic and aromatic amines, and strong oxidizing substances.

Monomers used for the synthesis of fluorinated polymers can be briefly classified into two categories, base monomers and special monomers, the former represented by those monomers that constitute the basic structure of modern fluoropolymers and the latter by those other monomers that add specially desired characteristics to match specialty application requirements.

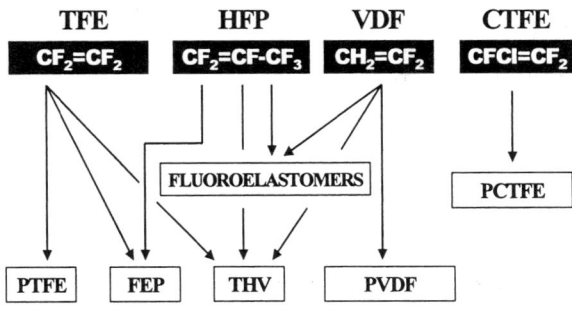

FIGURE 2. Constitution of base fluorinated homo- and copolymers.

Within this scheme, base fluoromonomers are tetrafluoroethylene (TFE), hexafluoropropylene (HFP), vinylidenefluoride (VDF), and chlorotrifluoroethylene (CTFE). Proper combination of these monomers yields homo- or copolymers with the most diverse characteristics: PTFE, FEP, fluoroelastomers, PVDF, PCTFE, and THV (see FIGURE 2). Hydrogenated monomers (e.g., ethylene) can be included in the structure (e.g., ECTFE, VDF/HFP/ethylene rubbers). In the field of films and membranes the base polymers mentioned above find wide application: just to mention some examples, films for anticorrosion (PTFE and FEP), films for protective packagings (PCTFE), microfiltration and ultrafiltration membranes (PVDF and PTFE), vapor permeable clothes and shoes (PTFE), and separators for lithium ion batteries (PVDF).

Special monomers produced on an industrial scale by Solexis are reported in FIGURE 3. These are perfluorovinylethers, such as perfluoromethylvinylether (MVE), perfluoropropylvinylether (PVE) and perfluorosulfonylfluoridevinylether (SFVE); perfluorinated cyclic monomers, such as 2,2,4-trifluoro-5 trifluoromethoxy-1,3 dioxole (TTD); or fluorinated diolefines, such as dodecafluoro-1,9 decadiene. MVE and PVE are used as perfluorinated modifiers both in semi-crystalline melt processable perfluoropolymers (e.g., MFA and PFA) and ultra-high resistance rubbery polymers (perfluoroelastomers). The main use of diolefine is in peroxide-curable fluoroelastomers and perfluoroelastomers to control the microstructure of the polymer chain and realize special polymer architectures by the so called "branching and pseudoliving" technology.[2] The two special monomers, TTD and SFVE reported in FIGURE 3 are those that find the most advanced application in the field of films and membranes. TFE–TTD copolymers are used to produce amorphous pellicles for photomask protection in far-UV microlithography and highly hydrophobic membranes, whereas TFE–SFVE copolymers have been employed for the preparation of perfluorinated high ionic conductance hydrophylic membranes. These two membrane typologies,

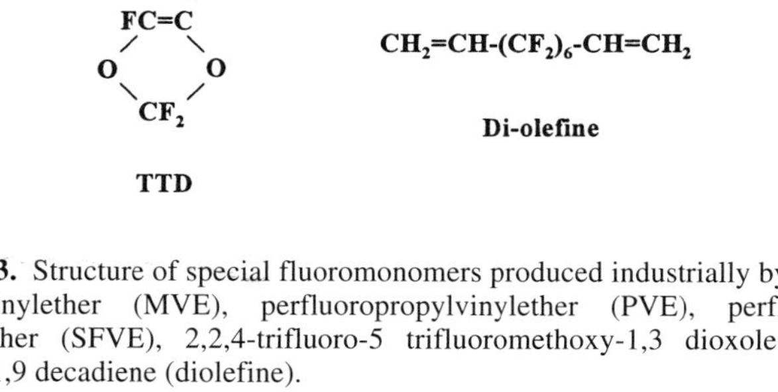

$CF_2=CF$ $CF_2=CF$ $CF_2=CF$
| | |
OCF_3 $OCF_2CF_2CF_3$ $OCF_2CF_2SO_2F$

MVE   PVE   SFVE

$$\begin{array}{c} OCF_3 \\ / \\ FC=C \\ / \quad \backslash \\ O \quad \quad O \\ \backslash \quad / \\ CF_2 \end{array}$$

$CH_2=CH-(CF_2)_6-CH=CH_2$

TTD   Di-olefine

FIGURE 3. Structure of special fluoromonomers produced industrially by Solexis: perfluoromethylvinylether (MVE), perfluoropropylvinylether (PVE), perfluorosulfonylfluoridevinylether (SFVE), 2,2,4-trifluoro-5 trifluoromethoxy-1,3 dioxole (TTD), and dodecafluoro-1,9 decadiene (diolefine).

that is, hydrophobic and hydrophylic, based on TFE–TTD and TFE–SFVE copolymers, respectively, are discussed in two separate sections.

HYDROPHOBIC AMORPHOUS PERFLUOROPOLYMER MEMBRANES

TFE–TTD copolymers, known commercially as Hyflon® AD, are amorphous perfluoropolymers with glass transition temperatures (T_g) greater than room temperature.[3] They show a thermal decomposition temperature exceeding 400°C and are highly transparent to light from far UV to near infrared.

TFE–TTD copolymers are synthesized by free radical polymerization. The structure of the resulting Hyflon AD polymer is depicted in FIGURE 4. Amorphous polymers are obtained when m/n is less than about 4; that is, when the TTD molar content is higher than about 20%. Due to the cyclic structure and the effective steric hindrance of the side group, chain motion is severely hindered and high T_g glassy polymers result. When $m = 0$, that is for the case of TTD homopolymer, $T_g = 170°C$. At increasing m/n values (decreasing TTD content), T_g decreases. In TABLE 1 we show T_g values for various TTD contents. In this table, T_g values for two other amorphous glassy polymers, commercially available from Du Pont (Teflon® AF) and Asahi Glass (Cytop®), are included for comparison.[4,5] The structure of these two polymers is also shown in FIGURE 4. Teflon AF is obtained by the copolymerization of perfluoro-2,2 dimethyldioxole (PDD) with TFE. Cytop is obtained by the cyclopolymerization of perfluoro-butenylvinylether (BVE), which offers a controlled alternate structure but fixes the content of the cyclic portion of the chain to 50%, thus limiting the T_g of the polymer to 108°C.

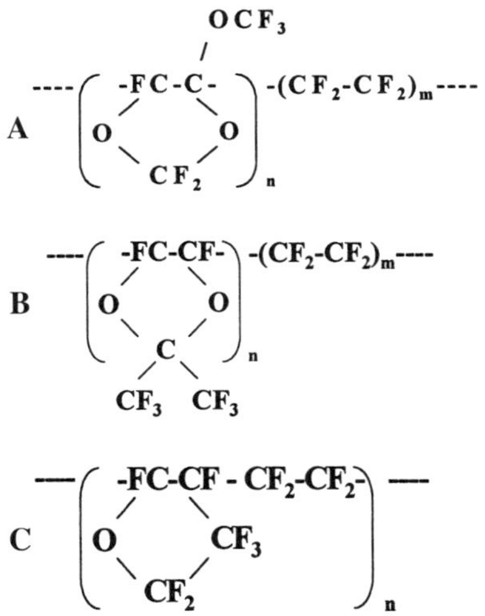

FIGURE 4. Structure of three commercial perfluorinated amorphous glassy polymers: **(A)** Hyflon AD, **(B)** Teflon AF, and **(C)** Cytop.

TABLE 1. Glass transition temperature (T_g) for amorphous perfluoropolymers at various compositions

Polymer	TFE	TTD	PDD	BVE	T_g
	(% mol)				(°C)
TTD homopolymer	—	100	—	—	170
Hyflon AD 80	15	85	—	—	135
Hyflon AD 60	40	60	—	—	110
PDD homopolymer	—	—	100	—	335
Teflon AF 2400	13	—	87	—	240
Teflon AF 1600	35	—	65	—	160
Cytop	—	—	—	100	108

TFE–TTD copolymers exhibit very high thermal stabilities, as shown by the weight loss at increasing temperatures obtained by TGA. One-percent weight losses are measured at temperatures as high as 400°C, demonstrating thermal stabilities typical of perfluoropolymers (see FIGURE 5).

Two characteristics of TFE–TTD copolymers that make them very attractive for membrane preparation are their high solubilities in perfluorinated solvents and their low solution viscosities.[6] Both features allow great flexibility in the selection of proper conditions for the preparation of membranes with various structures. Moreover, low solution viscosities imply greater ease of purification. This aspect is very important, since it is often crucial in the polymer and polymeric solutions to avoid the presence of both suspended and dissolved contaminants, such as dust or dissolved organic molecules. In fact contaminants can substantially affect properties, such as light transmittance and signal attenuation of optical devices. This is of

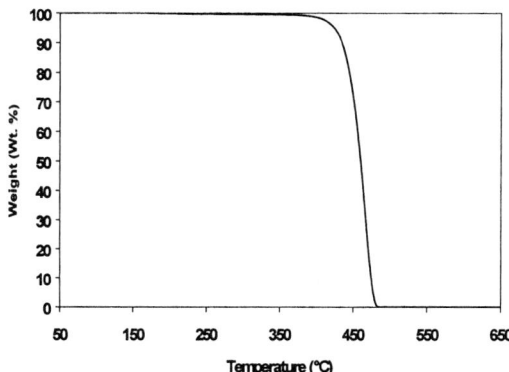

FIGURE 5. Dynamic TGA (in air at 10°C/min) of a Hyflon AD 80 membrane.

capital importance in the preparation of composite membranes, where the permselective layer is extremely thin and no defects are allowed.

Various procedures have been tuned to prepare membrane typologies to match application requirements. Examples of preparation methods and conditions for obtaining various membrane structures are reported in TABLE 2. In these preparation examples, Galden® HT 110 and Galden HT 55 are perfluopolyether oils with the following structure:

$$-(CF_2-CF_2O)_n-(CF_2O)_m-.$$

Normal boiling points are equal to 110°C and 55°C, respectively.

Flat sheet porous and non porous symmetric membranes can be obtained by following the evaporation method.[7] Laboratory membranes can be prepared by using a Braive Instruments knife to cast a 10% wt/wt polymeric solution on glass plates with various initial thicknesses and evaporation temperatures. Flat sheet laboratory asymmetric membranes can be prepared by following the dry–wet phase inversion method, with 10% wt/wt polymeric solutions.[7,8] The solvent can be Galden HT, the nonsolvent n-pentane, also using in this case a Braive Instruments knife to cast the polymeric film on glass plates at various initial thicknesses and coagulation bath temperatures. Flat sheet laboratory porous and non porous composite membranes of 1 μm thick films on PVDF inert supports were prepared by the spin coating process.[7,8]

Protective Pellicles for Microlithography

Symmetric dense films of Hyflon AD find application as UV resistant protective films for the semiconductor industry because of their high transparency at the far-UV wavelength, where conventional nitrocellulose pellicles are degraded due to high energy absorption. In this application, transparent pellicles are used in microlithography to cover the photomask and avoid dust depositing on it and disturbing the preparation of the photoresist. The protective film puts the dust particles out of focus (see FIGURE 6). The increase of information density requires the use of microlithography

TABLE 2. Examples of preparation procedures for various flat sheet membrane types obtained from Hyflon AD 60

Membrane type	symmetric	asymmetric	composite
Method	solution casting	dry–wet phase inversion	spin coating
Solvent	Galden HT 110	Galden HT 55	C_7F_{16}
Solution concentration	10% w/w	10% w/w	1% w/w
Temperature for solvent elimination	50°C	25°C	25°C
Non-solvent	—	n-C_5H_{12}	—
Temperature of coagulation bath	—	12°C	—
Support	glass	glass	PVDF

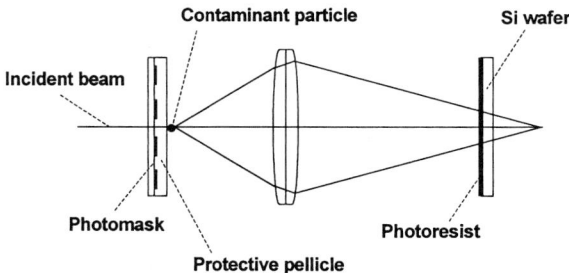

FIGURE 6. Use of the protective transparent pellicle in the microlithography process.

at ever lower wavelengths. Therefore, pellicles with high transparency at wavelengths as low as 193 nm, with a future target of 157 nm, are needed. FIGURE 7 shows the transmittance of Hyflon AD in the UV region.

Gas Separation Membranes

Hyflon AD membranes show values of permeability and selectivity to gases that make them interesting for gas separation applications. Asymmetric Hyflon AD 60 membranes give, under steady-state conditions, selectivity values for O_2/N_2 and CO_2/N_2 that are about 3 and 8, respectively, in the pressure range 5–7.5 bar. TABLE 3 gives mean permeance values to N_2, CO_2, and O_2 for these asymmetric membranes.

Permeability and selectivity values found on composite Hyflon AD membranes are in good agreement with values found on asymmetric membranes. Permeance values to N_2, CO_2, O_2, H_2, and CH_4 of composite membranes are reported in TABLE 4. Since the permselective Hyflon AD layer in this case is 1 μm thick, the permeances expressed in GPU coincide numerically with permeabilities expressed in Barrer. In TABLE 4, permeability values for Teflon AF are reported for comparison,[4] together

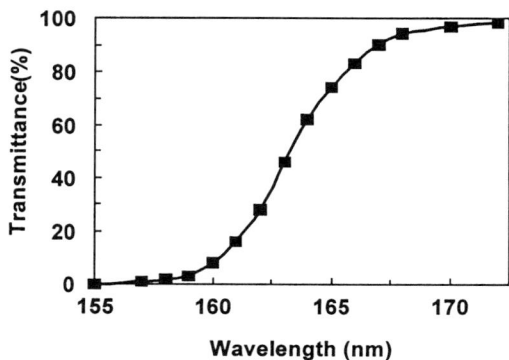

FIGURE 7. Light transmittance in the UV wavelength region for Hyflon AD.

TABLE 3. Gas permeances of Hyflon AD 60 asymmetric membranes as a function of differential pressure

Δp (bar)	Permeance (GPU)		
	N_2	CO_2	O_2
1	—	15.3	6
5	1.7	14	5.3
7.5	1.7	14.7	5.3
10	1.8	15.3	—

$1\,\text{GPU} = 10^{-6}\,\text{cm}^3(\text{STP})/\text{cm}^2\,\text{sec}\,\text{cmHg}$.

with the T_g values of the various polymers. Examination of these data shows that, independently of the fluoropolymer type and gas, a linear correlation exists between permeability and T_g (see FIGURE 8, where data for N_2, O_2, and H_2 are plotted). In this respect, gas permeation is often attributed to the presence of voids at the molecular scale and to their size distribution. Amorphous perfluoropolymers show an experimental density that is lower than that expected theoretically. Considering the chemical structure of amorphous perfluoropolymers, the low values of the experimental density can be ascribed, at least qualitatively, to differences in chain packing resulting from the low chain mobility and steric hindrance offered by the large dioxole groups. In other words, the high chain stiffness leads to difficulties in the close packing of chain segments and thus to "nanovoids". Since, in amorphous perfluoropolymers, T_g is mainly related to the macromolecular chain stiffness, which controls the dimension of nanosize holes, it follows that the higher the T_g value, the higher is gas permeation.

Hydrophobic Porous Membranes

Symmetric porous membranes with pore size between 30 and 80 nm have shown no permeation to water at pressures as high as 10 bar. On the other hand, high permeation to gases, such as O_2, N_2, and CO_2 were observed. Composite membranes prepared by casting the polymeric solution (1% weight) on PVDF supports have shown a unique combination of features:

TABLE 4. Gas permeability data (expressed in Barrer) of amorphous perfluoropolymers with various glass transition temperatures

Polymer	T_g (°C)	$P(N_2)$	$P(CO_2)$	$P(O_2)$	$P(H_2)$	$P(CH_4)$
Hyflon AD 80	135	77	473	194	563	49
Hyflon AD 60	110	17	124	51	202	8
Hyflon AD 40	91	8	64	26	—	—
Teflon AF 2400	240	490	2,800	990	2,200	340
Teflon AF 1600	160	—	—	340	—	—

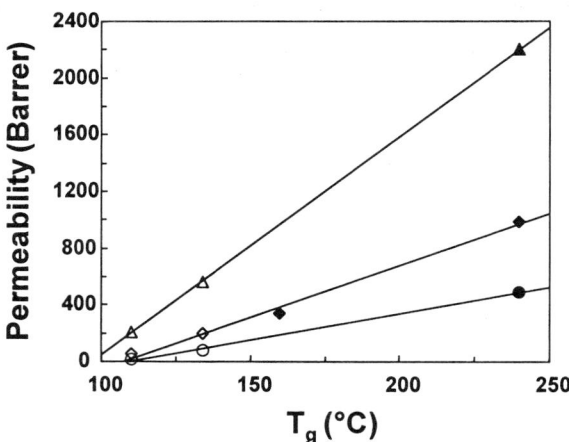

FIGURE 8. Correlation between ◇ oxygen, ○ nitrogen, and △ hydrogen gas permeability and T_g for amorphous perfluoropolymers.

- excellent organophobicity and hydrophobicity;
- gas fluxes many times those of commercial gas separation membranes; and
- chemical resistance and wear resistance.

Measures of contact angle to distilled water, compared with experimental results obtained on self-supported membranes and other commercial membranes, are reported in TABLE 5. Contact angles of 120° and above for the amorphous perfluoropolymer membranes demonstrate the extremely hydrophobic character of these membranes.

To test the organophobic character of the membranes, contact angle to hexadecane was measured. In this case the contact angle is about 65°, which is a high value

TABLE 5. Contact angle to water for various membranes

Membrane Polymer	Contact Angle to H_2O
PMMA	76
PEEK-WC-NO2	80
PES	82
PVDF	90
PE	96
PTFE	118
Hyflon AD 60	120
Hyflon AD 60 on PVDF	122

TABLE 6. Contact angle to hexadecane for various membranes

Membrane Polymer	Contact Angle to $C_{16}H_{34}$
PS	1
PTFE	39
CMS-7 on PP[a]	57
Hyflon AD 60	60
CMS-7 on PS[a]	63
Hyflon AD 60 on PVDF	65

[a]Compact Membrane Systems, Inc.

compared to other values reported in the literature (see TABLE 6). This result confirms the strong organophobicity of the membranes, which corresponds to excellent fouling resistance and inertness.

Oxygen permeation data through Hyflon AD composite membranes, in comparison to commercial gas separation membranes, are reported in TABLE 7.

In consideration of the above properties a particularly promising application can be found in the field of membrane contactors, that is, equipment in which membranes are used to improve mass transfer coefficients compared to traditional extraction and absorption processes.

HYDROPHYLIC PERFLUOROPOLYMER MEMBRANES

TFE and perfluorosufonylfluoridevinylether (SFVE) are copolymerized by free radical polymerization to obtain the polymer depicted in FIGURE 9A (Hyflon Ion). Amorphous polymers are obtained when *m/n* is less than about 4; that is, when the SFVE molar content is higher than about 20%. The glass transition temperature of the polymer is a function of polymer composition, but increases typically from about

TABLE 7. Oxygen permeances through various membrane types

Membrane Polymer	Permeance to O_2 (GPU)
Generon polycarbonate[a]	7
Permea polysulfone[a]	15
Dow 4-methylpentene-1[a]	90
Ethyl cellulose[a]	100
Gore-TEX	350 (5 bar, 16°C)
Silicone Rubber[a]	500
Hyflon AD 60 on PVDF	700 (5 bar, 27°C)

[a]Compact Membrane Systems, Inc.

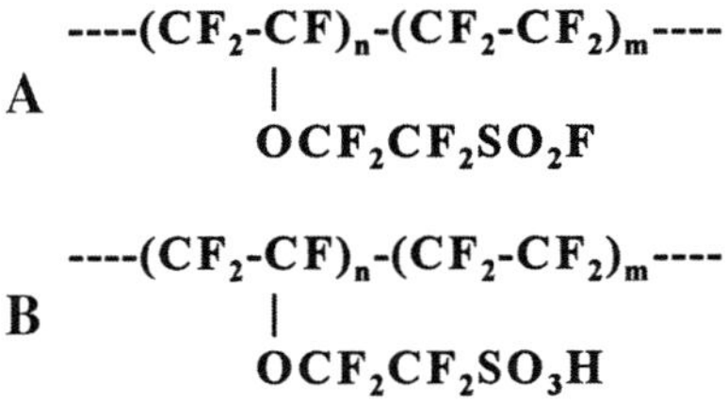

A
----(CF2-CF)n-(CF2-CF2)m----
OCF2CF2SO2F
B
----(CF2-CF)n-(CF2-CF2)m----
OCF2CF2SO3H

FIGURE 9. Structure of (**A**) Hyflon Ion and (**B**) Hyflon Ion H.

5°C to 50°C for SFVE content decreasing from 30% to 10% (molar). SFVE contents yielding amorphous polymers correspond to T_g values below ambient temperature, therefore amorphous Hyflon Ion polymers are rubbers at room temperature. These can be dissolved in a variety of perfluorinated or partially fluorinated solvents. On the other hand, when a crystalline phase appears, TFE-SFVE copolymers become scarcely soluble in any solvent. In this case, films can be prepared taking advantage of the melt-processability of the polymer.

Thermal stability of the Hyflon Ion polymers is very high, because of their perfluorinated nature. TGA shows 1% weight losses at temperatures as high as 420°C (see FIGURE 10).

After synthesis of the polymer in the sulfonyl fluoride form shown in FIGURE 9 A, the polymer is transformed into an ionomer (i.e., an ion-containing polymer) by conversion of the $-SO_2F$ group to $-SO_3X$, where X is a metal or hydrogen atom. This conversion is typically carried out in alkaline aqueous solutions at medium temperature (e.g., 80°C). The polymer is finally treated with a strong acid solution if the $-SO_3H$ form of the functional group is required in the application. Therefore, the

Weight (Wt. %)
100
90
80
70
60
50
40
30
20
10
0
50
150
250
350
450
550
650
Temperature (°C)

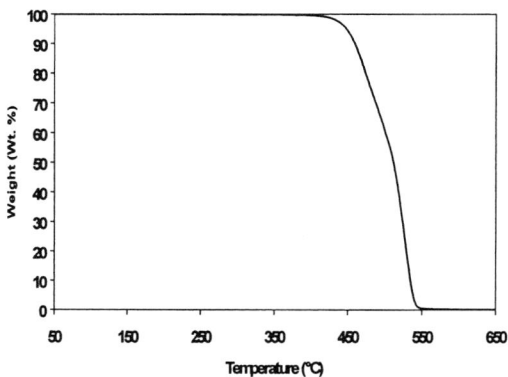

FIGURE 10. Dynamic TGA (in air at 10°C/min) of Hyflon Ion ($-SO_2F$ precursor) with EW = 1,000 g/mol.

sulfonyl fluoride form of the polymer can be considered the precursor form of the ionomer in the salt or acid form (FIG. 9B, Hyflon Ion H). Transformation of the precursor into an ionomer dramatically changes its properties.[9] This is due to strong coulombic associations that lead to the formation of ionic regions, commonly referred to as clusters. TFE rich ionomers can, therefore, be thought of as consisting of three different phases: an amorphous phase, a crystalline phase, and an ionic phase.

Perfluorinated sulfonate ionomers find effective or potential application in the form of membranes in a wide variety of fields, ranging from electrochemical electrolyzers (chloralkali[10] and HCl electrolysis[11]), to fuel cells (proton exchange membrane fuel cells[10] and direct methanol fuel cells[12]), energy storage and delivery devices (lithium ion batteries[12] and Br_2/S batteries[13]), microfiltration,[14] reverse osmosis and ultrafiltration,[12] pervaporation,[12] electrodialysis,[15] diffusion dialysis,[15] and membrane catalytic reactors.[16]

Short-Side-Chain (SSC) Ionomers and Related Membranes

Conventional ion-exchange perfluoropolymer membranes such as Nafion® (Du Pont), Aciplex® (Asahi Chemical), Flemion® (Asahi Glass), and Gore-Select™ (Gore and Associates) are based on the so-called long-side-chain (LSC) polymers.[9,17,18] Compared to Hyflon Ion (SSC) these polymers have a longer pendant group carrying the ionic functionality. The most extensively used and studied LSC ionomer is Du Pont's Nafion, which was developed in the late 1960s as polymer electrolyte for a GE fuel cell designed for NASA spacecraft missions. Since then, Nafion and polymers of the same family have found wide application in the chloralkali industry, due to their very high chemical inertness. The very high chemical stability of Nafion has been demonstrated in fuel cell applications by operating lifetimes in excess of 57,000 hours.

In the mid 1980s, Ballard Power Systems showed significant improvements in fuel cell performance using SSC ionomer membranes obtained from Dow Chemical.[19] The chemical structure of the Dow membrane was the same as Hyflon Ion (FIG. 9); that is, the Dow polymer was obtained by copolymerizing TFE with SFVE, shown in FIGURE 3. Six-cell stacks giving four times the power obtained with a standard Nafion membrane were demonstrated.

As well as improved power output, another important aspect of SSC ionomers compared to LSC ionomers is their different behavior with temperature. SSC ionomers in the protonic form ($-SO_3H$) present a primary transition, defined as the α-transition, at about 160°C, whereas LSC ionomers show this transition at about 110°C.[20] This difference is very important when the use of the membrane in a fuel-cell system is considered. First, the fact that SSC ionomers present the α-transition at 160°C implies that one necessary condition for the membrane to operate up to such a high temperature is ensured. An increase of the fuel-cell temperature is highly desirable since this means a reduction of the complexity of the system, both in terms of cooling and fuel pre-processing (CO content reduction), and consequent lower cost. In direct methanol fuel cells, higher temperatures are also required to increase the fuel oxydation kinetics. Second, if the fuel cell system for any reason goes out of control and the temperature of the stack rises locally or on the whole, the SSC ionomer membrane has a good capability for recovery. In fact, if the temperature exceeds the α-transition temperature, a catastrophic change in the membrane structure can result.[21] The higher the transition temperature, the lower is the probability for this to happen. Last, but not least, SSC ionomers, as a consequence of their lower molecular weight pendant group, show a crystallinity content that is higher than corresponding LSC ionomers of the same equivalent weight (EW).[9,22] Therefore, lower EW (i.e., higher ionic content) membranes can be prepared with same crystallinity (i.e., similar mechanical properties) or same EW membranes with higher crystallinity (i.e., higher mechanical properties). This last possibility is quite important because higher mechanical properties make it possible to achieve membranes with lower thickness, which in turn means higher membrane conductance and peak power. These are highly desirable especially in automotive applications.

A New Technology for SSC Ionomers and Membranes

A number of patents and papers on the Dow polymer and membrane appeared during the 1980s and the immediately following years.[9,19,20,22–29] Since then, interest in SSC membranes seemed to reduce. No industrial development and commercialization of these very promising experimental membranes followed.

The process developed by Dow[30] for the synthesis of the SSC monomer is illustrated in FIGURE 11. As it can be seen, the scheme involves a large number of steps. This complexity was probably one of the main obstacles to the industrial development of these ionomers and membranes, although very interesting fuel cell membrane properties were envisaged.

Recently, Solexis has applied its proprietary fluorovinylether process to the production of the SSC monomer on an industrial scale. This process, which is extremely simple compared to the old Dow process, is outlined in FIGURE 12. Within this process, the fluorination and fluoro-olefine addition steps are, in practice, a single reaction step.

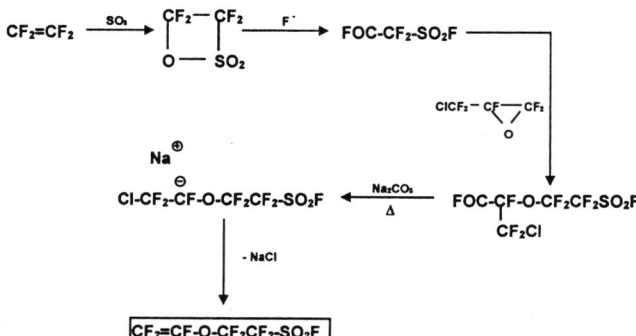

FIGURE 11. Dow route for the synthesis of the SSC sulfonylfluoridevinylether monomer.

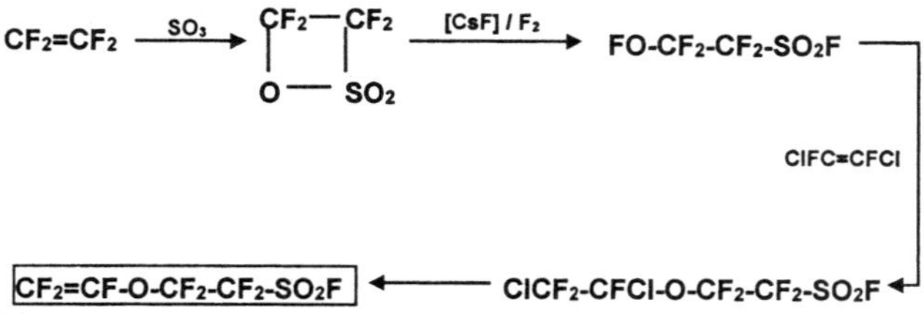

CF2=CF2
SO3
CF2—CF2
O — SO2
[CsF] / F2
FO-CF2-CF2-SO2F
ClFC=CFCl
CF2=CF-O-CF2-CF2-SO2F
ClCF2-CFCl-O-CF2-CF2-SO2F

FIGURE 12. Solexis route for the synthesis of the SSC sulfonylfluoridevinylether monomer.

Hyflon Ion-FM 900 Membranes

Based on its capability for producing this monomer on an industrial scale, Solexis has started a research and development project to create new ionomer membranes for fuel cells and other applications. Starting from the monomers, ionomers are synthesized by taking advantage of a proprietary microemulsion polymerization process.[31] This technology, broadly applied to the polymerization of other fluoropolymers, is able to give very high polymerization kinetics and high molecular weight polymers with accurate control of the molecular structure.

Many Hyflon Ion ionomers were polymerized on a pilot scale in a broad range of molecular weights and EWs, from amorphous soluble ionomers to highly crystalline ionomers. Many different membranes have been prepared with these ionomers.

Self-supported crosslinked membranes have been prepared with EW from 500 to 700 g/mol and thickness from 100 to 300 μm.[32] These membranes have shown very

1
0.9
0.8
0.7
0.6
0.5
0.4
0.3
0.2
0.1
0
Voltage (V)
0
0.2
0.4
0.6
0.8
1
1.2
1.4
1.6
Current density (A/cm2)

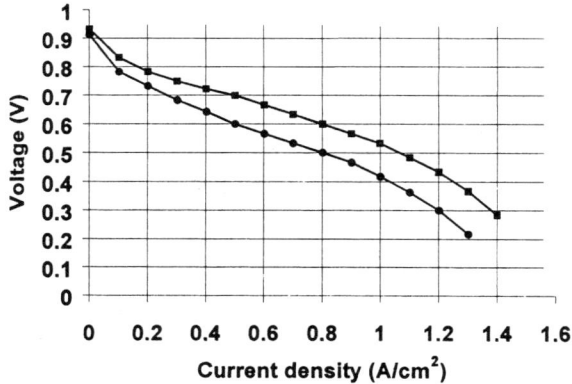

FIGURE 13. Comparison between polarization curves obtained with (■) a self-supported (melt extruded) and (●) supported (on expanded PTFE) membrane. Both membranes have the same thickness (20 μm) and were prepared with the same ionomer (Hyflon Ion, EW = 900).

high average conductivities, up to 10^{-1} S/cm under fuel cell operation, which implies high conductances (exceeding 3 S/cm^2) despite their thickness. However, the mechanical properties are insufficient, probably due to their excessive hydration. In addition, the compression molding process, used at the laboratory scale for crosslinked membrane formation, appears to be not easily industrially viable.

Composite membranes have also been prepared with EW from 750 to 1,100 g/mol and thickness from 20 to 80 μm. The ionomer in the acid form, dissolved at ambient[33] or high[34] temperature in water–ethanol mixtures, has been used to impregnate perfluorinated porous supports.[35] Dissolution at low temperature is assisted by addition of a small amount of fluoropolyether in the solvent mixture.[33] These membranes have also shown high conductance (up to 4 S/cm^2) as expected considering the low thickness. They possess also good mechanical properties. However the casting method, used on a laboratory scale to prepare small membrane samples, appears to be quite complex and is especially susceptible to inconsistency, so also not easily industrially viable. Moreover, for the more crystalline higher EW polymers, the limited gain in mechanical stability due to the support is countered by a loss in conductivity due to the presence of a fraction of volume that is occupied by the non-conducting material (see FIGURE 13).

Finally, self-supported semicrystalline membranes have been prepared with EW higher than 750 g/mol and thickness as low as 15 μm. These membranes show very good conductance and also very good mechanical properties; they are easy to handle, even at extremely low thickness. Furthermore, the ionomer, tuned with the appropriate molecular weight distribution, gave high quality and consistent membranes by film extrusion. This last process appeared to be the most viable and low cost industrial process. An activity was therefore carried out to select the ionomer that gave the best combination of processability, conductance, mechanical properties, and dimensional stability, while also being able to guarantee extremely low membrane thickness and high duration.

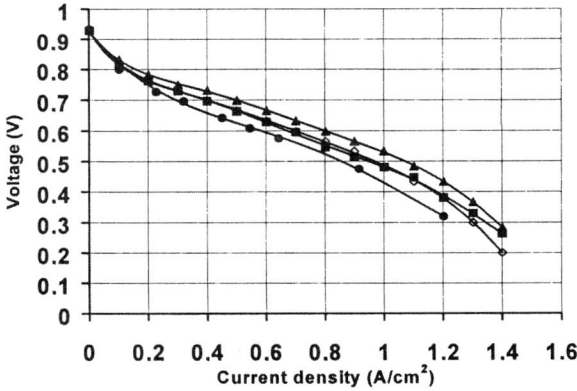

FIGURE 14. Comparison between polarization curves obtained with Hyflon Ion–FM EW = 900 membranes of various thicknesses (● 65 μm, ■ 40 μm, and ▲ 20 μm) and (○) a Gore-Select membrane.

TABLE 8. Stress and strain at break for Hyflon Ion–FM membranes and other commercial membranes, measured at RH = 50% and 23°C

Membrane	Stress (MPa)		Strain (%)	
	MD	TD	MD	TD
Hyflon Ion–FM EW = 800	23	20	130	120
Hyflon Ion–FM EW = 900	30	23	90	140
Nafion 115	30	26	119	188
Gore-Select[36]	34	24	—	—

Membranes conditioned under the measurement conditions: MD, machine direction; TD, transverse direction.

The resulting fuel cell membranes were perfluorosulphonic acid (PFSA) Hyflon Ion–FM membranes with EW = 900 g/mol. Membrane properties, in comparison with commercial Nafion 115, Nafion 117, and Gore-Select membranes, are reported in FIGURE 14 and in TABLES 8–10.

CONCLUSIONS

Fluoropolymer membranes offer, in many cases, the key to processes not easily achievable by conventional routes. Hydrophilic membranes have been used for many years in the chlorine and caustic electrolytic process as the sole ecologic alternative to mercury and asbestos diaphragm processes; they have also been used in fuel cells developed in the 1960s for the NASA space program. More recently, the use of hydrophobic membranes is being developed for applications, such as gas purification and gas–liquid contactors.

The chemistry that leads to such membranes is accessible to few companies in the world. A process developed by Solexis is particularly suitable for producing special fluoromonomers of crucial importance to the development of advanced

TABLE 9. Stress and strain at break for Hyflon Ion–FM membranes and other commercial membranes, measured at RH = 50% and 23°C

Membrane	Stress (MPa)		Strain (%)	
	MD	TD	MD	TD
Hyflon Ion–FM EW = 800	14	15	100	125
Hyflon Ion–FM EW = 900	22	17	70	110
Nafion 115	23	18	85	110
Gore-Select	32	17	—	—

Membranes pretreated by soaking in water at 100°C for 30 min. MD, machine direction; TD, transverse direction.

TABLE 10. Dimensional increase (%) in the plane directions from the dehydrated to the hydrated state (water soaking at 25°C and 100°C) for Hyflon Ion–FM membranes and other commercial membranes

Membrane	25°C		100°C	
	MD	TD	MD	TD
Hyflon Ion–FM EW = 800	7	8	18	20
Hyflon Ion–FM EW = 900	2	5	6	16
Nafion 117	8	10	10	12
Nafion 115	2	10	4	20
Gore-Select	—	—	3	3

fluoropolymer membranes. In particular, two special monomers produced on an industrial scale are the basis for the two classes of hydrophobic and hydrophilic membranes.

A cyclic monomer (2,2,4-trifluoro-5 trifluoromethoxy-1,3 dioxole) is the basis for the production of amorphous perfluoropolymers known commercially as Hyflon AD, showing T_g higher than room temperature, thermal decomposition temperature exceeding 400°C, and high transparency to light from far UV to near infrared. These polymers have been used to prepare protective pellicles for microlithography that are transparent in the far UV, gas separation membranes that show an optimum compromise between permeability and selectivity, and hydrophobic porous membranes that show excellent hydrophobicity and organophobicity and allow gas fluxes many times those of commercial gas separation membranes. A special sulfonyl fluoride monomer and a proprietary microemulsion polymerization process available from the Solexis technology are the basis for the production of short-side-chain ionic perfluoropolymers called Hyflon Ion. Based on these new ionomers, self-supported and composite sulfonic acid membranes for fuel cells are currently under development. Compared to presently commercial membranes, self-supported Hyflon Ion membranes show high conductivity, superior mechanical properties, and a higher ionic T_g, which is a necessary condition for running the cell at a higher temperature without damaging the membrane. The chemical stability has not yet been proved in long life tests, but the perfluorinated chemical structure guarantees the success of the tests currently being performed.

REFERENCES

1. SMART, B.E. 1986. *In* Molecular Structure and Energetics, Vol. 3, Ch. 4. J.F. Liebman & A. Greenberg, Eds. VCH Publishers, Deerfield Beach.
2. ARRIGONI, S., M. APOSTOLO & V. ARCELLA. 2003. Fluoropolymers architecture design: the branching and pseudo-living technology. Curr. Trend. Polym. Sci. In press.
3. NAVARRINI, W., V. TORTELLI, P. COLAIANNA & J. ABUSLEME, Inventors; Ausimont S.p.A., Assignee. 1997. European patent specification 0 633 257 B1. Date of application: June 24, 1994.

4. RESNICK, P.R. & W.H. BUCK. 1997. Teflon® AF amorphous fluoropolymers. *In* Modern Fluoropolymers. High Performance Polymers for Diverse Applications. J. Scheirs, Ed.: 397–419. Wiley, Chichester.
5. SUGIYAMA, N. 1997. Perfluoropolymers obtained by cyclopolymerization and their applications. *In* Modern Fluoropolymers. High Performance Polymers for Diverse Applications. J. Scheirs, Ed.: 541–556. Wiley, Chichester.
6. COLAIANNA, P., G. BRINATI & V. ARCELLA, Inventors; Ausimont S.p.A., Assignee. 1999. US patent 5,883,177. Date of application: April 21, 1997.
7. GORDANO, A., *et al.* 1999. Hydrophobic membranes of tetrafluoroethylene and 2,2,4 trifluoro 5 trifluorometoxy 1,3 dioxole. Korean Membr. J. **1**(1): 50–58.
8. ARCELLA, V., *et al.* 1999. A study on a perfluoropolymer purification and its application to membrane formation. J. Membr. Sci. **163**: 203–209.
9. TANT, M.R., K.P. DARST, K.D. LEE & C.W. MARTIN. 1989. Structure and properties of short-side-chain perfluorosulfonate ionomers. *In* Multiphase Polymers: Blends and Ionomers, ACS Symp. Ser. 395. L.A. Utracki & R.A. Weiss, Eds.: 370–400. ACS, Washington, DC.
10. EISENBERG, A. & H.L. YAEGER, Eds. 1982. Perfluorinated Ionomer Membranes, ACS Symp. Ser. 180. ACS. Washington, DC.
11. TRAINHAM, III, J.A., *et al.*, Inventors; E.I. Du Pont de Nemours and Company, Assignee. 1995. US patent 5,411,641. Date of application: November 22, 1993.
12. YEAGER, H.L. & A.A. GRONOWSKI. 1997. Membrane applications. *In* Ionomers. Synthesis, Structure, Properties and Applications. M.R. Tant, K.A. Mauritz & G.L. Wilkes, Eds.: 333–364. Blackie Academic & Professional, London.
13. COOLEY, G.E. & V.F. D'AGOSTINO, Inventors; National Power PLC, Assignee. 1997. US patent 5,626,731. Date of application: April 7, 1995.
14. MOYA, W., Inventor; Millipore Corporation, Assignee. 2001. US patent 6,179,132. Date of application: March 13, 1998.
15. NAYLOR, T. DE V. 1996. Polymer membranes—materials, structures and separation performance. Rapra Rev. Rep. **8**(5): 39–45.
16. OLAH, G.A., P.S. IYER & G.K.S. PRAKASH. 1986. Perfluorinated resinsulfonic acid (Nafion-H®) catalysis in synthesis. Synthesis **7**: 513–531.
17. FIELDING, H.C. 1992. Fluoropolymer membranes. *In* Fluoropolymers 1992. Synthesis, Properties & Commercial Applications: paper 5. UMIST, Manchester, 6–8 January.
18. HEITNER-WIRGUIN, C. 1996. Recent advances in perfluorinated ionomer membranes: structure, properties and applications. J. Membr. Sci. **120**: 1–33.
19. PRATER, K. 1990. The renaissance of the solid polymer fuel cell. J. Power Sources **29**: 239–250.
20. EISMAN, G.A. 1986. The physical and mechanical properties of a new perfluorosulfonic acid ionomer for use as a separator/membrane in proton exchange processes. *In* Proceedings of the 168th Electrochemical Society Meeting, Vol. 83-13: 156–171.
21. ZAWODZINSKI, JR., T.A. 1993. Water uptake by and transport through Nafion® 117 membranes. J. Electrochem. Soc. **140**(4): 1041–1047.
22. MOORE, III, R.B. & C.R. MARTIN. 1989. Morphology and chemical properties of the Dow perfluorosulfonate ionomers. Macromolecules **22**: 3594–3599.
23. EZZELL, B.R., W.P. CARL & W.A. MOD, Inventors; The Dow Chemical Company, Assignee. 1982. US patent 4,330,654. Date of application: June 11, 1980.
24. EZZELL, B.R., W.P. CARL & W.A. MOD, Inventors; The Dow Chemical Company, Assignee. 1982. US patent 4,358,545. Date of application: June 11, 1980.
25. ALEXANDER, L.E. & G.A. EISMAN, Inventors; The Dow Chemical Company, Assignee. 1993. European patent specification 0 221 178 B1. Date of application: May 15, 1986.
26. EZZELL, B.R., W.P. CARL & W.A. MOD. 1986. Ion exchange membranes for the chloralkali industry. *In* Industrial Membrane Processes. AIChE Symp. Ser. 248. R.E. White & P.N. Pintauro, Eds. **82**: 45–50. AIChE, New York.
27. EZZELL, B.R. & W.P. CARL, Inventors; The Dow Chemical Company, Assignee. 1990. US patent 4,940,525. Date of application: May 8, 1987.
28. CARL, W.P., *et al.*, Inventors; The Dow Chemical Company, Assignee. 1992. European patent application 0 498 076 A1. Date of application: December 20, 1991.

29. TSOU, Y.-M., M.C. KIMBLE & R.E. WHITE. 1992. Hydrogen diffusion, solubility and water uptake in Dow's short-side-chain perfluorocarbon membranes. J. Electrochem. Soc. **139**(7): 1913–1917.
30. EZZELL, B.R., W.P. CARL & W.A. MOD, Inventors; The Dow Chemical Company, Assignee. 1982. US patent 4,358,412. Date of application: June 11, 1980.
31. APOSTOLO, M. & V. ARCELLA, Inventors; Ausimont S.p.A., Assignee. 2002. European patent application 1 172 382 A2. Date of application: June 19, 2001.
32. ARCELLA, V., et al., Inventors; Ausimont S.p.A., Assignee. 2002. European patent application 1 167 400 A1. Date of application: June 19, 2001.
33. MACCONE, P. & A. ZOMPATORI, Inventors; Ausimont S.p.A., A. 2001. US patent 6,197,903. Date of application: March 31, 1999.
34. MARTIN, C.R., T.A. RHOADES & J.A. FERGUSON. 1982. Dissolution of perfluorinated ion containing polymers. Anal. Chem. **54:** 1639–1641.
35. BAHAR, B., et al., Inventors; W.L. Gore & Associates, Assignee. 1996. US patent 5,547,551. Date of application: March 15, 1995.
36. KOLDE, J.A., et al. 1995. Advanced composite polymer electrolyte fuel cell membranes. Proc. Electrochem. Soc. **95**(23): 193–201.

Novel Charge-Mosaic Membranes

BENHUI SUN AND SHUYING CHENG

*School of Materials Science and Engineering,
Beijing University of Chemical Technology, Beijing, China*

ABSTRACT: Novel charge-mosaic composite membranes composed of cation and anion permeable domains were prepared. Poly(4-methylstyrene–isoprene–styrene–isoprene–4-methylstyrene) pentablock copolymer (MsISIMs), poly(isoprene–4-methylstyrene) diblock copolymer (IMs), and poly(isoprene–styrene–isoprene) triblock copolymer (ISI), used as the precursor of the charge-mosaic membranes, were polymerized from styrene (S), isoprene (I), and 4-methylstyrene (Ms). MsISIMs and a blend of ISI and IMs were then dissolved in benzene and cast on a polypropylene microporous supporting membrane. After chemical modification, the MsISIMs and IMs/ISI charge-mosaic composite membranes were obtained. Our experimental results show that the IMs/ISI charge-mosaic composite membrane is similar in membrane structure and properties to the MsISIMs charge-mosaic composite membrane.

KEYWORDS: charge-mosaic membrane

INTRODUCTION

A method for preparing charge-mosaic membranes from a multiblock copolymer was first suggested by Fujimoto.[1–3] According to this method, an MsISIMs charge-mosaic membrane[4] is prepared from the pentablock copolymer, poly(4-methylstyrene–isoprene–styrene–isoprene–4-methylstyrene), synthesized by a sequential anionic polymerization method. However, the preparation of MsISIMs pentablock copolymer is very difficult. Therefore, its application is limited. To overcome this limitation, IMs/ISI blend charge-mosaic membranes were studied and successfully substituted for MsISIMs membranes.

Our results indicate that the structure and properties of IMs/ISI membranes are quite similar to those of MsISIMs membranes, but IMs and ISI membranes are more readily available at lower cost than MsISIMs.

EXPERIMENTAL

Materials Polymerization

The polymers used in this work were manufactured by a living anionic polymerization. The initiator, naphthoic lithium, was prepared by the reaction of naphthalene with lithium in tetrahydrofuran. Another initiator, *n*-butyllithium (*n*-BuLi), was prepared by the reaction of *n*-butyl with lithium in hexane. Isoprene was dried

---

Address for correspondence: Benhui Sun, School of Materials Science and Engineering, Beijing University of Chemical Technology, Beijing 100029, China.

over calcium hydride under reduced pressure and over sodium metal. Styrene and 4-methylstyrene were each dried over calcium hydride under reduced pressure. The cyclohexane used for polymerization and the hexane and tetrahydrofuran used for the initiator solution were purified by distillation.

Polymerization of pentablock copolymer MsISIMs, diblock copolymer IMs, and triblock copolymer ISI were carried out in sealed glass apparatus under nitrogen. MsISIMs was prepared by a living anion polymerization via a three-stage sequential addition of styrene, isoprene, and 4-methylstyrene in cyclohexane, using naphthoic lithium as the initiator. The anion polymerization of triblock copolymer ISI was carried out by a two-stage sequential addition of styrene and isoprene in cyclohexane, using naphthnic lithium as the initiator. The copolymer IMs was prepared by a two-stage sequential addition of isoprene and 4-methylstyrene.

The polymerization steps for MsISIMs were as follows:

Initiation

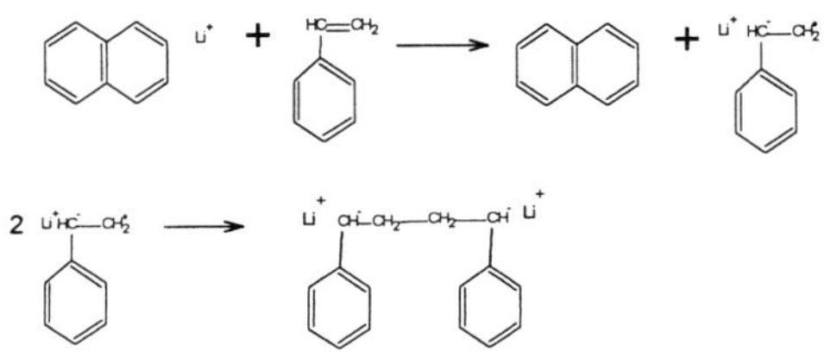

Propagation

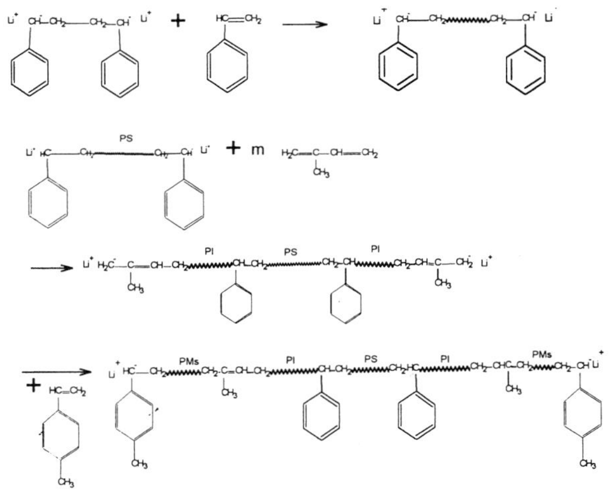

Termination

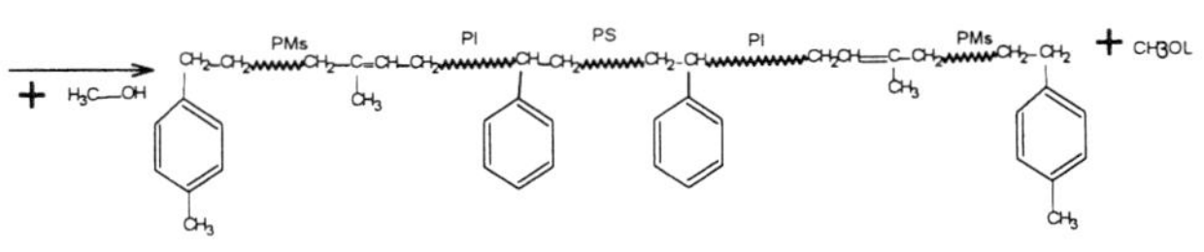

The polymerization steps of ISI were as follows:

Initiation

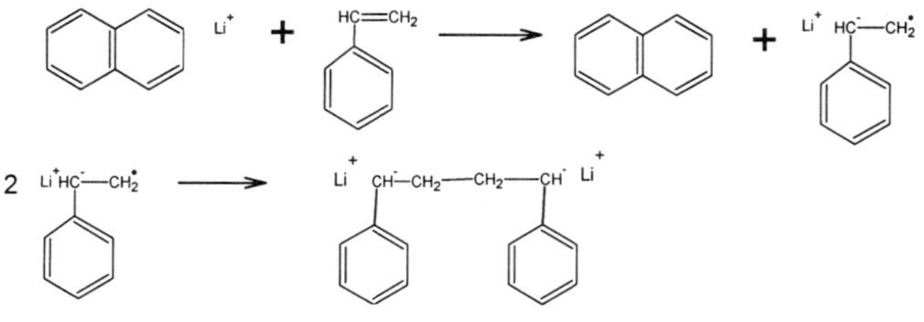

Propagation

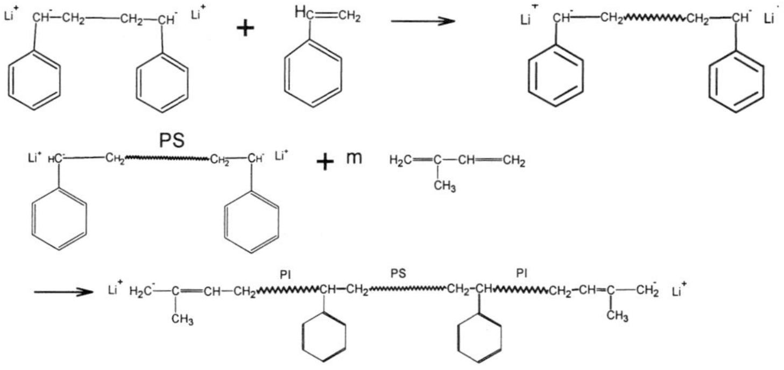

Termination

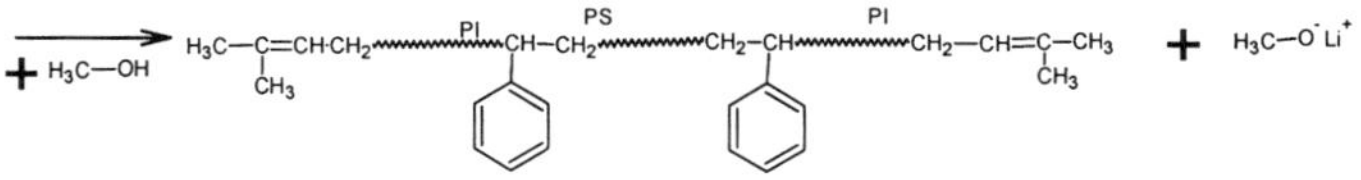

The polymerization steps of IMs were as follows:

Initiation

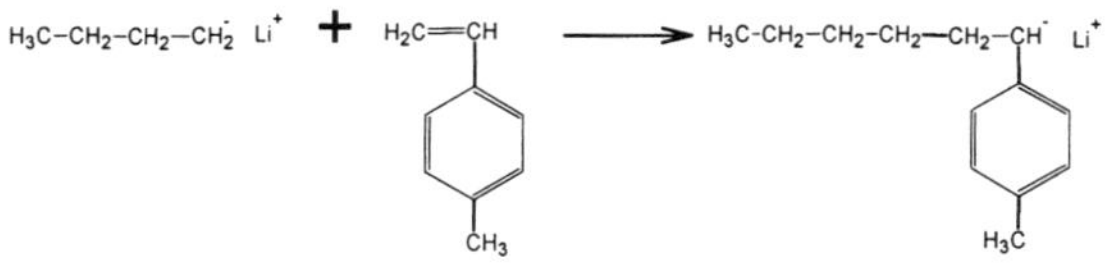

Propagation

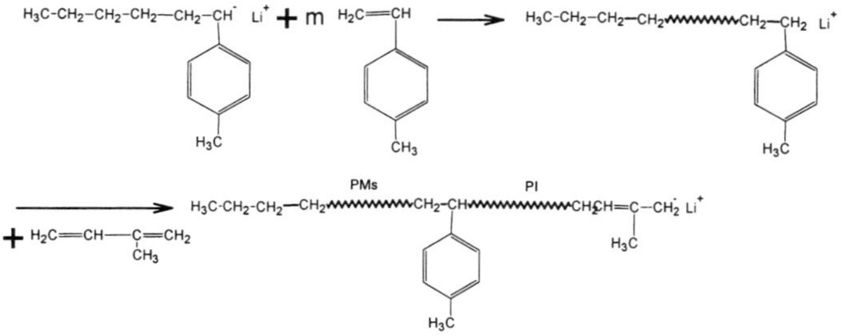

Termination

$$+ \; H_3C\text{-}OH \longrightarrow H_3C\text{-}CH_2\text{-}CH_2\text{-}CH_2\overset{PMs}{\sim\sim\sim\sim\sim}CH_2\text{-}HC\overset{PI}{\sim\sim\sim\sim\sim\sim}CH_2\text{-}CH=\underset{H_3C}{\overset{|}{C}}\text{-}CH_3 \; + \; H_3C\text{-}O^-\;Li^+$$

(with pendant C_6H_4-CH_3 group on the PMs segment)

Analysis of Copolymers

TEM Analysis

The block copolymers MsISIMs, ISI, IMs, and the IMs/ISI blends were each dissolved in the dissolvant benzene. The samples were prepared for transmission electron microscopy by coasting the solutions on a water surface and then staining with OsO_4. The TEM micrographs of these block copolymers were taken by JEM-100cx TEM.

DSC Analysis

The glass transition temperatures of the block copolymers MsISIMs, ISI, IMs, and the IMs/ISI blend were tested by using a Perkin-Elmer DSC-II, at an increasing temperature rate of 20°C/min and a temperature range between –70°C and 150°C.

Preparation for Composite Charge-Mosaic Membranes

Films of Pentablock MsISIMs and IMs/ISI Blend

PP MF membranes were manufactured and supplied by the Institute of Chemistry, the Chinese Academy of Science. The pentablock copolymer MSISIMs and the blend of ISI and IMs in a certain ratio were dissolved in the solvent benzene. The solutions were cast on the PP MF membrane to form an ultrathin layer and then they were dried at room temperature for chemical modification.

Chemical Modification Reactions

To prepare charge-mosaic membranes from films of the pentablock copolymer and the blend of IMs/ISI, crosslinking of the I block, chloridation and quaternization of the Ms block, and sulfonation of the S block were carried out sequentially. A nitromethane solution of sulfur monochloride (S_2Cl_2) was chosen as the crosslinking reagent for the I block; the chloridation of Ms block was carried out in a solution of sulfonyl chloride/tetrachloride with BPO, then quaternization in trimethylamine aqueous solution; and finally sulfonation of the S block was carried out in chlorosulfonic acid/trichloromethane solution.

Characterization of Charge-Mosaic Membranes

Infrared Spectrum Analysis

The blend charge-mosaic membrane was tested by using a 60SXB Fourier transform infrared spectrum instrument.

Measurement of Resistance of Membrane Surface

The resistance of membrane was investigated by the use of a cell consisting of conductive electrode, a DDS-II conductor, and a DT-300 digital multimeter. The membrane was submerged in a solution of 0.5 N NaCl with conductive electrode and nylon cloth. The resistance of nylon–membrane–nylon layer and nylon–nylon layer were each measured. The resistance of the membrane surface was determined from the difference between these two layers and the effective area of the membrane.

Measurement of the Flowing Electric Potential

The measuring instrument for flowing electric potential consists of two compartments, I and II, with Ag–AgCl electrodes connected by a DT-830 digital multimeter and separated by the membrane. The bath compartments were filled with a 0.1 mol/L solution of KCl. The potentials varied with a change in pressure in compartments I and II.

RESULTS AND DISCUSSION

Microphase Separation of Block Copolymers

Micro-Morphologies

The samples of IMs/ISI and MsISIMs had micro-separation morphologies as shown in FIGURE 1. The black domains represent I blocks and the white domains represent S and Ms blocks. This suggests that the fine microphase separations of the pentablock copolymer and IMs/ISI blend are very similar.

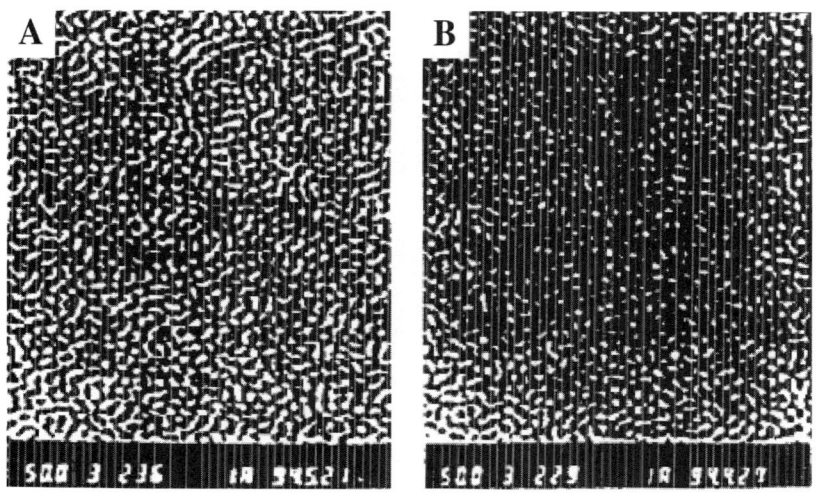

FIGURE 1. TEM micrographs of **(A)** IMs/ISI blend and **(B)** MsISIMs pentablock copolymer.

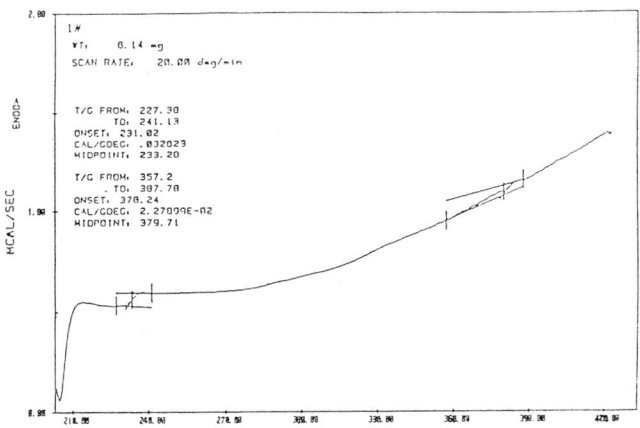

FIGURE 2. DSC graph for IMs.

DSC Analysis of Block Copolymers

In the multiphase systems of microphase separation of block polymers, there are at least two glassy transform regions. In the block copolymers (ISI, IMs, and MsISIMs) the I block is the soft block, S and Ms blocks are the hard blocks. The internal effects of soft and hard blocks caused a change in their T_g to medium values. The DSC spectra of IMs, ISI, IMs/ISI, and MsISIMs are shown in FIGURES 2, 3, 4, and 5, respectively. The individual blocks of IMs/ISI and MsISIMs had similar glass

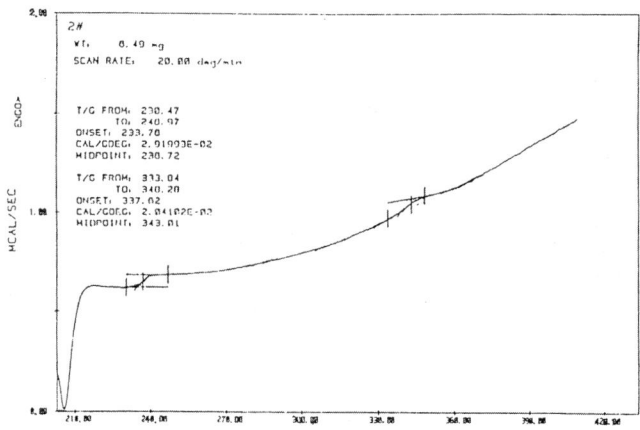

FIGURE 3. DSC graph for ISI.

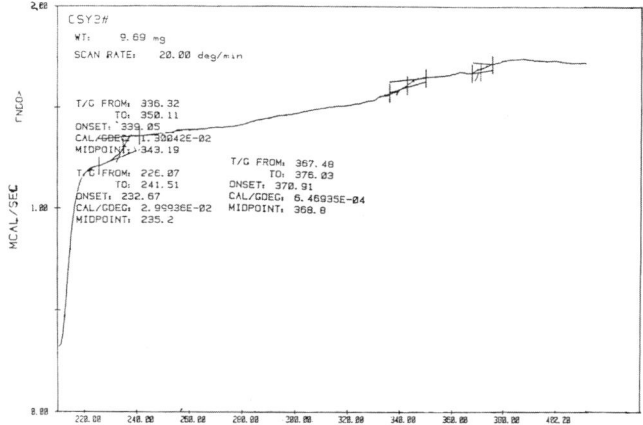

FIGURE 4. DSC graph for IMs/ISI.

transition temperatures, as is shown in TABLE 1. This illustrates that IMs/ISI and MsISIMs exhibit microseparation morphologies.

Infrared Spectrum Analysis of Blend Charge-Mosaic Membranes

FIGURE 6 shows the infrared spectrum for the modified blend charge-mosaic composite membrane. The band at $1,636\,cm^{-1}$ is due to the C–N bond in the quaternary group, the bands at $1,191\,cm^{-1}$ and $1,040\,cm^{-1}$ are due to the group $-SO_3H$, and the band at $631\,cm^{-1}$ is due to C–S bond formed on crosslinking double bonds. The

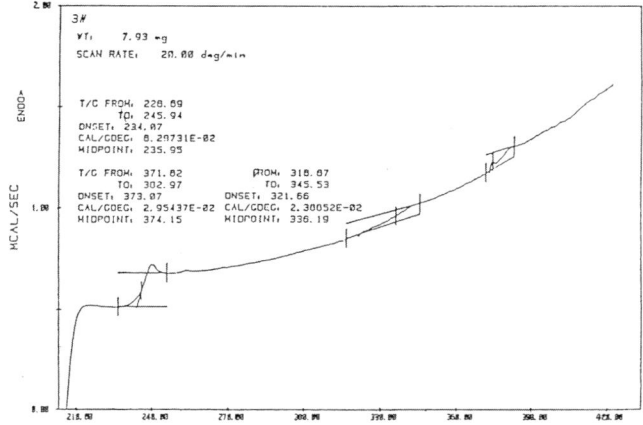

FIGURE 5. DSC graph for MsISIMs.

TABLE 1. T_g for block copolymers and polymers

Copolymer	Block I T_g (°C)	Block S T_g (°C)	Block Ms T_g (°C)
I-S-I	−34.4	69.9	—
I-Ms	−38.8	—	106.6
MsISIMs	−38.8	63.0	101.0
IMs/ISI	−37.9	70.0	102.8
Monopolymer	−42.0	88.0	112.0

infrared spectra results indicate that there are ionic exchange groups $-CH_2N^+(CH_3)_3$ and cationic exchange groups $-SO_3^-$ on the charge-mosaic membrane, and that crosslinking occurred at the C=C double bond.

Resistance of Membrane Surface and Its Effects[5]

The electric charge conductivity of a membrane depends on the inner network structure, exchange groups, and quantity of electric charge (exchange capacities). The electric conductance of a membrane is determined by the speed of transference of charges with different polarity. In the 1–1 type electrolyte solution, the specific conductance was calculated as follows:

$$X = 10^{-3}(\Lambda_+ C_+ + \Lambda_- C_-),$$

where Λ_+ and Λ_- are the equivalent conductances of ions in the membrane, and C_+ and C_- are the concentrations of ions. Specific conductance improves with increasing ionic concentration in the membrane. The resistance of the membrane was calculated from

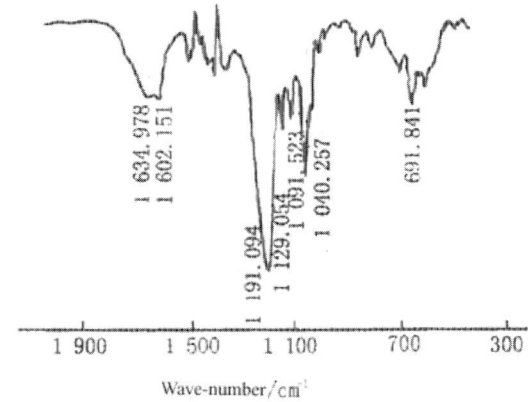

FIGURE 6. Infrared spectrum for a charge-mosaic membrane after removing the effect of PP.

$$R = \frac{L}{AX},$$

where L is the distance between electrodes and A is the area of electrode sections. The resistance of the membrane surface was calculated from

$$R_s = R \times A.$$

The value of R_s increases with the decreasing specific conductance.

The resistance values for blend charge-mosaic membrane and pentablock charge-mosaic membranes investigated in 0.1%mol/L KCl solution were $3.5 \times 10^2 \, \Omega \cdot cm^2$ and $970 \times 10^2 \, \Omega \cdot cm^2$, respectively, at a temperature of 18–21°C.

In general, the membrane resistance is used to estimate desalination of the membrane; the resistance of charge membrane was between 10^2 and $10^3 \, \Omega \cdot cm^2$. The resistances of the blend charge-mosaic membrane and pentablock membrane were $350 \, \Omega \cdot cm^2$ and $970 \, \Omega \cdot cm^2$, respectively, within this scale. Thus, it was concluded that good desalination was achieved.

Flowing Electric Potential of Charge-Mosaic Membranes[6]

The flowing electric potential of blend charge-mosaic membranes under the various pressure deviations was measured in 0.1 mol/L KCl solution and compared with pentablock charge-mosaic membranes. The flowing electric potentials of charge-mosaic membranes as functions of the pressure deviation are shown in FIGURE 7.

The slope of the curve in FIGURE 3 is the voltage permeation coefficient, β,[7]

$$\beta_I = 8.0 \times 10^{-9} \text{ V/Pa}$$

$$\beta_{II} = 6.1 \times 10^{-9} \text{ V/Pa}.$$

If the curve goes up, β is negative. Negative β indicates that there is a positive charge in the membrane and anions pass through the membrane; if down, β is positive, and cations pass through the membrane.

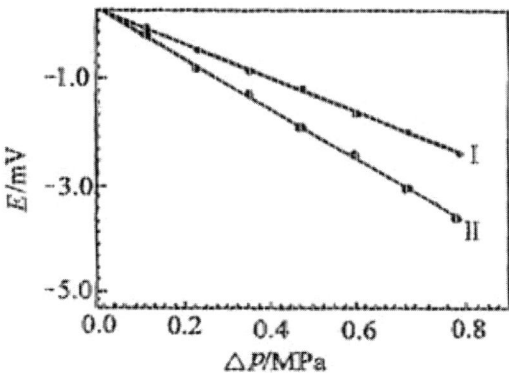

FIGURE 7. Flowing electric potential, E, versus pressure deviation, ΔP, of solution for (I) pentablock and (II) blend charge-mosaic membranes.

TABLE 2. Comparison of the electric properties of two kinds of charge-mosaic membrane

Type of Membrane	Resistance of Membrane ($\Omega \cdot cm^2$)	Potential Permeable Coefficient ($V \cdot Pa^{-1}$)
Blend type membrane	350	8.0×10^{-9}
Pentablock type membrane	970	6.1×10^{-9}

The curves in FIGURE 7 slope downward, indicating that the number of sulfonic acid groups exceeds that of quaternary groups. That was caused for the following reasons:
1. it was very difficult to make the equivalent weights of polystyrene and polymethylstyrene exactly equal;
2. when the membrane was modified, it was difficult to control the equal equivalent weights of sulfonic acid group and quaternary group; and
3. the I block was affected by the modifying agent when the membrane was modified.

Comparison of the Electric Properties of Blend and Pentablock Charge-Mosaic Membranes

TABLE 2 shows that the values of the resistances of these two membranes have the same order of magnitude and that both of them have good desalination efficiency. The potential permeable coefficients are close.

CONCLUSIONS

1. The pentablock MsISIMs and the blend of IMs/ISI were qualified for the preparation of charge-mosaic membranes because of their fine microphase separation.
2. The infrared spectra illustrated that the modified membranes had anionic exchange groups and cationic exchange groups and that the I block was crosslinked at the double bonds.
3. The resistance values for the blend and pentablock types of charge-mosaic membranes were $350\Omega \cdot cm^2$ and $970\Omega \cdot cm^2$, respectively; that is, they had the same order of magnitude. Both of these two types of membranes have good desalination capacity. The potential permeable coefficients were, respectively, $8.0 \times 10^{-9} V \cdot Pa^{-1}$ and $6.1 \times 10^{-9} V \cdot Pa^{-1}$ (similar values). Thus, it was concluded that the pentablock charge-mosaic membrane can be replaced by the blend type.

REFERENCES

1. FUNABASHI, H., Y. MIYAMOTO, T. FUJIMOTO, et al. 1983. Preparation and characterization of a pentablock copolymer of the ABACA type. Macromolecules **16:** 1–4.
2. ISONO, Y., H. TANISUGI, T. FUJIMOTO, et al. 1983. Morphological and mechanical properties of multiblock copolymer. Macromolecules **16:** 5–9.
3. FUJIMOTO, T. 1984. Artificial membrane from multiblock copolymer, 3. preparation and characterization of charge-mosaic membrane. Macromolecules **17:** 2331–2336.

4. SUN, B.H., J.H. HU, S.Y. CHENG, *et al.* 1998. Preparation of charge-mosaic composite membrane from poly(4-methylstyrene–isoprene–styrene–isoprene–4-methylstyrene) pentablock copolymer. J. Membr. Sci. Tech. **18**(3): 40–42.
5. MUO, J.X. & S.M. LIU. 1986. Testing of resistance of ionic exchange membrane using conductance electrode, Tech. Water Treatment **12**(2): 84–86.
6. WANG, Z.K. 1986. Ionic exchange membrane—manufacturing, properties and application. Chem. Ind. Express, Beijing.
7. MUO, J.X. & S.M. LIU. 1991. Theory and testing method of flowing electrical potential. Tech. Water Treatment **17**(3): 153–160.

Material Transport through Charged Mosaic Membrane

AKIRA YAMAUCHI AND TAKASHI FUKUDA

Chemistry Division, Graduate School of Science, Kyushu University, Higashi-ku, Fukuoka, Japan

ABSTRACT: Several characteristics of a charged mosaic membrane with parallel array of negative and positive charges were investigated by using transport studies and the related analysis. From an analysis of the volume flux and salt flux based on irreversible thermodynamics, preferential salt transport across the charged mosaic membrane was clearly demonstrated. Additionally, transport properties of amino acids and sucrose through the charged mosaic membrane were estimated, relatively, on the basis of KCl transport. As a result, amino acid transport depends largely on the charged states and molecular weight; however, for sucrose transport, non-electrolyte was rejected under all experimental conditions.

KEYWORDS: charged mosaic membrane; reflection coefficient; separation technology; amino acid transport; salt enrichment

INTRODUCTION

Our goal was to elucidate the material transport mechanism across a charged mosaic membrane. Since K. Sollner foresightedly proposed a model for a charged mosaic membrane,[1–4] several investigations have been presented, but unfortunately a functional membrane for practical applications has not been obtained to date.[5–19] If such a charged membrane were available, new membrane technology for the effective production of salts the from ocean would be possible. Conventional techniques that use membranes, such as electrodialysis methods, need to be improved to become more effective because present systems cannot provide highly concentrated seawater. This may be the result of limited performance in the electrodialysis method. As one of the potential techniques for this purpose, pressure dialysis using a charged mosaic membrane is proposed.[18,19] Charged mosaic membranes are well known as membranes containing two different fixed charges within the matrix. The two kind of charges, anion and cation exchange groups, are arranged parallel with each other inside the membrane and the array of charge groups links continuously from one membrane surface to the other. The structure of the charges, it is supposed, induce concurrent migrations of cations and anions along the respective fixed charges. Thus, preferential salt fluxes and a resultant separation between electrolyte and nonelectrolyte can be expected. This characteristic feature of the mosaic membrane is very fascinating in the field of membrane technology.

Address for correspondence: Akira Yamauchi, Chemistry Division, Graduate School of Science, Kyushu University, 6-10-1 Hakozaki, Higashi-ku, Fukuoka 812-8581, Japan. Voice/fax: +81-92-642-2595.
a.yamscc@mbox.nc.kyushu-u.ac.jp

Recently, Nakamura *et al.* developed a novel charged mosaic membrane using microsphere gels.[20,21] In previous studies, the fundamental properties of the membranes were also reported by our group.[22–25] In this paper, transport phenomena across the membrane and other properties were examined and the transport mechanism, from basic viewpoint, was characterized.[26]

EXPERIMENTAL

Membrane

The preparation and characteristics of charged mosaic membranes in this paper are described in detail elsewhere.[20,21] The membranes were kindly supplied by Dainichi Seika Co. Ltd. To prepare the membrane, identical quantities of exchangeable anions and cations were introduced into the membrane matrix and the cation exchange capacities were determined by acid–base titration and are given together with other membrane properties in TABLE 1. As indicated, the cation exchange capacity was 0.945 meq/g, the membrane thickness 50 μm, and the water content 31–42%. The membranes were stored in $0.1\,\text{mol}\,\text{dm}^{-3}$ KCl solution before experimental use.

For comparative study, the conventional cation and anion exchange membranes, CMV and AMV, which are commercially available from Asahi Glass Co., were used.

Reagents

General chemicals, such as simple electrolytes or sucrose, were procured from KATAYAMA Chemical Co. Ltd. and were used without further purification. Amino acids, sodium glutamate (GluNa H_2O, MW = 187), arginine (Arg, MW = 174), and alanine (Ala, MW = 89.1) were kindly supplied as ultrapure samples by Ajinomoto Co. Inc. Doubly distilled water was used to prepare aqueous electrolyte solutions.

Transport Studies

The cell for our experiments consisted of two half-glass cells; the charged mosaic membrane was tightly clamped between the two cells. The temperature was kept at 25°C by circulating water constantly around the two cells during the experiment.

TABLE 1. Membrane characteristics

Membrane thickness	50 μm
Anion exchange site (+)	$-C_5H_5^+CH_3$
Cation exchange site (−)	$-C_6H_5SO_3^-$
Water content	
water	40.9%
0.1 M KCl	37.6%
0.5 M KCl	31.2%
Ion exchange C. (−)	$0.945\,\text{meqg}^{-1}$

Two kinds of measurements, volume changes and salt concentration changes in either cell, were performed as functions of time by using a graduated capillary and an electrode type conductivity meter, respectively.[27–30]

RESULTS AND DISCUSSION

Membrane Parameters

According to irreversible thermodynamics as it relates to transport phenomena, the practical phenomenologic equations are as follows:[26]

$$J_v = L_p(\Delta P - \sigma \Delta \Pi) \quad (1)$$

$$J_s = C_s(1-\sigma)J_v + \omega \Delta \Pi. \quad (2)$$

Under the appropriate experimental conditions, the characteristic parameters of the membrane, such as filtration coefficient, L_p, reflection coefficient, σ, and solute permeability, ω, are

$$L_p = \left(\frac{J_v}{\Delta \Pi s}\right)_{\Delta P = 0, \sigma = 1}, \quad (3)$$

$$\sigma = -\left(\frac{1}{L_p}\right)\left(\frac{J_v}{\Delta \Pi s}\right)_{\Delta P = 0}, \quad (4)$$

$$\omega = \left(\frac{J_v}{\Delta \Pi s}\right)_{J_v = 0}, \quad (5)$$

where J_v and J_s are the volume flux and salt flux, which can be obtained from the change in volume and change in salt concentration over time, respectively. $\Delta \Pi s$ is the osmotic pressure calculated from the concentration differences of salt or sucrose across the membrane. It should be noted that J_v has positive sign for the transport direction from solution to pure water. Under the conditions indicated by each subscript, one can obtain the membrane parameters. Membrane performance can be discussed in terms of the parameters. That is, L_p represents water transport index across membrane, σ is the separation index between water (solvent) and solute for the membrane, and ω is the solute transport index across the membrane.

Fundamental Studies

Prior to our transport study, the separation efficiency of the charged mosaic membrane for electrolyte and nonelectrolyte mixture was examined. The results indicated that the separation in mixed solution of glucose and KCl was satisfactory and the glucose molecules were almost completely rejected by the membrane (see Figure 1 in Ref. 22).

Based on the above fact, it is expected that sucrose, a disaccharide molecule, is sufficiently rejected by the membrane because the molecule is nonelectrolyte and is larger than glucose. This leads to a sufficiently high osmotic pressure in the water/sucrose system. As typical example, the volume change driven by the osmotic pressure when 0.5 mol/dm³ sucrose is put into one side of the cell, are plotted as a function of time in FIGURE 1, where the decreased volume changes were taken as

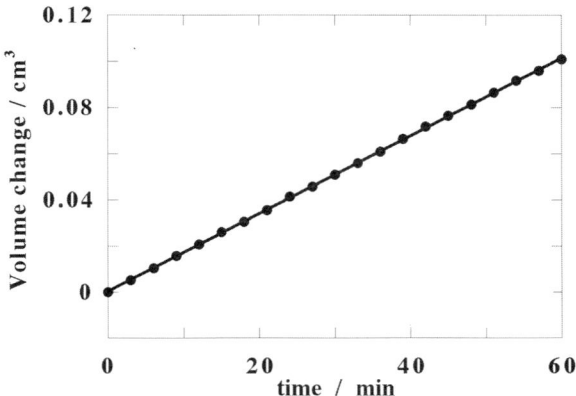

FIGURE 1. Volume changes with time in a water/0.5 M sucrose system. The increased volume in sucrose solution is taken as positive.

positive values. The relation turned out to be linear, within the range of time examined, and suggests a steady state. Accordingly volume flux, J_v was deduced from the slope of the linear relation by taking account of the effective membrane area, $3.14 \times 10^{-4} m^2$.[18,19,24] Experiments in a similar system that contained identical KCl concentrations in both solutions across the membrane, were carried out in order to determine the influence of electrolyte on the volume flux. Inserting the volume flux into Equation (3), one can obtain the filtration coefficient, L_p, that permits solvent (water) permeability through the membrane. The L_p values in the presence of KCl are summarized in FIGURE 2. As can be seen in FIGURE 2, the values were almost independent of KCl concentration in the range from zero to $0.5 \, mol/dm^3$. In this case, the presence of KCl did not affect water permeability through the membrane and $7.07 \times 10^{-14} m^3 N^{-1} sec^{-1}$ for L_p was determined as an average value. For comparative study, the same experiments on conventional cation (CMV) and anion (AMV) exchange membranes were carried out; the results are included in FIGURE 2.

In the water/KCl system, where only KCl instead of sucrose was put into the half cell, KCl flow due to the concentration gradient takes place across the membrane. This results in a decrease in osmotic pressure, due to the KCl concentration difference across the membrane. As a result, the volume flux of solvent is reduced. From viewpoint of separation between solute and solvent, an increase of KCl flux means that the separation index, σ, becomes less than unity. Provided that water permeability is constant and independent of the presence of KCl, σ can be estimated from Equation (4) by assuming $L_p = 7.07 \times 10^{-14} m^3 N^{-1} sec^{-1}$. As expected from the KCl flux values, negative values of σ appeared within range of KCl concentrations examined as shown in FIGURE 3.

In comparison with conventional charged membranes, the filtration coefficient in the charged mosaic membrane was almost the same order as those of ion exchange membranes.[7,19,22–25,26–32] Interestingly, the reflection coefficient turned out to be negative in contrast to that of an ordinary ion exchange membranes and the

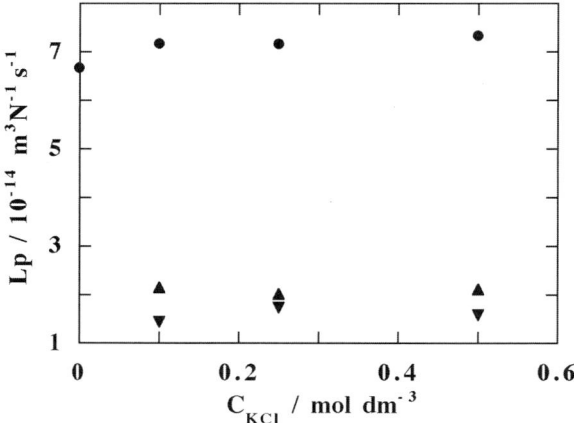

FIGURE 2. Dependence of L_p on KCl concentration. L_p values were estimated from the volume flux in water/sucrose or KCl/KCl + 0.5 mol dm^{-3} sucrose systems: ●, charged mosaic membrane; ▲, cation exchange membrane; and ▼, anion exchange membrane.

dependence on the electrolyte concentration was immediately apparent. It is suggested by irreversible thermodynamics that the salts can be preferentially transferred across the membrane.

As mentioned above, the fact that σ in the water/KCl system becomes less than unity or negative is attributed to the generation of salt flux across the membrane. The solute flux in water/KCl at several concentrations was also observed experimentally

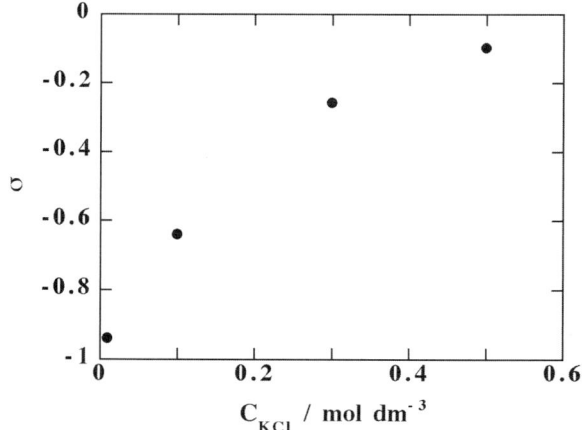

FIGURE 3. Dependence of σ on KCl concentration: the σ values were estimated from volume flux in water/KCl system by using the average L_p value 7.07×10^{-14} m^{3}N^{-1}sec^{-1}.

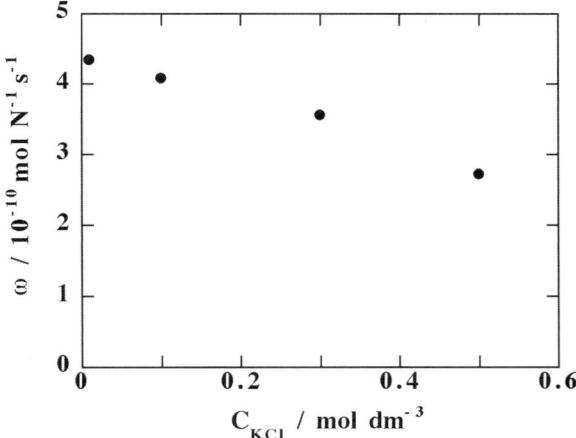

FIGURE 4. Dependence of ω on KCl concentration: ω values were estimated from the solute flux in the water/KCl system.

and the results indicated that KCl concentrations in water phase increased linearly with the elapsed time. The solute flux, J_s, was calculated from the slope of straight line, in the same manner as the volume flux. Furthermore, the salt permeability coefficient, ω, was obtained in terms of Equation (5). The values of ω are displayed as a function of KCl concentration in FIGURE 4. The solute permeability increased with a decrease in σ. Both ω and σ obviously imply the existence of KCl transport through the charged mosaic membrane. The ω value indicating that KCl transport was 100-fold more than that of the transport through an ordinary charged membrane[7] and this result supports the interpretation mentioned concerning σ. Thermodynamic parameters, such as σ and ω, suggest that application of the charged mosaic membrane offers promise for salt enrichment or for getting pure water from the ocean.

Amino Acid Transport

Based on information concerning the simple electrolytes mentioned above, the transport behavior of an amino acid through the charged mosaic membrane was examined. Three kinds of amino acids, known as typical, neutral, and basic compounds, were chosen to investigate the transport phenomena within a charged mosaic membrane. From the volume flux obtained using the system mentioned above, values for the reflection coefficient for the amino acids were obtained and are shown in FIGURE 5 together with those of KCl and LiCl for comparison. The data for amino acids have provided interesting results: the reflection coefficients are larger than for the simple salts but are less than unity. This means that the amino acids can permeate through the charged mosaic membrane and can be recovered from aqueous solution by a diffusion dialysis method using this type of membrane. Among these amino acids, GluNa showed the largest transport, perhaps because it completely dissociated under the pH condition. The difference for KCl or LiCl may be attributed to their ionic size. The pH values of aqueous Arg and Ala solutions prepared without any

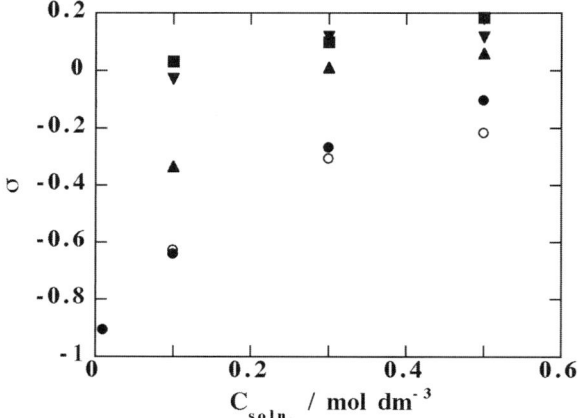

FIGURE 5. Dependence of σ on corresponding amino acid concentration: ▲, sodium glutamate; ▼, arginine; and ■, alanine. For comparison, σ values for KCl (●) and LiCl (○) are included.

salts for a buffer adjustment were 11.0 and 6.2, respectively, over amino acid concentration examined in this study. When the pH values are taken into consideration, together with the corresponding pK values, part of the Arg and Ala concentration is supposedly undissociated in aqueous solution. This means that there is no apparent net charge on these species, but it does not mean that the given molecules behave as a nonelectrolyte. We imagine that Arg and Ala molecules permeate, while forming a kind of ion complex, through the charged mosaic membrane. This could be one of the reasons why the reflection coefficients of amino acids are larger than those of simple salts, but less than that of the nonelectrolyte. The molecular size should also be considered for the transport of amino acids.

Preferential transport of amino acid across a charged mosaic membrane was suggested on the basis of the small reflection coefficient. To verify this speculation, a direct observation of the electrolyte flux was made using the same system as for the volume flux measurement. Using the same procedure as reported in a previous study,[19] the permeability coefficient was obtained from the solute flux. The ω values of two amino acids, GluNa and Arg, are given, together with the ω values of the simple salts, in FIGURE 6. Unfortunately, the electrical conductance of the Ala solution is too small to detect reliable change, and thus a reasonable concentration change across the membrane could not be obtained. As can be seen in FIGURE 6, an obvious difference in the permeabilities between simple salts and amino acids was observed, and a difference between GluNa and Arg was also recognized. The value of ω increases together with a decrease in solute concentration, indicating a correlation with σ (FIG. 5). The decrease in σ and the increase in ω, with a decrease in the solute concentration, may take place as a result of the balance between the osmotic flow and the diffusion flow. The difference between GluNa and Arg could be caused by the ionic state and a sieve effect recognizing ion size, as discussed above.

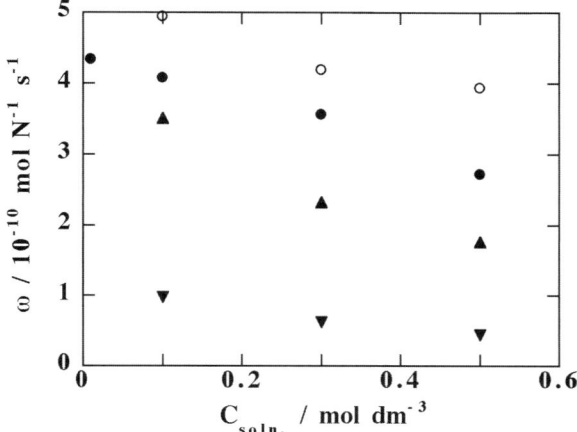

FIGURE 6. Dependence of ω on corresponding amino acid concentration: ▲, sodium glutamate; ▼, arginine; ●, KCl; and ○, LiCl.

Electrolyte Transport in the Presence of Non-Electrolyte

In this section material transport under various experimental conditions was examined. That is, the electrolyte transport in the presence of non-electrolyte was investigated. The experimental system is shown in FIGURE 7. As indicated, KCl was put on one side and sucrose was put on the other side of the membrane. This kind of

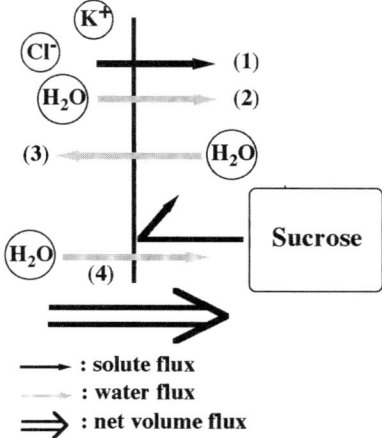

FIGURE 7. Schematic model for solute and solvent permeation across charged mosaic membrane in the KCl/sucrose system: (1) diffusion of ions, (2) water transport accompanied by (1), (3) osmotic flow to KCl solution side, and (4) osmotic flow to sucrose solution side.

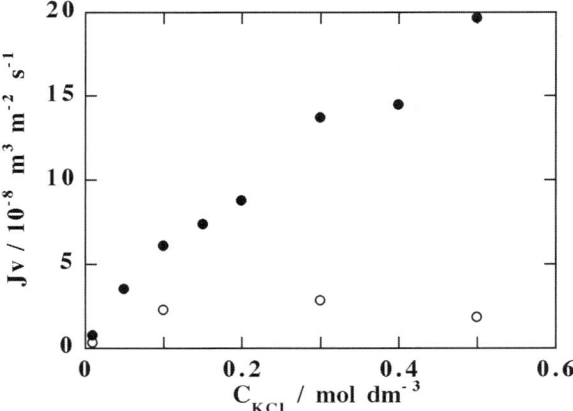

FIGURE 8. Relation between volume flux, J_v, and KCl concentration: ○, KCl/water, ●, KCl/sucrose.

system would be important in treating solute separation from a mixed solution. The equivalent concentrations of solutes in both phases were kept equal, so that initially the osmotic flow originating from the solutes canceled each other. The volume changes observed as a function of time and the volume flux for various KCl concentrations are given in FIGURE 8, together with results for the KCl/water system. Interestingly, the apparent volume fluxes indicate more negative osmotic flow than that of simple KCl/water system. When the effect of sucrose was subtracted from total flux, the reflection coefficient values were almost the same as those in simple salt

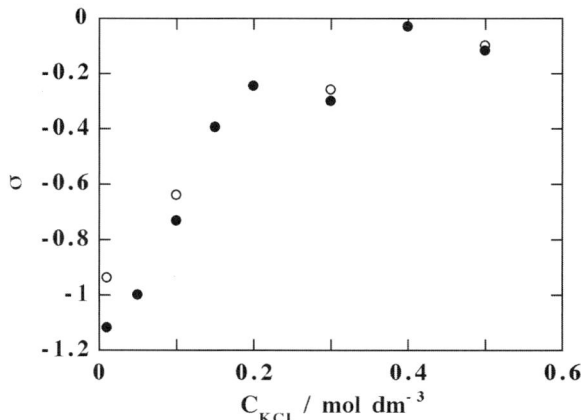

FIGURE 9. The reflection coefficient, σ, versus KCl concentration in KCl/sucrose (●) and KCl/water (○).

system, as indicated in FIGURE 9. This means that electrolyte transport across the charged mosaic membrane is independent of the presence of non-electrolyte. In other words, water transport based on electrolyte and non-electrolyte turns out to be additive.

CONCLUSIONS

In this work, the filtration coefficient for water transport across the membrane was of the same order as that for ordinary charged membrane, $7.07 \times 10^{-14} m^3 N^{-1} sec^{-1}$. Under certain KCl concentration conditions, the membrane indicated a negative reflection coefficient; that is, negative σ values, which suggests preferential salt transport compared to solvent transport. This interpretation was also supported by salt flow measurements. Three different amino acids indicated transport behavior corresponding to the ionic forms in solution. The order of transport was KCl, LiCl > GluNa > Arg, Ala > glucose, sucrose. Water transport due to electrolytes and nonelectrolytes through the charged mosaic membrane appear to be independent of each other. As a result, the charged mosaic membrane is promising for enrichment of salt or desalination from seawater and for separation technology between electrolyte and nonelectrolyte.

ACKNOWLEDGMENTS

Financial support aided by The Salt Science Research Foundation "0117" and supplies of variable samples, of charged mosaic membranes and amino acids made by Dainichi Seika Co. Ltd. and Ajinomoto Co. Inc. are highly appreciated and acknowledged.

REFERENCES

1. SOLLNER, K. 1932. Biochem. Z. **244:** 370.
2. NEIHOF, R. & K. SOLLNER. 1950. J. Phys. Colloid Chem. **54:** 157.
3. SOLLNER, K. 1955. Arch. Biochem. Biophys. **54:** 129.
4. NEIHOF, R. & K. SOLLNER. 1955. J. Gen. Physiol. **38:** 613.
5. MIZUTANI, Y., et al. 1963. Bull. Chem. Soc. Jpn. **36:** 361.
6. YAMANE, R., et al. 1965. Denki Kagaku **33:** 589.
7. WEINSTEIN, J.N. & S.R. CAPLAN. 1968. Science **61:** 70.
8. WEINSTEIN, J.N., B.J. BUNOW & S.R. CAPLAN. 1972. Desalination **11:** 341.
9. WEINSTEIN, J.N., et al. 1973. Desalination **12:** 1.
10. GARDNER, C.R., J.N. WEINSTEIN & S.R. CAPLAN. 1973. Desalination **12:** 19.
11. LEITZ, F.B. & W.A. MACRAE. 1972. Desalination **10:** 293.
12. LEITZ, F.B. 1973. Desalination **13:** 373.
13. SHORR, J. & F.B. LEITZ. 1974. Desalination **14:** 11.
14. YAMABE, T., et al. 1974. Desalination **15:** 127.
15. MATSUSHITA, Y., et al. 1980. Macromolecules **13:** 1053.
16. FUJIMOTO, T., et al. 1984. J. Membr. Sci. **20:** 313.
17. ISONO, Y., et al. 1989. J. Membr. Sci. **43:** 205.
18. YAMAUCHI, A. & Y. TANAKA. 1993. Effective Membrane Processes. R. Paterson, Ed.: 179–185. BHR Group Limited, England.
19. YAMAUCHI, A. 1993. Bull. Soc. Sea Water Sci. Jpn. **47:** 227–232.

20. NAKAMURA, M., *et al.* USP-5543045.
21. NAKAMURA, M., *et al.* Japan Kokai 10-87855.
22. YAMAUCHI, A., *et al.* 2000. J. Memb. Sci. **173:** 275–280.
23. YAMAUCHI, A. 2000. Eighth World Salt Symposium. 653–657.
24. FUKUDA, T. & A. YAMAUCHI. 2000. Bull. Chem. Soc. Jpn. **73:** 2729–2732.
25. FUKUDA, T. & A. YAMAUCHI. 2000. Bull. Soc. Sea Water Sci. Jpn. **56:** 33–38.
26. KATCHALSKY, A. & P.F. CURRAN. 1965. Nonequilibrium Thermodynamics in Biophysics. Harvard University Press.
27. YAMAUCHI, A., *et al.* 1998. J. Memb. Sci. **148:** 139–146.
28. YAMAUCHI, A., *et al.* 1999. J. Memb. Sci. **163:** 297–305.
29. YAMAUCHI, A., *et al.* 2000. J. Memb. Sci. **163:** 289–296.
30. YAMAUCHI, A., *et al.* 2000. J. Memb. Sci. **170:** 1–7.
31. YAMAUCHI, A., *et al.* 1990. Desalination **80:** 61–71.
32. HIRATA, Y. & A. YAMAUCHI. 1990. J. Memb. Sci. **48:** 25–31.

Effect of Compatibility of PVC/P$_2$ Alloy System on Membrane Structure and Performance

PATRICIA B. SUN[a] AND BENHUI SUN[b]

[a]*Wayne State University, Detroit, Michigan, USA*
[b]*Beijing University of Chemical Technology, Beijing, China*

ABSTRACT: The effects of the secondary polymer component (P$_2$) on poly(vinyl chloride) (PVC)/P$_2$ alloy membrane structure and performance were systematically investigated. A series of P$_2$ with varying compatibility with PVC used in this study included vinyl chloride–vinyl acetate copolymer (VC-co-VAc); copolymer of vinyl chloride, vinyl acetate–maleic anhydride copolymer (VC-co-VAc-co-MAL); isobutylene–maleic anhydride copolymer (IB-co-MAL); poly(methyl methacrylate) (PMMA); poly(vinylidene dichloride) (PVDC); and styrene–acrylonitrile copolymer (SAN). Alloy membranes were prepared by means of solution blending—phase inversion technique. On the basis of our experimental results, the compatibility of PVC/P$_2$ was proved to be the most critical factor affecting the alloy membrane structure and performance. Systems with good compatibility, such as PVC/VC-co-VAc, are more suitable for preparing membranes with small pore size; whereas systems with partial compatibility, such as PVC/PMMA, are more favored for the formation of large-pore membranes.

KEYWORDS: membrane; polymer alloy; PVC; compatibility

INTRODUCTION

Intensive efforts on the research of membrane separation technology have focused on two aspects: membrane materials with excellent properties and separation membranes with high efficiency. However, material and structural limitations cause commercial separation membranes fabricated from a single material to be unable to simultaneously meet various demands for applications. To solve this problem, many modification methods that can endow existing membranes with new features or functions have been proposed Among them, the polymer alloying method is gaining more and more attention because of its obvious modification effects, simple manufacturing process, low cost and industrial feasibility.

Alloys are made by mixing different kinds of metals together. Similarly, composite polymer systems composed of various kinds of polymers that exhibit physical or chemical interactions, are named as *polymer alloys*. Polymeric membranes made from polymer alloy materials are called *polymer alloy membranes*. According to their composition, polymer alloys can be separated into two main categories:

Address for correspondence: Benhui Sun, Beijing University of Chemical Technology, Beijing 100029, China. Fax: 86-10-84615790.
sunbenhui@hotmail.com

polymer blends and copolymers.[1] All of the alloy membranes investigated in this study were made by a physical blending method.

The earliest research on alloy membranes dates back to the 1960s. In 1960, the first isotropic cellulose nitrate (CN)/cellulose acetate (CA) microfiltration (MF) membrane was prepared by Goeth.[2] Ten years later, manufacture of a CA/cellulose triacetate (CTA) reverse osmosis (RO) membrane initialized the commercialization of alloy membranes.[3] Subsequently, asymmetric alloy membranes consisting of CA/poly-4-vinylpyridine or CA/aromatic polyphosphate[4–7] were used in RO desalination and azeotropic liquid separation with satisfactory results. In the 1980s, Kesting *et al.* did systematic research on blending systems of CA and Nylon RO membranes[8–10] and obtained important results. During the past two decades, studies on novel alloy membrane systems have been widespread,[11–20] all of them indicating that alloy membranes provide a very effective and promising approach to modifying both the chemical and the physical structure of membranes, as well as controlling membrane performance. Consequently, many membrane properties can now be significantly improved, including permeability, selectivity, hydrophilicity, fouling resistance, bacteriological resistance, and mechanical strength.

In this study, a series of PVC/P_2 alloy MF and ultrafiltration (UF) membranes were prepared by solution blending—phase inversion method, and the effects of P_2 on PVC/P_2 alloy membrane structure and performance were thoroughly investigated.

PRINCIPLE OF MATERIALS DESIGN OF THE SECOND POLYMER ADDITIVE

The purposes for introducing P_2 are generally: making up for the shortcomings of the primary polymer additive (P_1); and achieving control of membrane structure and improving membrane properties by means of altering the compatibility between P_1 and P_2.

Kesting[21] pointed out that, whereas complete compatible systems might be a requirement for membranes employed in gas separation and hyperfiltration, it was less likely to be necessary for UF and MF membranes. Less compatible blends than mixture systems with a single T_g were preferred in the separation of larger solutes or suspended particles. As a result, the compatibility between polymer components of alloy membranes can dramatically influence membrane structure, especially the skin-layer pore size and distribution, and, thus, affect their performance.

For a PVC/P_2 system, three aspects should be taken into account when selecting the second polymer additive. First, since PVC has poor hydrophilicity, leading to low water flux and low fouling resistance of a PVC membrane, P_2 should possess good hydrophilicity. Second, to assure the stability of the casting solution system, P_2 should at least be partially compatible with PVC. Finally, differences in gelation rate should exist between PVC and P_2 so that microscopic phase separation can occur during the precipitation process, and then be used to alter the membrane pore structure, as well as performance, by blending. On the basis of these principles, VC-co-VAc, PMMA, PVDC, SAN, and IB-co-MAL were chosen as P_2.

EXPERIMENT

Membrane Preparation

The PVC (S-800) resin used in our experiments was provided by Qilu Petrochemical Corp. (SINOPEC). The secondary polymer P_2 included VC-co-VAC (Beijing Second Chemical Industry Company, China), PMMA (Xinhua Chemical Industry Factory, China), PVDC (Quzhou Chemical Industry Ltd., China), SAN (Lanzhou Petrochemical Corp., China), and IB-co-MAL synthesized by Beijing University of Chemical Technology, China.

The polymer blend solution was prepared by mixing a certain amount of PVC with P_2 in an appropriate solvent system. The concentration of the polymer blend solution was kept at 13 wt%. Alloy membrane was obtained from this casting solution by the Loeb–Sourirajan (L–S) phase inversion method.

Measurement of Pure Water Flux

Water flux was characterized by the quality of deionized water permeating through the membrane under an operating pressure of 0.3 MPa (UF)/0.1 MPa (MF) and at 25°C. This is defined by

$$F = \frac{Q}{A \times T},$$

where F is water flux, Q is the quantity of permeated water, A is the effective area of membrane, and T is the time during which the water permeated.

Measurement of Retention

Retention rate was measured with 0.1% bovine serum albumin in deionized water as feed solution. The UV spectrophotometer was adjusted so that the absorbance of the initial feed solution and water at $\lambda = 280$ nm were 100% and 0%, respectively. By filtering the feed solution through a membrane in an ultrafiltration cell at 0.3 Mpa and measuring the absorbance of permeate (C), the membrane retention (R) can be expressed as follows:

$$R = 100 - C.$$

RESULTS AND DISCUSSION

The Prediction of PVC/P_2 System Compatibility

From a thermodynamic point of view, in accord with Hildebrand's solubility parameter principle,[22,23]

$$\Delta h_m = (\delta_1 - \delta_2)^2 \Phi_1 \Phi_2,$$

where $\Delta h_m = \Delta H_m/\Delta V_m$ is the change in enthalpy per unit volume; δ_i and Φ_i ($i = 1, 2$) are the solubility parameter and mixing volume fraction, respectively, of component i. It can be seen from the equation that the closer δ_1 and δ_2 are in value, the smaller is the value of ΔH_m. With an increase in $\Delta\delta$, more and more energy is required to

TABLE 1. Alloy system thermodynamic compatibility according to the solubility parameter rule

Polymer	δ (cal/cm^3)$^{1/2}$	$\Delta\delta = |\delta_{PVC}| - \delta_{P_2}$ (cal/cm^3)$^{1/2}$	Anticipated Thermodynamic Compatibility
PVC	9.6	—	—
VC-co-Vac	VC block 9.6 VAc block 9.6	0	best
PMMA	9.3	0.3	good
PVDC	10.1	0.5	good
SAN	S block 9.4 AN block 12.6	3	poor, AN block is not compatible with PVC
IB-co-MAL	8.0	1.6	bad

blend the mixture into one phase because of the increase of Δh_m. Generally, when $\Delta\delta$ is zero, the system is thermodynamically completely compatible, whereas when $\Delta\delta$ exceeds a threshold value, phase separation occurs. Therefore, the value of $\Delta\delta$ reflects the compatibility of the mixture system to some extent.

Solubility parameters of the polymers involved in this study are listed in TABLE 1;[22] according to these data, the compatibility between VC-co-VAc and PVC should be the best; IB-co-MAL should have the worst compatibility with PVC. However, since the actual compatibility of a blend system depends on many factors, such as solvent, temperature, additives, and so forth, solubility parameters can only be used as reference instead of a quantitative criterion to determine the system compatibility.

Effect of P_2 on the Compatibility of a Polymer Blend in the Casting Solution

The compatibility of alloy polymers in casting solutions is represented by phase difference micrographs of their casting films. Casting films of polymer blends were prepared by putting drops of casting solution on a small piece of glass plate, casting it to let the solvent gradually evaporate until the film started to solidify, and then putting it into vacuum oven for 24 hours until fully dried. The polymer blend morphology was then observed by using a Nikon's Blopont 1500X phase contrast microscope.

Phase difference micrographs of the blend systems investigated in this study are shown in FIGURE 1. Almost no phase separation was observed for the PVC/VC-co-VAc system. Slight phase separation was obtained from both PVC/PMMA and PVC/PVDC alloy systems, in which the size of each phase is even and very small. This indicated relatively good compatibility between PVC and P_2 in all of the three systems mentioned above. Obvious phase separation could be seen in PVC/SAN and PVC/IB-co-MAL solutions, where each component tends to self-associate. This is caused by the relatively bad compatibility of PVC/P_2, which agrees well with the thermodynamic predictions shown in TABLE 1.

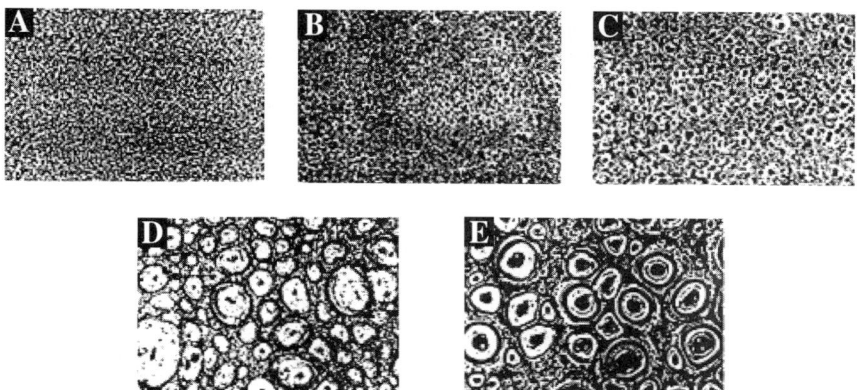

FIGURE 1. Phase difference micrographs of casting solutions for five alloy systems: **(A)** PVC/VC-co-VAc; **(B)** PVC/PMMA; **(C)** PVC/PVDC; **(D)** PVC/SAN; and **(E)** PVC/IB-co-MAL.

Effect of P_2 on the Compatibility of Polymer Blends in Alloy Membranes

Measuring alloy polymer glass transition temperatures (T_g) by thermal analysis method, such as differential scanning calorimeter (DSC), is by far the most widely used approach in characterizing their compatibility. The glass transition zones of a polymer blend are useful in judging whether it is homogenous or heterogeneous. For a one-phase system, only one T_g can be seen; two or more T_g peaks indicate that the system is multiphasic.

DSC measurements for three PVC/P_2 alloy membranes (P_1:P_2 = 7:3) with varying compatibility are shown in TABLE 2. A Perkin-Elmer DSC-2C was employed in the tests and the rate of temperature increase was 20°C/min. Only one glass transition region with T_g = 79.63°C was obtained for PVC/VC-co-VAc system. In addition, the T_g of PVC shifted towards the VAC segment T_g value, 28°C. Therefore, this system has the best compatibility among the three. For the other two systems, two glass transition regions were observed, indicating that they both contain heterogeneous phases. For the PVC/PMMA system, T_g shifted toward higher temperatures because of the intermolecular interaction between PVC and PMMA (the T_g of PMMA is 106°C); thus, this system is partially compatible. Although for PVC/SAN system both PS and PAN block T_g values are higher than PVC (S block, T_g = 100°C;

TABLE 2. T_g and glass transition region for PVC and its alloy systems

Polymer System	PVC	PVC/VC-co-VAc	PVC/PMMA	PVC/SAN
T_g, PVC (°C)	84.42	79.63	87.68	84.48
Glass transition region (°C)	68–94	72–92	78–95	74–90
			98–120	97–111

AN block, $T_g = 105°C$), no shift of the PVC T_g values was observed, which indicated that compatibility between PVC and SAN is the worst. All of the DSC results agree well with the phase difference microscopy results.

Effect of P_2 on Alloy Membrane Cross-Section Structure

The morphology of the alloy membrane cross-section was determined by employing a Cambridge 250MK3 scanning electron microscope (SEM). The specimens were gold–palladium coated for 60 to 70 seconds to have a coating thickness of 50 to 100 nm. Cross sections of the membranes were obtained by freeze fracturing the sample under liquid nitrogen.

The SEM micrographs of the cross-section structure of PVC, PVC/VC-co-VAc, and PVC/PMMA alloy membranes (see FIGURE 2), prepared under the same conditions, showed that the better the system compatibility, the smaller the pore size in both skin and bottom layers, and the denser the cross section. Three different kinds of pores can be found on single component membranes formed by the phase inversion method; in the order of small to large, they are: network pores, aggregate pores, and liquid–liquid phase separation pores.[24] For polymer blends, a fourth kind of pore can be observed on the skin-layer, which is caused by the phase separation of various kinds of polymers.

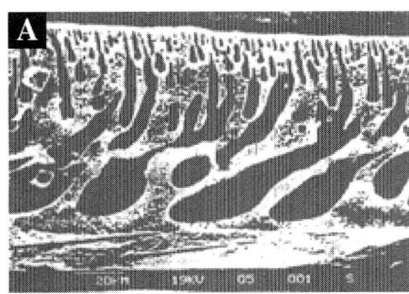
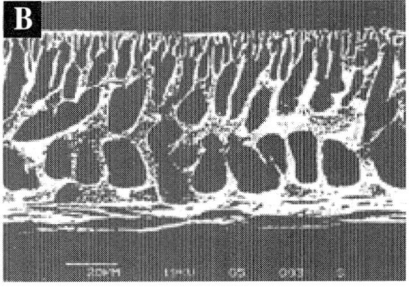
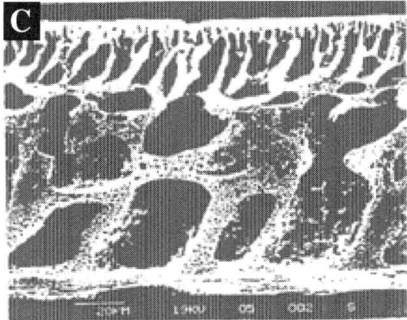

FIGURE 2. SEM micrographs of alloy membrane cross-section structure: **(A)** PVC; **(B)** PVC/VC-co-VAc; and **(C)** PVC/PMMA.

A typical finger-like structure could be seen on PVC membrane, where both the skin layer and the transition layer are densely covered with fine pores. Finger-like pores with evenly distributed diameters are connected in a vertical direction and go through the membrane cross-section. In the second system, interaction between PVC and VC-co-VAc is strong, the two polymers disperse well in the casting solution and, even when the solvent is gradually removed from the system, phase separation is not obvious. Thus, similar to the PVC membrane, the pores on skin layer of the alloy membrane are mainly composed of network and aggregate pores that cause the skin layer to appear dense. The dense layer in turn acts as a barrier to the quinch medium, which slows down the exchange rate of water and solvent. Therefore, a sponge-like structure is formed. Compared with this system, the PVC/PMMA system compatibility is poor. In addition, the difference of gelation rate between PMMA and PVC is larger than that for PVC and VC-co-VAc (the coagulation value[25] for PVC in water is 1.43 mL, for PMMA is 2.64 mL, and for VC-co-VAc is 1.67 mL; i.e., among the three polymers, the gelation rate for PVC is the highest in water and that of PMMA is the smallest). For the two reasons mentioned above the obvious phase separation region is more likely to be formed for PVC/PMMA system during gelation. As a result, phase separation pores are dominant in both skin and bottom layers and a loose membrane structure can be obtained.

Effect of P_2 on PVC/P_2 Alloy Membrane Hydrophilicity

Hydrophilic materials usually have high water flux and low wet strength, whereas hydrophobic materials possess high wet strength, low flux, and fouling resistance. Hence, blending polymers with various hydrophilicity may alter the membrane flux and fouling resistance. Despite its many advantages, PVC is not suitable for the preparation of UF and MF membranes that require high water flux because of its poor hydrophilicity. Therefore, good hydrophilicity is required from P_2 to make PVC alloy macroporous membrane with excellent water permeability.

Measurements of contact angles (θ) of water on various membrane surfaces were conducted to characterize the alloy membrane hydrophilicity. Literally, the smaller θ, the better the hydrophilicity. Alloy membrane θ values and flux rates are listed in TABLE 3.

The addition of hydrophilic P_2 causes a decrease in θ value and increase in membrane hydrophilicity, as well as water flux. Based on the θ value data, VC-co-VAc is more hydrophilic than PMMA and, hence, is more useful in improving PVC membrane hydrophilicity. However, the PVC/VC-co-VAc alloy membrane pure water

TABLE 3. Effect of P_2 on alloy membrane hydrophilicity and pure water flux

Membrane Material	Contact Angle θ (deg)	Water Flux F L/(m^{2}h)
PVC	66	61
PVC/PMMA	62	770
PVC/VC-co-VAc	57	482
PVC/VC-co-VAc-co-MAL	54	603

TABLE 4. Effect of P_2 on alloy membrane performance

Polymer System	Compatibility	Water Flux F L/(m^{2}h)	Rejection Rate R %
PVC	—	51	98
PVC/VC-co-VAc	best	506	87
PVC/PMMA	good	827	79
PVC/PVDC	good	773	78
PVC/IB-co-MAL	bad	1.2×10^4	0

flux is not as large as that of the PVC/PMMA system. This can be explained as a result of PVC/P_2 compatibility. Since the PVC/PMMA system compatibility is not as good as that of PVC/VC-co-VAc, more phase separation pores exist in its membrane structure and contribute greatly to the enhancement of alloy membrane water flux.

Although the compatibility of PVC/VC-co-VAc and PVC/VC-co-VAc-co-MAL are close to each other, the water flux of the latter system is higher than that of the former system. This is attributed to the highly hydrophilic maleic anhydride functional groups it contains.

Effect of P_2 on Alloy Membrane Performance

Permeability and selectivity are the most important properties for membrane performance evaluation. Membrane selectivity is principally determined by skin layer pore size, whereas permeability depends on membrane materials, cross-section, and skin layer structure.

It is difficult to improve permeability and selectivity at the same time for a monomaterial membrane, whereas it can be easily achieved in alloy membranes because of the cooperation between the two components.

The effect of P_2 on alloy membrane performance can be summarized as follows. First, the interaction between PVC and P_2 weakened the cohesive energy of PVC, disrupted the relatively regular order of PVC chains, and thus decreased the density of network and increased the content of aggregate in casting film. Second, the different solubility of PVC and P_2 in solvent resulted in the phase separation and promoted the formation of large polymer–polymer phase separation pores. The larger the pores, the worse the selectivity. Moreover, the difference in their gelation rates enhanced phase separation during the precipitation procedure. As a result, porous cross-section structure ready formed. Last, the presence of hydrophilic P_2 is helpful in improving the membrane material hydrophilicity. High productivity of water can be obtained from membranes with good hydrophilicity and porous cross-section.

It can be seen from TABLE 4 that when fixing the ratio of P_1/P_2 at 7/3, with the decrease of alloy system compatibility, the membrane water flux increases and rejection decreases. The worse the system compatibility, the more severe the phase separation between PVC and P_2, and the more phase separation pores form in the membrane. Consequently, membranes with higher flux and lower rejection are obtained.

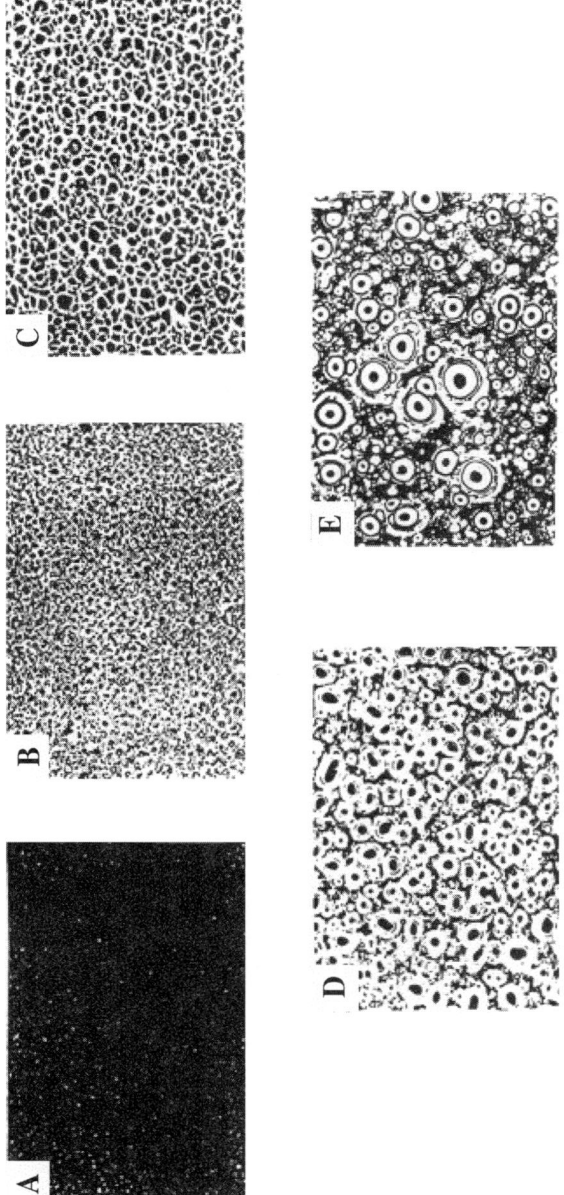

FIGURE 3. Phase difference micrographs for PVC/IB-co-MAL casting film with P_1/P_2 mixture ratio (**A**) 10:0, (**B**) 9:1, (**C**) 8:2, (**D**) 7:3, and (**E**) 5:5.

TABLE 5. Pure water contact angle for PVC/IB-co-MAL alloy materials with various mixture ratios

Mixture Ratio	Contact Angle θ (deg)
10/0	68.5
9/1	60.8
8/2	56.5
7/3	52.7
6/4	49.9
5/5	47.8
3/7	42.5
0/10	37.8

Effect of P_1/P_2 Ratio on the Compatibility of Polymer Blends and Alloy Membrane Performance

For the PVC/IB-co-MAL system, which has rather poor compatibility, according to the calculation of mixing enthalpy versus mixture ratio within the range 10:0 to 6:4, ΔH_m increased steadily with an increasing amount of P_2. This inevitably caused the aggravation of P–P phase separation. This was proved by the phase contrast micrographs shown in FIGURE 3.

The effect of PVC/IB-co-MAL ratio on the contact angle with pure water and alloy membrane performance is shown in TABLE 5 and FIGURE 4, respectively. Obviously, the greater the P_2 content, the worse the compatibility of the system and the better the hydrophilicity of the membrane. As a result, both the membrane pore size and flux increase.

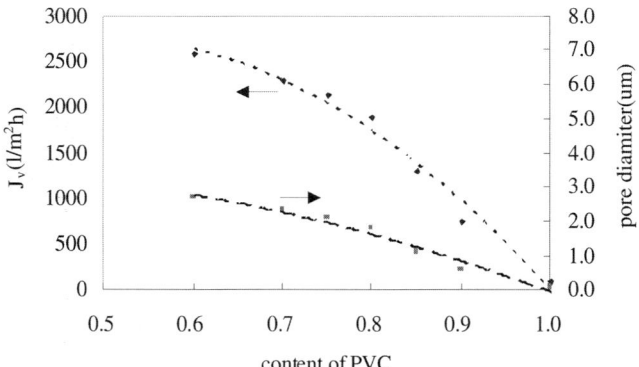

FIGURE 4. Effect of PVC/IB-co-MAL mixture ratio (wt:wt) on alloy membrane performance.

CONCLUSIONS

The compatibility of the secondary polymeric component with PVC significantly affects the resulting alloy membrane structure and performance. Generally, better compatibility of PVC/P_2 causes the formation of membrane with relatively dense and even structure, whose water flux is low and rejection is high; whereas for alloy systems whose compatibility between polymer components is bad, strong phase separation gives rise to loose structure with large pores. Therefore, the membrane rejection decreases and flux increases.

REFERENCES

1. JIANG, M. 1988. *In* Physical Chemistry of Polymer Alloys. Sichuan Education Publishing Company, Chengdu, China.
2. GOETZ, A., Inventor and Assignee. 1960. U.S. Patent 2,926,204. Date of application: Feb. 23.
3. CANNON, C., Inventor; Aerojet-General Corp., Assignee. 1970. U.S. Patent 3,497,072. Date of application: Feb. 24.
4. CABASSO, I., *et al.* 1974. Polymeric alloys of polyphosphonates and acetyl cellulose I. sorption and diffusion of benzene and cyclohexane. J. Appl. Polym. Sci. **18:** 2117–2136.
5. CABASSO, I., *et al.* 1974. A study of permeation of organic solvents through polymeric membranes based on polymeric alloys of polyphosphonates and acetyl cellulose II. separation of benzene, cyclohexene and cyclohexene and cyclohexane. J. Appl. Polym. Sci. **18:** 2137–2147.
6. CABASSO I. & C.N. TRAN. 1979. Polymer alloy membrane I. cellulose acetate—poly(bromophenylene oxide phosphonate) dense and asymmetric membrane. J. Appl. Polym. Sci. **23:** 2967–2988.
7. APTEL, P. & I. CABASSO. 1980. Novel polymer alloy membranes composed of poly(4-vinyl pyridine) and cellulose acetate I. asymmetric membranes. J. Appl. Polym. Sci. **25:** 1969–1989.
8. KESTING, R.E., Inventor; Nuclepore Corp., Assignee. 1980. U.S. Patent 4,220,477. Date of application: September 2.
9. KESTING, R.E., Inventor; Puropore, Inc., Assignee. 1982. U.S. Patent 4,333,972. Date of application: June 8.
10. KESTING, R.E., *et al.* 1983. Dry-RO membranes from blends of ionogenic and nonionogenic polymers. Desalination **46:** 343–348.
11. URAGAMI, T., *et al.* 1980. Studies on syntheses and permeabilities of special polymer membranes. 28. Permeation characteristics and structure of interpolymer membranes from poly(vinylidene fluoride) and poly(styrene sulfonic acid). Desalination **34:** 311–323.
12. KANG, Y., K. ARAKI & K. IWAMOTO. 1982. Preparation and gas permeability of polymer blend membranes of polystyrene and poly(1,1,1–tris(trimethylsiloxane) methacrylate propylsilane). J. Appl. Polym. Sci. **27:** 2025–2032.
13. KASI, M. & N. KOYAMA, Inventors; Terumo Kabushiki Kaisha T/A Terumo Corp., Assignee. 1988. U.S. Patent 4,772,440. Date of application: September 20.
14. HUANG, R.Y.M., Inventor; University of Waterloo, Assignee. 1990. U.S. Patent 4,892,661. Date of application: January 9.
15. MALINI, B. & R.Y.M. HUANG. 1991. Pervaporation separation of pentane-alcohol mixtures through ionically crosslinked blended I. thermal measurements, electron microscopy, tensile studies. Angew. Makromol. Chem. **191:** 36–69.
16. TOSHIKI, A., *et al.* 1993. Temperature-sensitive ethanol permselectivity of poly(dimethylsiloxane) membrane by the modification of its surface with copoly(*N*-isopropylacrylamide/1H,1H,2H,2H–perfluorododecylacrylate). Polymer **34**(7): 1538–1543.

17. CHERDRON, H., *et al.* 1994. Misible blends of polybenzimidazole and polyacramides with polyvinylpyrrolidone. J. Appl. Polym. Sci. **53:** 507–512.
18. SUN, P.B., *et al.* 1995. The study of PVC/VC-co-VAc alloy membrane. *In* Symposium of the 6th National Congress on Applied Chemistry. 177–185. Chinese Chem. Soc. Hai Nan, China.
19. SCHAUER, J. & W. ALBRECHT. 2001. Microporous membranes prepared from blends of polysulfone and sulfonated poly(2,6-dimethyl-1,4-phenylene oxide). J. Appl. Polym. Sci. **81:** 134–142.
20. YANG, G. & L. ZHANG. 2001. Cellulose/casein blend membranes from NaOH/urea solution. J. Appl. Polym. Sci. **81:** 3260–3267.
21. KESTING, R.E. 1985. Polymer Solutions. *In* Synthetic Polymeric Membranes, 2nd edit. 186–223. John Wiley & Sons, New York.
22. SUN, B.H. 1983. The calculation and measurement of solubility parameters. Special Rubber Products (Chinese) **4:** 47–59.
23. VAN KREVELEN, D.W. 1976. Cohesive energy density and solubility. *In* Properties of Polymers—Their Estimation and Correlation with Chemical Structure, Chap. 7. 95–119. Elsevier Scientific Publishing Company, Amsterdam-Oxford-New York.
24. ZHU, Z.X. 1991. The discussion of surface pore formation mechanism for reverse osmosis, ultrafiltration and microfiltration membranes made by L-S phase inversion method. *In* Symposium of the First National Membrane and Membrane Processing Conference. 39–42. Da Lian, China.
25. SUN, B.H. 1993. Characterization of the relationship between gelation rate and membrane structure in asymmetric membrane formation via phase inversion. Technol. Water Treatment (Chinese) **6:** 313–318.

Stable Liquid Membranes

Recent Developments and Future Directions

A. SARMA KOVVALI[a] AND KAMALESH K. SIRKAR[b]

[a]*Baker Petrolite, Sugar Land, Texas, USA*

[b]*Department of Chemical Engineering, New Jersey Institute of Technology, Newark, New Jersey, USA*

ABSTRACT: Immobilized liquid membranes offer the advantages of high selectivity and species flux in separation of species from either gaseous or liquid streams. Despite these advantages, they are not yet commercially viable because of their instabilities and limited lifetime. This paper presents recent advances made in developing stable immobilized liquid membranes for gas separation with high fluxes and selectivities for the species of interest. Some of the recent approaches being studied for development of stable liquid membranes for liquid separation are also discussed.

KEYWORDS: liquid membrane; gas separation; liquid separation; membrane stability; recent developments; future directions

INTRODUCTION

All mass transfer operations in chemical engineering aim to achieve the transfer of one or more species from one phase to another in the most efficient manner possible. Transfer of solute(s) is implemented to (1) purify the fluid stream of undesirable impurities, (2) separate a mixture of solutes to result in fluid streams with increased concentrations of species, or (3) increase the concentration of a solute so that the product stream is more amenable to further treatment/processing. Examples of such operations are abundant in chemical engineering practice, ranging from pollution control to enrichment of dilute streams of valuable compounds. In all cases, the emphasis is on achieving the separation in a cost effective and, if possible, elegant manner. Various mass transfer operations achieve this goal by creating a substantial interface between two or more phases involved in a variety of phase contacting devices. Examples include separation of gas mixtures via absorption and stripping, adsorption, separation of liquid mixtures by extraction, adsorption, and so forth.

Membrane processes often provide a compact and efficient structural platform for contacting two fluid phases. Membrane-based absorption/desorption, solvent extraction/back extraction are some of the phase-contacting applications in which microporous membranes provide the contact area for phase interface immobilization.

Address for correspondence: Kamalesh K. Sirkar, Department of Chemical Engineering, New Jersey Institute of Technology, Newark, NJ 07102, USA. Voice: 973-596-8447; fax: 973-642-4854.

Sirkar@adm.njit.edu

Such membrane devices are called *membrane contactors* to highlight their important function. Excellent reviews of the advantages and unique features of membrane contactors are available in the literature.[1–4]

The general scheme of such membrane contactors is that one membrane unit is employed for solute absorption/extraction from the feed stream into a fluid (generally a liquid containing absorbents or extractants); a separate membrane unit is used to desorb/back extract the solute from the liquid medium into another fluid medium. This allows the absorbing/extracting liquid to be regenerated and fed back to the absorbing/extracting membrane unit. It also facilitates collection of solute on the strip side of the desorbing unit at a much higher concentration than originally present in the feed stream. Sometimes, the desorbing unit (mainly in gaseous solute transfer) is a conventional unit operation, such as vacuum or thermal desorption. However, the fact remains that for recovery of solute from the feed stream in higher concentrations, two separate units are needed.

It is safe to generalize in that for most applications of membrane contactors the main resistance to mass transfer lies in the liquid phase. If the liquid phase occupies the pores of the membranes, it contributes significantly to the overall resistance to mass transfer. This is particularly prominent in membrane-based back extraction or gas-liquid desorption. Mass transfer resistance by the pore fluid can often be reduced by reducing the membrane thickness, filling the pores with a fluid of lower mass transfer resistance, among other strategies. In almost all situations, the governing strategy is to minimize the liquid film thickness (in the pore and on the outside of the membrane).

IMMOBILIZED LIQUID MEMBRANES

It is possible to combine the absorption/extraction and desorption/back extraction membrane units into a single membrane unit. Such a membrane process is often referred to as immobilized (or supported) liquid membrane. Immobilized liquid membranes (ILMs) contain a liquid solution immobilized in the pores of the polymeric or ceramic substrate by physical forces of capillarity/wetting. They are also referred to as supported liquid membranes (SLMs), particularly when the feed and sweep sides are also liquid streams. For convenience, they are all referred to as ILMs in this paper. FIGURE 1 shows schematically the immobilized liquid membrane. The ILM configuration can be visualized as an absorption unit on the feed side, and a desorption unit on the sweep/strip stream side. The liquid membrane is located in the pores of the membrane, forming a thin partition between the feed and sweep phases. Because the thickness of the membranes is often low, ranging from a few microns to a few hundred microns, the diffusion resistance for a species/species-carrier complex is often minimized, increasing the species flux across the liquid membrane.

The liquid solution in the pores usually consists of a carrier and a solvent. The carrier reacts reversibly with the species of interest. ILMs can potentially provide the highest fluxes and selectivities for reacting species, such as carbon dioxide and olefins, especially at low concentrations in gas mixtures. Similarly, ILMs have demonstrated high selectivities and fluxes for separation of metals from aqueous streams.[5]

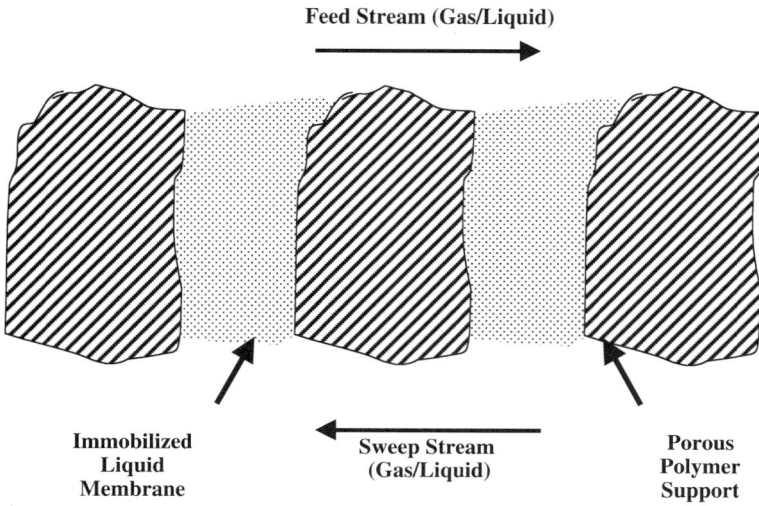

FIGURE 1. Schematic of an immobilized liquid membrane.

STABILITY OF ILMs

Despite the obvious advantages offered by the ILMs, commercialization of these membranes has not taken place due to the inherent limitation of stability of the liquid membranes. The main reasons for the instability of the ILMs both for gas separation and liquid separation are discussed extensively elsewhere.[6,7] Some of the main reasons for ILM instability are:

- absence of any chemical bonding of the carrier/solvent to the substrate matrix,
- evaporation/dissolution of the carrier species and/or the solvent liquid into the feed/strip phases during the operation,
- lower breakthrough pressures for the pore liquids in most membranes,
- displacement of pore liquid by the aqueous phases,
- emulsion formation of the pore liquid induced by lateral shear forces, and
- presence of osmotic pressure gradient.

Depending on the application, some of these factors can be dominant over others in determining the overall stability of the ILM. For example, the ILMs for gas separation mainly suffer from the loss of solvent and carrier due to evaporation, whereas the main routes of ILM degradation in liquid separation have been the solubility of the carrier and/or the solvent in the aqueous phases and progressive wetting of the membrane by the feed and strip phases. To date, there has not been a systematic study delineating these factors in ILM instability encountered in gas separation.

Role of Solvent in the Stability of ILMs

Solvent is the major component in most of the liquid membranes studied in literature. Water has been the solvent of choice for most ILMs studied in literature because of its unique properties, such as solubility of various carriers and ease of wetting. Many organic solvents are studied for recovery of metals from aqueous streams. Vapor pressure of the solvent has the highest impact on the stability of the ILMs in gas separation. However, due to the high vapor pressure of water and its presence in small quantities in the pores, loss of water from the membranes is rapid, leading to failure of the ILM in a short span of time, usually hours. Approaches adapted in literature for improving the lifetime of the ILMs include:

- complete humidification of feed and sweep streams, minimizing the loss of water[8] and
- use of low-volatility and hygroscopic solvents like polyethylene glycol for preparation of the ILM.[9]

However, both these approaches suffer from various limitations[6] and have not found commercial acceptance.

Recently, efforts to improve the stability of ILMs for gas separation have been successful in providing highly stable and selective liquid membranes for carbon dioxide separation and olefin–paraffin separation.[6,10–12] These efforts can be categorized as follows:

- use of carriers that do not require a solvent;
- use of glycerol as a solvent in place of water, which can accommodate carrier species for efficient transport of reacting gas species; and
- use of non-volatile solvent/carrier, which may not require the presence of water for selectivity.

Each of these approaches has its unique characteristics. The experimental procedures followed in these recent efforts are substantially different from those used for water-based liquid membranes. The distinct features of these experiments are:

- Use of a dry sweep gas. This feature ensures that vacuum operations are possible with these liquid membranes. Compared to this, all water-based liquid membranes need humidified feed and sweep gas streams for maintaining stable operation.
- Deliberate oscillations in feed gas humidity conditions, ranging from completely humid to completely dry. This feature was employed to test the *ruggedness* of these ILMs under practical operating conditions, where spikes in feed gas humidity are a reality.
- Most of these ILMs were studied for at least 1–3 months without any regeneration of the liquid membrane. Due to the inherent features of the solvents and carriers used, much longer lifetimes for these liquid membranes are anticipated.

Glycerol was found to be an excellent alternative solvent in place of water for dissolving various salts for CO_2 separation[6,13–15] and olefin–paraffin separation.[11]

Glycerol is a viscous liquid with extremely low vapor pressure and high hygroscopicity. It is an environmentally benign chemical with reasonable solubility for water-soluble carriers. TABLE 1 presents the physical properties of glycerol together with other solvents used for preparing liquid membranes.

Glycerol-based ILMs provided high fluxes and selectivities for the desired species (e.g., CO_2 and olefins) while offering exceptional stability under severe operating conditions. One unique feature of glycerol-based ILMs was their resilience with changing feed side relative humidities and dry sweep gas stream. Under completely humidified conditions on the feed side and dry sweep gas, the glycerol-based ILMs achieved CO_2/N_2 selectivities as high as 7,000 at low CO_2 partial pressures.[13] When the feed gas was made completely dry, the CO_2 and other gas fluxes became negligible. There are two reasons for this: first, the facilitated transport of carbon dioxide requires the presence of water; second, in the absence of water, viscosity of glycerol shoots up drastically and cuts down the permeation of all gases. Once the feed side humidification was restarted, the CO_2 permeance and CO_2/N_2 selectivities were restored to their original levels. This revival of ILM performance is not feasible for water-based ILMs. TABLE 2 presents the performance of the liquid membranes studied recently under completely dry and humidified feed gas conditions and compares them to water-based liquid membranes.

The second category of liquid membranes studied recently are dendrimer-based ILMs.[10,15] Polyamidoamine (PAMAM) dendrimer liquid of generation 0 was used to form the ILM, where it acted both as the carrier for CO_2 separation and solvent as well. The pure dendrimer liquid membranes provided exceptional selectivities (α_{CO_2/N_2} of 18,000 and 700 for CO_2 partial pressures of 0.5 cmHg and 30 cmHg, respectively) for carbon dioxide; the selectivity for the high CO_2 partial pressure is comparable to the highest selectivities reported for polyelectrolyte membranes without the associated problems of membrane formation.[16] It was also found that addition of a small amount of glycerol to dendrimer solutions improved their performance over a broader range of feed side relative humidities.[10,15]

The ILMs we are studying need to be distinguished from multilayer composite (MLC) membranes studied at Air Products.[17] The MLC membrane consists of a thin film of an active transport polyelectrolyte, for example, poly(diallyldimethylammonium fluoride) (DADMAF) contained/sandwiched between two nonporous, highly permeable nonselective support polymeric films of poly(dimethylsiloxane) or

TABLE 1. Comparison of physical properties of various solvents used in ILMs

ILM Solvent	Mol. Wt.	Hygroscopicity	Density (g/cc)	Viscosity (cP)	Vapor Pressure (mmHg)
Water	18	—	1	1	21 (at 23°C)
Glycerol	92	high	1.25	1410	7.5×10^{-5}
PAMAM dendrimer	518	moderate	—	—	—
Glycerol carbonate	118	moderate	1.4	62	137–140°C[a]
PEG 400	400	low	1.124	100	—

[a]Boiling point at 0.5 mm Hg.

poly(trimethylsilylpropyne). These membranes exhibited good selectivity for CO_2 over H_2 (about 50–80) at CO_2 feed pressures in the range 40–85 psi and were stable for a period of three weeks at a transmembrane pressure of 90 psig.

The third category of liquid membranes we have studied is based on glycerol carbonate.[12] Glycerol carbonate is an experimental compound that satisfies most of the requirements for an ideal solvent for carbon dioxide separation. It is non-toxic and non-volatile with moderate viscosity. Glycerol carbonate liquid membranes were found to have CO_2/N_2 selectivities that are independent of the substrate thickness, CO_2 partial pressure, and humidity conditions of the feed and sweep gas streams. Essentially, glycerol carbonate seems to function as a physical solvent with a high affinity for carbon dioxide. These features of glycerol carbonate are important in designing ultrathin liquid membranes with good CO_2 flux and selectivity.

These recent efforts have resulted in improving the stability of ILMs for gas separation tremendously; the ILMs formed were tested for a few months without any loss of stability or performance. Similarly, the approach of using non-volatile solvents/carriers was equally successful in developing highly VOC-selective ILMs for VOC removal from gas streams[18] and aqueous streams.[19] For example, trichloroethylene (TCE) was removed from aqueous streams using hexadecane immobilized liquid membrane in a membrane-based pervaporation technique using an ultrathin silicone rubber-coated microporous hollow fiber membrane. The TCE and water fluxes were studied for a period of four months and were essentially stable.[19]

Role of Carriers in ILM Stability

The selection of a suitable carrier for ILM is a critical aspect of successful operation of an ILM. Most carriers used in ILMs transport the selected species (e.g., CO_2, olefins, and metals) via facilitated transport.

TABLE 2. Performance of various ILMs under varying feed side relative humidities for CO_2–N_2 separation

	Feed RH = 100%			Feed RH = 0%		
	$Q_{CO_2}{}^a$	$Q_{N_2}{}^a$	α_{CO_2/N_2}	$Q_{CO_2}{}^a$	$Q_{N_2}{}^a$	α_{CO_2/N_2}
Glycerol[b]	110	2.85	39	5.6	3.8	1.5
75% Dendrimer–glycerol[c]	3,200	0.19	16,000	5,800	860	7
Pure dendrimer[c]	3,600	0.2	18,000	4,700	930	5
Glycerol carbonate[d]	380	3.6	105	100	1	100
Na_2CO_3–glycerol[b]	1,800	1.6	1,140	2.8	1.2	2.3
Water[b]	2,100	50	42	unstable due to evaporation		

[a] Q_i in units of $10^{-10} cm^2$ (STP)·cm/cm^2·sec·cmHg.
[b] For details on feed and sweep gas compositions see Reference 6.
[c] For details on feed and sweep gas compositions see Reference 10.
[d] For details on feed and sweep gas compositions see Reference 12.

Although most carriers used for gas separation are non-volatile salts, some of the highly promising carriers (particularly for carbon dioxide separation) are volatile amine liquids, such as monoethanolamine (MEA), diethanolamine (DEA), and ethylenediamine (EDA). Our recent efforts focused on using non-volatile amines (such as dendrimers) and salts (such as sodium carbonate, sodium glycinate, and silver nitrate). This approach eliminated one of the main causes of ILM degradation—the loss of carrier due to volatility. The solubility of these salts in the non-aqueous solvents of interest has been reasonable and adequate. For example, sodium carbonate and sodium glycinate are soluble in glycerol up to 2.5 M and 5 M, respectively. However, due to a salting out effect, the permeance of the active species (CO_2 with these carriers) passes through a maximum at salt concentrations below their solubility limits. The role of glycerol carbonate as a solvent for facilitated transport is being studied. It appears to have little solubility for sodium carbonate. However, solutions containing 0.1 M sodium glycinate or 0.127 M PAMAM dendrimer were easily prepared and used. The effect of addition of small amounts of glycerol on the dissolution of salts in glycerol carbonate, if any, is not known.

Role of Substrate in ILM Stability

The substrate employed for immobilizing the liquid membrane has a very important role to play in the stability of the ILM. The pore size and pore size distribution of the substrate determine the critical displacement pressure of the ILM and possibly in progressive wetting of the pores by the feed/strip streams. In general, the ideal substrate for an ILM should have high porosity, smaller pore size and narrow pore size distribution. The chemical nature of the substrate (hydrophilic/hydrophobic and its solvent resistance) is also a determining factor in the ILM stability. As a rule of thumb, most ILMs for gas separation require a hydrophilic substrate, whereas most ILMs for liquid separation employ a hydrophobic substrate. Wetting of the pore surface by the liquid is important for the prevention of gas leakage. However, systematic studies on the role of substrate in the stability of ILMs appear to be scant.[5] Regardless of these factors, chemical stability and absence of swelling by the solvents are prerequisites for successful ILM operation.

COMMERCIAL SUPPORTED LIQUID MEMBRANES

Although this article focuses mainly on the use of liquid membranes for gas separations, supported (immobilized) liquid membranes have been studied extensively for recovering/removing various species of interest from liquid streams. However, the ILMs for liquid separation are inherently more unstable in comparison with gas separation ILMs. Various possible mechanisms were proposed in literature for the causes of SLM instability[7,20] and approaches to optimize the solvent/carrier selection.[21,22] Similarly, efforts were also made to improve the lifetime of ILMs.[23,24] Two recent approaches in preparing stable SLMs for liquid separation should be mentioned for their commercial viability and novelty.

The first approach employs the strip phase dispersed as drops in the extraction phase on one side of the membrane.[25] The solution containing the solvent and carrier fills the membrane substrate pores as well as the strip side. The feed extracting solvent interface is immobilized at the feed side of the membrane via a pressure differential as recommended by Sirkar and coworkers.[26] This concept was demonstrated commercially for extraction of chromium from aqueous waste streams. The second approach, by Facilichem, mentions a proprietary design that allows continual replenishment of liquid membrane for indefinite life.[27] However, few details are available about the design or the mechanism for extended stability of their SLMs.

SLM STABILIZATION AND REGENERATION

One common problem with all liquid membranes, for either gas separation or liquid separation, has been their short lifetime. This often creates the need to provide some type of resistance for the carriers/solvents from leaving the membrane pores, and/or effective methods of regenerating the liquid membranes for their continued use.

Most efforts in stabilizing the SLMs were in applications dealing with liquid separations, where the instability problems are more severe. These efforts can be broadly categorized as follows:

- formation of gel structures with carrier and solvent trapped within the membrane pores,[28] and
- formation of barrier layers on substrate membrane surfaces, either by physical deposition[29] or by interfacial polymerization.[30–32]

These approaches suffer from a number of severe limitations to their commercially viability. ILMs with gel structure may not be a general approach because they are limited by the compatibility of the gel with fluid streams. ILMs prepared with modified substrates by forming barrier layers (e.g., interfacial polymerization) are promising but there are practical issues that need to be addressed before they can be commercially viable.[33] Our group has been working on improving the stability of supported liquid membranes by using various approaches. Our efforts will be discussed in future communications.

FUTURE DIRECTIONS

For immobilized liquid membranes to achieve commercial viability, their stability and lifetime need to be improved significantly. Recent developments in forming stable liquid membranes are highly promising. Additional efforts should include research activity on new solvent and/or carriers for ILM preparation, a fundamental understanding of the role of substrate on ILM stability, and novel ILMs with modified substrate structures.

REFERENCES

1. SIRKAR, K.K. 1992. Other new membrane processes. *In* Membrane Handbook. W.S.W. Ho & K.K. Sirkar, Eds. Chapman and Hall, New York.

2. REED, B.W., M.J. SEMMENS & E.L. CUSSLER. 1995. Membrane contactors. In Membrane Separations Technology, Principles and Applications. R.D. Noble & S.A. Stern, Eds.: 468–498. Elsevier Science, Amsterdam.
3. GABLEMAN, A. & S.T. HWANG. 1999. Hollow fiber membrane contactors. J. Membr. Sci. **159:** 61–106.
4. KOVVALI, A.S. & K.K. SIRKAR. 2003. Membrane contactors: recent developments. In New Insights into Membrane Science and Technology: Polymeric, Inorganic, and Biofunctional Membranes. D. Bhattacharyya & A. Butterfield, Eds. Elsevier Science, Amsterdam. In Press.
5. TAKIGAWA, D.Y. 1992. The effect of porous support composition and operating parameters on the performance of supported, liquid membranes. Sep. Sci. Technol. **27:** 325–339.
6. CHEN, H., A.S. KOVVALI, S. MAJUMDAR & K.K. SIRKAR. 1999. Selective CO_2 separation from CO_2–N_2 mixtures by immobilized carbonate–glycerol membranes. Ind. Eng. Chem. Res. **38:** 3489–3498.
7. KEMPERMANN, A.J.B., D. BARGEMAN, TH. VAN DEN BOOMGAARD & H. STRATHMANN. 1996. Stability of supported liquid membranes: state of the art. Sep. Sci. Technol. **31:** 2733–2762.
8. TERAMOTO, M., K. NAKAI, N. OHNISHI, et al. 1996. Facilitated transport of carbon dioxide through supported liquid membranes of aqueous amine solutions. Ind. Eng. Chem. Res. **35:** 538–545.
9. MELDON, J.H., A. PABOOJIAN & G. RAJANGAM. 1986. Selective CO_2 permeation in immobilized liquid membranes. AIChE Symp. Ser. **248:** 114.
10. KOVVALI, A.S. & K.K. SIRKAR. 2001. Dendrimer liquid membranes: CO_2 separation from gas mixtures. Ind. Eng. Chem. Res. **40:** 2502–2511.
11. KOVVALI, A.S., H. CHEN & K.K. SIRKAR. 2002. Glycerol-based immobilized liquid membranes for olefin–paraffin separation. Ind. Eng. Chem. Res. **41:** 347–356.
12. KOVVALI, A.S. & K.K. SIRKAR. 2002. Carbon dioxide separation with novel solvents as liquid membranes. Ind. Eng. Chem. Res. **41:** 2287–2295.
13. CHEN, H., A.S. KOVVALI & K.K. SIRKAR. 2000. Selective CO_2 separation from CO_2–N_2 mixtures by immobilized glycine–Na–glycerol membranes. Ind. Eng. Chem. Res. **39:** 2447–2458.
14. CHEN, H., G. OBUSKOVIC, S. MAJUMDAR & K.K. SIRKAR. 2001. Immobilized glycerol-based liquid membranes in hollow fibers for selective CO_2 separation from CO_2–N_2 mixtures. J. Membr. Sci. **283:** 75–88.
15. KOVVALI, A.S., H. CHEN & K.K. SIRKAR. 2000. Dendrimer membranes: a CO_2-selective molecular gate. J. Am. Chem. Soc. **122:** 7594–7595.
16. QUINN, R. 1998. A repair technique for acid gas selective polyelectrolyte membranes. J. Membr. Sci. **139:** 97–102.
17. LACIAK, D.V. 1994. Development of novel active transport membrane device. USDOE/ID/12779-3. Report on Contract FC36-89ID12799 by Air Products and Chemicals, Allentown, PA.
18. OBUSKOVIC, G., S. MAJUMDAR & K.K. SIRKAR. 2003. Highly VOC-selective hollow fiber membranes for separation by vapor permeation. J. Membr. Sci. In press.
19. QIN, Y.J., J.P. SHETH & K.K. SIRKAR. 2002. Supported liquid membrane-based pervaporation for VOC removal from water. Ind. Eng. Chem. Res. **41:** 3413–3428.
20. DREHER, T.M. & G.W. STEVENS. 1998. Instability mechanisms of supported liquid membranes. Sep. Sci. Technol. **33:** 835–853.
21. DEBLAY, P., S. DELEPINE, M. MINIER & H. RENON. 1991. Selection of organic phases for optimal stability and efficiency of flat-sheet supported liquid membranes. Sep. Sci. Technol. **26:** 97–116.
22. BARNES, D.E., G.D. MARSHALL & J.F. VAN STADEN. 1995. Rapid optimization of chemical parameters affecting supported liquid membranes. Sep. Sci. Technol. **30:** 751–776.
23. YANG, X.-J. & A. FANE. 1997. Effect of membrane preparation on the lifetime of supported liquid membranes. J. Membr. Sci. **133:** 269–273.
24. NAKANO, M., K. TAKAHASHI & H. TAKEUCHI. 1987. A method for continuous operation of supported liquid membranes. J. Chem. Eng. Jpn. **20:** 326–328.

25. HO, W.S.W. & T.K. PODDAR. 2001. New membrane technology for removal and recovery of chromium from waste waters. Env. Progr. **20**(1): 44–52.
26. KIANI, A.K., R.R. BHAVE & K.K. SIRKAR. 1984. Solvent extraction with immobilized interfaces in a microporous hydrophobic membrane. J. Membr. Sci. **20**: 125.
27. FACILICHEM. 2002. <www.facilichem.com>.
28. NAPLENBROEK, A.M., D. BARGEMAN & C.A. SMOLDERS. 1990. The stability of supported liquid membranes. Desalination **79**: 303–312.
29. WIJERS, M.C., M. JIN, M. WESSLING & H. STRATHMANN. 1998. Supported liquid membranes modification with sulfonated poly(ether ether ketone). Permeability, selectivity and stability. J. Membr. Sci. **147**: 117–130.
30. WANG, Y., Y.S. THIO & F.M. DOYLE. 1998. Formation of semi-permeable polyamide skin layers on the surface of supported liquid membranes. J. Membr. Sci. **147**: 109–116.
31. WANG, Y. & F.M. DOYLE. 1999. Formation of epoxy skin layers on the surface of supported liquid membranes containing polyamines. J. Membr. Sci. **159**: 167–175.
32. KEMPERMANN, A.J.B., H.H.M. ROLEVINK, D. BARGEMAN, *et al.* 1998. Stabilization of supported liquid membranes by interfacial polymerization top layers. J. Membr. Sci. **138**: 43–55.
33. KOVVALI, A.S. & K.K. SIRKAR. 2002. Unpublished work.

The Use of Conducting Polymers in Membrane-Based Separations

A Review and Recent Developments

JOHN PELLEGRINO

Santa Fe Science and Technology, Inc., Santa Fe, New Mexico, USA

ABSTRACT: As a material family, π-conjugated polymers (also known as intrinsically conductive polymers) elicit the possibility of both exploiting the chemical and physical attributes of the polymer for membrane-based separations and incorporating its electronic and electrochemical properties to enhance the separation figures-of-merit. This review article, although by no means comprehensive, provides a current snapshot of the investigations from many research laboratories in the use of conducting polymers for membrane-based separations. The review focuses primarily on polyaniline, polypyrrole, and substituted-polythiophene and includes applications in gas separations, liquid (and/or vapor) separations, and ion separations. Additionally, we discuss the broad challenges and accomplishments in membrane formation from conducting polymers.

KEYWORDS: conducting polymers; membranes; polyaniline; polypyrrole; substituted-polythiophene; separations

INTRODUCTION

Currently, we are not aware of any commercial membrane technology products that use π-conjugated polymers (conducting polymers). Nonetheless, during the past 10–15 years there have been many research studies that have explored the potential for exploiting the unique chemical, physical, and electrical properties of the conducting polymers, polyaniline (PANI), polypyrrole (PPy), and polythiophene (PT) and their derivatives (note: only substituted PTs are stable after doping). These conducting polymers are environmentally robust and have the lowest potential production costs (polyphenylene vinylene is also stable but is also more costly). Although very novel separations uses have been explored and exciting results have been obtained, commercial applications have been slowed for at least two reasons: (1) in general, conducting polymers are difficult to process into useful forms for membranes with common organic solvents and (2) there are few sources of large quantities of reproducible quality polymer suitable for membrane formation. Most membrane formation and subsequent separation studies have been performed with materials synthesized in the laboratory of the individual investigator, which tends to

Address for correspondence: John Pellegrino, Santa Fe Science and Technology, Inc., 3216 Richards Lane, Santa Fe, NM 87507, USA. Voice: 505-474-6656; fax: 505-474-9489.
jjp@sfst.net

cause poorly-reproducible results. In this review we present developments that address both of these obstacles.

BACKGROUND

Electrically conductive polymers and their derivatives have the capacity to intrinsically carry and transport electric charge in the semiconducting to metallic regime.[1] FIGURE 1 presents the nominal repeat unit for representative conducting polymers. All of these monomers may also be derived with a side group that can alter the solubility, chemical, and electronic properties of the resulting polymer. Their intrinsically conductive nature arises from a unique bonding structure along the polymer backbone consisting of alternating double (π) and single (σ) bonds. If an electron is added to (via chemical reduction or n-type doping) or removed from (via chemical oxidation or p-type doping) the conjugated polymer backbone in the chemical or electrochemical *doping* process, and if the polymer conjugation length is sufficiently long, then the charge can freely travel down these conjugation paths when an electrical potential is applied. When doped, the electrical conductivity can vary from semiconducting (10^{-5}–$10^{-1}\,\Omega^{-1}\mathrm{cm}^{-1}$) to metallic ($10^{0}$–$10^{5}\,\Omega^{-1}\mathrm{cm}^{-1}$), depending on the extent of doping. The conductivity achieved is very dependent on the type of dopant, the polymer (such as, the specific repeat unit, molecular mass, polydispersity, and chain defects like branching and chemical heterogeneity), and how it was processed. For example, stretching doped conducting polymer films and fibers can increase their conductivity by two orders of magnitude due to anisotropic alignment of the polymer chains.

For separations uses, it is important to bear in mind that the dopant counter-ion is closely bound to the ionic repeat unit sites of the doped polymer by electrostatic forces that serve to maintain overall charge neutrality of the system. Thus, the volume conductivity is truly electronic in nature not simply ionic, although ionic conductivity may also be implicated in some transport mechanisms. With the notable exception of PANI, all the conjugated polymers require redox (n- or p-type chemical) doping to activate the electrical conductivity pathways. Polyaniline has three oxidation states, and only the emeraldine base (EB) form (shown in FIG. 1) can be chemically doped by a protonic acid (PANI-EB is weakly basic, $pK_b = 8.6$) to become electrically conductive without a change in its oxidation state. The acid polymer becomes undoped by exposure to aqueous base (pH>3) to form the insulating emeraldine base. During this undoping process, the dopant acid anion is removed from the polymer, unless the dopant is immobilized in the polymer matrix. Nevertheless, the doped EB form of PANI can be electrochemically doped and dedoped by electrochemical charge injection and removal with techniques, such as cyclic voltammetry.[2]

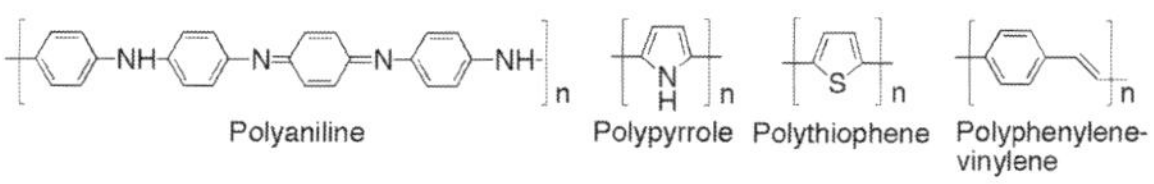

FIGURE 1. Repeat unit of several pertinent conducting polymers.

GAS SEPARATION

PANI

For polymers to show high selectivity for different gases, they must be fabricated into fully dense, nonporous films. This prevents many conducting polymers from being considered for gas separations due to their open fibrillar morphology. On the other hand, it has been shown that dense PANI membranes can be fabricated by dissolving the polymer in its emeraldine base oxidation state and casting the solution to remove the solvent. The resulting free-standing EB films can be subsequently doped with a wide variety of organic and inorganic acids. Earliest work required that the films be at least 10μm thick to avoid micropore defects during formation that lead to Knudsen-type diffusion.[3–6] (When Knudsen diffusion dominates the transport mechanism of the gases through the membrane, the separation factor is equal to the inverse square root of the molecular masses of the two gases.)

Interest in PANI membranes for gas separation stems from the work by Anderson et al.,[3,4,7,8] where the authors reported some of the highest ideal separation factors for O_2/N_2 ($\alpha = 30$), H_2/N_2 ($\alpha = 3,590$), and CO_2/CH_4 ($\alpha = 336$) ever achieved with polymer films. They showed that the permeability of gases through doped PANI membranes depends on the size of the penetrating gas based on the kinetic diameter, with values ranging between 20 Barrers (1 Barrer = 10^{-10} cm^3(STP)/cm·sec·cmHg) for the smallest gas, He (2.6 Å), to 0.03 Barrers for the largest gas tested, CH_4 (3.8 Å). The only exception from this trend was oxygen, since its permeability was actually higher than Ar even though O_2 (3.46 Å) is a slightly larger molecule than Ar (3.4 Å). This was later shown to be due to a relative high solubility of O_2 in PANI films, although O_2 and Ar have similar solubilities in PANI powder.[9]

The enhanced permeation of O_2 over N_2 suggests that PANI membranes could be effective in separating O_2 from air, which is especially challenging since N_2 (3.64 Å) is only slightly larger than O_2. The as-cast PANI-EB (not doped) film had an ideal O_2/N_2 selectivity of 9, which is comparable to commercially available polysulfone and polyimide membranes ($4 < \alpha < 8$). After the membrane was fully doped with HCl, the O_2 and N_2 permeability of the membrane is lowered. This is presumably due to dopant molecules changing the free volume distribution of the polymer. After dedoping (removal of the acid dopant molecules) by treatment with base, higher permeation was obtained for both N_2 and O_2. However, the O_2/N_2 ideal permselectivity was identical as to that obtained for the as-cast membrane. Partial redoping of the membrane with HCl lowered the permeability of both gases; however the N_2 permeability decreased more than did the O_2 permeability and, consequently, a selectivity of 30 was observed.

Mattes et al.[10] suggested that this phenomenon is related to selective transport of para-magnetic oxygen with the free electrons in the doped polymer. It was latter shown by Rebattet et al.,[11,12] from electron spin resonance (ESR) studies, that the polaron-O_2 interactions are the dominating mechanism for O_2 permeability in the HCl-redoped membrane. For a PANI membrane that was doped with HCl, dedoped with ammonium hydroxide and redoped with HCl, in accord with Reference 3, the permeability of O_2, H_2, and CO_2 increased by approximately 15%, whereas a 45% decrease was observed in the permeability of N_2 and CH_4.

Like Rebattet et al.,[11,12] Kang et al.[13] also reported EPR measurements of O_2 transport through an HCl-doped PANI film. They observed that the intensity of the polaron spectra decreased when O_2 was exposed to the film confirming that its paramagnetic property was lost due to a specific interaction with O_2. However, this interaction is completely reversible—thus supporting a contention of facilitated diffusion. They measured the pure gas permeabilities using the constant-volume, time lag method and confirmed the O_2 and N_2 permeation results reported by Anderson et al.[3,4] The highest O_2/N_2 ideal selectivity reported by Kang et al.[13] was 12 (the O_2 permeability was about 0.03 Barrers) for a dopant concentration 0.015 M (at higher dopant levels the N_2 permeability was below their detection limits). They concluded that since O_2 permeability decreased with increasing dopant (and consequently polaron) density, the facilitated transport mechanism was not a significant contributor to the high O_2/N_2 selectivity. When results of X-ray diffraction d-spacing measurements were considered, they argued that the high O_2/N_2 (and other reported gas pairs) selectivity is a result of very precise structural changes upon redoping and not specific interactions.

The results of Anderson et al.[3] and Rebattet et al.[12] were later questioned by Illing et al.,[6] who examined the permeability of several gas pairs through dense PANI membranes prepared from commercially available Ormecon PANI dispersed in NMP (this is a suspension made by emulsion polymerization). The separation factors based on these PANI membranes were independent of the membrane thickness once Knudsen diffusion due to the presence of micropores in thin membranes (less than 10 µm thick) was eliminated. However, they found separation factors for H_2/N_2 (151), CO_2/N_2 (17), and O_2/N_2 (5) to be significantly lower than previously reported, but acknowledged that the membrane fabrication process may have caused these discrepancies. Kaner[14] later commented that the lower separation factors for the dedoped membrane may be associated with extended curing times and high temperatures in the membrane formation process. Excessive heating leads to crosslinking between the PANI chains that will result in minimal expansion in the free volume with doping and dedoping, as confirmed recently by solid state NMR spectroscopy. The as-cast membranes prepared by Anderson et al.[3,4] were dried for one to three hours at 135°C, whereas the cast membranes prepared by Illing et al.[6] were dried at 80°C for 12 hours and subsequently annealed at 125°C for one hour.

Many researchers have attempted to reproduce the high O_2/N_2 selectivity factor reported by Anderson et al.,[3] and at the same time produce highly permeable membranes. The most common approach is to deposit a highly selective thin PANI film onto a highly permeable support membrane. This approach was first used by Kuwabata and Martin,[5] where the authors prepared thin PANI films on alumina support membranes that showed a remarkable increase in the total gas permeation rate. The thickness of the as-cast emeraldine base films was shown to make a difference on the diffusion behavior of gases through the membrane. O_2/N_2 selectivity values less than 1 were obtained when the thickness of the emeraldine base film was less than 1.5 µm, which indicates Knudsen-type diffusion that arises from microporous defects in the membrane. The selectivity factor increased to 9.5 when the thickness of the neutral PANI-EB exceeded 3 µm, which is the same value that was reported by Anderson et al.[3] Acid doping of the composite membrane with aqueous HCl resulted in decreased permeation of O_2 and N_2 through the membrane. The selectivity increased from 9.5 for the as-cast membrane to 15.2 for the fully doped membrane.

Other researchers have also used composite membranes with a PANI layer to improve gas permeability through the membrane. For example, Lee et al.[15] used a porous nylon support membrane to improve the permeation of both N_2 and O_2 through a partially redoped PANI membrane with an ideal separation factor of 28. Alternatively, Li et al.[17] fabricated multilayer composite membranes by blending ethylcellulose with either poly(o-toluidine) or poly(aniline-co-o-toluidine) and casting the solution onto traditional porous membrane supports. The introduction of the methyl side chain to the phenylene ring in PANI enhances the polymer interchain free volume and the resulting membrane exhibited higher O_2 permeation rates. The O_2 permeation through this poly(o-toluidine)/ethylcellulose composite membrane was 27 times larger than for the PANI membranes reported by Anderson et al.,[3,4] but the latter had an ideal selectivity factor 1.5 times larger than the composite membrane. (The higher permeance for the composite was likely due to the lower thickness of its permselective layer.)

The effect of the dopant size on the gas permeation rates were initially studied by Anderson et al.[3] for the halogen acid series HF, HCl, HBr, and HI. The authors doped the as-cast membrane with these acids, followed by subsequent dedoping with ammonium hydroxide to remove the dopant anion. It was found that the permeation of several gases (He, H_2, Ar, O_2, and N_2) decreased from HF to HCl to HBr to HI. Although counterintuitive to the actual size of the anion in a crystal lattice, this trend is consistent with the size of the solvated halogen acids, in which HF is actually the largest solvated acid in this series. The use of other dopants can also lead to high selectivities.[5] For example, a 36 mass% sulfuric acid doped PANI membrane possessed an O_2/N_2 selectivity of 13.4, whereas a membrane doped with p-toluenesulfonic acid had an O_2/N_2 selectivity of 14.9. However, it was found by Lee et al.[16] that doping the membrane with substantially larger dopant molecules, such as dedocylbenzenesulfonic acid, increased permeability of the membrane but the selectivity decreased.

In contrast to doping the PANI membrane with small molecule acids, PANI membranes for O_2/N_2 separation have also been prepared in which the polymer is doped using a polymeric dopant.[7] The polymeric dopant used in this study was poly(amic acid) (a precursor to polyimides), which has been extensively studied for gas separations. Relative to a pure PANI membrane, gas permeability increased for the polymer blend membranes but maintained (and in some cases increased) the selectivity from that of the polyimide alone.

Although most of the emphasis in PANI membranes for gas separation has focused on the inconsistent O_2/N_2 ideal selectivity results, most reports do acknowledge that for H_2 separations, PANI films outperform (by several-fold) conventional membrane materials.[6]

PPy

The use of PPy membranes for gas separation was investigated soon after the initial publications about PANI membranes. Because PPy is insoluble in all organic solvents, PPy membranes are either prepared by electrochemical polymerization of the monomer onto a large platinum or stainless steel electrode or by growing the polymer chemically on an existing membrane substrate. The latter approach was used by Liang and Martin[18] to deposit a thin PPy coating onto an alumina support membrane using an interfacial polymerization technique with iron (III) nitrate as the oxidant. Their doped PPy was highly porous, and consequently, Knudsen diffusion was observed with an O_2/N_2 selectivity of 0.94. However, using a different pyrrole monomer, a poly(N-methylpyrrole) film deposited onto the alumina support membrane using the same technique possessed good permeation rates and selectivity. The selectivity for O_2/N_2 was 7.9 for the doped polymer, and this decreased slightly to 6.2 for the undoped polymer. The undoped polymer showed significantly higher permeability, and the selectivity factor for CO_2/CH_4 doubled as the poly(N-methylpyyrole) was chemically reduced to its neutral form. These results indicate that permeability and selectivity of poly(N-methylpyyrole) membranes can also be altered by doping and undoping. Similar results were obtained by Kamada et al.[19,20] using a ceramic filter (Vycor glass oxide) coated with poly(N-methylpyyrole). However, the O_2/N_2 separation factor was only 3.9 for the doped poly(N-methylpyyrole) membrane they prepared.

Subsequent work by Martin et al.[21] showed that the low gas-transport selectivity of PPy membranes is attributed to the microporous morphology obtained when thin films of this polymer are deposited onto the support membrane by conventional polymerization methods. Dense PPy membranes can be prepared by slowing down the rate of the polymerization reaction by adding iron (II) chloride to the reaction medium. Dense PPy films were deposited onto a polycarbonate support membrane and subsequently treated with base to yield a partially oxidized PPy film. This membrane showed extraordinary gas-transport properties with an O_2/N_2 selectivity of 18. When the partial pressure of O_2 in the feed was 0.1% (99.9% N_2), the O_2/N_2 selectivity coefficient increased to 92, which is the highest value reported in the literature for polymeric materials. Unfortunately, this work did not report flow rates (only compositions) so the permeance of the membrane materials were unavailable.

Andreeva et al.[22] recently reported on the fabrication of PPy-coated sulfonated poly(phenylene oxide) membranes. A dense PPy layer was formed by saturating the support membrane with pyrrole in the vapor phase, followed by polymerization of the sorbed pyrrole in an aqueous solution of the oxidant. The O_2/N_2 selectivity coefficient of the PPy layer was 15, which is one of the highest values for fully-doped PPy found in the literature. Unfortunately, the two layer composite membranes had lower selectivities (α_{O_2/N_2} = 8.1 with P_{O_2} = 0.42 Barrers). Yilmaz and coworkers[23] have tried to exploit the apparent intrinsic gas transport properties of PPy by fabricating mixed-matrix membranes—these mixed-matrix membranes were fabricated as two interspersed matrices of different materials—based on PPy interspersed in a polycarbonate continuum. Their best results (α_{CO_2/N_2} = 5.9, α_{O_2/N_2} = 4.2, α_{H_2/CH_4} = 5.2 and permeabilities for CO_2, O_2, and H_2 of 11.6, 8.2, and 19.5 B, respectively) were for a formulation with p-toluenesulfonic acid as the dopant anion.

Substituted PT

Although the majority of conducting polymer membranes for gas separation have been fabricated from either PANI or PPy, there have been a few scattered reports using other conducting polymers. As mentioned already, there are difficulties with obtaining fully dense layers from some conducting polymers because of the tendency to form a porous morphology that leads to Knudsen diffusion. For example, Musselman et al.[24,25] studied gas transport through poly(3-substituted thiophenes) membranes. Although solution cast poly(3-alkylthiophenes) membranes, such as poly(3-octylthiophene) and poly(3-dodecylthiophene), exhibit high permeability, these membranes suffer from low selectivity values. For example, a dense poly (3-dodecylthiophene) membrane cast from the polymer in tetrachloroethylene, possessed an O_2/N_2 selectivity factor of 2.2, with an O_2 permeability of 20.2 Barrers. Doping the membrane with $SbCl_3$ resulted in the polymer adopting a more rigid structure, and consequently, the permeability significantly decreased and the selectivity increased slightly (P_{O_2} = 10.5 Barrers and α_{O_2/N_2} = 3.0.) Subsequent work[25] has shown that the selectivity of poly(3-substituted thiophene) membranes can be enhanced by using an alkylester substituent. The O_2/N_2 selectivity of a poly(3-(2-acetyloxyethyl) thiophene) membrane was 5.1, but this membrane exhibited two orders of magnitude lower permeation rates than the poly(alkylthiophene)-based films. Hydrolysis of the ester side chain resulted in a further decrease in permeability, but this was accompanied by a dramatic increase in the selectivity of the membrane. The O_2/N_2 selectivity increased to 12.9 for the base-treated membrane and to 11.7 for the acid-treated one.

LIQUID–LIQUID SEPARATIONS

Since most liquids are composed of molecules with similar dimensions to gaseous molecules (2–10 Å), both PANI and PPy membranes have been studied for liquid–liquid separations using pervaporation. In pervaporation applications, it is important to remember that PANI membranes can be tailored after fabrication by doping (undoping) with a wide variety of acids (bases), whereas PPy membranes are only tailored by oxidizing or reducing the polymer using an appropriate electrical potential. Thus, when PANI membranes in the emeraldine base oxidation state are used in the presence of acids, they partially dope the polymer, with the degree of doping depending on the pH of the feed solution. However, when a predoped PANI membrane is used, dopant exchange can occur based on the relative interactions (including acidity and anion mobility) between the PANI and the species present.

PANI

The separation of carboxylic acids from aqueous solutions using PANI membranes has been extensively studied by Kaner et al.[26–28] Their results indicate that fully doping the PANI membrane with HCl has a dramatic effect on their selectivity. For example, undoped PANI displays relatively poor selectivities for separating a mixture of 50/50 (g/g) acetic acid–water (α = 2.4), but fully HCl-doped PANI selectively permeates water over acetic acid with a separation factor of 1,300. This places the doped PANI among the most selective membranes reported for separating acetic acid/water mixtures. Preferential water transport through doped PANI is most likely associated with a combination of favorable diffusion, due to the small size of water, and favorable solubility, due to the hydrophilicity of the PANI membrane induced by doping. Whereas formic acid (solvated diameter 3.5 Å) can permeate through a doped PANI membrane, larger carboxylic acids, such as acetic acid (4.5 Å), have an exceedingly low rate of permeation or can be virtually excluded, as was observed for propionic acid (5.5 Å). Although the water permeation rates from acetic acid/water feeds through dense, as-cast PANI membranes are relatively low, a two-fold or more increase in the permeation rate can be obtained by using asymmetric membranes of approximately the same overall thickness. Increasing the temperature of the feed solution can also lead to large increases in permeation rate. For example, the permeation rates from a 50/50 (g/g) acetic acid–water feed solution through doped PANI membranes increased two-fold on going from room temperature to 85°C.

One of the major drawbacks encountered with HCl-doped PANI membranes is that their performance deteriorates over time, which is associated with the HCl dopant molecules leaching out of the membrane into the feed solution due to prolonged exposure to water. This phenomena has also been observed in the separation of alcohol/water mixtures.[27] When the feed solution is not liable to solvate the HCl molecules (e.g., pure ethanol), the permeability of the membrane remains constant. To overcome this dopant-leaching problem, the PANI membrane can be doped with a polymeric dopant that cannot leach out of the membrane into the feed solution despite prolonged exposure to water. For example, Lee et al.[29] used a PANI membrane doped with poly(acrylic acid) for the separation of isopropanol from water, and Ball et al.[27] used a PANI membrane doped with either poly(acrylic acid) or poly(amic acid) for separation of primary and secondary alcohols. Although the permeation through these membranes with polymeric dopants is higher, their selectivity is lower, which is a common trade-off in membrane science.

PPy

In contrast to the work published on PANI membranes for pervaporation, there have been only few scattered reports on using PPy membranes. Sarrazin et al.[30,31] prepared polypyrrole membranes doped with either hexafluorophosphate or p-toluenesulfonate by electrochemical deposition of the polymer onto stainless steel mesh. Pervaporation studies using these PPy membranes revealed that high pervaporation rates could be obtained for organic pairs based on differences in polarity, hydrophilicity, and molecular size. In particular, they focused on the extraction of methanol from a wide variety of organic solvents including toluene, isopropanol, methyl tert-butyl ether and acetonitrile. Both types of PPy membranes displayed preferential permeation of methanol in all of these organic pairs, with the highest selectivity ($\alpha = 590$) obtained for the methanol/toluene pair using the hexafluorophosphate-doped PPy membrane. Subsequent work[30] involved using the hexafluorophosphate-doped PPy membrane for the separation of ethanol–cyclohexane by pervaporation. Ethanol preferentially permeates, but the permeation fluxes and selectivity are lower when the polymer is electrochemically reduced to its neutral form. It is postulated that the small primary alcohols are transported through the charged sites of the membrane by a push–pull mechanism due to an electron exchange between the molecules and the polarons on the polymer chain. The larger, less polar organic molecules are transported through the amorphous regions of the membrane according to the solution–diffusion mechanism.

Instead of using flat PPy membranes, Martin et al.[32] developed novel hollow fiber composite membranes for pervaporation in which PPy, poly(N-methylpyyrole), and PANI are chemically polymerized onto a microporous cellulose hollow fiber membrane. Pervaporation data for the separation of methanol from methyl *tert*-butyl ether indicates that the selectivity of the uncoated, microporous cellulose hollow fiber membrane is 1.06. This is attributed to the more polar methanol being more soluble in the cellulose membrane than methyl *tert*-butyl ether. The PPy-coated hollow fiber showed slightly enhanced transport of methanol versus that of the uncoated fiber (the selectivity increased to 1.4) probably limited by the highly porous nature of the PPy film. In contrast, a dense poly(*N*-methylpyyrole) coating on the hollow fiber improved the selectivity to 3.9, whereas the highest selectivity was obtained for the PANI-coated hollow fiber since it showed a selectivity of 4.9.

SELECTIVE ION TRANSPORT

The most widely studied application of conducting polymer membranes is their use for selective ion transport because of the ability to dynamically control ion mobility through the membrane by applying an appropriate electrical signal. With the application of various electrical potential waveforms, the permeation and selectivity of the membrane can be dynamically changed. Furthermore, the electroactivity and selective nature of these membranes can be tailored by synthesizing the polymer with different anions. The majority of ion transport membranes fabricated from conducting polymer have used either PPy or PANI, since they are relatively easy to synthesize. Unlike PANI membranes that can only function in acidic solutions (pH < 3), PPy membranes can operate in neutral, acidic, and basic solutions. The electronic transport of molecules through the conducting polymer membrane involves the incorporation/expulsion of both anions and cations because the membrane is electrochemically altered between its reduced neutral state and its doped oxidized state. The relative amount of anions and cations transported through the membrane depends on the size and mobility of the dopant anion initially incorporated in the conducting polymer membrane. As this dopant anion becomes more immobile, the movement of anions through the membrane decreases and cation transport is observed. In this situation, the immobile dopant molecules are not expelled from the membrane into the electrolyte upon polymer reduction, and consequently, cations from the electrolyte are incorporated into the membrane to maintain charge neutrality.

PPy

Although a wide range of anions can, in principle, be used as the dopant anion for the fabrication of PPy membranes, the required electroactivity and mechanical properties of the resulting free standing membrane severally limits the choice of anions for electrochemical synthesis. For example, the dodecylsulfate anion has excellent cation exchangeability with minimal leaching when incorporated into a PPy membrane,[33] but the electrochemically deposited dodecylsulfate-doped PPy membrane has poor mechanical properties.[34] A common approach to overcoming this problem is to deposit the PPy coating on a microporous substrate. For example, PPy has been chemically deposited on a wide range of polymeric supports including polyethylene,[35] polycarbonate,[36,37] nylon,[37] and poly(ethylene terephthalate)[38] microporous membranes or commercially available ion exchange membranes such as Nafion[39] and Neosepta.[40,41] Alternatively, PPy has been electrochemically deposited on gold or platinum-sputter-coated polymeric membranes.[42–47]

The concept of selective ion transport through conducting polymer membranes was initially reported by Burgmayer and Murray,[48,49] where they demonstrated that a PPy-coated mircroporous gold membrane could act as an ion gate. The PPy was electrochemically deposited onto the gold membrane, and the oxidation state of the PPy film was electrochemically altered before placing the membrane in a transport cell. They demonstrated that altering the oxidation state of the membrane influences the permeability of several ionic species through this membrane. For example, the flux of chloride ions was 1,000 times greater for the doped polymer than for the reduced neutral polymer. It was also observed that, compared to the transport of the chloride ion, permeation of the potassium ion through the doped PPy membrane was negligible. Subsequent work by Wang et al.[50] has demonstrated cation exchange through a conducting polymer membrane. Using cyclic voltammetry, they studied the movement of different cations through PPy membranes between two electrolyte solutions and found that the ion transport depends on the size and charge of the cation. Subsequent work by Wallace and coworkers[34,43,47,51–55] expanded this concept by demonstrating that the most efficient method for inducing ion transport through conducting polymer membranes is to apply a square wave potential to the membrane, which repeatedly reduces and oxidizes the membrane. The length of time (pulse width) at which the potential is held at a particular value must be sufficient to allow the ion species to be incorporated and expelled.

Although the majority of the research into ion-exchange membranes is focused on membrane fabrication and transport of particular ions through the conducting polymer membrane, recent studies have demonstrated that these membranes can be used for selective ion transport. Using a two compartment stirred cell, Mirmohseni et al.[51] examined the separation of sodium and potassium ions from an electrolyte solution with an electrochemically deposited PPy membrane doped with p-toluenesulfonate using different electrical stimuli. The p-toluenesulfonate dopant anion has restricted mobility within the PPy membrane. When no potential was applied to the membrane, no transport of the sodium and potassium ions was detected. Application of negative potentials to the membrane resulted in cation transport being instigated. Although both cations competed for the transport sites in the membrane, the potassium ions had a higher flux through the membrane. The separation is based on the fact that the sodium and potassium ions are incorporated and/or expelled at slightly different potentials, as was observed for the peak currents in their respective cyclic voltammograms. By optimizing the pulse width (20 sec) and the pulse potential range (-0.5 to -1 V), a selectivity factor of 9.1 was obtained for the separation.

It was later commented by Partridge[56] that a disadvantage of this approach is that the ions are selectively absorbed from the feed solution, but simultaneously desorbed into both compartments, which limits the flux of ions through the conducting polymer membrane. To overcome this problem, Partridge[56] achieved mono-directional flow of cations through a p-toluenesulfonate doped PPy membrane by applying a potential gradient across the membrane. The gradient was achieved by using a three-electrode electrochemical cell in both the feed and permeate compartments, with the PPy membrane as the common working electrode that separates both compartments.

This configuration showed a higher flux for both potassium and sodium ions than the results published by Mirmohseni et al.,[51] but the selectivity was reduced to 3.5.

Another approach developed by Zhao et al.[34] was the use of an asymmetric PPy membrane in a stirred electrochemical cell. A bilayer of various PPy films (p-toluenesulfonate-doped polypyrrole and docedylsulfate-doped polypyrrole) was electrochemically deposited onto a platinum electrode. For the separation of sodium and potassium ions by applying square wave pulses to the membrane, this highly asymmetric membrane had a ratio of up to 35:1 in terms of the flux of these cations in one direction as compared with the other. This asymmetric transport of cations arises from the different surface chemistry, electroactivity and morphology of the various PPy films. Using this membrane a reasonable selectivity factor of up to 4.5:1 was obtained for the separation of potassium from sodium ions.

To overcome possible boundary-layer, mass transfer resistances in stirred electrochemical cells, Davey et al.[43] used a crossflow cell in which the feed and permeate solutions were continually cycled by pumping between the cells and larger solution reservoirs. The mechanical properties of electrochemically-deposited PPy membranes, co-doped with poly(styrenesulfonate) and dodecylbenzenesulfonate, were not sufficient to withstand the applied hydrodynamic forces and ruptured. Consequently, they deposited the doped-PPy film onto a platinum-sputter-coated, polyvinylidenefluoride (PVDF) microporous membrane. Compared to the stirred cell with a pulsed electrical potential protocol, a three- to four-fold increase in the flux rates of sodium, potassium and calcium ions was observed in a crossflow configuration (with a constant potential gradient applied across the membrane.)

An interesting approach developed by Misoska et al.[47] for improving the selectivity of the membrane is to dope the PPy membrane with chelating ligands. For example, a doped-PPy film containing the chelating ligand bathocuproinedisulfonic acid, that was electrochemically deposited onto platinum-sputter-coated PVDF membranes, proved to be permeable to alkali metal ions, alkali earth metal ions, and transition metal ions under the influence of an applied electrical potential. The flux of transition metal ions, such as copper, across the membrane was 50-fold higher than for PPy membranes doped with non-chelating acids.

In contrast to the number of studies on selective cation separations, research involving selective anion separation through conducting polymer membranes is fairly limited. Using a p-toluenesulfonate-doped PPy membrane, Mirmohseni et al.[52] showed selective transport of functionalized sulfonated aromatics across the membrane. A square wave potential was used to first incorporate the sulfonated aromatics from the feed solution when the polymer is in its oxidized state, and release the molecules into the permeate solution upon reduction of the PPy membrane. High selectivity factors could be obtained for this membrane for certain pairs of sulfonated aromatics; for example, a selectivity of 27 for the separation of p-toluenesulfonate from 3-sulfobenzoate was obtained.

Separation of proteins using a PPy-coated platinized PVDF substrates was demonstrated by Zhou et al.[46] The PPy membrane doped with p-toluenesulfonate faced the feed solution and platinum was sputter-coated on the other side of the PVDF support. When a constant current was applied across the membrane this configuration acted as an electrophoretic system with a high electric field. Using this configuration with a negative current applied to the feed side of the membrane (the feed solution was negatively-polarized), separation of human serum albumin from myoglobin was accomplished in an aqueous solution at pH 6.5 (at this pH human serum albumin is negatively charged and myoglobin is positively charged)—the human serum albumin transported through the membrane to the permeate solution while retaining the myoglobin in the feed solution. A selectivity exceeding 6 was obtained for this separation. The PPy coating also facilitates different selectivities by incorporating different functional groups into the polymer and changing the hydrophobicity of the surface layer. Using this concept, Lee et al.[36] separated amino acids using a microporous polycarbonate membrane coated on both sides using PPy doped with p-toluenesulfonate. By applying a negative potential across the membrane, positively charged L-lysine and negatively charged L-aspartic acid were separated from their binary mixture at pH 4.7.

PANI

Laboratory studies have usually been performed on dense PANI film due to its good mechanical properties. PANI membranes are generally fabricated by dissolving the polymer in N-methyl-2-pyyrolidone[51,57,58] and casting the solution onto a glass substrate. The resulting freestanding PANI membrane is peeled off the glass plate and subsequently doped in an acidic solution. Selective ion transport through PANI membranes is much less studied than for PPy membranes. Mirmohseni et al.[51] observed that proton transport through PANI membranes in the emeraldine oxidation state from acidic solutions is highly dependent on the nature of the acid employed. Large fluxes through the membrane were observed for sulfuric and nitric, whereas the transport of perchloric acid was negligible. Reducing the polymer to its leucomeraldine oxidation state (that is, it is fully reduced) by applying a negative potential (-0.1 V) to the membrane, proton transport through the membrane was completely halted. On application of a positive potential ($+0.65$ V) that reoxidizes the polymer beyond the emeraldine oxidation state, higher fluxes were observed than when the polymer was in the emeraldine oxidation state. The reason for this increase in permeability of the membrane arises from the larger number of imine nitrogens present in the polymer that can interact with the acid molecules. Similar work by Wen and Kocherginsky[57,58] shows that doped PANI membranes become anion selective with larger permeabilities observed for monovalent ions than for multivalent anions. This selectivity is presumably related to the hydration volume of the anion in the feed solution. However, the effect of electrical stimuli on the selectivity of PANI membranes has not yet been considered.

ADVANCES IN POLYANILINE MEMBRANE FORMATION

Clearly, one of the major limitations for exploiting the possibilities of conducting polymer membranes is the low permeability of free-standing dense films. The challenge of developing asymmetric membranes for improving gas permeability through polyaniline membranes was initially addressed by Mattes et al.[59] In this preliminary study, asymmetric, hollow polyaniline fibers were prepared in an attempt to increase the permeance of oxygen through the membrane by decreasing the thickness of the selective skin.

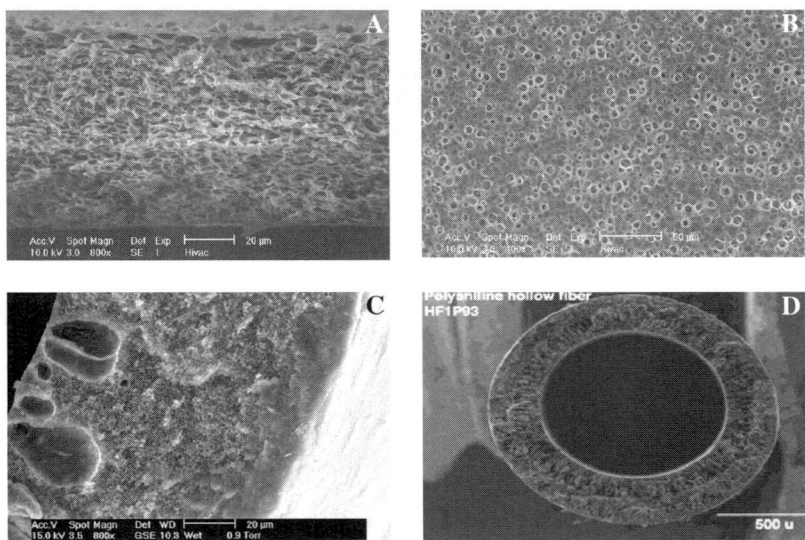

FIGURE 2. Membrane structures from PANI: **(A)** flat sheet microfilter, cross-sectional view; **(B)** flat sheet microfilter, top view; **(C)** integrally-skinned asymmetric membrane (ISAM) hollow fiber, close up of inner surface; **(D)** ISAM, full cross-sectional view.

Building on this initial work, their approach has focused on the commercial-scale processing of polyaniline into porous membrane materials for gas and liquid separations. In particular, the solution processing of polyaniline into integrally-skinned asymmetric membrane structures (ISAMs) is challenging due to unique issues associated with the control of the phase inversion process for creating porous structures from concentrated polyaniline solutions. Concentrated polyaniline solutions are inherently difficult to process, due to their rapid gelation when the concentration exceeds about 6% (mass). However, they have demonstrated that addition of gel inhibitors to concentrated polyaniline solutions can stabilize these solutions for more than 30h.[60–63] This increased processing window enables the large scale processing of polyaniline into both ISAM hollow fiber and flat sheet membranes.

The mesoscopic morphology of immersion-precipitated polyaniline membranes determines both their mechanical properties and usefulness for molecular discrimination. The presence of macrovoids caused by fast precipitation kinetics in the coagulation bath lowers the mechanical properties of the membrane and may lead to defects in the selective layer (which renders the membranes unviable for gas and many vapor separations applications). However, for some applications, a small amount of macrovoids developed under the skin in asymmetric membranes can enhance their permeability without sacrificing the selectivity of the membrane. By investigating the effect of a wide variety of processing parameters, such as solution composition, coagulants, coagulation temperatures, and spinning conditions on the morphology of immersion-precipitated polyaniline membranes, they have been able to control the extent of macrovoid formation. For the fabrication of polyaniline

ISAMs, the thickness and porosity of the *skin layer* can be controlled to provide a range of morphologies from a highly permeable interface to a *barrier layer.* FIGURE 2 illustrates the range of porous structures for both flat sheet and hollow fiber membranes that have been developed in our laboratory. For the first time, asymmetric polyaniline membranes with an ultrathin, defect-free skin layer are being produced on a continuous basis.

ACKNOWLEDGMENT

This work was supported by funds provided by DAPRPA/DSO under contract NBCHC020069 for which the author is thankful.

REFERENCES

1. SKOTHEIM, T.A., R.L. ELSENBAUMER & J.R. REYNOLDS. 1998. Handbook of Conducting Polymers. Marcel Dekker Inc.
2. LU, W., A.G. FADEEV, B. QI, *et al.* 2002. Use of ionic liquids for conjugated polymer electrochemical devices. Science **297:** 983–987.
3. ANDERSON, M.R., B.R. MATTES, H. REISS, *et al.* 1991. Conjugated polymer films for gas separations. Science **252:** 1412–1415.
4. ANDERSON, M.R., B.R. MATTES, H. REISS, *et al.* 1991. Gas separation membranes. A novel application for conducting polymers. Synthetic Metals **41:** 1151–1154.
5. KUWABATA, S. & C.R. MARTIN. 1994. Investigation of the gas-transport properties of polyaniline. J. Membr. Sci. **91:** 1–12.
6. ILLING, G., K. HELLGARDT, R.J. WAKEMAN, *et al.* 2001. Preparation and characterization of polyaniline based membranes for gas separation. J. Membr. Sci. **184:** 69–78.
7. MATTES, B.R., M.R. ANDERSON, J.A. CONKLIN, *et al.* 1993. Morphological modification of polyaniline films for the separation of gases. Synthetic Metals **55–57:** 3655–3660.
8. CONKLIN, J.A., M.R. ANDERSON, H. REISS, *et al.* 1996. Anhydrous halogen acid interaction with polyaniline membranes: a gas permeability study. J. Phys. Chem. **100:** 8425–8429.
9. PELLEGRINO, J.J., R. RADEBAUGH & B.R. MATTES. 1996. Gas sorption in polyaniline. 1. Emeraldine base. Macromolecules **29:** 4985–4991.
10. MATTES, B.R., M.R. ANDERSON, H. REISS, *et al.* 1992. The separation of gases using conducting polymer films. *In* Intrinsically Conducting Polymers: an Emerging Technology. M. Aldissi, Ed. Kluwer Academic Publishers. The Netherlands.
11. REBATTET, L., E.M. GENIES, J.-J. ALLEGRAUD, *et al.* 1993. Evidence of oxygen-polaron interactions to explain the high selectivity in oxygen/nitrogen gas permeation experiments. Polym. Adv. Technol. **4:** 32–37.
12. REBATTET, L., M. ESCOUBES, M. PINERI, *et al.* 1995. Gas sorption in polyaniline powders and gas permeation in polyaniline films. Synthetic Metals **71:** 2133–2137.
13. KANG, Y.S., H.J. LEE, J. NAMGOONG, *et al.* 1997. Oxygen transport through electronically conductive polyanilines. Polym. Mater. Sci. Eng. **76:** 437–438.
14. KANER, R.B. 2002. Gas, liquid and enantiomeric separations using polyaniline. Synthetic Metals **125:** 65–71.
15. LEE, Y.M., S.Y. HA, Y.K. LEE, *et al.* 1999. Gas separation through conductive polymer membranes. 2. Polyaniline membranes with high oxygen selectivity. Ind. Eng. Chem. Res. **38:** 1917–1924.
16. LEE, Y.K., S.Y. HA, Y.M. LEE, *et al.* 1996. Gas separation through conductive polymer membranes. 1. Effect of dopants on properties and gas separation of polyanilines. Membr. J. **6:** 258–264.

17. LI, X.-G., M.-R. HUANG, G.-F. GU, et al. 2000. Actual air separation through poly(aniline-co-toluidine)/ethylcelluidine blend thin-film composite membranes. J. Appl. Polym. Sci. **75:** 458–463.
18. LIANG, W. & C.R. MARTIN. 1991. Gas transport in electronically conductive polymers. Chem. Mater. **3:** 390–391.
19. KAMADA, K., J. KAMO, A. MOTONAGA, et al. 1994. Gas permeation properties of conducting polymer/porous media composite membranes. I. Polym. J. **26:** 141–149.
20. KAMADA, K., J. KAMO, A. MOTONAGA, et al. 1994. Gas permeation properties of conducting polymer/porous media composite membranes. II. Polym. J. **26:** 833–839.
21. PARTHASARATHY, R.V., V.P. MENON & C.R. MARTIN. 1997. Unusual gas-transport selectivity in a partially oxidized form of the conductive polymer polypyrrole. Chem. Mater. **9:** 560–566.
22. ANDREEVA, D.V., Z. PIENTKA, L. BROZOVA, et al. 2002. Effect of polymerization conditions of pyrrole on formation, structure and properties of high gas separation thin polypyrrole films. Thin Solid Films **406:** 54–63.
23. GULSEN, D., P. HACARLOGLU, L. TOPPARE, et al. 2001. Effect of preparation parameters on the performance of conductive composite gas separation membranes. J. Membr. Sci. **182:** 29–39.
24. MUSSELMAN, I.H., L. LI, L. WASHMON, et al. 1999. Poly(3-dodecylthiophene) membranes for gas separations. J. Membr. Sci. **152:** 1–18.
25. REID, B.D., V.H.M. EBRON, I.H. MUSSELMAN, et al. 2002. Enhanced gas selectivity in thin film composite membranes of poly(3-(2-acetoxyethyl)thiophene). J. Membr. Sci. **195:** 181–192.
26. BALL, I.J., S.-C. HUANG, T.M. SU, et al. 1997. Permselectivity and temperature-dependent permeability of polyaniline membranes. Synthetic Metals **84:** 799–800.
27. BALL, I.J., S.-C. HUANG, R.A. WOLF, et al. 2000. Pervaporation studies with polyaniline membranes and blends. J. Membr. Sci. **174:** 161–176.
28. HUANG, S.-C., I.J. BALL & R.B. KANER. 1998. Polyaniline membranes for pervaporation of carboxylic acids and water. Macromolecules **31:** 5456–5464.
29. LEE, Y.M., S.Y. NAM & S.Y. HA. 1999. Pervaporation of water/isopropanol mixtures through polyaniline membranes doped with poly(acrylic acid). J. Membr. Sci. **159:** 41–46.
30. ZHOU, M., M. PERSIN, W. KUJAWSKI, et al. 1995. Electrochemical preparation of polypyrrole membranes and their application in ethanol-cyclohexane separation by pervaporation. J. Membr. Sci. **108:** 89–96.
31. ZHOU, M., M. PERSIN & J. SARRAZIN. 1996. Electrodeposition of membrane-oriented conducting poly(pyrrole, thiophene) on stainless steel meshes. J. Membr. Sci. **117:** 303–309.
32. MARTIN, C.R., W. LIANG, V. MENON, et al. 1993. Electronically conductive polymers as chemically-selective layers for membrane-based separations. Synthetic Metals **55–57:** 3766–3773.
33. EHRENBECK, C. & K. JÜTTNER. 1996. Ion conductivity and permselectivity measurements of polypyrrole membranes at variable states of oxidation. Electrochimica Acta **41:** 1815–1823.
34. ZHAO, H., W.E. PRICE & G.G. WALLACE. 1998. Synthesis, characterization and transport properties of layered conducting electroactive polypyrrole membranes. J. Membr. Sci. **148:** 161–172.
35. BLEHA, M., V. KDELA, E.Y. ROSOVA, et al. 1999. Synthesis and characterization of thin polypyrrole layers on polyethylene microporous films. Eur. Polym. J. **35:** 613–620.
36. LEE, H.S. & J. HONG. 2000. Electrokinetic separation of lysine and aspartic acid using polypyrrole-coated stacked membrane system. J. Membr. Sci. **169:** 2000.
37. LEE, H.S. & J. HONG. 2000. Chemical synthesis and characterization of polypyrrole coated on porous membranes and its electrochemical stability. Synthetic Metals **113:** 115–119.
38. ERMOLAEV, S.V., N. JITARIOUK & A.L. MOËL. 2001. Polymerization of pyrrole onto "track-etch" membranes. Nucl. Instr. Meth. Phys. Res. Sect. B: Beam Interact. Mater. Atoms **185:** 184–191.

39. SCHWITZGEBEL, G. & F. ENDRES. 1995. The determination of the apparent diffusion coefficient of HCl in Nafion-117 and polypyrrole + Nafion-117 by simple potential measurements. J. Electroanal. Chem. **386:** 11–16.
40. SATA, T., T. YAMAGUCHI & K. MATSUSAKI. 1996. Preparation and properties of composite membranes composed of anion-exchange membranes and polypyrrole. J. Phys. Chem. **100:** 16633–16640.
41. SATA, T., T. FUNAKOSHI & K. AKAI. 1996. Preparation and transport properties of composite membranes composed of cation exchange membranes and polypyrrole. Macromolecules **29:** 4029–4035.
42. EHRENBECK, C. & K. JÜTTNER. 1996. Development of an anion/cation permeable freestanding membrane based on electrochemical switching of polypyrrole. Electrochimica Acta **41:** 511–518.
43. DAVEY, J.M., S.F. RALPH, C.O. TOO, et al. 2001. Electrochemically controlled transport of metal ions across polypyrrole membranes using a flow-through cell. React. Funct. Polym. **49:** 87–98.
44. KONTTURI, K., P. PENTTI & G. SUNDHOLM. 1998. Polypyrrole as a model membrane for drug delivery. J. Electroanal. Chem. **453:** 231–238.
45. KONTTURI, K., L. MURTOMAKI, P. PENTTI, et al. 1998. Preparation and properties of a pyrrole-based ion-gate membrane as studied by the EQCM. Synthetic Metals **92:** 179–185.
46. ZHOU, D., C.O. TOO, G.G. WALLACE, et al. 2000. Protein transport and separation using polypyrrole coated, platinised polyvinylidene fluoride membranes. React. Funct. Polym. **45:** 217–226.
47. MISOSKA, V., J. DING, J.M. DAVEY, et al. 2001. Polypyrrole membranes containing chelating ligands: synthesis, characterisation and transport studies. Polymer **42:** 8571–8579.
48. BURGMAYER, P. & R.W. MURRAY. 1982. An ion gate membrane: electrochemical control of ion permeability through a membrane with an embedded electrode. J.A.C.S. **104:** 6139–6140.
49. BURGMAYER, P. & R.W. MURRAY. 1984. Ion gate electrodes. Polypyrrole as a switchable ion conductor membrane. J. Phys. Chem. **88:** 2515–2521.
50. WANG, E., Y. LIU, S. DONG, et al. 1990. Electrochemical study of ion transport across the polypyrrole membrane between two electrolyte solutions. J. Chem. Soc. Faraday Trans. **86:** 2243–2247.
51. MIRMOHSENI, A., W.E. PRICE, G.G. WALLACE, et al. 1993. Adaptive membrane systems based on conductive electroactive polymers. J. Intel. Mater. Syst. Struct. **4:** 43–49.
52. MIRMOHSENI, A., W.E. PRICE & G.G. WALLACE. 1995. Electrochemically controlled transport of small charged organic molecules across conducting polymer membranes. J. Membr. Sci. **100:** 239–248.
53. ZHOU, D., H. ZHAO, W.E. PRICE, et al. 1995. Electrochemically controlled transport in a dual conducting polymer membrane system. J. Membr. Sci. **98:** 173–176.
54. ZHAO, H., W.E. PRICE, C.O. TOO, et al. 1996. Parameters influencing transport across conducting electroactive polymer membranes. J. Membr. Sci. **119:** 199–212.
55. PRICE, W.E., C.O. TOO, G.G. WALLACE, et al. 1999. Development of membrane systems based on conducting polymers. Synthetic Metals **102:** 1338–1341.
56. PARTRIDGE, A.C. 1995. Ion transport through conducting polymer membranes. Electrochim. Acta **40:** 1199–1202.
57. WEN, L. & N.M. KOCHERGINSKY. 1999. Doping-dependent ion selectivity of polyaniline membranes. Synthetic Metals **106:** 19–27.
58. WEN, L. & N.M. KOCHERGINSKY. 2000. Coupled H^+/anion transport through polyaniline membranes. J. Membr. Sci. **167:** 135–146.
59. MATTES, B.R. 1996. Transport and other physicochemical properties of polyaniline. Proceedings of the 8th Annual Meeeting of the North American Membrane Society, Ottawa, Canada.
60. MATTES, B.R., H.L. WANG, D. YANG, et al. 1997. Formation of conductive polyaniline fibers derived from highly concentrated emeraldine base solutions. Synthetic Metals **84:** 45–49.

61. WANG, H.-L., R.J. ROMERO, B.R. MATTES, *et al.* 2000. Effect of processing conditions on the properties of high molecular weight conductive polyaniline fiber. J. Polym. Sci. Part B: Polym. Phys. **38:** 194–204.
62. MATTES, B.R., H.L. WANG & D. YANG. 1997. Electrically conductive polyaniline fibers prepared by dry-wet spinning techniques. Society of Plastic Engineers Annual Technical Conference Proceedings **2:** 1463–1467.
63. YANG, D., A. FADEEV, P.N. ADAMS, *et al.* 2001. Controlling macrovoid formation in wet-spun polyaniline fibers. Proc. SPIE. **4239:** 59–71.

Natural Gas Cleanup by Means of Membranes

KLAUS OHLROGGE AND TORSTEN BRINKMANN

GKSS-Forschungszentrum Geesthacht GmbH, Institut für Chemie, Geesthacht, Germany

ABSTRACT: This paper deals with the use of membranes for hydrocarbon dewpointing and dehydration of natural gas. Based on experience gained from membrane applications in separating organic vapors from off-gas and process streams, as well as the dehydration of compressed air, membranes have been developed and tested for use in high pressure applications. Membranes and membrane modules have been modified to withstand the high operating pressure. Calculation programs were developed to understand the separation performance and to provide the necessary information for optimizing membrane design. A real challenge was the introduction of the vacuum mode dehydration operation in order to achieve the highest possible dewpoint reduction with minimum methane loss.

KEYWORDS: natural gas; hydrocarbon dewpointing; dehydration; gas permeation

INTRODUCTION

The capability of rubbery polymers to separate or to concentrate hydrocarbons has been known for many years. Data on the permeability of various rubbery polymers were published in the 1940s by Amerongen and Barrer *et al.*[1,2] In 1939 and 1952, two patents were granted for processes to concentrate hydrocarbons or to separate hydrocarbons from non-hydrocarbons by diffusion.[3,4] In 1983, a patent was granted to upgrade the fuel gas of a gas engine by means of membrane separation.[5]

Introduction of membrane processes on a commercial scale is based on the development of thin film composite membranes with sufficient selectivities and permeabilities. The first applications to treat off-gases were commissioned in the late 1980s in the US, Japan, and Germany.[6–8] Membrane based separation and recovery of organic vapor is now an accepted technology.

Nearly all kinds of membrane polymers have a high permeability for water vapor. When membranes were applied to CO_2 separation from natural gas, water vapor dewpoint reduction was a beneficial side effect. An application with an increasing market share is the dehydration of compressed air. In the framework of a project supported by the European Community and several industrial projects, membranes have been tested for natural gas applications. The potential advantages of the use of membrane separation are:

Address for correspondence: Klaus Ohlrogge, GKSS-Forschungszentrum Geesthacht GmbH, Institut für Chemie, Max-Planck-Str., 21502 Geesthacht, Germany. Voice: +49-4152-87-2400; fax: +49-4152-87-2444.
Klaus.Ohlrogge@gkss.de

- low space and weight requirements;
- ease of operation;
- no need of additional separation agents and the associated regeneration of those agents; and
- the inherent gas pressure provides the driving force for gas and vapor permeation.

MEMBRANE, MEMBRANE MODULE, AND PROCESS REQUIREMENTS

The challenges to the use of membranes in high-pressure applications are:

- *Membrane stability:* High pressure requires membranes, and especially substructures, that are able to withstand mechanical stress.
- *Real gas behavior:* This has to be taken into account at elevated pressures because it reduces the driving force of some compounds to be separated.
- *Membrane swelling:* Natural gas contains swelling compounds that cause an increase in permeation rates through the membrane.
- *Concentration polarization:* A boundary layer is formed adjacent to the surface of the membrane and may govern the separation selectivity. The gas composition of the laminar layer is depleted from "fast" permeating compounds while the "slow" compounds are enriched. This leads to a reduction of driving force and separation selectivity.
- *Joule-Thomson effect:* The pressure difference between the high pressure (feed) and the low pressure (permeate) side causes an adiabatic throttling of the permeating molecules.[9,10] This behavior causes a cooling down of the gas that is described by the Joule–Thomson effect. Since the permeances are functions of temperature, the Joule–Thomson effect influences separation.
- *Use of highly selective membranes:* To reduce product losses, highly selective membranes ($\alpha > 1,000$) are required. Highly selective membranes require the application of an adjusted pressure ratio (feed pressure over permeate pressure) to achieve the highest possible separation performance. This favors the introduction of vacuum mode separation, which is novel in membrane based natural gas applications.

Membrane Swelling

Hydrocarbon vapors cause a swelling of elastomeric membrane polymers. The polymer chains become more flexible and the distance between them is increased. The "free volume" is enlarged by hydrocarbons penetrating into the selective layer. According to Fujitu[11] this increase in free volume facilitates the diffusion of gases through the polymer. The permeance of all compounds is then increased, as experimentally demonstrated by Stern and Fang.[12]

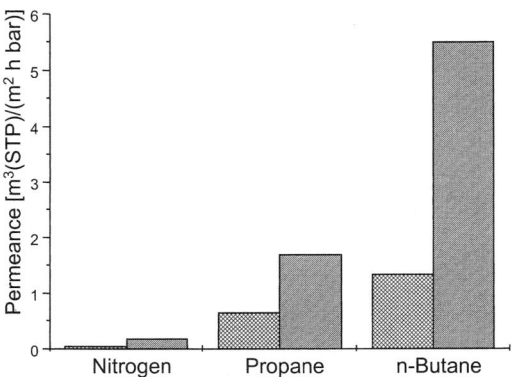

FIGURE 1. Increase in permeabilities due to swelling caused by hydrocarbon vapors: before and after 10 days exposition.

This behavior was investigated experimentally by exposing a hydrocarbon selective membrane based on an elastomeric polymer in "rich" gas, that is, gas that was collected directly at the well head, at 80 bar, during 10 days. FIGURE 1 shows the pure component permeances of nitrogen, propane, and n-butane, measured before and after exposition. It is apparent that the pure component permeances drastically increase because of swelling, whereas the selectivity, if based on these pure component permeances, remains at the same value.[13]

Membrane Stability

The pore structure of the supported membrane needs to be as open as possible to provide enough space for unrestricted permeate removal. On the other hand the structure has to be stable against elevated pressures, so that no compaction or collapse of pores occurs. Membrane casting procedures must be optimized to achieve a pore structure appropriate for the membrane process operating pressure and membrane permeance.

Real Gas Behavior

The permeance through a membrane is defined by

$$L_i = \frac{\dot{V}^P_{STP,j}}{A_M \cdot (p_{i,H} - p_{i,L})}, \tag{1}$$

where L_i is the permeance, $\dot{V}^P_{STP,j}$ is the permeate volume flowrate of component i at standard state, A_M is the membrane area, $p_{i,H}$ and $p_{i,L}$ are the partial pressures of component i on the high pressure and the low pressure sides, respectively, of the membrane. The partial pressures should be replaced by fugacities in case of high-pressure applications.

At pressures exceeding 10 bar the driving force of higher hydrocarbons to be separated is reduced by real gas behavior. The difference in partial pressures then had to be replaced by the difference in fugacities:

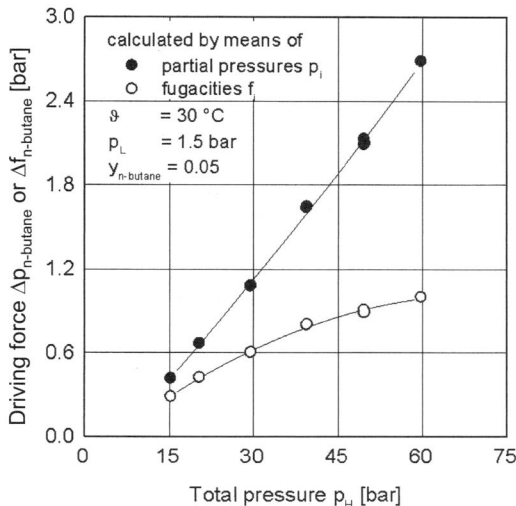

FIGURE 2. Calculated ideal and real driving force of n-butane in a binary n-butane/methane mixture.

$$L_i = \frac{\dot{V}_{STP,j}^P}{A_M \cdot (\varphi_{i,H} \cdot y_{i,H} \cdot p_H - \varphi_{i,L} \cdot y_{i,L} \cdot p_L)}, \quad (2)$$

where y_i is the mole fraction of component i and φ_i is the fugacity coefficient. In FIGURE 2 the ideal driving force is compared with the real driving force for the permeation of n-butane/methane mixtures. The fugacity coefficients were calculated on the basis of the Redlich–Kwong–Soave equation.[14] Although the ideal driving force steadily increases with increasing pressure according to Dalton's law, the slope of the real driving force decrease depends on pressure and approaches a limiting value. It is obvious that for high-pressures no additional increase in effective driving force can be obtained by further increasing the pressure. In FIGURE 2 the driving force over total pressure is plotted with and without consideration of real gas behavior under the given process conditions.

The dependence of fugacities of some natural gas compounds on pressure is shown in FIGURE 3. It is obvious that the influence of non-ideal gas behavior cannot be neglected for processes with operating pressures in excess of 10 bar.

Concentration Polarization

A composite membrane, consisting of a selective non-porous layer, a porous substructure and a non-woven fleece needs to be designed with the conditions of the membrane separation process in mind. The thickness of the selective layer is chosen in accord with the flux densities and selectivities of the feed compounds, as well as the possible Reynolds number of the feed flow over the membrane surface. The Reynolds number directly controls the influence of the boundary layer between the bulk gas and the membrane surface; the thickness of the selective layer has a reciprocal

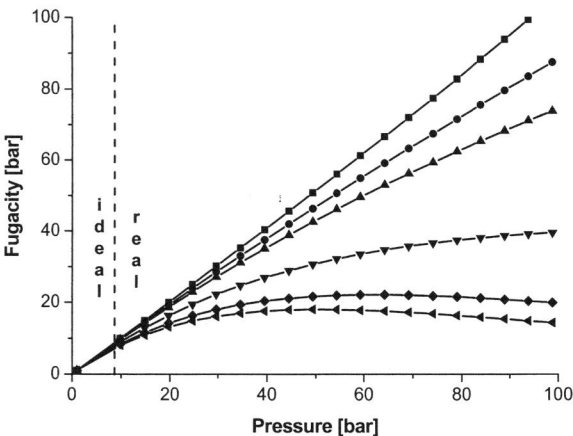

FIGURE 3. Influence of pressure on real gas behavior of natural gas compounds: ■ N_2, ● CH_4, ▲ CO_2, ▼ n-butane, ◆ toluene, and ◀ n-octane.

dependency on the flux density. The substructure must be as open as possible to provide a nearly unrestrained permeation of all compounds that still provides the required mechanical strength. Boundary layer, selective layer, and porous support constitute the major transport resistances in high pressure membrane separations. The first two of these conditions are interdependent: a faster membrane will increase the importance of the boundary layer and, hence, cause concentration polarization to exert a larger influence on the overall separation performance. Concentration polarization causes the concentration of a fast permeating compound to decrease on the membrane surface. This in turn causes the concentration of a slow component to increase. Hence, the driving force, Equation (1), is decreased for the component to be separated whereas it is increased for the component to be retained. FIGURE 4 shows the effects described for a binary system.

Transport through the boundary layer is of diffusive nature. Hence, concentration polarization is of increased importance for high pressure applications since the diffusion coefficient is inversely proportional to the pressure. Mathematically, concentration polarization can be described by an appropriate mass transfer correlation or the Maxwell–Stefan theory.[15] For a binary mixture the transport equation is

$$\dot{n}_i^P = \dot{V}_{STP,j}^P \cdot \frac{p_s}{RT_s} = A_M\left(\frac{\beta_i^*}{RT}(y_{i,H} \cdot p_H - y_{i,HM} \cdot p_H) + y_{i,H} \sum_{j=1}^{n} \dot{n}_j^P\right). \quad (3)$$

In Equation (3), $\dot{n}_i^P$ is the molar transmembrane flow of component i, p_s and T_s are standard pressure and temperature, respectively, β_i^* is the mass transfer coefficient, and the subscript HM indicates conditions on the membrane surface.

The interdependence of permeance and concentration polarization is illustrated in FIGURE 5.[16] Here two membranes consisting of the same polymer but of different thickness are exposed to a feed flow of varying Reynolds numbers. Membrane 1 is approximately six times as fast as membrane 2. If concentration polarization (i.e.,

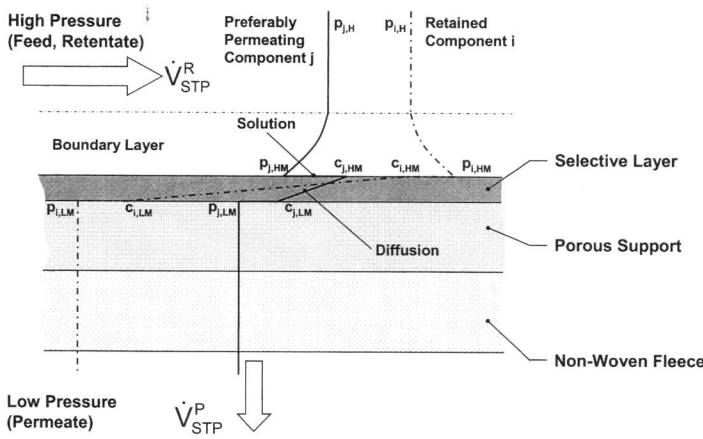

FIGURE 4. Transport through polymeric membranes: p, partial pressure in gas phase; c, concentration in selective layer; and V, volumetric flowrate.

boundary layer mass transfer restriction) was of no effect, the average butane flux density would be equal to the intrinsic permeance of the membrane. This is clearly not the case. Instead, the transport resistance of the boundary layer and that of the selective layer need to be treated akin to a network of electrical resistances. For the faster membrane 1, transport is clearly controlled by the boundary layer whereas the average butane flux density of the slower membrane 2 is much closer to its intrinsic permeance. This effect also illustrated by the selectivities: the selectivity values of membrane 2 are much closer to its intrinsic value and increase with increasing Reynolds number, that is, a diminished effect of concentration polarization. The

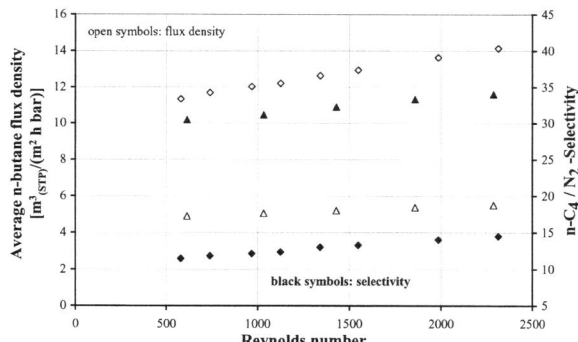

FIGURE 5. Interdependence of concentration polarization and membrane thickness: ◇, membrane 1; △, membrane 2. Flux densities, m^3_{STP}/m^2 h bar: membrane 1, $L_{N_2} = 0.93$, $L_{n\text{-}C_4} = 41.4$; membrane 2, $L_{N_2} = 0.16$, $L_{n\text{-}C_4} = 6.9$. Feed concentration, 2.1 vol% $n\text{-}C_4$, 97.9 vol% N_2. Pressure, $p_{feed} = 3.5$ bar, $p_{permeate} = 0.3$ bar. Temperature, 295 K.

observed selectivity of membrane 1 is lower, indicating a decreased butane concentration (and hence, increased nitrogen concentration) at the membrane surface.

Joule–Thomson Effect

The Joule–Thomson effect is of importance in gas permeation if the trans-membrane pressure difference is high.[9] In natural gas processing this is clearly the case. The temperature change can be viewed as caused by two effects:

- adiabatic throttling of the permeating gas, and
- heat transfer from the high to the low pressure side of the membrane.

The Joule–Thomson effect is especially important for applications with high stage cuts, that is, high permeate flow rates. In this situation it is insufficient to describe a gas permeation process solely in terms of the material balances, permeation equations, and mass transfer kinetics. The energy balances also need to be solved. The temperature change along the membrane module has direct influence on the permeances, since they are functions of temperature. Depending on the component-membrane material combination the permeance can increase or decrease with increasing temperature. The former is true if the permeation mechanism is diffusion controlled, whereas the latter is valid for solution controlled scenarios.[17]

Use of Highly Selective Membranes

An important application of membrane separators in natural gas processing is dehydration. Because natural gas is typically saturated with water it must be dehydrated in order to meet pipeline specifications. During this process step it is important to minimize product (i.e., methane) loss, which in turn requires membranes highly selective for water. The application of a vacuum to the permeate side allows a high water driving force to be to maintained across the membrane, even for highly selective membranes, and hence decreases the required membrane area. A decrease in selectivity and hence a higher membrane copermeation serves as an "internal" purge and, hence, results in lower required membrane areas or operation at lower permeate pressures.

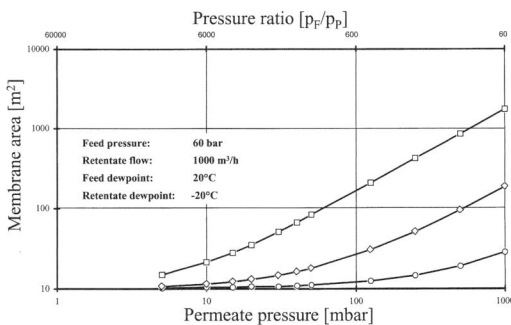

FIGURE 6. Membrane area versus permeate pressure for selectivity 100 (○), 1,000 (◇), and 10,000 (□).

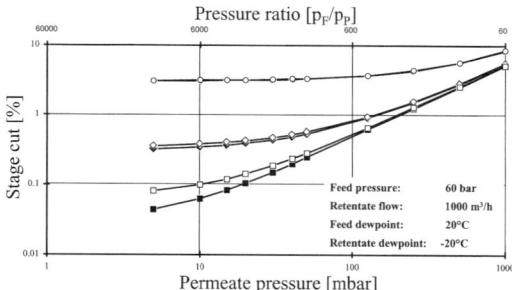

FIGURE 7. Stage cut versus permeate pressure for selectivity 100 (○, without H_2O ●), 1,000 (◇, without H_2O ◆), and 10,000 (□, without H_2O ■).

FIGURES 6 and 7 show the required membrane area and the achieved stage cut, respectively, as functions of applied permeate pressures at various selectivities. The calculations were based on the following conditions:

- binary mixture of methane and water vapor;
- the water vapor permeance is kept constant, the methane permeance varies;
- feed pressure, 60 bar (882 psi);
- retentate volume flow, $1{,}000\,m^3_{STP}/h$ (35,314 scf/h);
- feed dewpoint, 20°C (68°F); and
- retentate dewpoint, $-20°C$ ($-4°F$).

It is apparent that theoretically product losses of less than 0.1% can be achieved at high pressure ratios with highly selective membranes.

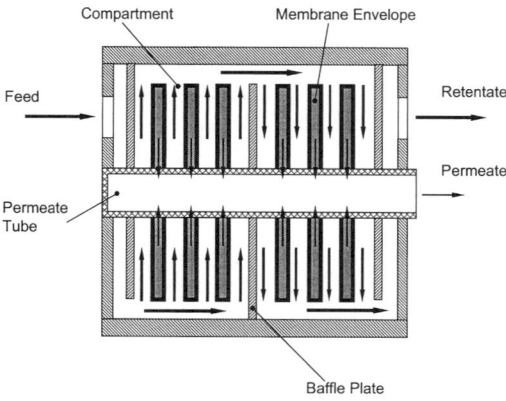

FIGURE 8. Envelope type membrane module.

MODULE DESIGN

The challenges for the use of membrane processes in high pressure applications require a well designed membrane module. The membrane module developed is of the envelope type. A membrane envelope consists of two membrane sheets that are thermally welded at their edge. Between the membrane sheets, spacer material is located, facilitating the permeate withdrawal in a radial direction toward a centrally located permeate tube. Membrane envelopes and permeate tube form a membrane stack, which is subdivided into compartments by means of baffle plates and mounted inside a pressure vessel. The purpose of the subdivision is to keep the feed flow velocity constant and, hence, minimize the effects of concentration velocity. Depending on the application, the subdivision can be asymmetric in order to account for the reduction of feed flowrate by permeation. FIGURE 8 shows the schematic design of the envelope type module.

MEMBRANE PROCESS SIMULATION

Simulation is a tool that is gaining importance in modern process design. Typically, commercial process simulators are employed for this purpose. However, the libraries of these simulators lack models for technologically newer processes, such as membrane processes. Hence, it is necessary to develop simulation models that are compatible with commercial packages. A model for high pressure gas permeation applications was developed using the equation based process simulator Aspen Custom Modeler™. The model takes account of the various physical phenomena that are important to gas permeation at high pressures, as described above. It has direct access to a physical property package and, hence, can perform rigorous thermodynamic calculations, such as fugacity, dewpoint, and enthalpy calculations. The model is capable of performing *forward* and *backward* calculations; for example,

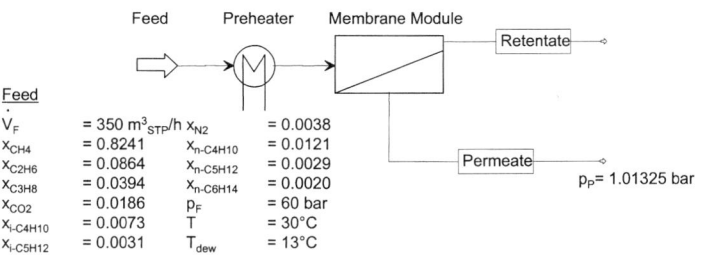

FIGURE 9. Hydrocarbon dewpointing example.

determining module performance from given feed conditions and determining the membrane area for a required product quality, respectively.

The model is well suited to illustrate the non-ideal effects that are important in natural gas processing. The example given here was derived from field tests carried out with a GKSS GS membrane module employed for hydrocarbon dewpointing. The employed membrane was a rubbery, silicone based membrane that is highly selective to higher hydrocarbons.[18] FIGURE 9 shows the process conditions. The preheater was employed to ensure that no condensation occurs on the membrane surface. The goal was to reduce the hydrocarbon dewpoint to −10°C in the retentate stream.

FIGURE 10 shows calculated concentration profiles of hexane along the membrane surface. Hexane was the highest hydrocarbon considered and its presence is of a major influence on the achievable dewpoint reduction. It is apparent that using an ideal simulation assumption, that is, employing Equation (1) to express the permeation behavior, leads to an extreme overestimation of the separation efficiency. Once the real gas behavior according to Equation (2) is accounted for, complete removal of hexane is not predicted, as it is in the ideal case. If concentration polarization by means of Equation (3) is considered, additional decrease in hexane removal is indicated.

FIGURE 11 shows the relative membrane areas required to achieve a retentate dewpoint of −10°C: 100% stands for the membrane area required if all non-ideal effects are taken into account. In the case of a ideal simulation assumption, the membrane area was underestimated by 72%. Not considering concentration polarization causes an underestimate of about 13%. In this example, the Joule–Thomson effect does not have a major influence on separation performance. This was caused by the fact that the permeabilities did not show a large temperature dependence in the investigated temperature range of 20°C to 30°C.

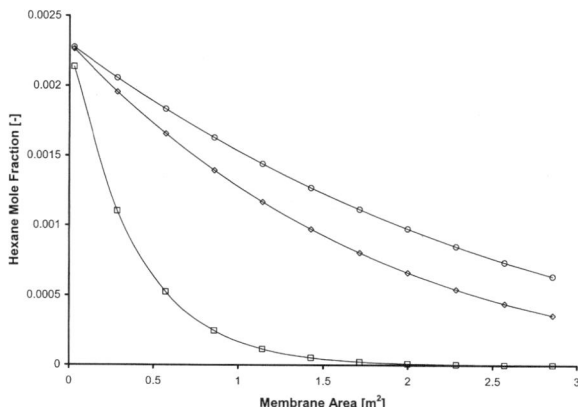

FIGURE 10. Hexane mole fraction versus membrane area: □ ideal; real gas, J–T effect without (◇) and with (○) concentration polarization.

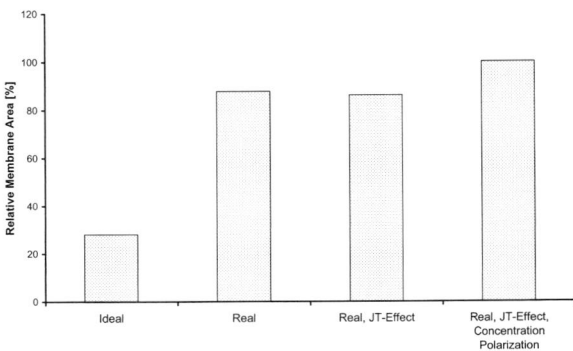

FIGURE 11. Membrane areas relative to the required membrane area.

CONCLUSIONS

Membrane processes offer alternatives to conventional technologies in natural gas processing. When gas permeation membrane processes are used in high pressure environments, the following effects are important:

- membrane stability,
- real gas behavior,
- membrane swelling,
- concentration polarization,
- Joule–Thomson effect, and
- application of highly selective membranes.

If real gas behavior and concentration polarization are not considered in the design stage of a process, a considerable underestimate of the required membrane area is the result. Membranes that are highly selective for water allow the design of dehydration processes with minimized product loss.

Process simulation is an important tool in modern plant design. A model describing the performance of a gas permeation membrane module in high pressure application was developed. This model can be used in combination with commercial process simulators to evaluate the performance of the membrane process.

REFERENCES

1. AMERONGEN, G.J. 1950. Influence of structure of elastomers on their permeability to gases. J. Appl. Poly. Sci. **5**.
2. BARRER, R.M. 1951. Diffusion In and Through Solids. Cambridge University Press, London.
3. FREY, F.E., Inventor; Phillips Petroleum Company, Assignee. 1939. Process for concentrating hydrocarbons. U.S. patent 2,159,434.
4. JONES, J.P., Inventor; Phillips Petroleum Company, Assignee. 1952. Separation of hydrocarbons from non-hydrocarbons by diffusion. U.S. patent 2,167,493.

5. FENSTERMAKER, R.W., Inventor; Phillips Petroleum Company, Assignee. 1983. Engine performance operating on field gas as engine fuel. U.S. patent 4,370,150.
6. BAKER, R.W., N. YOSHIOKA, J.M. MOHR & A.J. KAHN. 1987. Separation of organic vapors from air. J. Membr. Sci. **31**(2–3): 259–271.
7. KATOH, M. 1989. Hydrocarbon vapor recovery with membrane technology. NKK Tech. Rev. **56**.
8. OHLROGGE, K., J. BROCKMÖLLER, J. WIND & R.-D. BEHLING. 1993. Engineering aspects of the plant design to separate volatile hydrocarbons by vapor permeation. Sep. Sci. Tech. **28**(1–3): 227–240.
9. RAUTENBACH, R. 1997. Gaspermeation. *In* Membranverfahren. 312–354. Springer-Verlag, Berlin.
10. KEIL, B., J. WIND & K. OHLROGGE. 2001. Membrane simulation tools for flowsheeting programmes. *In* Proceedings ICheaP-5, Vol. 2. S. Pierucci, Ed. AIDIC. Milan.
11. FUJITA, H. 1961. Diffusion in polymer diluent systems. Fortschr. Hochpolym. Forsch. **3**: 1–47.
12. STERN, S.A. & S.M. FANG. 1972. Effect of pressure on gas permeability coefficients. A new application of "free volume" theory. J. Polym. Sci. **10**: 201–219.
13. ALPERS, A. 1997. Hochdruckpermeation mit selektiven Polymermembranen für die Separation gasförmiger Gemische. Ph.D. Thesis, University of Hannover, Germany.
14. SOAVE, G. 1972. Equilibrium constants from a modified Redlich-Kwong equation of state. Chem. Eng. Sci. **6**: 1197–1203.
15. BIRD, R.B., W.E. STEWART & E.N. LIGHTFOOT. 1960. Interphase transport in multicomponent systems. *In* Transport Phenomena. 636–684. John Wiley & Sons, New York.
16. LÜDTKE, O. 1997. Trennung gasförmiger Gemische mit Hilfe von Membranen. Ph.D. Thesis, University of Hannover, Germany.
17. SCHONERT, M. 1999. Charakterisierung einer Kompositmembran für die Konditionierung von Raffenerie- und Erdgas. Diploma Thesis, Fachhochschule Ostfriesland, Emden, Germany.
18. SCHULTZ, J. & K.V. PEINEMANN. 1996. Membranes for the separation of higher hydrocarbons from methane. J. Membr. Sci. **110**: 37–45.

Polysulfone Hollow Fiber Gas Separation Membranes Filled with Submicron Particles

V. BHARDWAJ, A. MACINTOSH, I.D. SHARPE, S.A. GORDEYEV, AND S.J. SHILTON

Department of Chemical and Process Engineering, University of Strathclyde, Glasgow, Scotland

ABSTRACT: Three different fillers, carbon black (CB), vapor grown carbon fibers (VGCF), and TiO_2, were incorporated into polysulfone spinning solutions with the intention of producing highly selective membranes with enhanced mechanical strength. The effect of filler presence on gas permeation characteristics, mechanical strength (bursting pressure), and morphology was investigated and compared to unfilled membranes. As well as studying filler types, the influence of CB filler concentration on membrane performance was also examined. For all filler types (at a concentration of 5%w/w), the pressure-normalized flux of O_2, N_2, and CH_4 was greater in the composite than in the unfilled membranes. The CO_2 pressure-normalized flux was only greater in the TiO_2 composite membranes. For CB and VGCF, the CO_2 pressure-normalized flux was reduced compared with unfilled membranes. Three CB concentrations were investigated (2, 5, and 10%w/w). For O_2, N_2, and CH_4, pressure-normalized flux peaked at 5%w/w CB. CO_2 exhibited the opposite trend, showing a minimum pressure-normalized flux at 5%w/w. Considering O_2/N_2 and CO_2/CH_4 gas pairs and the various filled membrane categories, only the O_2/N_2 selectivity of the 2%w/w CB filled membranes was higher than that of the unfilled fibers—all other selectivities were lower. In terms of CB concentration, selectivity was a minimum at the intermediate concentration of 5%w/w. All the filled membrane types exhibited greater mechanical strength (bursting pressure) than unfilled fibers apart from the 5%w/w VGCF composites. The 2%w/w CB composites were the strongest. Electron microscopy showed no visible differences in general morphology between the various filled and unfilled membranes.

KEYWORDS: submicron fillers; hollow fiber membranes; gas permeation; bursting pressure

NOMENCLATURE:
C_{filler} weight fraction of filler in the membrane, kg/kg
$M_{particle}$ filler particle weight, kg
$\rho_{polymer}$ polymer density, kg/m^3
ρ_{filler} filler density, kg/m^3
ρ_{comp} density of polymer/filler composite, kg/m^3
$n_{particle}$ number of particles per unit volume of composite material, m^{-3}
$X_{p-p,i}$ interparticle spacing for isotropic fillers, m
$X_{p-p,a}$ cross-sectional interparticle spacing for conterminous anisotropic fillers, m
$L_{particle}$ length of anisotropic filler particle, m

Address for correspondence: S.J. Shilton, Department of Chemical and Process Engineering, University of Strathclyde, James Weir Building, 75 Montrose Street, Glasgow G1 1XJ, Scotland. Voice: +44-141-548-2380.
simon.shilton@strath.ac.uk

M_w	molecular weight of membrane polymer, amu
M_{ws}	molecular weight of repeating unit in macromolecular chain, amu
X_s	length of repeating unit in macromolecular chain, m
$X_{polymer}$	size of polysulfone molecule, m
P	pressure-normalized flux $\times 10^6$, cm^3(STP)/(s cm^2 cmHg)

INTRODUCTION

Asymmetric gas separation hollow fibers with attractive separation properties have been developed in our group by controlling shear induced molecular orientation and the mechanism of skin formation in the membrane active layer.[1,2] Although these membranes have enhanced separation properties (selectivities often surpass the recognized intrinsic value for the polymer), they are mechanically weak. This lack of robustness is manifest through low bursting pressures, which limits their use in industrial gas separations.

Fillers are submicron solid inclusions capable of altering the physical and chemical properties of materials through their own physical characteristics or through their interaction with the host substrate. Fillers are commonly used to mechanically reinforce materials. A systematic classification of fillers and their applications has been documented by Wypych.[3]

In the work described here, various fillers were incorporated into polysulfone hollow fibers in the hope of producing highly selective and mechanically strong gas separation membranes. Such membranes, with high bursting pressures, would be highly attractive for industrial gas separation applications. Three different fillers, carbon black (CB), vapor grown carbon fibers (VGCF), and TiO_2, were selected.

CB is a fluffy powder of extreme fineness and high surface area, composed essentially of elemental carbon with well-developed graphite platelets.[4,5] It has been used extensively in the rubber and plastics industries and has earned the reputation of being an effective low-cost filler. VGCF is a relatively new class of reinforcing anisotropic materials consisting of entangled fibrils.[6,7] TiO_2 has been used to enhance the permselectivities of selected gas pairs and to improve the mechanical strength of poly(amide-imide) membranes.[8,9] The fillers, two from the carbon family (CB and VGCF) and one inorganic (TiO_2), have dissimilar particle sizes and shapes and represent a range of inclusion types.

Each filler was incorporated into the polysulfone spinning solution using the same mixing technique. However, achieving good dispersion of the filler is a challenge; primarily due to the formation of aggregates of CB and TiO_2 and the randomly entangled characteristics of VGCF. The spinning of the composite dopes was carried out using the same process used previously to produce highly selective unfilled polysulfone membranes.[1,2]

Recently, Gordeyev et al.[10] carried out a preliminary study of VGCF filled polysulfone membranes that examined fiber tensile strength as well as permeation properties. In the current study, bursting pressure tests, as opposed to tensile measurements, were employed since they give a more realistic reflection of hollow fiber mechanical strength in the context of gas separation. Scanning electron microscopy was also used to examine filled membrane morphology.

METHODS

Fillers and Pretreatment

The general properties of the fillers are given in TABLE 1. The fillers were oven dried at 120°C for 48 hours to remove moisture. Adsorbed water creates high-energy sites (–OH– groups) that cause polarity and hydrogen bonding that favors filler–filler interaction and hinders polymer–filler interaction, that is, dispersion. FIGURE 1 shows the general structure of the filler particles used in this work.

Spinning Dope Preparation

Polysulfone (Amoco Chemicals, Udel P1700, $M_w = 35,400$) was dissolved in a solvent/non-solvent cocktail to produce an optimized four-component dope[11] for use in spinning processes that involve a dry forced convection stage. The dope comprised 22%w/w polysulfone, 31.8%w/w N,N-dimethylacetamide, 31.8%w/w tetrahydrofuran, and 14.4%w/w ethanol. Five filled dopes were prepared by incorporating the filler particles into this standard dope composition. Three filled dopes were prepared to give CB concentrations in the resultant membranes of 2, 5, and 10%w/w. Two other dopes corresponding to membrane compositions of 5%w/w VGCF and 5%w/w TiO_2 were also produced.

The polymer was first dissolved in the solvents by stirring under total reflux at 70–75°C. The mixture was then allowed to cool to about 30°C while being stirred. Cooling increases the viscosity of the polymer solution. Filler, prewetted with a small amount of solvent, was added to the cooled solution, which was then stirred vigorously for eight hours without heating. It was thought that the higher shear stresses created during stirring in the higher viscosity cooled mixture would aid filler dispersion. The ethanol (non-solvent) was then added to the mixture to achieve the aforementioned dope composition. This was followed by prolonged mixing. The filled dope was then passed through a 150-micron filter to aid the dispersion of any remaining agglomerates and entangled particles. The filtered dope was finally transferred to the dope reservoir for spinning.

TABLE 1. Details of fillers

	Filler Type		
	Carbon Black	Titanium Dioxide	Vapor-Grown Carbon Fiber
Supplier	Degussa	Fisher	Pyrograf Products
Trade name	lamp black 101 powder	titanium(IV) oxide	PR-19-PS
Chemical composition	C	TiO_2	C
Density, g/cm^3	1.7–1.9	3.5	2.0
BET surface area, m^2/g	20	75	25–30
Average primary particle size, μm	0.095	0.15	$L > 100$ $D \approx 0.15$–0.2

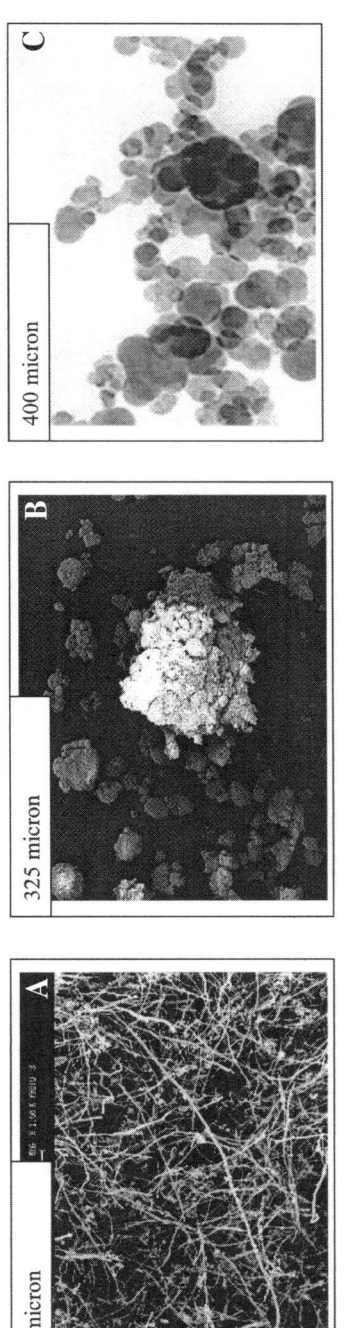

FIGURE 1. SEM photographs of fillers: **(A)** entangled vapor grown carbon fibers (photograph top edge is 75 μm), **(B)** TiO$_2$ aggregates (photograph top edge is 650 μm), and **(C)** carbon black aggregates (photograph top edge is 800 μm).

Membrane Spinning

Asymmetric filled hollow fiber membranes for gas separation were fabricated using a dry/wet spinning process with forced convection in the dry gap. The dope reservoir was at ambient temperature ($20 \pm 2°C$) during spinning. On extrusion from the spinneret (spinneret dimensions: OD 600μm, ID 330μm; dope extrusion rate: 2.5 cm^3/min), the fiber passed through a cylindrical forced convection chamber (diameter 5 cm, height 9 cm), which was flushed with 4 L/min of nitrogen gas. The nitrogen was introduced through a 0.25 inch (0.6 cm) tube that abutted upon the chamber normal to the surface at midheight. A 2-mm clearance existed between the top of forced convection chamber and the bottom of the spinneret and also between the bottom of the forced convection chamber and the water level in the first coagulation bath. Pure water at $14 \pm 0.5°C$ was used in the external coagulation baths. The bore coagulant was a 20%w/w solution of potassium acetate in water at ambient temperature, which equates to a water activity of 0.9.[12] The jet stretch ratio (windup speed/extrusion speed) was fixed at 1 throughout. The ratio of the dope extrusion rate to the bore fluid injection rate was 3. After spinning, the membranes were steeped in water and then dried using a methanol solvent exchange technique.[13]

Gas Permeation

The membranes were potted into modules and tested before and after coating with silicone (Sylgard 184, Dow Corning). Coating is a standard technique that attempts to repair any small defects or pores that exist in the membrane active layer.[14] The pressure-normalized fluxes of the fibers were measured for O_2, N_2, CO_2, and CH_4 at 25°C and at a pressure drop of 5 bar. Membrane selectivities were determined by taking the ratio of pressure-normalized fluxes.

Bursting Pressure

Bursting pressure is a measure of the robustness of the hollow fiber membrane to endure operating pressures in an industrial gas separation process. Hollow fibers are subjected to increasing pressures applied to the lumen in a stepped ramp until a certain pressure, called the bursting pressure, is reached where the gas flux becomes immeasurably high owing to a rupture on the surface wall and leakage of the permeate gas. Five-centimeter lengths of fiber were used in this technique and nitrogen was the test gas. The pressure was increased in increments of 2.5 bar initially and 1 bar after the flow rate of N_2 gas tended to rise swiftly. This was done to ensure test sensitivity near the bursting pressure. The experimental apparatus is shown in FIGURE 2.

Electron Microscopy

The structures of the unfilled and filled hollow fiber membranes were visually observed using a scanning electron microscope (JEOL SEM Model 840 A). The fibers were cut to lengths of 1–2 mm using a fresh razor blade and mounted on sample stubs. These were then sputter-coated with gold under vacuum before being viewed under the microscope. Micrographs of both the fiber cross-sections and surfaces were taken at various magnifications.

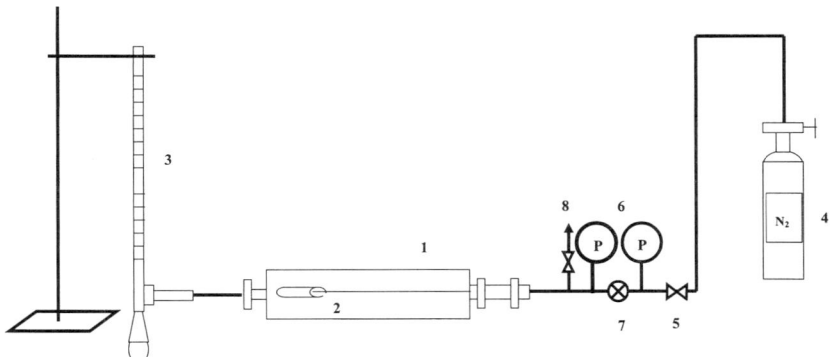

FIGURE 2. Bursting pressure apparatus: (1) test chamber, (2) membrane sample, (3) bubble flowmeter, (4) nitrogen cylinder, (5) ball valve, (6) pressure gauges, (7) pressure regulator, and (8) vent.

Filler Interparticle Spacing and Size of a Polymer Molecule

Given the weight fraction of filler in the membrane, C_{filler}, the filler particle weight, $M_{particle}$, and the densities of both the polymer, $\rho_{polymer}$, and filler, ρ_{filler}, the number of filler particles per unit volume of polymer matrix and, hence, an interparticle spacing, assuming uniform dispersion, can be calculated using simple geometry.

Density of polymer/filler composite:

$$\rho_{comp} = \frac{\rho_{filler}\rho_{polymer}}{\rho_{polymer}C_{filler} + \rho_{filler}(1 - C_{filler})}.$$

Number of particles per unit volume of composite material:

$$n_{particle} = \frac{\rho_{comp}C_{filler}}{M_{particle}}.$$

Interparticle spacing for isotropic fillers:

$$X_{p-p,\,i} = n_{particle}^{-1/3}.$$

Cross-sectional interparticle spacing for conterminous anisotropic fillers of length $L_{particle}$:

$$X_{p-p,\,a} = (L_{particle}n_{particle})^{-1/2}.$$

The interparticle spacing can then be compared to the diameter of the filler particle in order to gauge the proximity of the inclusions in the polymer.

Knowing the molecular weight of the membrane polymer ($M_w = 35,400$), and the weight ($M_{ws} = 442$) and approximate length ($X_s = 80\,\text{Å}$) of each repeating unit in the macromolecular chain, the size of the polysulfone molecules can be estimated,[15]

$$X_{polymer} = X_s\left(\frac{M_w}{M_{ws}}\right)^{1/2}.$$

TABLE 2. Permeation properties of membranes

Membrane Type	Uncoated			Coated				
	P_{O_2}	P_{N_2}	P_{O_2}/P_{N_2}	P_{O_2}	P_{N_2}	P_{O_2}/P_{N_2}	P_{CO_2}	P_{CH_4}	P_{CO_2}/P_{CH_4}
Polysulfone (PSF)	97.71	91.68	1.06	12.21	2.05	6.05	86.12	2.10	40.98
PSF + CB (2%)	292.84	172.61	1.76	12.65	1.94	6.53	76.25	2.16	35.40
PSF + CB (5%)	153.36	97.12	1.61	12.78	2.81	5.06	68.72	4.77	16.37
PSF + CB (10%)	46.775	34.35	1.43	11.89	2.25	5.47	75.13	2.44	33.73
PSF + VGCF (5%)	119.56	72.34	1.78	12.63	3.95	3.27	57.12	5.52	11.66
PSF + TiO$_2$ (5%)	119.52	123.00	0.97	14.70	3.17	4.66	89.56	3.67	24.51

NOTE: P is pressure-normalized flux $\times 10^6 \text{cm}^3(\text{STP})/(\text{s cm}^3 \text{cmHg})$, measured at 25°C and at a pressure differential of 5 bar.

RESULTS AND DISCUSSION

All filled dopes were spinnable. The gas permeation and bursting pressure results for the hollow fibers are given in TABLES 2 and 3, respectively. The dimensions of the filler particles and the interparticle spacings for the various membrane categories are given in TABLE 4 together with the estimated average size of a polysulfone macromolecule.

Considering coated membranes, TABLE 2 shows that for all filler types (at a fixed filler loading of 5%w/w in polysulfone), the pressure-normalized flux of O_2, N_2, and CH_4 was greater in the composite membranes than in the unfilled. The CO_2 pressure-normalized flux was only greater in the TiO_2 composite membranes. For CB and VGCF, the CO_2 pressure-normalized flux was actually reduced compared with unfilled membranes.

Thus, there seems to be a special interaction between the carbon fillers (CB and VGCF) and the carbon dioxide: the CO_2 pressure-normalized flux is lower in the carbon filled than in the unfilled membranes. The retardation of the carbon dioxide may relate to its sorption onto/into the carbon inclusions. For the other gas/filler combinations, the composite membranes tend to be more permeable than the unfilled membranes. TABLE 4 shows that the proximity and size of the inclusions relate to the same order of magnitude as the size of a polysulfone macromolecule and, hence, the fillers

TABLE 3. **Bursting pressure of membranes**

Membrane Type	Bursting Pressure, bar
Polysulfone (PSF)	18
PSF + CB (2%)	24
PSF + CB (5%)	22
PSF + CB (10%)	20
PSF + VGCF (5%)	9
PSF + TiO_2 (5%)	19

TABLE 4. **Filler interparticle spacing and size of a polymer molecule**

Membrane Type	Number of Particles per m^3	Interparticle Spacing, nm	Filler Particle Diameter, nm	Size of Polysulfone Molecule, nm
PSF + CB (2%)	3.09×10^{19}	319	95	71.6
PSF + CB (5%)	7.80×10^{19}	234	95	
PSF + CB (10%)	15.8×10^{19}	185	95	
PSF + VGCF (5%)	1.31×10^{16}	872	175 ($L = 100$ micron)	
PSF + TiO_2 (5%)	1.036×10^{19}	459	150	

could conceivably increase the free volume of the polymer matrix. Thus the greater gas transmission rates in the non-CO_2/carbon gas/filler combinations compared to the unfilled membranes may be dictated by increased polymer free volume in the composite material.

The effect of CB filler concentration was also investigated (2, 5, and 10%w/w CB). TABLE 2 shows that for O_2, N_2, and CH_4, pressure-normalized flux peaked at 5%w/w CB. CO_2 exhibits the opposite trend showing a minimum pressure-normalized flux at 5%w/w. This result is difficult to explain but does again demonstrate the peculiar behavior of the CO_2/carbon filler system.

Considering O_2/N_2 and CO_2/CH_4 gas pairs and the various filled membrane categories, only the O_2/N_2 selectivity of the 2%w/w CB filled membranes was higher than that of the unfilled fibers—all other selectivities were lower. In terms of CB concentration, selectivity experienced a minimum at the intermediate concentration of 5%w/w. The enhanced O_2/N_2 selectivity of the 2%w/w CB filled membranes may be due to subtle free volume effects, enabling the membrane to better discriminate between the two similarly sized penetrants.

All uncoated membrane samples exhibited very high pressure-normalized fluxes and Knudsen diffusion type selectivities, suggesting that they suffered from surface pores or imperfections.

As shown in TABLE 3, all the filled membrane types exhibited greater mechanical strength (bursting pressure) than unfilled fibers apart from the 5%w/w VGCF composites, which showed a reduction in strength. The 2%w/w CB composites were the strongest. The full bursting pressure test plot for each membrane type is given in FIGURE 3.

The mechanical strength of the VGCF-filled membranes is interesting. This filler produces the weakest membranes in terms of bursting pressure but has been shown

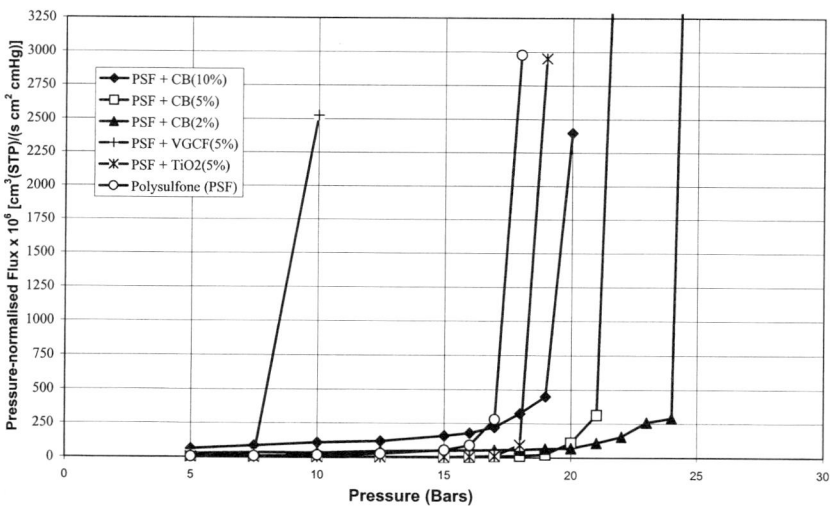

FIGURE 3. Bursting pressure plots.

to increase the tensile strength of membranes compared to unfilled samples.[10] The VGCF fibrils are aligned longitudinally along the hollow fiber length by the action of extrusion shear (micrographs) during membrane spinning. This axial alignment improves tensile strength, which acts in that direction. However, the axial orientation of fibrils undermines bursting pressure strength, which relies on structural integrity perpendicular to the fibrils. The radial stresses set up during the bursting pressure test work across the "grain" and are particularly destructive to the VGCF filled membranes. The membranes containing the isotropic fillers (CB and TiO_2) do not suffer from this direction dependency and, in fact, have greater bursting pressures than unfilled hollow fibers.

Electron micrographs given in FIGURE 4 show no visible differences in general morphology between the various filled and unfilled membranes.

CONCLUSIONS

The presence of fillers alters the permeation properties and bursting pressure strength of polysulfone gas separation hollow-fiber membranes. Both O_2/N_2 selectivity and bursting pressure were improved over unfilled membranes for composites containing 2%w/w carbon black. These results should be regarded as preliminary

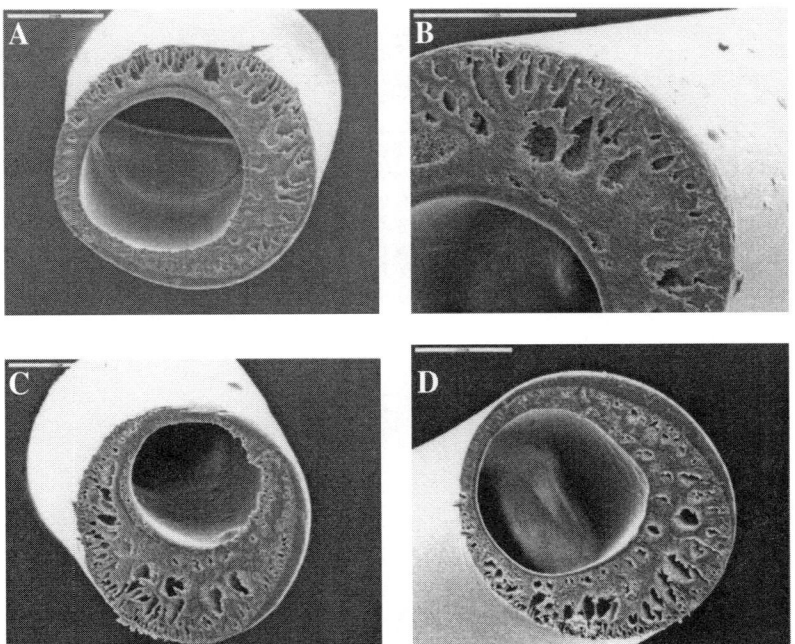

FIGURE 4. SEM photographs of membrane cross-sections: **(A)** unfilled polysulfone (scale bar, 200 μm), **(B)** CB (5%w/w) (scale bar, 200 μm), **(C)** VGCF (5%w/w) (scale bar, 200 μm), and **(D)** TiO_2 (5%w/w) (scale bar, 200 μm).

since additional work is required to improve filler dispersion. It is hoped that a more detailed understanding of filler–filler and filler–polymer interactions, and improved bursting pressures (approximately 60 bar for industrial membranes), will be achieved in the future.

REFERENCES

1. ISMAIL, A.F., I.R. DUNKIN, S.L. GALLIVAN & S.J. SHILTON. 1999. Production of super selective polysulfone hollow fiber membranes for gas separation. Polymer **40:** 6499.
2. SHARPE, I.D., A.F. ISMAIL & S.J. SHILTON. 1999. A study of extrusion shear and forced convection residence time in the spinning of polysulfone hollow fiber membranes for gas separation. Sep. Purification Tech. **17:** 101.
3. WYPYCH, G. 1999. Handbook of Fillers, Vol. 1. Chemical Technology Publishing.
4. MALLETTE, J.G., L.M. QUEJ, A. MARQUEZ & O. MANERO. 2001. Carbon black filled PET/HDPE blends: effect of the CB structure on rheological and electrical properties. J. Appl. Polym. Sci. **81:** 562.
5. JAGER, K.M. & D.H. MCQUEEN. 2001. Fractal agglomerates and electrical conductivity in carbon black polymer composites. Polymer **42:** 9575.
6. REYNAUD, E.T., T. JOUEN, C. GAUTHIER, *et al.* 2001. Nano fillers in polymer matrix: a study on silica reinforced PA6. Polymer **42:** 8759.
7. JANA, S.C. & S. JAIN. 2001. Dispersion of nanofillers in high performance polymers using reactive solvents as processing aids. Polymer **42:** 6897.
8. HU, A., E. MARAND, S. DHINGRA, *et al.* 1997. Poly(amide-imide)/TiO_2 nano-composite gas separation membranes: Fabrication and characterisation. J. Membr. Sci. **135:** 65.
9. TONG, Y., Y. LI, F. XIE & M. DING. 2000. Preparation and characteristics of polyimide–TiO_2 nanocomposite film. (Short Communication.) Polym. Intl. **49:** 1543.
10. GORDEYEV, S.A., I.D. SHARPE & S.J. SHILTON. 2001. Processing and properties of polysulfone hollow fibre membranes for gas separation filled with sub-micron carbon fibres. Macromol. Sympos. **170:** 273.
11. PESEK, S.C. & W.J. KOROS. 1993. Aqueous quenched asymmetric polysulfone membranes prepared by dry/wet phase separation. J. Membr. Sci. **81:** 71.
12. PESEK, S.C. & W.J. KOROS. 1994. Aqueous quenched asymmetric polysulfone hollow fibers prepared by dry/wet phase separation. J. Membr. Sci. **88:** 1.
13. MANOS, P. 1978. Solvent exchange drying of membranes for gas separation. U.S. Patent 4120098.
14. WARD, R.R., R.C. CHANG, J.C. DANOS & J.A. CARDEN, Monsanto Co. 1980. Processes for coating bundles of hollow fiber membranes. U.S. Patent 4214020.
15. STROBL, G. 1997. The Physics of Polymers, 2nd edit. Springer, Berlin.

Carbon Molecular Sieve Membranes
A Promising Alternative for Selected Industrial Applications

MAY-BRITT HÄGG, JON A. LIE, AND ARNE LINDBRÅTHEN

Norwegian University of Science and Technology, Chemical Engineering Department, Trondheim, Norway

ABSTRACT: Carbon molecular sieve (CMS) membranes (hollow fibers) have been studied for application as possible separation units for selected industrial gas streams. Gas streams at petrochemical plants (polypropene and polyethene) and upgrading of biogas to fuel specifications have been in focus. Gases present in biogas (N_2, CO_2, H_2O_{vap}, and CH_4) and gas streams at polyolefin plants (C_2H_4, C_3H_6, and C_3H_8) have been measured; both as pure gases and in mixtures. Aging of the CMS-membranes as a function of humidity and pore blocking is discussed; likewise, possible regeneration methods when flux decrease is experienced. Transport mechanisms depending on pore size and molecular properties are also discussed. Excellent separation properties were documented for these applications, but also the need for frequent regeneration of the membrane in order to maintain permeability flux. The mixed gas experiments documented clearly the need for careful pore tailoring in order to optimize selectivity when the membranes were used for alkane–alkene separation.

KEYWORDS: carbon membranes; microporous membranes; gas separation; transport mechanisms; aging; biogas; alkanes; alkenes

NOMENCLATURE:
C	concentration, mol m^{-3}
d	diameter, m
D	diffusion coefficient, $\text{m}^2 \text{sec}^{-1}$
e	electron charge, $1.602 \times 10^{-19} \text{C}$
E	energy, J
h	Planck constant, $6.626 \times 10^{-34} \text{J sec}$
J	flux, $\text{mol m}^{-2} \text{sec}^{-1}$
k	Boltzmann constant, 1.381×10^{-23}, J K^{-1}
l	effective membrane thickness, m
M	molecular mass, kg mol^{-1}
P	permeability, $\text{m}^3(\text{STP}) \text{m h}^{-1} \text{m}^{-2} \text{bar}^{-1}$
p	pressure, bar
q	volume flow rate, $\text{m}^3 \text{sec}^{-1}$
R	molar gas constant, $8.3145 \text{J K}^{-1} \text{mol}^{-1}$
S	entropy, J K^{-1}
T	temperature, K
x	distance coordinate, m
x	mole fraction in retentate
y	mole fraction in permeate

Address for correspondence: May-Britt Hägg, Norwegian University of Science and Technology, Chem. Eng. Dept., N-7491 Trondheim, Norway. Voice: +47-73594033; fax: + 47-73594080.

may-britt.hagg@chemeng.ntnu.no

Indices

a	activation
ads	adsorption
c	critical
d	diffusion
h	high
i	any species *i*
K	Knudsen
l	low
m	membrane
MS	molecular sieving
p	pore
S	surface diffusion

Greek letters

λ	mean free path, m
ν	velocity, m sec^{-1}
θ	stage cut

INTRODUCTION

Carbon molecular sieve (CMS) membranes are basically prepared in four different configurations: flat sheet, membrane supported on tube, capillary, and hollow fiber. This article will focus on carbon membranes prepared as hollow fibers since, in the opinion of the authors, it is this configuration that has the largest potential for becoming a successful separation unit used on an industrial scale for gas separation and purification. Although the concept of carbon membranes has been known for almost 30 years, it is only during the past 10 years that the potential for difficult gas separation has been fully acknowledged. A comprehensive review on development of the various types of carbon membranes is provided by Ismail and David.[1]

Although excellent separation properties have been reported for gas mixtures like hydrocarbons, sour gases, and air, commercial applications are still very limited. The most important reported are the production of high purity nitrogen from air and recovery of hydrogen from gas streams.[2,3] CMS-membranes have also been identified as having a very large potential for application in the petrochemical industry.[4] Carbon fibers still find their largest application as adsorption media (or molecular sieve) in separation processes like pressure swing adsorption (PSA).[5,6] The properties that place the carbon membranes among the most promising membrane materials are their high temperature resistance and excellent chemical resistance to acids, hot organic solvents, and alkaline baths. They are also rather easy to produce as fibers or flat sheets since much is known about how the pyrolysis conditions effect separation properties.[6–11] Thus, a carbon membrane can be tailored with a pore size giving excellent separation properties for a given gas mixture (high flux for permeating component and high selectivity for gas pairs). The membranes can be prepared as bundles of fibers, and thus, modules may have a high packing density (m^2/m^3) for commercial applications.

So why are these membranes that possess such potential still not used on a large-scale basis? First, the production processes for larger commercial units are challenging when the process environment is harsh and the process temperature high. Several producers are preparing carbon membranes as tubes,[1] but only a couple of producers

(Carbon Membranes Ltd., Ube Industries) have succeeded in making commercial fiber modules for selected applications. Sealing of fibers and selecting potting material that is thermal and chemical resistant has been difficult. Because some of the fibers may break during mounting, a technique for plugging broken fibers also needed to be developed.

The most serious disadvantages that have to be overcome or controlled are still that the fibers are vulnerable to (1) oxidizing agents (pore size may increase and selectivity decrease), (2) water vapor (giving performance loss), (3) pore blocking from contaminants, and (4) flux decline over time. When these effects are detected in membrane performance, it is important to know how to address the problem (or preferably how to avoid it). It is important to know how to "open" pores that are blocked, how to regenerate the membrane, and how to optimize the separation performance by controlling the process variables. If a basic knowledge about these membranes is not known to a user one may very quickly draw the conclusion that CMS-membranes are unstable and exhibit too large a decrease in performance over time, and will then probably judge them as "unsuitable" for gas separation. This may be very wrong.

This article addresses various aspects of how properties of a CMS–membrane may change over time and how important it is to find the optimum process conditions (temperature and pressure) for the gas mixture. Results are presented—examples from separation of hydrocarbons (alkanes–alkenes), upgrading of biogas, and the effect of humidity in the gas stream. Finally, the importance of regeneration is discussed.

TRANSPORT MECHANISMS

The ability of a microporous carbon fiber to separate gases depends on the pore size of the membrane, the physiochemical properties of the gases, and surface properties of the membrane pore. The pore size of a carbon fiber for gas separation is usually within the range 3.5–10 Å; depending on the conditions for preparation of the membrane during the pyrolysis or treatment afterwards (post oxidation or chemical vapor deposition).[6–11]

The transport mechanism takes place according to one of three mechanisms:[12–15]

1. Knudsen diffusion; hence, the square root of the ratio of the molecular weights gives the separation factor.
2. Selective surface diffusion governed by a selective adsorption of the larger nonideal components on the pore surface, hence retaining the smaller components from permeation.
3. Molecular sieving; hence, the smallest molecules permeate, the larger being retained.

A fourth mechanism is mentioned, but is not really considered when the carbon fiber operates as a membrane:

4. Capillary condensation; hence, the ability of the gas to condense, in addition to pore size, determines which component will permeate.

The mechanisms are illustrated in FIGURE 1.[16]

Knudsen diffusion dominates for the largest pores, molecular sieving for the smallest. Molecular sieving is often referred to as a configurational diffusion, and it

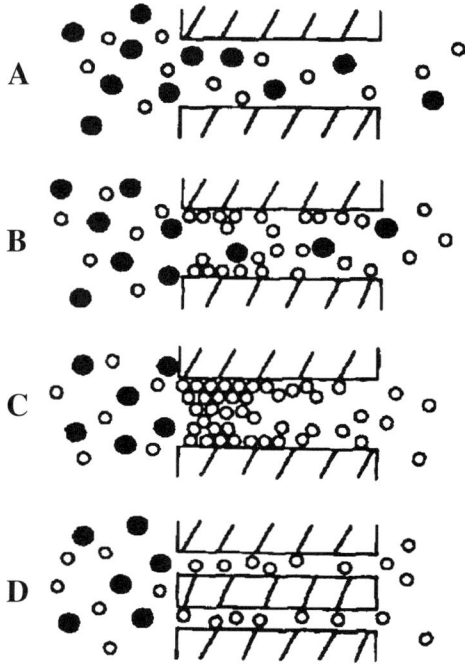

FIGURE 1. Mechanisms of mass transfer through microporous membranes: (**A**) Knudsen diffusion, (**B**) surface diffusion, (**C**) capillary condensation, and (**D**) molecular sieving.[16]

is an activated diffusion like surface selective flow (SSF). For each of the mechanisms, the molar flux through the pore is defined by the following equation:

$$J_i = -D_i \frac{\partial C_i}{\partial x}, \qquad (1)$$

where D_i is the diffusion coefficient and $\partial C_i/\partial x$ is the driving force.

Knudsen Diffusion

For Knudsen diffusion to take place, the lower limit for pore diameter is usually set to $d_p > 20\,\text{Å}$.[13,17] Gilron and Soffer[14] have, however, recently discussed thoroughly how Knudsen diffusion may contribute to transport even in smaller pores. Using a model and considering pore structure, they showed that contributions to transport may both come from activated transport and Knudsen through one specific fiber. Experiments referred to here appear to document and confirm the theory of Gilron and Soffer. Thus, it may be difficult to know exactly when transport due to Knudsen diffusion is taking place. One way to approach this problem is to calculate the Knudsen number, $N_{Kn} = \lambda/d_p$, for the system, where λ is the mean free path for the molecule. This means that λ, which is temperature dependent, must be calculated first, and Knudsen number checked. If $N_{Kn} \geq 10$, then the separation can be assumed to take place according to Knudsen diffusion.[17] The Knudsen diffusivity of a component A, $D_{K,A}$, is independent of pressure, and can be calculated from

$$D_{K,A} = \frac{d_p}{3}\bar{v}_A = \frac{d_p}{3}\sqrt{\frac{8RT}{\pi M_A}} = 48.5 d_p \sqrt{\frac{T}{M_A}}, \qquad (2)$$

where d_p is the average pore diameter, $\bar{v}_A$ is the average molecular velocity, M_A is the molecular weight, and T is the temperature.

The flux can be calculated according to

$$J_K = \frac{D_K \cdot \Delta p}{RT \cdot l}, \qquad (3)$$

where l is the effective membrane thickness.

Selective Surface Flow

The driving force for separation according to a SSF is basically the difference in the concentration of the adsorbed phase of the diffusing components. This means that a large driving force can be attained even with a small partial pressure difference for the permeating component. The activation energy for surface diffusion may also be significantly lower than, for instance, diffusion through a polymeric membrane,[18] hence the flux may be increased compared to a dense membrane. The larger (more condensable) molecules in a gas mixture will be selectively adsorbed, hence, the smaller molecules will be retained due to reduced pore size. The pore size region in which SSF is expected to take place is about $5\,\text{Å} < d_p < 10\,\text{Å}$; or up to three times the diameter of the molecule.[13,15]

Activated diffusion can be described by an Arrhenius type of equation,[19]

$$D_a = D_0 \exp\left(\frac{-E_d}{RT}\right), \qquad (4)$$

where E_d is the activation energy for diffusion. If Henry's law is assumed to apply, the integrated flux equation may be written as

$$J_a = \frac{\Delta p}{RTl} D_0 \exp\left(\frac{-[E_{a,S} - E_{ads}]}{RT}\right) = \frac{\Delta p}{RTl} D_0 \exp\left(\frac{-E_S}{RT}\right), \qquad (5)$$

where ΔE_S, the difference in transport activation energy and adsorption energy, may be positive or negative. When $\Delta E_S < 0$, transport due to SSF increases with decreasing temperature; for $\Delta E_S > 0$ it decreases. Plainly stated; adsorption, and hence selectivity, increase with decreasing temperature. This is the opposite of the temperature influence for molecular sieving separation (see below).

Molecular Sieving

Molecular sieving is the dominating transport mechanism where carbon membranes are applied; this has also given the name to these membranes, CMS. The pore size is usually within the range of a few Ångstrøm, (3-5)Å. The dimensions of a molecule are usually described either with a Lennard–Jones radius or a Van der Waals radius. For separation by molecular sieving, this is not really a satisfactory way for giving molecular size; a shape factor should be included.[20] This can be easily understood from the illustration in FIGURE 2, which shows the shapes of an oxygen and a nitrogen molecule, together with a slit in a zeolite crystal that has to be passed in order to permeate.[21] Sorption selectivity has little influence on the separation when molecular sieving is considered. Equation (4) remains valid for activated transport, but now attention should be drawn to the preexponential term, D_0. From the transition state theory, this factor may be expressed by[22]

$$D_0 = e\lambda^2 \frac{kT}{h} \exp\left(\frac{S_{a,d}}{R}\right), \quad (6)$$

where k and h are Boltzmann's and Planck's constants, respectively, and $S_{a,d}$ is the activation entropy for diffusion. This means that a change in entropy will give a significant contribution to the increase in selectivity when molecular sieving is considered. Singh and Koros have discussed this thoroughly.[21] Flux may now be described as follows:

$$J_a = \frac{\Delta p}{RTl} D_0 \exp\left(\frac{-E_{a,MS}}{RT}\right), \quad (7)$$

where $E_{a,MS}$ is activation energy for diffusion in the molecular sieving process. The sorption will, in this case, have little influence and thus the selectivity for separation will increase with increasing temperature due to increased diffusion rate of the permeating component. Likewise, it will become more difficult for the larger molecule to pass the narrow slit when temperature increases (see FIG. 2).

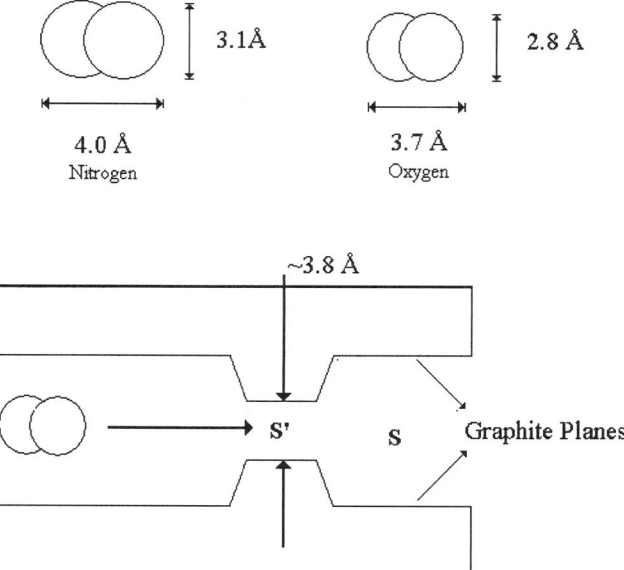

FIGURE 2. Idealized schematic cross section of 4 Å zeolite crystals and micropores in CMS. Average pore diameter (S) is 11.2 Å; constricted window (S') is about 3.8 Å.[21]

Capillary Condensation

For condensation to take place the pores must be in the mesopore range ($20 < d_p < 30$ Å).[23] Very high selectivity can be obtained by removing an easily condensable component from the gas stream and controlling the pressure. The drawback, however, is that in order to keep this as a continuous process, the carbon has to be "emptied" or regenerated quite often by pressure or temperature change. Separation according to this mechanism is basically what is taking place when activated carbon is used as a "bed" in a pressure or temperature swing adsorption (PSA or TSA) system and not really as a membrane. Further discussion on this mechanism is, therefore, not given here.

PRINCIPLES FOR THE PERMEABILITY MEASUREMENTS

For the results reported, permeation was measured the standard way, as described by Mulder[19] and calculations performed according to Equations (**8**) (for pure gas measurements) and (**9**) (for mixed gas measurements). For details about the specific experimental setup, we refer to Hägg.[24] The flux, J, was recorded at the given pressure difference, Δp ($= p_h - p_l$). The carbon membrane module (pyrolyzed cellulose) consisted of fibers mounted in a steel housing with the gas being fed into the fibers. The diameter of the fibers was approximately 165 µm and the effective permeation area was about 180 cm². The thickness of the selective membrane, l, was either known from SEM-analysis or given by the producer of the carbon fibers. The experimental set-up was mounted in an insulated cabinet with temperature control.

Permeation of the pure gas, i, was measured according to

$$J_i = \frac{P_i(p_{0,j} - p_{1,j})}{l} = \frac{P_i}{l}\Delta p_i. \qquad (8)$$

Mixed gas permeation measurements were performed according to

$$J_i = \frac{q_i}{A_m} = \frac{q_p y_{p,j}}{A_m} = \frac{P_i}{l}(p_h x_{0,j} - p_l y_{p,j}). \qquad (9)$$

Since the thickness of the selective membrane, l, was known, P_i could be calculated when composition of retentate and permeate ($x_{0,i}$ and $y_{p,i}$) were known.

An attached GC was available for online measurements of the composition of both the retentate and permeate stream, x_0 and y_p, with great accuracy. The flow of feed was measured (L/min) with a flowmeter and the retentate stream (L/min) was kept constant with a flow controller.

AGING OF THE CMS MEMBRANES TESTED

Aging is defined as the change in membrane performance caused by time and environment. The effects of oxygen, water, hydrocarbons, and/or carbon dioxide in the environment are here in focus because they are all related to the two major potential applications discussed in the paper (biogas separation and alkane/alkene separation). The aspects discussed are also of general interest when the change in performance for a CMS membrane is discussed.

The Influence of Oxygen on Performance

Oxygen is one of the most detrimental species for CMS membranes. However, oxygen may be used for postoxidation at elevated temperatures to increase the micropore volume and, thereby, the permeability flux.[25] As a result, the selectivity may be damaged, but only if the pore size distribution broadens.

When carbon materials are exposed to air at room temperature, irreversible chemisorption of oxygen may take place and C–O surface groups are formed.[26] These groups also provide sites for adsorption of H_2O. Both phenomena slightly reduce the effective size of micropores. The chemically bonded oxygen is only completely removed (as CO and CO_2) by heating the sample to temperatures as high as 700–800°C with an inert gas.

The Influence of Water Vapor on Membrane Performance

The effect of water vapor on carbon membrane performance has been studied by several authors.[27–30] It has been clearly documented that the adsorption of water is strongly influenced by both the nature of the carrier gas from which the adsorption take place, the humidity level in the gas, and of course, temperature. Adsorption kinetics varies for the different components; and when water vapor starts to fill the pores, other adsorbed components will be more or less easily removed. Adsorbed oxygen, for instance, is removed more easily than nitrogen.[27] This explains how water uptake from various gases with same humidity level may differ greatly. It has also been documented that water uptake from methane is much greater than from nitrogen.[28] Different sorption mechanisms are involved depending on vapor concentration, according to Gawryz et al.[29] At low relative humidities, only active polar centers seem to be involved, and this adsorption is so weak that the negative effect on separation can easily be managed. When relative humidity is too high (greater that 25%), the negative effect on separation may be substantial. Hydrogen bonding between neighboring water molecules leads to clusters of adsorbed water, which may trap the carrier gas molecules and form new sites for adsorption. This has also been documented by Jones and Koros[30] studying humid oxygen. With oxygen present, C–O surface groups may form and H_2O adsorbs more easily.[26]

The results from a permeability experiment with humidified nitrogen in the current study are shown in FIGURE 3A. The relative humidity was controlled by leading the feed gas (pressure of 2 bar and ambient temperature) through a solution of $MgCl_2$. The permeability flux of pure nitrogen is reduced by 25%. The membrane was then regenerated before permeability measurements were performed for humidified oxygen. A similar curve as for nitrogen was obtained (FIG. 3B). From these figures it may be documented that a decrease in O_2/N_2 selectivity from $\alpha = 10.5$ after 20 hours to $\alpha = 8.5$ after 80 hours occurred.

The Influence of Carbon Dioxide on Membrane Performance

CMS-membranes have been evaluated for removal of CO_2 from gas streams; in the current study removal from biogas is in focus. CO_2 is an easily condensable gas having a critical temperature (T_c) of 304 K, and critical pressure (p_c) of 73.9 bar. Its kinetic L–J diameter is equal to 3.9 Å.[31] This indicates that CO_2 should be removed fairly easy from a gas mixture of CO_2/CH_4, with CO_2 permeating according to the SSF mechanism and CH_4 retained if the pore size is correctly tailored. CO_2 does, however, also have a tendency to plug pores over time and, hence, reduce the permeability flux. In the current study, this was documented by exposing a new CMS membrane to pure CO_2 at ambient temperature over time (see FIGURE 4).

The membrane was regenerated (see next section) with an N_2 purge at 200°C after 165 and 225 hours of exposure. Different feed pressures (1.3 and 3.0 bar) were used, with the low-pressure side being constant at atmospheric pressure. Also, different regeneration times (1 h and 2 h) were tested. FIGURE 4 shows that the recovery of flux through the regenerations is limited. Increasing the time to 2 h indeed has a positive effect. The flux increase at first regeneration is 32%, whereas at second regeneration

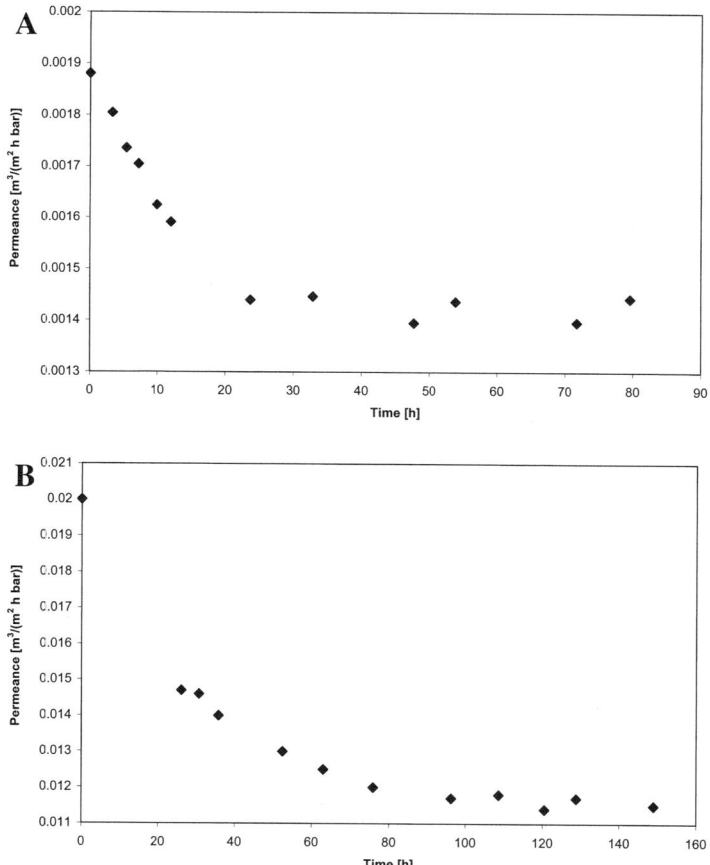

FIGURE 3. Influence of water vapor on **(A)** nitrogen permeance and **(B)** oxygen permeance. Humidified feed, nitrogen or oxygen with relative humidity about 25%.

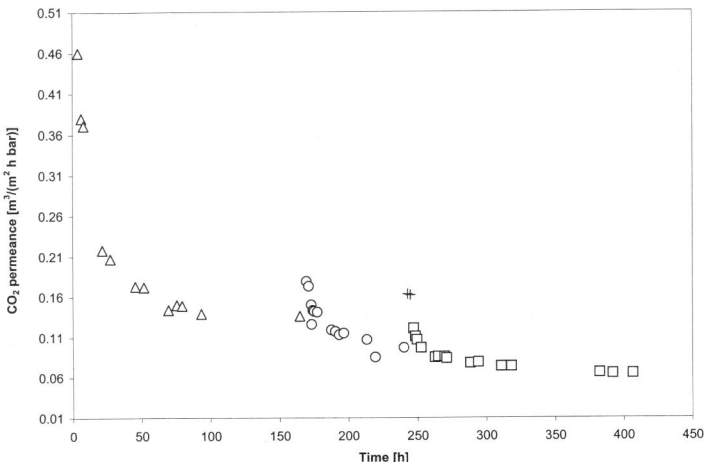

FIGURE 4. Permeance of CO_2 as a function of exposure time: (△) new membrane, feed pressure 1.3 bar; (○) membrane regenerated 1 h, feed pressure 1.3 bar; (+) membrane regenerated 2 h, feed pressure 1.3 bar; and (□) membrane regenerated 2 h, feed pressure 3.0 bar.

it is 68%. Increasing the feed pressure of CO_2 to 3 bar results in a more rapid decrease in permeability flux.

Approaches to Membrane Regeneration

To recover a decreased membrane flux, two main approaches are reported:

1. The membrane may be treated at elevated temperatures, at least 200°C, under vacuum or inert atmosphere. Regeneration with N_2 at 200°C was used in the current experiments (see previous section). Menendez and Fuertes[32] report only partially restored flux after regeneration of an air separating membrane at 600°C and vacuum (one hour). This could be the result of incomplete removal of C–O surface groups.
2. If exposed to organics, treatment of the membrane with propene may offer a good solution. Jones and Koros[33] found propene to be very effective in removing sorbed organics. In some cases the flux was completely restored. The effect of propene was also noted in the current experiments (see next section).

SEPARATION OF GASES WITH CMS MEMBRANES: TWO EXAMPLES

CMS Membrane Used for Alkane/Alkene Separation

Background

At petrochemical plants there are numerous gas streams within the process that contain valuable components which need to be recovered and reused; typically, non-reacted monomers, by-products from reactor, inerts, and solvents, carrier gas, polymers as dust particles, and so forth. The traditional separation technology in use is expensive and complicated with compressors, refrigeration systems, and columns. In the current study, integration of membranes has been evaluated for replacement of columns at a polyethene (PE) and polypropene (PP) plant. The most challenging separations are the alkane–alkene separations; for example propane–propene or propane–ethene mixtures. Chemically and physically the alkanes and alkenes are very similar compounds with almost identical critical properties[31] and molecular weight that differs by only 2 g/mole. However, the Lennard–Jones diameter varies from 4.7 Å to 5.1 Å for propene and propane, respectively. A regular dense membrane would, thus, give very low selectivity for the alkane/alkene pair. On the other hand, a carbon membrane operating in accord with the molecular sieving mechanism could theoretically obtain a much higher selectivity.

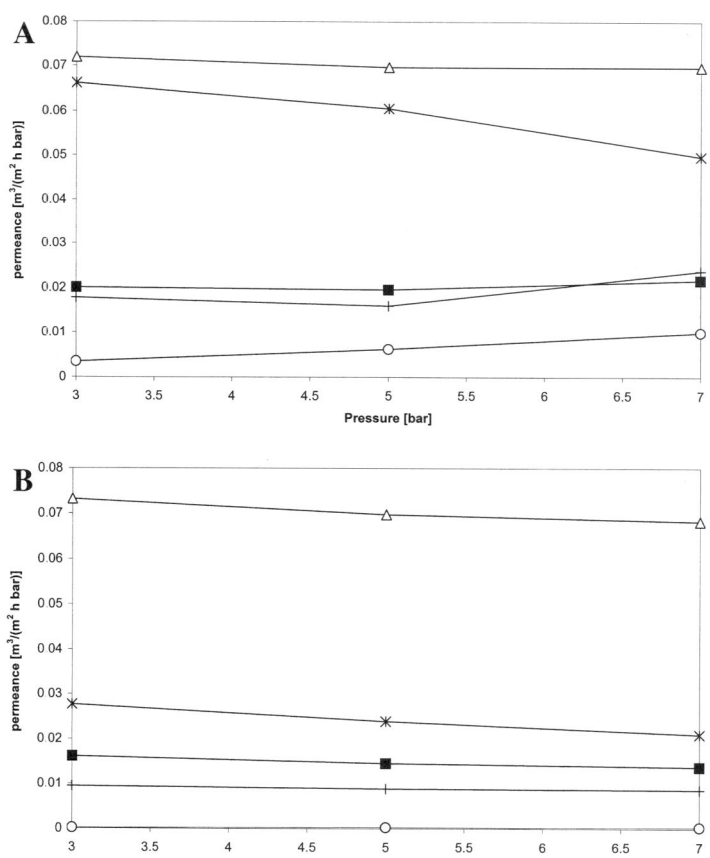

FIGURE 5. Permeances at **(A)** 30°C and **(B)** 50°C: ✻, nitrogen; ✕, ethene; △, propene, ▽, ethane, and ○, propane.

Pure Gas Permeation Tests

Pure gas permeation tests were performed as explained in PRINCIPLES FOR THE PERMEABILITY MEASUREMENTS. The gases tested were nitrogen, ethane, ethene, propane, and propene at 20°C to 80°C and 2 to 7 bar. FIGURE 5 A and B show the pressure dependence of the permeability for the tested gases at 30 and 50°C, respectively.

The nitrogen and ethene show no decrease, or a weak decrease, in permeability as a function of pressure at 30 and 50°C. Comparing FIGURE 5 A and B, for ethene, it is quite evident that we are moving from one flow regime into another in this temperature–pressure region—this was confirmed in later experiments (to be published). FIGURE 6 shows the temperature dependence of propane permeability for other samples of CMS membranes, hence pore size varies.

The minima in the permeability for propane can be explained by a combination of the selective surface flow and the molecular sieve mechanisms. The SSF dominates at 30°C according to the description of the mechanism given previously, whereas molecular sieving is dominant at 80°C. This leaves 50°C in the "middle" of the two mechanisms and, thus, at the minimum. This is in partial agreement with what other investigations have found.[34] The deviation may be caused by different mean pore size.

Mixed Gas Tests

The mixed gas permeation tests were performed as described in PRINCIPLES FOR THE PERMEABILITY MEASUREMENTS. Based on the pure gas experiments the highest propene/propane selectivity is expected at 3 bar and 50°C (the ideal selectivity from pure gases is 86). This assumption is used in the mixed gas experiments to investigate if this is achievable for a mixture. A slight decrease in selectivity was, however, recorded (about 50 for an equimolar mixture). Additionally for an equimolar mixture

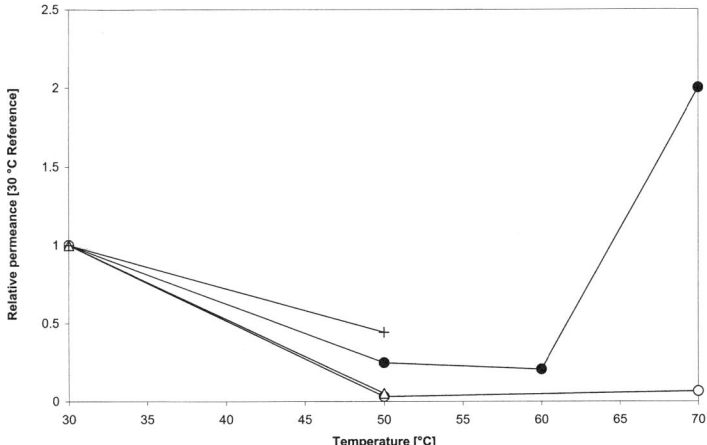

FIGURE 6. Temperature dependence for the propane permeance measured at 3 bar in various carbon membranes: ✻, Sample 1 (same as in FIGURE 5); △, Sample 2, first run; ○, Sample 2, second run; and ▽, Sample 3.

of ethene and propane (under the same process conditions) a selectivity of only 45 was obtained (the ideal selectivity calculated from pure gas experiments is 148).[35,36] These results may be difficult to understand but are clearly related both to the fact that the components have different temperature–pressure dependence and the possibility of aging of the membrane over time.

The experiments were performed within the temperature–pressure region where the alkenes (propene and ethene) still should permeate, according to the SSF mechanism, whereas propane has moved into another flow regime with a very low (and almost constant) flux.[35,36] However, as can be seen from FIGURE 5, the flux of ethene steadily decreased in this temperature–pressure range, but the change in permeation may also be influenced by changing pore size (as documented for propane in FIG. 6). Knowing that propene is acknowledged as an efficient gas for regeneration, the membrane may be continuously regenerated while running the mixed gas experiments, and propane may no longer block the pores but, rather, result in an increased flux. The selectivity for propene–propane will then decrease. The same effect may apply to the ethene–propane mixture. These effects are currently being studied.

CMS Membranes Used for Separation of Biogas

Background

In this context, biogas is the gaseous mixture evolved from microbial digestion of organic waste in absence of molecular oxygen. The main components are methane (CH_4) and carbon dioxide (CO_2). Biogas is a renewable energy resource that at present is exploited only to a small extent. The use of biogas gives no net contribution of CO_2 to the atmosphere, provided that the production and distribution is based on biofuels. In addition, a reduction in the greenhouse effect is achieved by collecting and burning the CH_4 gas.

Upgraded biogas is well suited for vehicle fuel. Compared to engines using conventional diesel and gasoline, exhaust emissions (especially NO_x and particles), and noise is reduced. Maltesson[37] has summarized the specification of biogas meant for vehicle engines: content of CH_4 must be 96–100 vol% and maximum content of H_2O 32 mg/Nm^3.

The composition of the feed gas is important for CO_2/CH_4 selectivity. Ogawa and Nakano[38] report an increase in selectivity as the feed CO_2 concentration increases. This may be explained in terms of the hindrance of CH_4 permeation by CO_2 molecules existing inside the micropores.

Two Examples from Pilot Plant Tests

Example 1: At a municipal waste and wastewater treatment plant in Southern Norway, biogas is produced. The company wanted to evaluate the possibility of producing upgraded biogas for fuel to vehicles. A CMS module was tested on-site for 12 hours. The gas was dried with silica gel before entering the membrane module. The results obtained were in accord with laboratory results obtained with synthetic biogas (mixtures of CO_2 and CH_4). The selectivity of CO_2 over CH_4 was found to be about 50 at a temperature of 50°C. The total flux of permeating gas was 0.027 m^3(STP)/(m^{2}h). For the gas stream in question this would require a membrane area of 21,000 m^2, and give a recovery of 99% for methane (very low cut rate). If a

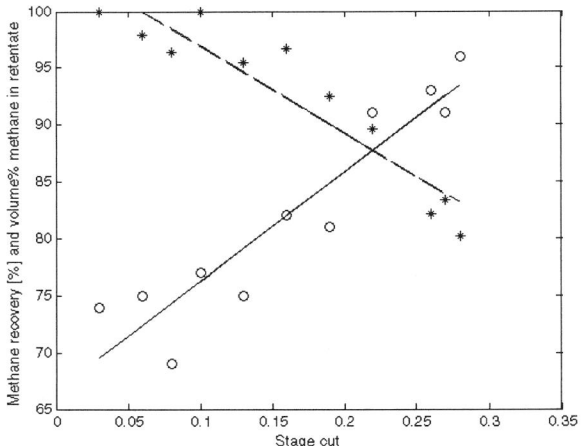

FIGURE 7. Concentration of CH_4 (○) and percent CH_4 recovery (✶) in retentate product stream as a function of stage cut.

recovery of only 70% methane could be accepted, the required area would be significantly reduced.

Example 2: Tests for upgrading biogas with CMS-membranes were performed on pilot scale at a commercial site for biogas production in Denmark. The results clearly documented the importance of operating with an optimized stage cut, θ (the ratio between permeate and feed volumetric flow rate), in order to both meet the specifications for the gas and recover maximum amount of methane. FIGURE 7 illustrates the trade-off that was found between the purity of the retentate (gas product) and the recovery of methane.

To reach the required concentration of CH_4 (min, 96 vol%), it is recommended that θ is in the range 0.3–0.5. A value of θ << 0.5 indicates too small membrane area for the given feed gas flow, whereas θ > 0.5 indicates an over dimensioned area.

Discussion

The effect of humidity on carbon membranes has been broadly studied (as reported already) and an effect of significantly reduced permeance (about 25%) was also documented in the current study for gas streams with high relative humidity (see FIG. 3). Likewise, a reduction in selectivity for N_2/O_2 was expected from theory. Water vapor will always be present in a biogas stream but, then, usually at a fairly low humidity level (below 10%) in the system. The effect of flux reduction due to humidity could only be detected at one of the pilot plant modules tested, and permeance was fairly easy to recover by periodic regeneration of the membrane with dry air at 70°C. The regeneration was performed after every third run, which meant about once a week and then over a period of 1–2 hours. The oxygen present in the air may of course also contribute to pore opening by oxidation or formation of C–O surface groups at this temperature.

The tendency of CO_2 to block the pores, hence resulting in reduced flux, was documented in FIGURE 4—the membrane was then exposed to pure CO_2. The same figure also clearly shows how flux is recovered with regeneration of the membrane, but only to a certain level depending on the type of regeneration used. For better recovery of flux, a higher temperature should be used. This indicates that for a carbon membrane used for biogas separation, the process conditions should be carefully controlled and regular membrane regeneration integrated as part of the process procedure. The decrease in permeability flux is not so severe when the CO_2 concentration is about 30% and in a mixture with CH_4 and N_2 as it was in the two pilot plant tests. Hence, the CMS membrane may still be a good choice for this separation. More optimal conditions for regenerations may also be found.

Membranes with an activated transport mechanism is probably a good option for separation of alkane/alkene mixtures, as reported here for propane/propene/ethene. Tailoring of correct pore size is extremely important in order to obtain the best trade-off between selectivity and permeability; likewise to set the optimum process conditions with respect to temperature and pressure. Depending on these parameters, one may obtain higher or lower selectivity of a mixed gas compared to the pure gas experiments. In the reported mixed gas results we were not able to document increased selectivity for equimolar propene–propane or ethene–propane gas mixtures. There may be several reasons for this. For both mixtures the alkenes may act as a regeneration gas by opening the pores, hence reducing selectivity, but more likely the propane is hindering the permeating component (here propene or ethene). Additional documentation is needed to fully understand the observed effect. Experiments should be carried out with carbon membranes of varying pore size for the same gas mixture.

CONCLUSIONS

The current study has shown promising results for separation of components in a mixture such as propane–propene–ethene using CMS-fiber membranes. The importance of correct pore tailoring and process separation conditions have also been proved, likewise the need for regular regeneration of the membrane when exposed to humidity and CO_2. Assuming this is taken into account, CMS membranes may be a good choice for upgrading of biogas to fuel specifications. The carbon membranes are, however, judged as not suitable for gas streams with high relative humidity. Oxygen may be used for restoring pore size, but may also ruin the membrane if exposure is carried out at higher temperature (hence, oxidation of the membrane).

ACKNOWLEDGMENTS

The authors thank Carbon Membranes Ltd. in Israel for supplying test modules of CMS-membranes.

REFERENCES

1. ISMAIL, A.F. & L.I. DAVID. 2001. A review of the latest development of carbon membranes for gas separation. J. Membr. Sci. **193:** 1–18.
2. RAO, M.B. & S. SIRCAR. 1993. Nanoporous carbon membranes for separation of gas mixtures by selective surface flow. J. Membr. Sci. **85:** 253–264.
3. FUERTES, A.B. & T.A. CENTENO. 1999. Preparation of supported carbon molecular sieve membrane. Carbon **37:** 679.
4. SUDA, H. & K. HARAYA. 1997. Alkene/alkane permselectivities of carbon molecular sieve membrane. J. Chem. Soc., Chem. Commun. **1:** 93.
5. Ullmann's Encyclopedia of Industrial Chemistry. 1988. Pressure swing adsorption. Vol. B-3: Section 9, 35–36. VCH Weinheim.
6. MOREIRA, R.F.P.M., H.J. JOSE & A.E. RODRIGUES. 2001. Modification of pore size in activated carbon by polymer deposition and its effect on molecular sieve selectivity. Carbon **39:** 2269–2276.
7. SOFFER, A., J. GILRON, S. SAGUEE, *et al.*, Inventors. 1995. European Patent 95103272.1. Date of publication of application: September 13.
8. KORESH, J.E. & A. SOFFER. 1980. Study of molecular sieve carbons. Part 1. Pore structure, gradual pore opening and mechanism of molecular sieving. J. Chem. Soc., Faraday Trans. **76:** 2457.
9. KORESH, J.E. & A. SOFFER. 1980. Molecular sieving range of pore diameters of adsorbents. J. Chem. Soc., Faraday Trans. **76:** 2507.
10. FUERTES, A.B. & T.A. CENTENO. 1998. Carbon molecular sieve membranes from polyetherimide. Micropor. Mesopor. Mater. **26:** 23.
11. GEIZLER, V. & W.J. KOROS. 1996. Effects of polyimide pyrolysis conditions on carbon molecular sieve membrane properties. Ind. Eng. Chem. Res. **35:** 2999–3003.
12. MACELROY, J.M.D., S.P. FRIEDMAN & N.A. SEATON. 1999. On the origin of transport resistance within carbon molecular sieves. Chem. Eng. Sci. **54:** 1015–1027.
13. RAO, M.B. & S. SIRCAR. 1996. Performance and pore characterization of nanoporous carbon membranes for gas separation. J. Membr. Sci. **110:** 109–118.
14. GILRON, J. & A. SOFFER. 2003. Knudsen diffusion in microporous carbon membranes with molecular sieving character. J. Membr. Sci. **209:** 339–352.
15. BHAVE, R.R. 1991. Inorganic Membranes; Synthesis, Characteristics and Applications, Chapter 6. Van Nostrand Reinhold, New York.
16. FLEMING, H.L. 1986. Membrane Technology/Planning Conference Proceedings. Business Communications Co., Cambridge, MA.
17. GEANKOPLIS, C.J. 1993. Transport Processes and Unit Operations, 3rd edit. Chapter 7. Prentice-Hall, Englewood Cliffs.
18. KESTING, R.E. 1985. Synthetic Polymeric Membranes: A Structural Perspective, 2nd edit. 1–18. J. Wiley & Sons, New York.
19. MULDER, M. 1996. Basic Principles of Membrane Technology, 2nd edit. Kluwer Academic Publishers, Dordrecht.
20. BURGGRAAF, A.J. 1999. Single gas permeation of thin zeolite (MFI) membranes: theory and analysis of experimental observations. J. Membr. Sci. **155:** 45.
21. SINGH, A. & W.J. KOROS. 1996. Significance of entropic selectivity for advanced gas separation membranes. Ind. Eng. Chem. Res. **35:** 1231–1234.
22. GLASSTONE, S., K.J. LAIDLER & H. EYRING. 1941. The Theory of Rate Processes, 1st edit. McGraw-Hill Book Co., New York.
23. SAKATA, J. & M. YAMAMOTO, Inventors; Kabushiki Kaisha Toyota Chuo Kenkyusho, Assignee. 1986. U.S. patent 4,583,996. Date of application: Nov. 2. 24.
24. HÄGG, M.-B. 2000. Membrane Purification of Chlorine Gas. Dissertation, Norwegian University of Science and Technology, Trondheim.
25. KUSAKABE, K., *et al.* 1998. Gas permeation and micropore structure of carbon molecular sieving membranes modified by oxidation. J. Membr. Sci. **149:** 59–67.
26. MATTSON, J.S. & H.B. MARK. 1971. Activated Carbon. Marcel Dekker, New York.
27. IFAT, A., A. DANON & J.E. KORESH. 1999. Water coabsorption effect on the physical adsorption of N_2 and O_2 at room temperature on CMS-fibers. Phys. Chem. Chem. Phys. **1:** 479.

28. GROSZEK, A.J. 2001. Heats of water adsorption on microporous carbons from nitrogen and methane carriers. Carbon **39:** 1857–1862.
29. GAWRYS, M., P. FASTYN, J. GAWLOWSKI, et al. 2001. Prevention of water vapor adsorption by carbon molecular sieves in sampling humid gases. J. Chrom. A. **933:** 107–116.
30. JONES, C.W. & W.J. KOROS. 1995. Characterization of ultramicroporous carbon membranes with humidified feeds. Ind. Eng. Chem. Res. **34:** 158–163.
31. REID, R.C. & T.K. SHERWOOD. 1966. The Properties of Gases and Liquids. McGraw-Hill, New York.
32. MENENDEZ, I. & A.B. FUERTES. 2001. Aging of carbon membranes under different environments. Carbon **39:** 733–740.
33. JONES, C.W. & W.J. KOROS. 1994. Carbon molecular sieve gas separation membranes—II. Regeneration following organic exposure. Carbon **32:** 1427–1432.
34. YAMAMOTO, M., K. KUSAKABE, J. HAYASHI & S. MOROOKA.1997. Carbon molecular sieve membrane formed by oxidative carbonization of a copolyimide film coated on a porous support tube. J. Membr. Sci. **133:** 195–205.
35. HÄGG, M.-B. & B. HALVORSEN. 2000. Membranes in petrochemical industry I: the recovery of propene/propane from gas streams. Presented at the Euromembrane 2000 conference. Jerusalem, Israel, September 27.
36. HÄGG, M.-B., A. LINDBRÅTHEN & J.A. LIE. 2001. Membranes in petrochemical industry II: a comparison of integrated membrane modules with traditional separation technology. Presented at the UEF-Conference on Advanced Membrane Technology. Barga, Italy, October 16.
37. MALTESSON, H.Å. 1997. Biogas for vehicles—quality spesifications (in Swedish). Kommunikationsforsknings-beredningen (KFB4). Stockholm.
38. OGAWA, M. & Y. NAKANO. 2000. Separation of CO2/CH4 mixture through carbonized membrane prepared by gel modification. J. Membr. Sci. **173:** 123–132.

Thin Composite Palladium and Palladium/Alloy Membranes for Hydrogen Separation

YI HUA MA, IVAN P. MARDILOVICH, AND ERIK E. ENGWALL

Center for Inorganic Membrane Studies, Department of Chemical Engineering, Worcester Polytechnic Institute, Worcester, Massachusetts, USA

> ABSTRACT: Dense composite Pd and Pd/alloy membranes are currently being extensively investigated. The synthesis and characterization of these membranes, with a special emphasis on Pd/alloy membranes, are reviewed in this paper. Experimental results on Pd/Cu membranes supported on porous stainless steel exhibited good thermal stability and reasonable hydrogen flux. Furthermore, optical micrographs showed the formation of the dense palladium layer was unaffected by the topological features of the porous stainless steel, although the surface of the support directs the topology of the final Pd layer.
>
> KEYWORDS: thin composites; palladium membrane; palladium/alloy membrane; hydrogen separation

INTRODUCTION

The twenty-first century has been termed by some to be the century of the global hydrogen economy. Hydrogen is becoming one of the key energy resources of the future and the focal point of strong global concern. At the heart of this global hydrogen economy is worldwide research and development for low-cost production of hydrogen from hydrocarbons or water. One of the most effective ways to produce pure hydrogen is to use inorganic membranes for hydrogen separation and membrane reactor applications. Dense composite Pd and Pd/alloy membranes supported on a porous substrate, in particular, porous stainless steel (PSS), are especially suited to these applications. This paper provides a review on the preparation and characterization of thin composite Pd and Pd/alloy membranes and discusses the effects of the substrate properties on the Pd layer thickness and permeance.

COMPOSITE PALLADIUM MEMBRANES

Early work on hydrogen permeation in palladium was summarized in detail by Lewis[1] and most work involved the use of palladium foils. For example, measurements of hydrogen permeability in α-palladium carried out by Koffler *et al.*[2] using a low temperature technique under conditions such that the bulk diffusion was the

---

Address for correspondence: Yi Hua Ma, Center for Inorganic Membrane Studies, Department of Chemical Engineering, Worcester Polytechnic Institute, Worcester, MA 01609, USA.

rate-limiting step, demonstrated a square root pressure dependence of the hydrogen permeability in the temperature range 300 to 709 K and pressure range 2.9×10^{-5} to 5.0×10^{-3} cmHg. Using their permeability values in conjunction with the hydrogen solubility data of Favreau et al.,[3] Koffler et al.[2] concluded that the diffusion coefficient was not affected by substructural defects in the palladium bulk layer. Balovnev[4] also found a square root pressure dependence in the ultra-high vacuum region under catalytic diffusion pumping conditions. He further concluded that sorption of hydrogen in palladium obeyed Henry's law, under low pressure conditions.

As pointed out later in this paper, there are a number of advantages to using composite palladium membranes supported on porous substrates over palladium foils and tubes. The porous supports used by various investigators comprised porous alumina and glass, and porous metals including porous Ni and porous stainless steel. Some of the techniques used for the deposition of Pd or its alloys on a support were the spray pyrolysis method to deposit Pd–Ag alloy on an alumina support,[5] the metal–organic chemical vapor deposition (MOCVD) technique by decomposing palladium (II) acetate in argon under a reduced pressure to form a thin palladium membrane inside the porous wall of an α-alumina tube,[6,7] the supercritical fluid transport-chemical deposition (SFTCD) method using the metal β-diketonate complex, (2,2,7-trimethyl-3,5-octanedionato) palladium (II), to pyrolytically deposit a thin 1–2 μm Pd layer,[8] the sputter-deposition technique to deposit an ultrathin Pd layer on polymeric membranes,[9] porous alumina,[10,11] anodic alumina,[12,13] and Vycor glass,[14] magnetic sputtering to deposit Pd and Pd-alloys on polymer membranes, porous stainless steel, and oxide plates,[15] and electron beam evaporation and ion-beam sputtering to deposit Pd on the surface of tantalum foil.[16] Some other examples of the application of these techniques, as well as physical vapor deposition and electroplating, are briefly discussed in the review by Shu et al.[17] The major drawbacks of all these powerful techniques, that are especially useful for the deposition of alloys, are low area of the prepared membranes and/or high cost of the necessary equipment. Therefore, the practical application of these techniques requires further investigation.

As pointed out by Mardilovich et al.,[18] electroless plating is quite attractive due to the possibility of uniform deposition on complex shapes and large substrate areas, hardness of the deposited film, and very simple equipment. Electroless plating has been used to deposit a Pd film on a wide variety of supports, including tantalum and niobium tubes,[19,20] porous silver,[21] porous glass,[22,23] porous alumina,[24–28] and porous stainless steel,[18,29,30,31] Pd-alumina,[32] Pd-Vycor,[33] and Pd/PSS[34] composite membranes have been prepared by electroless plating using osmotic pressure to manipulate the microstructure, porosity, and thickness of the deposited metal. A controlled in situ oxidation of the porous stainless steel to produce an oxide layer to minimize the intermetallic diffusion was developed by Ma et al.[35] The Pd membrane produced by this method has been shown to be stable for more than 6,000 hours between 350 and 450°C.[30,31]

The main advantages of porous stainless steel (PSS) supports over porous ceramics and Vycor glass are resistance to cracking and simplicity of module construction. Composite Pd/PSS membranes, welded from both ends with non-porous stainless steel tubes, can be very easily assembled. Additionally, the thermal expansion

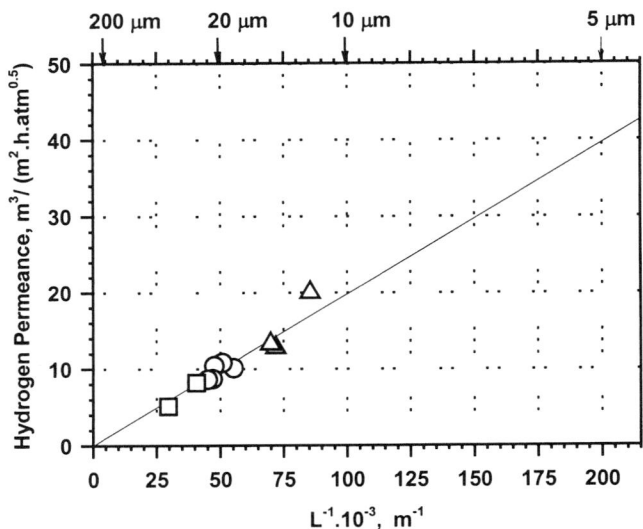

FIGURE 1. Hydrogen permeance versus reciprocal palladium thickness for a pressure difference of 1 atm and a temperature of 350°C. The Pd layer thickness is indicated along the top. The grade of the supports were: □, 0.5 μm; ○, 0.2 μm; △, 0.1 μm.

coefficient of stainless steel is almost identical to that of palladium, insuring good mechanical properties of the composite membrane during temperature cycling.

There are few studies dealing with the inverse relationship between the permeance and the thickness of the palladium layer. Koffler et al.[2] measured the specific permeability of hydrogen in Pd foils with three different thicknesses and found an inverse relationship between specific permeability and thickness. The experimental results[36] from a systematic study of composite Pd/PSS membranes with various Pd layer thicknesses showed an inverse relationship between the hydrogen permeance and Pd layer thickness at 350°C. As shown in FIGURE 1, these experimental results covered the range of Pd layer thickness from 11.7 μm to 33.8 μm, an almost threefold difference in membrane thickness. This inverse relationship is additional evidence that, even at such small membrane thicknesses, the permeation is still controlled by hydrogen diffusion through the Pd bulk without any influence from mass transport at the membrane surface. If the permeation were controlled by mass transport at the surface, the permeance would be affected by the surface area and independent of the membrane thickness.

PALLADIUM ALLOYS

It has long been recognized that Pd alloys may have advantages over Pd for use in hydrogen separation and membrane reactor applications. In general Pd alloys have a reduced critical temperature for the α–β phase transition.[1] Therefore, alloy membranes may be operated in the presence of H_2 at temperatures below 300°C without

TABLE 1. Improvement in hydrogen permeability for various binary Pd alloys at 350°C (after Knapton[37])

Alloy Metal	Weight Percentage for Maximum Permeability	Normalized Permeability (P_{alloy}/P_{Pd})
Y	10	3.8
Ag	23	1.7
Ce	7.7	1.6
Cu	40	1.1
Au	5	1.1
Pure Pd	—	1.0

the hydrogen embrittlement observed for pure Pd membranes.[17] In some cases, the hydrogen permeability of alloys is greater than that of Pd. This is true for binary palladium alloys with yttrium, cerium, silver, copper, and gold for the appropriate compositions, as shown in TABLE 1 based on data from Knapton.[37] Several tertiary Pd alloys, such as Pd–Ag–Au and Pd–In–Ru, are also reported to have high hydrogen permeability.[17,38] Finally, some Pd alloys may produce membranes with enhanced chemical resistance. It has been suggested that both the Pd–Cu and Pd–Au systems can be used to produce membranes with enhanced resistance to H_2S.[39,40]

Of the metals listed above for binary palladium alloys, Cu and Ag represent the least expensive alternatives from the standpoint of materials costs. Palladium–silver alloys are attractive because they have good hydrogen permeability relative to pure palladium. In the case of Pd–Ag alloys, the solubility of hydrogen increases with increasing Ag content reaching a maximum at an Ag content between 20–40 wt%.[1] However, the diffusivity of hydrogen in Pd–Ag alloys decreases with increasing silver content. The simultaneous changes in solubility and diffusivity result in a maximum in the hydrogen permeability of 1.7 times the permeability of pure palladium, at a silver content of 23 wt% and a temperature of 350°C.[37]

Palladium–copper alloys are attractive from the standpoint of relatively high permeabilities coupled with their enhanced sulfur resistance. In the case of Pd–Cu alloys, the formation of a solid phase with a BCC structure at 40 wt% copper, leads to an increase in the hydrogen diffusivity of two orders of magnitude, and thus a sharp maximum in permeability for this composition.[17] Knapton[37] reports a hydrogen permeability for the palladium alloy with 40 wt% copper to be slightly higher than that of pure palladium at 350°C. However, McKinley et al.[40] report hydrogen permeabilities in the range of 1.25–1.5 times that of pure palladium for the same alloy.

COMPOSITE Pd–ALLOY MEMBRANES

Most of the fundamental work concerning the interaction of hydrogen with palladium and palladium alloys has been accomplished with foils or tubes or other bulk samples. The industrial use of self-supporting Pd membranes has been limited for economic reasons. During the more recent past, efforts to produce economically viable Pd membranes have focused on supported composite Pd membranes. Composite membranes use a thinner dense Pd layer applied to a porous or otherwise permeable structural support. The advantages of composite membranes are increased hydrogen flux and reduced consumption of noble metals. In the case of palladium alloy membranes, the challenges of applying a thin dense layer to a support are augmented by the difficulties of forming this layer from two or more metals and obtaining a uniform alloy throughout the layer.

The supports and metal deposition methods that have been used for the preparation of composite palladium–alloy membranes are largely drawn from the same list as those used for pure palladium composite membranes. However, in reviewing the synthesis of composite palladium–alloy membranes, it quickly becomes apparent that the formation of a dense homogeneous alloy layer is the critical step. Therefore, this discussion is organized based on the means by which the various authors have attempted to form true alloys. The approaches of greatest interest from the standpoint of producing large membrane modules involve aqueous plating methods. Work in this area may be categorized as efforts at the deposition of successive layers of metals followed by a thermal treatment, or attempts to codeposit two metals simultaneously.

COATING AND DIFFUSION

The method of forming an alloy membrane by thermal treatment of discrete metal layers has been referred to as "coating and diffusion treatment".[23] This technique has been employed extensively to prepare Pd–Ag composite membranes on various supports.[23–25,41–43] It was also used to prepare Pd–Cu composite membranes by Uemiya et al.,[23] Nam et al.,[44] and the present authors. The synthesis procedures, porous supports used, and other basic characteristics of membranes produced in these studies are summarized in TABLE 2. For most of these studies, electroless plating was used to deposit a single layer of Pd on the support, followed by a single layer of Ag or Cu. After plating, the membranes were formed by heating to temperatures (T) and times (t) as indicated. However, Nam et al.[44] and the present authors have explored the formation of supported membranes from multiple layers of each alloy component for Pd–Cu/PSS membranes.

All of these studies achieved some degree of alloy formation, as indicated by an increase in the hydrogen permeation following the heat treatment. For the Pd–Ag system, the formation of membranes with uniform compositions throughout the layer was only conclusively demonstrated by Uemiya et al.[24] and Kikuchi et al.[25] on a porous alumina support. For the former case, the Pd and Ag layers were deposited on an asymmetric porous alumina tube. The average pore size of these supports on the side to which the metals were applied was 200 nm. The diffusion treatment was accomplished at a temperature of 900°C for a period of 12 hours under an argon atmosphere. For the Pd–Ag membranes summarized in TABLE 2, this represents the highest diffusion treatment temperature. Over an Ag composition range of 0–30.5 wt%, this approach resulted in uniform Pd and Ag compositions throughout the membrane, as shown by WDX-EPMA.[25]

TABLE 2. Synthesis conditions for Pd–Ag and Pd–Cu membranes prepared by the coating and diffusion treatment method[a]

Alloy	Wt%	Support	δ (μm)	T (°C)	t (h)	Reference
Ag	7.0	PG	21.6	500	12	Uemiya et al.[23]
Ag	0–31	Al_2O_3	4.5–6.4	900	12	Uemiya et al.[24]
Ag	5.1	PSS	10.3	550	5[b]	Shu et al.[41]
Ag	26	PSS[c]	16.8	500–800	5[b]	Shu et al.[42]
Ag	20–25	Al_2O_3	1.4–2.2	500–600	varied	Keuler et al.[43]
Cu	37	PSS[c]	< 2	500	5	Nam et al.[44]
Cu	22–26	PSS[c]	33.8–35.2	600–675	16	this work
Cu	6.2	PG	18.9	500	12	Uemiya et al.[23]

[a]Membrane thickness (δ), treatment temperature (T), and treatment time (t).
[b]Treated under H_2.
[c]Diffusion barrier applied to support.

From an industrial perspective, alumina supports have disadvantages in terms of the ease of sealing and the fabrication of large modules. However, the stability of the these supports at high diffusion treatment temperatures enabled a very thorough evaluation of the influence of membrane composition for the Pd–Ag system. Therefore, the work by Uemiya et al.[24] and Kikuchi et al.[25] provides the best basis for comparison of Pd–Ag alloy properties, between a self consistent set of composite membranes and previous work on unsupported tube and foil membranes. Uemiya et al.[24] and Kikuchi et al.[25] found the hydrogen permeation coefficient at 400°C on a series of Pd–Ag alloy membranes with similar thickness, to pass through a maximum at a composition of 23 wt% Ag. The permeation coefficient of the 23%-Ag membrane was about double the value for pure Pd. These results are in good agreement with the properties of the Pd–Ag alloy as reviewed by Shu et al.[17] and Gryaznov.[38] Additionally, Uemiya et al.[24] and Kikuchi et al.[25] found that membranes with 23–30.5 wt% Ag maintained high selectivity down to a temperature of 200°C. This limiting temperature increased to 250°C for 11% Ag and to 325°C for pure Pd. These results provide clear evidence of suppression of the α–β phase transformation for these membranes.

The work of Uemiya et al.[24] and Kikuchi et al.[25] with alumina supports provides a good benchmark for the anticipated performance from composite Pd–Ag membranes. However, the literature indicates that it is considerably more difficult to form Pd–Ag alloy membranes by coating and diffusion treatment on other supports. In the case of porous glass (PG) supports, Uemiya et al.[23] were able to prepare a dense Pd–Ag membrane having good hydrogen selectivity in the temperature range 200–300°C. This membrane had a silver composition of 7 wt%. The porous glass supports had an average pore size of 300 nm. The diffusion treatment for this membrane was carried out at 500°C for 12 hours. The pore structure of the PG support collapses in the temperature range 550–600°C, placing a limitation on the temperature of the diffusion treatment with PG supports.

The formation of an alloy was indicated by XRD results and by an increase in hydrogen permeation of samples plated with both Pd and Ag, following their heat treatments. However, the membranes had lower permeability than similar membranes prepared using Pd only. If the uniform distribution of Ag at a composition of 7 wt% had been achieved, the Pd–Ag membrane should have a greater permeability than the Pd only membrane. Therefore this result indicates a higher Ag composition on the surface. EPMA results after heating confirmed this showing that the membrane consisted of a layer of pure Pd next to the support, with a Pd–Ag alloy region on top of it. The Ag composition of the alloy region increased towards the outer surface of the membrane.

Pd–Ag alloys formed by coating and diffusion treatments were evaluated on porous stainless steel supports (PSS) by Shu et al.[41,42] Membranes were prepared on the outside of 0.2 μm grade, 0.95 cm outside diameter PSS tubes manufactured by Mott Metallurgical. These membranes had the relatively low Ag composition of 5.1 wt%. A treatment temperature of 550°C was used for five hours under a hydrogen atmosphere.[41] For Pd–Ag alloy membranes in the presence of hydrogen, Pd and Ag have been shown to segregate over a wide range in operating temperatures.[45] The surface facing the highest hydrogen partial pressure becomes Pd enriched and the surface facing the lowest partial pressure becomes enriched in Ag. Therefore, the application of hydrogen over the Ag plated side of the membrane during initial heating may aid the diffusion treatment by drawing more Pd towards the exposed surface and drawing more Ag towards the membrane/support interface. However, Shu et al.[41] note that a similar treatment at the slightly lower temperature of 500°C produced a membrane in which the silver only partially diffused into the palladium and no quantitative characterization of the alloy composition was provided for the 550°C treatment.

Diffusion treatment at 600°C for five hours under hydrogen for Pd–Ag/PSS membranes using the 0.2 μm grade Mott support resulted in significant diffusion of Fe from the support into the membrane, as shown by Auger electron depth profiling.[42] It should be added that a deterioration in hydrogen permeability and an increase in inert gas flux were observed at temperatures above 550°C for a pure Pd membrane on a Mott PSS support without a diffusion barrier.[18] These results indicate the onset of intermetalic diffusion between the membrane and the PSS support between 550 and 600°C. Therefore, 550–600°C should be regarded as the reasonable limit for both operation and alloy formation if PSS supports are used without a diffusion barrier.

The application of a 0.1-μm titanium nitride diffusional barrier to the outer surface of a PSS support prior to plating was explored by Shu et al.[42] This was evaluated for diffusion treatments at 500 to 800°C for five hours under hydrogen. The thicknesses of the Pd and Ag layers prior to heat treatment were 14.8 μm and about 2 μm, respectively. Auger electron depth profiling showed that the dispersion of Ag across the entire Pd layer requires a treatment temperature of 700°C. However, the Ag composition was not completely uniform and was lower closest to the edge of the diffusion barrier. For the membrane treated at 800°C, the Auger results show significant diffusion of Fe into the membrane. Unfortunately, this study did not provide hydrogen permeation data for the membranes.

Application of the coating and diffusion treatment method to form Pd–Cu composite membranes has been far more limited than work with the Pd–Ag system.

Uemiya et al.[23] produced a Pd–Cu membrane with 6.2 wt% Cu on a porous glass support. The heat treatment in this case was 12 hours at 500°C under argon. Good hydrogen selectivity was observed in the temperature range 200–300°C. The resulting membrane had a lower hydrogen permeability than similar membranes prepared with either a Pd–Ag alloy or pure Pd on porous glass. This is not surprising because higher permeability for Pd–Cu alloys is only expected in a narrow range of composition near 40 wt% Cu.[17,37,39,40]

Recently a study of Pd–Cu/PSS membranes has appeared.[44] This work differs from the others in this section because the Pd and Cu layers were applied by vacuum electrodeposition, which has been described elsewhere.[46] The PSS supports were grade 0.5 µm obtained from Mott Metallurgical Corporation. Prior to deposition of the metals, the support was prepared by application of a submicron nickel powder on the surface followed by sintering at 800°C for five hours under high vacuum. The support was then coated on the same side with a thin intermediate layer of silica using the sol–gel method. Pd and then Cu were successively applied to the support for a total of two layers of each, with the first layer being Pd and the final layer being Cu. Heat treatment was carried out at 500°C for five hours under nitrogen. The final membrane was reported to be about 2 µm thick and had a Cu composition of 37 wt%. This indicates that individual metal layers were about 0.5 µm thick prior to heat treatment.

Prior to heat treatment the XRD patterns for these membranes had lines for both the Pd and Cu phases. After the treatment, only the lines corresponding to a Pd–Cu alloy were present. Additionally, there was no sign of diffusion of metals from the support into the membrane. Hydrogen and nitrogen permeation experiments were performed at 350, 400, and 450°C. The H_2/N_2 selectivity was measured at all temperatures with a transmembrane pressure difference of 10 psi, and found to be greater than 10,000 in all cases. The long term stability of the membrane was tested by cycling the membrane 18 times between hydrogen and nitrogen permeation during a 40-day period at a temperature of 450°C. Both the hydrogen and nitrogen permeance values remained constant over this period. EPMA was performed on a cross section after 40 days of testing and showed that both Cu and Pd were very uniformly distributed throughout the membrane.

Recently, the preparation of Pd–Cu/PSS composite membranes has been accomplished by the present authors using electroless plating with the coating and diffusion method. The supports for this work were 0.2 µm and 0.5 µm grade, 0.5 inch outside diameter (OD) tubes of 316L stainless steel, provided by Mott Metallurgical. Support modules were prepared by welding sections of 0.5 inch OD 316L stainless steel tubing to both ends of a porous section that was six inches in length. An oxide diffusion barrier was formed according to the procedure described by Ma et al.[35] The supports were then cleaned and activated as described by Mardilovich et al.[18] Following this, the supports were plated with successive layers of Pd and Cu. The electroless plating procedure used for Pd is described by Mardilovich et al.[18] The procedure for electroless plating of Cu was adapted from those described by Vansovskaya.[47] The details of the Cu plating bath solution used are given in TABLE 3. An aqueous $CuSO_4$ solution was prepared using Na_2EDTA as a complexing agent and several stabilizers. Sodium hydroxide was added as needed to obtain a pH of 12.6–12.8. The aqueous formaldehyde reducing agent was added just before

TABLE 3. Copper plating bath composition

Component	Amount
$CuSO_4 \cdot 5H_2O$	30 g/L
Na_2EDTA	60 g/L
NaOH	18–20 g/L
Formaldehyde (37 wt% aq)[a]	25 mL/L
$C_5H_{10}NS_2Na \cdot 3H_2O$	1.7 mg/L
$NH_2CH_2CH_2NH_2$	33.3 mg/L
$K_4Fe(CN)_6$	11.7 mg/L

NOTE: Stabilizers and reducing agent were added just prior to use.
[a]Added just prior to use.

plating. The Cu plating bath was held at room temperature. These conditions were found to produce an average Cu deposition rate of 2.5 µm per hour.

The deposition of Pd on a Cu surface was found to be difficult because Cu on the surface partially or completely dissolves in the Pd plating bath. Therefore, the first metal layer applied to the support was always Pd, then a Cu layer was applied on top of the Pd layer. The membrane was then heat treated to form an alloy. Then a second set of Pd and Cu layers was applied followed by a second heat treatment. Following the second heat treatment, a final layer of Pd was applied and a final heat treatment was performed. The temperature for the heat treatments was 600°C for the membrane on the 0.5 µm support and 675°C for the membrane on the 0.2 µm grade support. In all cases the time of heating was 16 hours and the treatments were performed under hydrogen. The thickness of individual metal layers varied during plating, but the thickness of the membranes after final heat treatment was in the range 33.6 to 35.2 µm. The composition of the heat treated membranes was in the range 22–26 wt% Cu. It was found during the preparation of these membranes that individual Cu layers with a thickness greater than 7 µm developed cracks during sintering.

Hydrogen permeation data are given for two of these membranes in FIGURES 2 and 3. FIGURE 2 shows the pressure dependence of the hydrogen flux for a Pd–Cu/PSS membrane with a thickness of 33.6 µm at 350°C. It is clear that the hydrogen flux exhibits a Sievert's law type of dependence on the difference in the square roots of shell and tube side pressures. Therefore, it can be concluded that the bulk diffusion of hydrogen in the membrane was the rate limiting step. The hydrogen selectivity of this membrane was about 100, based on helium and hydrogen flux data taken at 350°C and a pressure difference of 1 atm.

FIGURE 3 shows the stability of the hydrogen flux for a Pd–Cu/PSS membrane with a thickness of 35.2 µm at 350°C and a pressure difference of 1 atm. The steady state flux for this membrane was nearly constant at 1.27–1.31 m^3/m^{2}h for more than 50 hours. Mardilovich et al.[36] reported a steady state hydrogen flux of about 2 m^3/m^{2}h for a pure Pd membrane with a thickness of 33.8 µm prepared from an identical support to that used for FIGURE 3 and tested under the same conditions. The flux of the Pd–Cu/PSS membrane in this case was about 65% of the flux for the pure

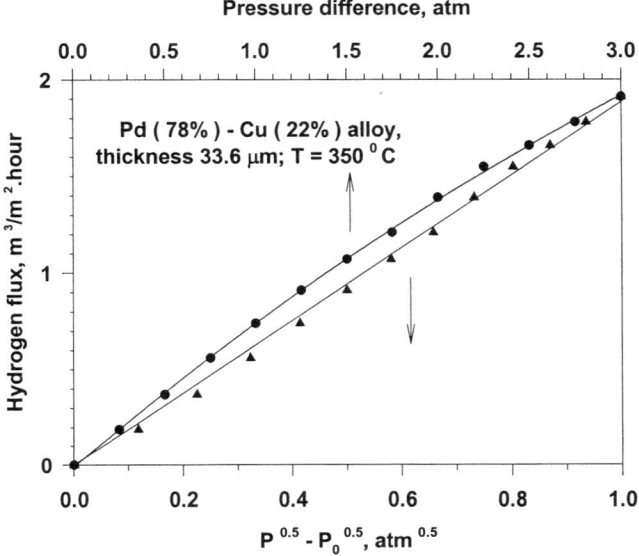

FIGURE 2. Pressure dependence of the hydrogen flux at 350°C for a Pd–Cu/PSS composite membrane. Pd, 78 wt%; Cu, 22 wt%; thickness, 33.6 μm; support, Mott PSS grade 0.2 μm.

Pd/PSS membrane. It is likely that the flux could have been improved if a composition closer to 40 wt% Cu had been obtained. These membranes were not analyzed by EPMA to determine if the alloy was uniform.

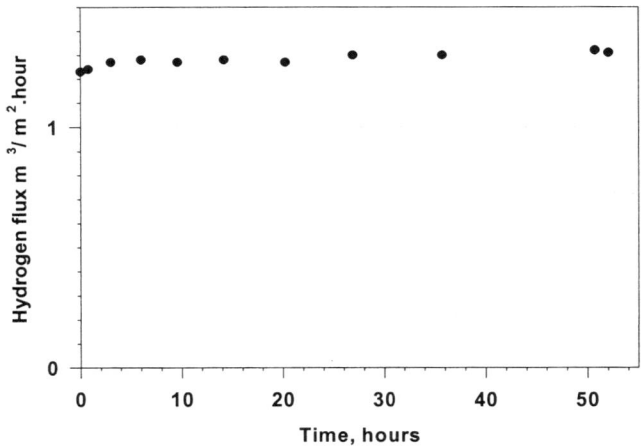

FIGURE 3. Hydrogen permeation over time at 350°C under $\Delta P = 1$ atm for Pd–Cu/PSS membrane. Pd, 74 wt%; Cu, 26 wt%; thickness, 35.2 m; support, Mott PSS grade 0.5 μm.

SIMULTANEOUS DEPOSITION OF ALLOY MEMBRANES

The simultaneous deposition of silver and palladium from a single electroless plating bath was attempted by Shu et al.[29] and by Cheng et al.[48] for porous stainless steel and porous glass supports, respectively. It has been shown that both metals can be plated simultaneously. However, two phases are formed as revealed by XRD, and therefore a high temperature treatment is needed. It is likely that the codeposition methods facilitate diffusion and ultimate formation of the alloy. This would be the case if the individual Ag and Pd grains were well intermingled and small compared to the thickness of metal layers used in the coating and diffusion treatment method.

Cheng et al.[48] characterized a series of unheated codeposited Pd–Ag surfaces after codeposition by electroless plating using SEM. Their results showed that the codeposited Pd and Ag films had an average grain size of 0.7 µm at the end of 60 minutes of plating. The membranes were annealed under a hydrogen atmosphere at 400°C for eight hours and then at 500°C for an additional eight hours. The resulting membranes had a hydrogen permeation rate 1.4–1.7 times higher than a pure Pd membrane. The magnitude of this improvement strongly suggests that the membrane was comprised of a uniform alloy. However, no EPMA results were provided on a cross section of this membrane after testing. Direct comparison of this result to the Pd–Ag/PG membranes produced by coating and diffusion treatment method[23] would suggest that for similar annealing temperatures, codeposition produces a more uniform alloy layer. This is probably the result of the intermingling of the small Ag and Pd grains achieved in the codeposition, prior to annealing.

MEMBRANE TOPOLOGY

It has already been noted that the electroless plating technique is useful in applying uniform coatings to surfaces of varied geometry. It has been the experience of the present authors that the electroless plating of Pd is very effective in covering regions of the support surface that have variations in topology, or topological defects. FIGURES 4, 5, and 6 show optical micrographs taken from a single dense Pd/PSS membrane, using Nikon SMZ800 stereoscopic zoom microscope with a SPOT Insight Camera. This membrane was produced using a 0.2 µm grade PSS tube with a 1-inch OD, obtained from Mott Metallurgical. The support module was prepared by welding sections of non-porous 1-inch OD 316L stainless steel tube to each end of the porous section. Cleaning, activation, and plating were performed to obtain a dense membrane, as described by Mardilovich et al.[18]

FIGURE 4 shows the weld area where the porous support is connected to the non-porous tube. FIGURES 5 and 6 show an area where there was a small cavity in the support surface at the start of plating with an opening of about 0.6–0.7 mm wide. The membrane was dense at the time these micrographs were taken. This indicates that the electroless plating methods used were effective in covering transitional regions in the substrate, such as the weld, or defects in the surface, such as the cavity. This is particularly apparent in FIGURE 6, which shows the Pd layer covering the bottom of the cavity. In general, macroscopic defects, such as dents, tool marks, and abrasions, transitions in the substrate, such as the weld, and other topological features,

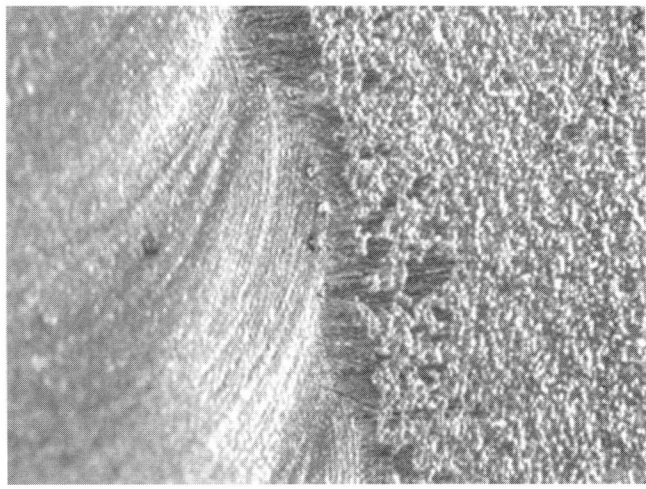

FIGURE 4. Optical micrograph of the weld region for a Pd/PSS membrane: magnification ×40 (the region shown is 2.8 mm wide).

such as the curvature of the support, have no influence on the ability to achieve a dense layer. The dense Pd layer is simply deposited so as to conform with the original surface topology. For Pd/PSS membranes, prepared by electroless plating, the thickness of the dense Pd layer is determined by the size of the largest pores in the

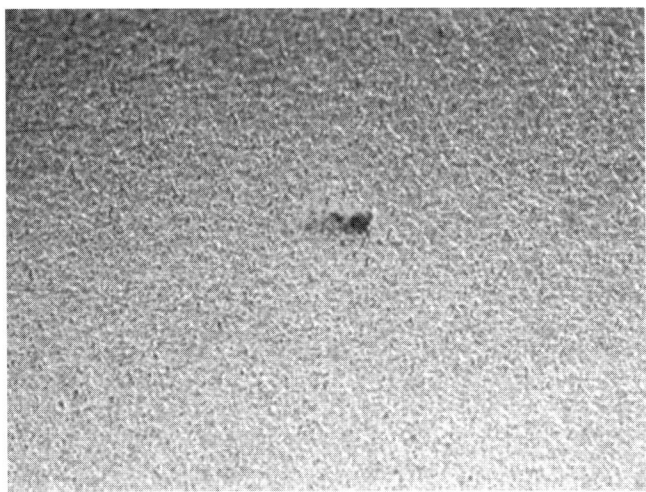

FIGURE 5. Optical micrograph of a cavity in the support surface (about 0.6–0.7 mm wide), after electroless plating to form a dense layer of Pd: magnification ×20 (the region shown is 5.6 mm wide).

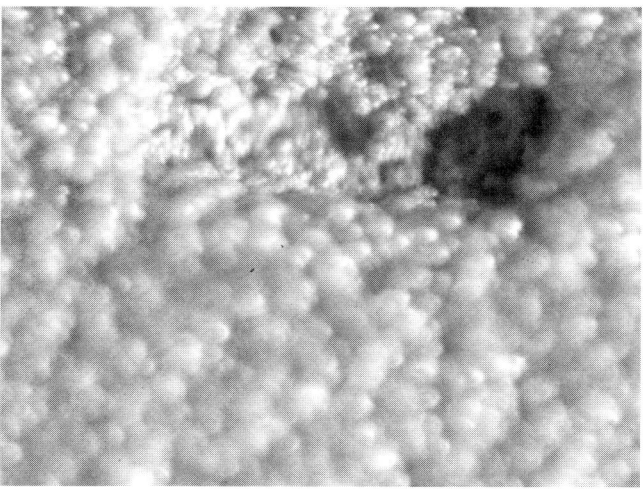

FIGURE 6. Optical micrograph showing Pd coverage at the bottom of the cavity shown in FIGURE 5: magnification ×120 (the region shown is 0.94 mm wide).

support and is unaffected by topological features that are several orders of magnitude larger in size than the pore size.[36,49]

CONCLUDING REMARKS

Because the demand for hydrogen will undoubtedly increase immensely in future, dense composite palladium, especially palladium/alloy, membranes will play an important and essential role in the production of pure hydrogen. The use of porous metal as substrates will facilitate the membrane module fabrication and process integration. There are, of course, many technical issues, such as producing thin membranes with good separation and long term thermal and mechanical stability, that still need to be developed. The use of palladium alloys to increase the hydrogen permeability and, at the same time, to enhance the thermal stability should be further explored.

REFERENCES

1. LEWIS, F.A. 1967. The Palladium Hydrogen System. Academic Press, London.
2. KOFFLER, S.A., J.B. HUSDON & G.S ASNELL. 1969. Hydrogen permeation through alpha-palladium. Trans. Metal. Soc. AIME. **245**: 1735–1740.
3. FAVREAU, R.L., R.E. PATTERSON, D. RANDALL & O.N. SALMON. 1954. Solubility of hydrogen and tritium in palladium black. USAEC Report. KAPL-1036. U.S. Atomic Energy Commission.
4. BALOVEV, YU.A. 1974. Russian J. Phys. Chem. **48**: 719–720.
5. LI, Z.Y., H. MAEDA, K. KUSAKABE, *et al.* 1993. Preparation of palladium–silver alloy membranes for hydrogen separation by the spray pyrolysis method. J. Membr. Sci. **78**: 247–254.

6. YAN, S., H. MAEDA, K. KUSAKABE & S. MOROOKA. 1994. Thin palladium membrane formed in support pores by metal–organic chemical vapor deposition method and application to hydrogen separation. Ind. Eng. Chem. Res. **33:** 616.
7. MOROOKA, S., S.Y. YAN & K. KUSAKABE. 1995. Palladium membrane formed in macropores of support tube by chemical vapor deposition with crossflow through a porous wall. Sep. Sci. Tech. **30:** 2877.
8. HYBERTSON, B.M., B.N. HANSEN, R.M. BARKLEY & R.E. SIEVERS. 1991. Deposition of palladium films by a novel supercritical fluid transport–chemical deposition process. Mater. Res. Bull. **26:** 1127.
9. ATHAYDE, A.L., R.W. BAKER & P. NGUYEN. 1994. Metal composite membranes for hydrogen separation. J. Membr. Sci. **94:** 299–311.
10. JAYARAMAN, V., Y.S. LIN, M. PAKALA & R.Y. LIN. 1995. Fabrication of ultrathin metallic membranes on ceramic supports by sputter deposition. J. Membr. Sci. **99:** 89–100.
11. JAYARAMAN, V. & Y.S. LIN. 1995. Synthesis and hydrogen permeation properties of ultrathin palladium–silver alloy membranes. J. Membr. Sci. **104:** 251–262.
12. KONNO, M., M. SHINDO, S. SUGAWARA & S. SAITO. 1988. A composite palladium and porous aluminum oxide membrane for hydrogen gas separation. J. Membr. Sci. **37:** 193–197.
13. MARDILOVICH P.P., P.V. KURMAN, A.N. GOVYADINOV, et al. 1996. Gas permeability of anodized alumina membranes with a palladium–ruthenium alloy layer. Russ. J. Phys. Chem. **70:** 514.
14. BRYDEN, K.J. & J.Y. YING. 1995. Nanostructured palladium membrane synthesis by magnetron sputtering. Mater. Sci. Eng. **A204:** 140.
15. GRYAZNOV, V.M., O.S. SEREBRYANNIKOVA, M.YU. SEROV, et al. 1993. Preparation and Catalysis over palladium composite membranes. Appl. Catal. A: General **96:** 15.
16. PEACHEY, N.M., R.C. SNOW & R.C. DYE. 1996. Composite Pd/Ta metal membranes for hydrogen separation. J. Membr. Sci. **111:** 123–133.
17. SHU, J., B.P.A. GRANDJEAN, A. VAN NESTE & S. KALIAGUINE. 1991. Catalytic palladium-based membrane reactors: a review. Can. J. Chem. Eng. **69:** 1036–1060.
18. MARDILOVICH, P.P., Y. SHE, Y.H. MA & M.H. REI. 1998. Defect-free palladium membranes on porous stainless-steel support. AIChE J. **44:** 310.
19. BUXBAUM, R.E. & T.L. MARKER. 1993. Hydrogen transport through non-porous membranes of palladium-coated niobium, tantalum and vanadium. J. Membr. Sci. **85:** 29–38.
20. BUXBAUM, R.E. & A.B. KINNEY. 1996. Hydrogen transport through tubular membranes of palladium-coated tantalum and niobium. Ind. Eng. Chem. Res. **35:** 530.
21. GOVIND, R., & D. ATNOOR. 1991. Development of a composite palladium membrane for selective hydrogen separation at high temperature. Ind. Eng. Chem. Res. **30:** 591.
22. UEMIYA, S., Y. KUDE, K. SUGINO, et al. 1988. A palladium/porous-glass composite membrane for hydrogen separation. Chem. Lett. **10:** 1687.
23. UEMIYA, S., N. SATO, H. ANDO, et al. 1991. Separation of hydrogen through palladium thin film supported on a porous glass tube. J. Membr. Sci. **56:** 303–313.
24. UEMIYA, S., T. MATSUDA & E. KIKUCHI. 1991. Hydrogen permeable palladium–silver alloy membrane supported on porous ceramics. J. Membr. Sci. **56:** 315–325.
25. KIKUCHI, E. & S. UEMIYA. 1991. Preparation of supported thin palladium-silver alloy membranes and their characteristics for hydrogen separation. Gas Sep. Purific. **5:** 261–266.
26. COLLINS, J.P. & J.D. WAY. 1993. Preparation and characterization of a composite palladium–ceramic membrane. Ind. Eng. Chem. Res. **32:** 3006–3013.
27. HUANG, T.C., M.C. WEI & H.I. CHEN. 2001. Permeation of hydrogen through palladium/alumina composite membranes. Sep. Sci. Tech. **36:** 199–222.
28. PAGLIERI, S.N., K.Y. FOO, J.D. WAY, et al. 1999. A new prepration technique for Pt/alumina membranes with enhanced high-temperature stability. Ind. Eng. Chem. Res. **38:** 1925–1936.
29. SHU, J., B.P.A. GRANDJEAN, E. GHALI & S. KALIAGUINE. 1993. Simultaneous deposition of Pd and Ag on porous stainless steel by electroless plating. J. Membr. Sci. **77:** 181–195.

30. MARDILOVICH, P.P., Y. SHE, M.-H. REI & Y.H. MA. 1996. Permeation characterization of defect free Pd/porous stainless steel membranes. Proceedings, ICIM4-96. 362–374.
31. MA, Y.H., P.P. MARDILOVICH & Y. SHE. 1998. Stability of hydrogen flux through Pd/porous stainless composite membranes. Proceedings, ICIM6. Nagoya, Japan. June 22–26.
32. YEUNG, K.L. & A. VARMA. 1995. Novel preparation techniques for thin metal-ceramic composite membranes. AIChE J. **41:** 2131.
33. YEUNG, K.L., S.C. CHRISTIANSEN & A. VARMA. 1999. Palladium composite membranes by electorless plating technique, relationship between plating kinetics, film microstructure and membrane performance. J. Membr. Sci. **159:** 107–123.
34. SOULEIMANOVA, R.S., A.S. MUKASYAN & A. VARMA. 2002. Pd membranes formed by electroless plating with osmosis: H_2 permeation studies. AIChE J. **48:** 262–268.
35. MA, Y.H., P.P. MARDILOVICH & Y. SHE. 2000. Hydrogen gas-extraction module and method of fabrication. U.S. Patent 6,152,987.
36. MARDILOVICH, I.P., E. ENGWALL & Y.H. MA. 2002. Dependence of hydrogen flux on the pore size and plating surface topology of asymmetric Pd-porous stainless steel membranes. Desalination **144:** 85–89.
37. KNAPTON, A.G. 1977. Palladium alloys for hydrogen diffusion membranes—a review of high permeability materials. Plat. Met. Rev. **21:** 44–50.
38. GRYAZNOV, V.M. 2000. Metal containing membranes for the production of ultrapure hydrogen and the recovery of hydrogen isotopes. Sep. Purific. Meth. **29:** 171–187.
39. MCKINLEY, D.L. & W.VA. NITRO. 1967. Metal alloy for hydrogen separation and purification. U.S. Patent 3,350,845.
40. MCKINLEY, D.L. & W.VA. NITRO. 1969. Method for hydrogen separation and purification. U.S. Patent 3,439,474.
41. SHU, J., B.P.A. GRANDJEAN & S. KALIAGUINE. 1995. Asymmetric Pd–Ag/stainless steel catalytic membranes for methane steam reforming. Catal. Today **25:** 327–332.
42. SHU, J., A. ADNOT, B.P.A. GRANDJEAN & S. KALIAGUINE. 1996. Structurally stable composite Pd–Ag alloy membranes: introduction of a diffusion barrier. Thin Solid Films **286:** 72–79.
43. KEULER, J.N. & L. LORENZEN. 2002. Developing a heating procedure to optimise hydrogen permeance through Pd–Ag membranes of thickness less than 2.2 mm. J. Membr. Sci. **195:** 203–213.
44. NAM, S.E. & K.H. LEE. 2001. Hydrogen separation by Pd alloy composite membranes: introduction of diffusion barrier. J. Membr. Sci. **192:** 177–185.
45. SHU, J., B.E.W. BONGONDO, B.P.A. GRANDJEAN, *et al.* 1993. Surface segregation of Pd–Ag membranes upon hydrogen permeation. Surf. Sci. **291:** 129–138.
46. NAM, S.E. & K.H. LEE. 2000. A study on the palladium/nickel composite membrane by vacuum electrodeposition. J. Membr. Sci. **170:** 91–99.
47. VANSOVKAYA, K.M. 1985. Electroless Metal Coatings. Mashinostroenie, Leningrad Otdelenie, Leningrad, USSR.
48. CHENG, Y.S. & K.L. YEUNG. 1999. Palladium-silver composite membranes by electroless plating technique. J. Membr. Sci. **158:** 127–141.
49. MA, Y.H., I.P. MARDILOVICH & P.P. MARDILOVICH. 2001. Effects of porosity and pore size distribution of the porous stainless steel on the thickness and hydrogen flux of palladium membranes. Preprints, American Chemical Society, Division of Petroleum Chemistry **46:** 154–156.

Mixed Matrix Membrane Development

SANTI KULPRATHIPANJA

UOP LLC, Des Plaines, Illinois, USA

ABSTRACT: Two types of mixed matrix membranes were developed by UOP in the late 1980s. The first type includes adsorbent polymers, such as silicalite–cellulose acetate (CA), NaX-CA, and AgX-CA mixed matrix membranes. The silicalite–CA has a CO_2/H_2 selectivity of 5.15 ± 2.2. In contrast, the CA membrane has a CO_2/H_2 selectivity of 0.77 ± 0.06. The second type of mixed matrix membrane is PEG–silicone rubber. The PEG–silicone rubber mixed matrix membrane has high selectivity for polar gases, such as SO_2, NH_3, and H_2S.

KEYWORDS: mixed matrix membrane

INTRODUCTION

Permeation of molecules through membranes is controlled by two major mechanisms: diffusivity (D) and solubility (S). Diffusivity is the mobility of individual molecules passing through the holes in a membrane material; solubility is the number of molecules dissolved in a membrane material. Permeability (P), defined in Equation (**1**), is a measure of the ability of molecules to permeate a membrane,

$$P = DS. \qquad (1)$$

The ability of a membrane to separate two molecules, for example, A and B, is the ratio of their permeabilities and is called the membrane selectivity, α_{AB},

$$\alpha_{AB} = \frac{P_A}{P_B}. \qquad (2)$$

Because P is the product of D and S, Equation (**2**) can be rewritten

$$\alpha_{AB} = \frac{D_A}{D_B} \cdot \frac{S_A}{S_B}. \qquad (3)$$

D_A/D_B is the ratio of the diffusion coefficients of the two molecules and can be viewed as the mobility or diffusivity selectivity, reflecting the different sizes of the two molecules. S_A/S_B is the ratio of the Henry's law sorption coefficients of the two molecules and can be viewed as the sorption or solubility selectivity of the two molecules. The balance between the solubility selectivity and the diffusivity selectivity determines whether a membrane material is selective for molecule A or molecule B in a feed mixture. Membranes with both high permeability and selectivity are desirable. The higher the permeability, the less the membrane area required to treat a given amount of gas; the higher the selectivity, the higher the purity of the product gas.

To increase membrane selectivity either the diffusivity or the solubility needs be enhanced. However, for a particular membrane polymer, these factors are fixed and

---

Address for correspondence: Santi Kulprathipanja, UOP LLC, Des Plaines, Illinois 60017, USA.

TABLE 1. Kinetic diameter for various molecules

Molecule	Kinetic Diameter (Å)
H_2	2.89
O_2	3.46
N_2	3.64
CO	3.76
CO_2	3.3
H_2O	2.65
NH_3	2.6
SO_2	3.6
CH_4	3.8
C_2H_2	3.3
C_2H_4	3.9

difficult to alter without chemically modifying the molecular structure. Chemical modification of a polymer membrane can be accomplished for gas separation. However, selectivity enhancement through gas diffusion mechanisms is difficult. This is due to the small difference in the kinetic diameters of each molecule to be separated (see TABLE 1).

To enhance the commercial applicability of membrane separation processes, our goal is to develop a novel type of membrane that can alter or enhance membrane solubility by physical modification in the polymer membrane phase. With this objective in mind, UOP LLC developed a mixed matrix membrane (MMM) that enables one to enhance membrane selectivity through gas solubility optimization. Two types of mixed matrix membranes (MMM) were developed in the early 1980s.[2-7] The first mixed matrix membrane is a membrane with adsorbent embedded in the polymer phase (MMM_{ADS}). The polymers can be cellulose acetate (CA), polysulfone, polyethersulfone, polyimide, or silicone rubber. The adsorbent can be zeolites, such as NaX, AgX, NaY, NaA, and silicalite, silica gel, alumina, or activated carbon. The second type of mixed matrix membrane is produced by casting polyethylene glycol (PEG) and silicone rubber on a porous polysulfone support (MMM_{PEG}). Both types of MMMs were evaluated for the separation of polar gas from non-polar gas, carbon dioxide from nitrogen and methane, and light paraffins from light olefins. The overall MMM development and permeation results are summarized in this paper.

ADSORBENT POLYMER MIXED MATRIX MEMBRANE (MMM_{ADS})

MMM_{ADS} Preparation

A typical MMM_{ADS} was prepared by stirring 3.75 grams of adsorbent powder (2–5 micron particle size) in 85 grams of acetone. Fifteen grams of cellulose acetate (acetyl content 39.8%) were subsequently added to the adsorbent–acetone suspension. A partial vacuum was applied for a brief duration to ensure the removal of air bubbles, while the suspension was stirred to obtain a suspended homogeneous solution. The solution was then coated on the top horizontal surface of a clean glass plate. A portion of the acetone was allowed to slowly evaporate until a film formed on the upper surface of the solution. The membrane was allowed to set for two minutes, followed by submersion in a water bath for two minutes. The membrane was removed from the ice water bath and soaked in a hot water bath at 90°C for one hour. The membrane was then dried. Three types of MMM_{ADS}, silicalite–CA, NaX–CA, and AgX–CA, were prepared. For comparison, a cellulose acetate reference membrane was also prepared and tested.

RESULTS AND DISCUSSION

To demonstrate the effect of silicalite in the polymer phase, the silicalite–CA MMM was tested for CO_2/H_2 separation.[2] In this study, a feed mixture of 50/50 CO_2/H_2 with a differential pressure of 50 psig was used. The calculated separation factor for CO_2/H_2 was found to be 5.15 ± 2.2. In comparison, a CO_2/H_2 separation factor of 0.77 ± 0.06 was found for cellulose acetate. This indicates that the silicalite in the membrane phase reversed the selectivity from H_2 to CO_2. It is known that silicalite is more selective toward CO_2 than H_2.

The same membrane, silicalite–CA, was also tested for O_2/N_2 separation. Using air as feed with a differential pressure of 150 psig, the calculated separation factor for O_2/N_2 was 3.63 ± 0.28. In comparison, the O_2/N_2 separation factor was 2.99 for the cellulose acetate membrane. This indicates that the O_2 permeation rate is enhanced by silicalite in the membrane phase. However, this enhancement of O_2/N_2 selectivity is not sufficient to ensure the commercial membrane requirement. Other membrane groups also attempted to formulate other types of MMM for O_2/N_2 separation. They included polyethersulfone–NaA and polyethersulfone–NaX,[8] silicalite-silicone rubber,[9,10] silicalite–ethylene–propylene rubber,[10] and polysulfone–NaX.[11] In general, a slight increase was observed in O_2/N_2 selectivity with zeolite addition. Despite the wide variety of combinations of zeolites with glassy and rubbery polymers, the MMM_{ADS} developed to date has remained short of the desired O_2/N_2 selectivity. Additional research and development is needed to properly select a suitable adsorbent and polymer combination for MMM formation with desirable performance characteristics.[12,13]

TABLE 2. Membranes and adsorbent selectivities for propylene/propane

Membrane	$P_{C_3H_6}/P_{C_3H_8}$	Zeolite	$P_{C_3H_6}/P_{C_3H_8}$
CA	1.77		
Sil-CA	2.28	silicalite	1.27
NaX-CA	0.58	NaX	16.23
AgX-CA	0.34	AgX	48.90

TABLE 3. Activities in the area of MMM_{ADS}

Adsorbent	Polymer Matrix	Gas Separation	Ref.
NaA	poly(vinyl chloride)	ethanol/water by pervaporation	16
Activated carbon	ethylene–propene rubber	toluene/ethanol by pervaporation	17
Silicalite, NaX, AgX	vitron, polyethylene–polypropylene	methanol/toluene, ethane/ethylene by pervaporation	18
HY	chitosan	ethanol/water by pervaporation	19
NaA	polyvinyl alcohol	dimethylformamide/water, alcohol/water by pervaporation	20
NaA, NaX	cellulose acetate	ethanol/water by pervaporation	21
NaA	cellulose	water/air—dehumidification	22
NaA, NaX, NaY, CaA,	polyimide	He, N_2, O_2, CO_2, CH_4	23
KA, NaA, CaA, NaX	polyethersulfone	CO_2/O_2, H_2/N_2, CO_2/N_2	24
Silicalite, NaX, CaA, KY, carbon molecular sieve	rubbery polymer	CO_2/CH_4, O_2/N_2	25
NaA	polyethersulfone	CO_2/CH_4, CO_2/air, H_2/CH_4	26
HZSM-5, CaA, NaZSM-5, NaA	polydimethylsiloxane	pentane/iso-pentane	27
Silicate	proton-conducting polymer	H_2 transport in fuel cell application	28
Silicalite	polydimethylsiloxane	O_2, N_2, CO_2	29
ZSM-5	polydimethylsiloxane	effect of zeolite particle size	30
NaA, NaX	polyethersulfone	N_2, O_2, Ar, CO_2, H_2	31
NaX, silicalite	vitron, polyethylene-polypropylene	methanol/toluene	32
Silicalite, dealuminized-Y	polydimethylsiloxane	ethanol, 1,1,1-trichloro-ethane, trichloroethylene	33
Silicalite, borosilicate, Y	polyimide	xylenes	34
Silicalite, KY	cellulose acetate, polyetherimide, poly(4-methyl-1-pentene)	CO_2/CH_4	35

The membranes, silicalite–CA, NaX–CA, and AgX–CA, were evaluated for propylene/propane separation.[14] In the gas permeation study, either propylene or propane was passed through the membrane under a pressure of 50–100 psig at ambient temperature. Each pure gas permeation rate was measured. The permeability of propylene and propane, as well as the propylene/propane selectivity, were then calculated. The results are summarized in TABLE 2.

$P_{C_3H_6}/P_{C_3H_8}$ selectivity decreases: silicalite–CA > CA > NaX–CA > AgX–CA. This indicates that silicalite enhances the propylene/propane selectivity. However, NaX and AgX reverse propylene/propane selectivity. To understand the effect of each adsorbent in the MMM to the propylene/propane selectivity, the adsorbent equilibrium selectivities of silicalite, NaX, and AgX for propylene/propane selectivity, $P_{C_3H_6}/P_{C_3H_8}$ values,[15] are summarized in TABLE 2. From the table, $P_{C_3H_6}/P_{C_3H_8}$ decreases: AgX > NaX > silicalite, which is the reverse of the MMM selectivity. One can speculate that the model of sorption mechanism might play a role in this observation. This indicates that careful selection of each of the components in the MMM, adsorbent, and polymer is critical to enhancing the potential benefits from this approach.

Interest in the area of adsorbent–polymer mixed matrix membrane has increased during the past few years, both in the fundamental and application aspects. TABLE 3 briefly summarizes the activities in the area.

POLYETHYLENE GLYCOL–SILICONE RUBBER MIXED MATRIX MEMBRANE (MMM$_{PEG}$)

MMM$_{PEG}$ Preparation

An MMM$_{PEG}$ membrane was prepared by dissolving 1.08 grams of a silicone rubber RTV-615A, 0.12 grams silicone rubber RTV-615B in 18.6 grams Freon TF or cyclohexane solvent. To this solution, 0.3 grams of 400–600 MW polyethylene glycol was added to form an emulsified solution. The emulsified solution was coated on a porous polysulfone under vacuum. After coating, the membrane was cured for a period of 30 minutes at a temperature of 82°C. A gly-MMM$_{PEG}$ membrane was also prepared by first soaking a wet porous polysulfone in a solution containing 15% glycerol by weight for a period of two hours, followed by drying for a period of 10 hours at room temperature. The treated polysulfone support was then coated with the emulsified solution described above.

Results and Discussion

Both MMM$_{PEG}$ and gly-MMM$_{PEG}$ were tested for the separation of gases, such as SO_2, NH_3, H_2S, CO_2, H_2, N_2, CH_4, C_2H_4, and C_2H_6. The gas to be tested was passed through the membrane under a pressure of 5 to 50 psig at ambient temperature. The permeation rate and selectivity for each pure gas, both polar and non-polar, was measured and calculated. For comparison, a reference membrane composed of silicone rubber coated on a porous polysulfone (Sil–PS) was also prepared and tested. The results are summarized in TABLES 4–8.

TABLE 4. Mixed matrix membrane for SO_2 separation

Membrane	P_{SO_2}/P_{N_2}	P_{SO_2}/P_{CO_2}	$P_{SO_2}/d \times 10^3$ cm^3(STP)/cm^2 sec cm(Hg)
MMM$_{PEG}$	4,350	74	10
Gly–MMM$_{PEG}$	30,000	300	42
Sil–PS	85	7.5	4.6

MMM$_{PEG}$, polyethylene glycol/silicone rubber coated on porous polysulfone.
Gly–MMM$_{PEG}$, polyethylene glycol/silicone rubber coated on glycerol treated porous polysulfone.
Sil–PS, silicone rubber coated on porous polysulfone.
d, thickness of the membrane.

TABLE 5. Mixed matrix membrane for NH_3 separation

Membrane	P_{NH_3}/P_{N_2}	P_{NH_3}/P_{H_2}	$P_{NH_3}/d \times 10^4$ cm^3(STP)/cm^2 sec cm(Hg)
MMM$_{PEG}$	1,270	80	2.25
Gly–MMM$_{PEG}$	7,200	510	2.89
Sil–PS	35	12	1.86

TABLE 6. Mixed matrix membrane for H_2S separation

Membrane	P_{H_2S}/P_{CH_4}	P_{H_2S}/P_{H_2}	P_{H_2S}/P_{CO_2}	$P_{H_2S}/d \times 10^5$ cm^3(STP)/cm^2 sec cm(Hg)
MMM$_{PEG}$	136	21	6.7	7.8
Gly–MMM$_{PEG}$	119	74	8.1	10.7
Sil–PS	10	9	2.3	14.0

TABLE 7. Mixed matrix membrane for CO_2 separation

Membrane	P_{CO_2}/P_{CH_4}	P_{CO_2}/P_{H_2}	P_{CO_2}/P_{N_2}	$P_{CO_2}/d \times 10^6$ cm^3(STP)/cm^2 sec cm(Hg)
MMM$_{PEG}$	24	4.8	67	13.6
Gly–MMM$_{PEG}$	22	9.1	100	13.2
Sil–PS	4.3	3.8	11.4	61

TABLE 8. Mixed matrix membrane for ethylene separation

Membrane	$P_{C_2H_4}/P_{C_2H_6}$	$P_{C_2H_4}/d \times 10^6$ cm³(STP)/cm² sec cm(Hg)	$P_{C_2H_6}/d \times 10^6$ cm³(STP)/cm² sec cm(Hg)
MMM$_{PEG}$	1.7	0.79 ± 0.010	0.47 ± 0.003
Gly–MMM$_{PEG}$	2.28	0.40 ± 0.020	0.17 ± 0.002
Sil–PS	0.86	7.19 ± 0.152	8.34 ± 0.070

The results in TABLES 4–8 indicate that PEG enhanced the permeation rates of SO_2 and NH_3. PEG, however, hindered the permeation rates of the other gases. The effect of PEG on lowering the gas permeation rates increased as follows: $H_2S < CO_2 < H_2 < CH_4 < N_2$. This enhanced the selectivity between polar gas to non-polar gas substantially. TABLE 8 also shows the effect of PEG on ethylene and ethane.[36] PEG slowed down the permeation rate of ethylene and ethane in the membrane phase. However, the effect is more prevalent for ethane than ethylene. This promotes ethylene/ethane separation.

The stability of MMM$_{PEG}$ was carried out at a temperature range of 40°C to 75°C. CO_2 and CH_4 gases were saturated with water and then were passed through the MMM$_{PEG}$ at pressures of 25 and 50 psi, respectively. The stability of MMM$_{PEG}$ was excellent over the period studied, namely 200 days on stream, with no apparent deterioration in CO_2 and CH_4 flux or CO_2/CH_4 selectivity.

A plot of selectivity versus critical temperature for these gases is presented in FIGURE 1. The plot shows that the permeabilities of the molecules increase as the critical temperature increases. This indicates that solubility is the controlling mechanism for mixed matrix membrane selectivity. However, for lighter gases, such as H_2 and N_2, the permeability decreases as the molecular weight increases. The inverse relationship between the diffusion coefficient and molecular weight implies the importance of molecular diffusion coefficients.

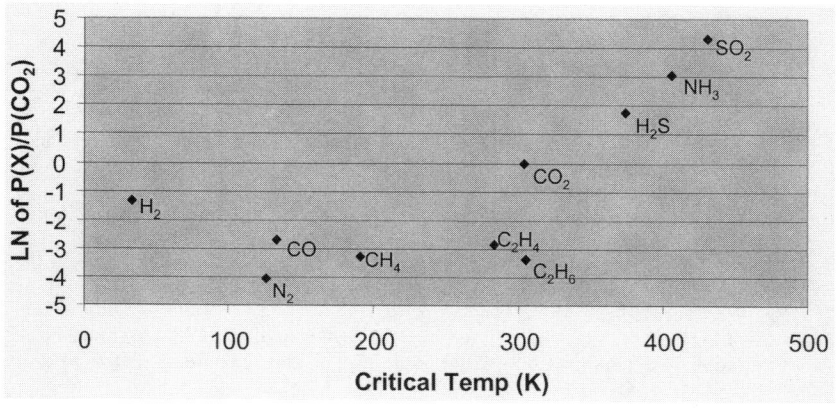

FIGURE 1. Critical temperature versus selectivity.

CONCLUSIONS

Two types of mixed matrix membranes, adsorbent–polymer and PEG–silicone rubber, have been developed. The concept of mixed matrix membrane enhancing the membrane solubility through adsorbent or PEG embedded in the polymer phase has been demonstrated. However, in the case of adsorbent–polymer mixed matrix membrane, a careful selection of adsorbent and polymer is critical to enhancing the potential benefits of this approach.

REFERENCES

1. BREAK, D.W. 1974. Zeolite Molecular Sieves. John Wiley & Sons, Inc., New York.
2. KULPRATHIPANJA, S., R.W. NEUZIL & N.N. LI. 1988. Separation of fluids by means of mixed matrix membranes. U.S. Patent 4,740,219, April 26.
3. KULPRATHIPANJA, S., R.W. NEUZIL & N.N. LI. 1988. Separation of gases by means of mixed matrix membranes. U.S. Patent 5,127,925, July 7.
4. KULPRATHIPANJA, S. 1986. Separation of polar gases from nonpolar gases. U.S. Patent 4,606,740, August 19.
5. KULPRATHIPANJA, S. & S.S. KULKARNI. 1986. Separation of polar gases from nonpolar gases. U.S. Patent 4,606,060, August 26.
6. KULPRATHIPANJA, S., S.S. KULKARNI & E.W. FUNK. 1988. Multicomponent membranes, U.S. Patent 4,737,165, April 12.
7. KULPRATHIPANJA, S., S.S. KULKARNI & E.W. FUNK. 1988. Preparation of gas selective membranes. U.S. Patent 4,751,102, June 14.
8. SÜER, M.G., N. BAC & L. YILMAZ. 1994. Gas permeation properties of polymer–zeolite mixed matrix membranes. J. Membr. Sci. **91:** 77.
9. JIA, M., K.-V. PEINEMANN & R.-D. BEHLING. 1991. Molecular sieving effect of the zeolite filled silicone rubber membranes in gas permeation. J. Membr. Sci. **57:** 289.
10. DUVAL, J.M., B. FOLKERS, M.H.V. MULDER, et al. 1993. Adsorbent filled membranes for gas separations. J. Membr. Sci. **80:** 189.
11. GÜR, T.M. 1994. Permselectivity of zeolite filled polysulfone gas separation membranes. J. Membr. Sci. **93:** 283.
12. ZIMMERMAN, C.M., A. SINGH & W.J. KOROS. 1997. Tailoring mixed matrix composite membranes for gas separations. J. Membr. Sci. **137:** 145.
13. ZIMMERMAN, C.M., R. MAHAJAN & W.J. KOROS. 1997. Fundamental and practical aspects of mixed matrix gas separation membranes. Polym. Mater. Sci. Eng. **77:** 328–329.
14. RATTANAWONG, W., S. OSUWAM, T. RISKSOMBOEN & S. KULPRATHIPANJA. 2001. Zeolite-cellulose acetate mixed membrane for olefin/paraffin separation. Am. Chem. Soc., Div. Pet. Chem. **46**(2): 166–167.
15. RATTANAWONG, W. 2001. Zeolite/Cellulose Acetate Mixed Matrix Membranes for Olefin/Paraffin Separations. M.S. Thesis, The Petroleum and Petrochemical College, Chulalongkovn University. Bangkok, Thailand.
16. GOLDMAN, M., D. FRAENKEL & G. LEVIN. 1989. A zeolite/polymer membrane for separation of ethanol–water azeotrope. J. Appl. Polym. Sci. **37**(7): 1791–1800.
17. DUVAL, J.M., B. FOLKERS, M.H.V. HULDER, et al. 1994. Separation of a toluene/ethanol mixture by pervaporation using active carbon-filled polymeric membranes. Sep. Sci. Tech. **29**(3): 357–373.
18. BOOM, J.P., D. BARGEMAN & H. STRATHMANN. 1994. Zeolite-filled membranes for gas separation and pervaporation. Stud. Surf. Sci. Catal. Zeolites Related Micropor. Mater. Part B. **84:** 1167–1174.
19. CHEN, X., H. YANG, Z. GU & Z. SHAO. 2001. Preparation and characterization of HY zeolite-filled chitosan membranes for pervaporation separation. J. Appl. Polym. Sci. **79**(6): 1144–1149.

20. SHAH, D., K. KISSICK, A. GHORPADE, et al. 2000. Pervaporation of alcohol-water and dimethylformamide–water mixtures using hydrophilic zeolite NaA membranes: mechanisms and experimental results. J. Membr. Sci. **179**(1–2): 185–205.
21. OKUMUS, E., T. GURKAN & L. YILMAZ. 1994. Development of a mixed matrix membrane for pervaporation. Sep. Sci. Tech. **29**(18): 2451–2473.
22. ITO, A., H. SASAKI & M. YONEKURA. 1998. Dehumidification of air by a zeolite-filled polymer membrane. Sekiyu Gakkaishi **41**(3): 216–221.
23. YONG, H.H., H.C. PARK, Y.S. KANG, et al. 2001. Zeolite-fillled polyimide membrane containing 2,4,6-triaminopyrimidine. J. Membr. Sci. **188**(2): 151–163.
24. SÜER, M.G., N. BAC, L. YILMAZ, et al. 1994. Gas separation with zeolite based polyethersulfone membranes. *In* Separation Technology. E.F. Vansant, Ed.: 661–669.
25. DUVAL, J.M., B. FOLERS, M.H.V. MULDER, et al. 1993. Adsorbent-filled membranes for gas separation. Part 1. Improvement of the gas separation properties of polymeric membranes by incorporation of microporous adsorbents. J. Membr. Sci. **80**: 189–198.
26. BUTTAL, T., N. BAC & L. YILMAZ. 1995. Effect of feed composition on the performance of polyer–zeolite mixed matrix gas separation membranes. Sep. Sci. Tech. **30**(11): 2365–2384.
27. TANTERKIN-ERSOLMAZ, S.B., L. SENORKYAN, N. KALAONRA, et al. 2001. *n*-Pentane/I-pentane separation by using zeolite-PDMS mixed matrix membranes. J. Membr. Sci. **189**(1): 59–67.
28. HAERING, T., J. KERRES & A. ULLRICH. 2000. Silicate-supported proton-conducting composite membranes containing polymeric ionomer blends. WO Patent 2000074827.
29. TANTEKIN-ERSOLMAZ, S.B., C. ATALAY-ORAL, M. TATHER, et al. 2000. Effect of zeolite particle size on the performance of polymer-zeolite mixed matrix membranes. J. Membr. Sci. **175**: 285–288.
30. TATLIER, M., S. TANTEKIN-ERSOLMAZ, S. BIRGUIL, et al. 2001. Power-law scaling behavior of membranes. J. Membr. Sci. 1821–1822, 183–193.
31. SÜER, M.G., N. BAC & L. YILMAZ. 1994. Gas permeation characteristics of polymer-zeolite mixed matrix membranes. J. Membr. Sci. **91**: 77–86.
32. BOOM, J.P., I.G.M. PUNT, H. ZWIJNENBERG, et al. 1998. Transport through zeolite filled polymeric membranes. J. Membr. Sci. **138**(2): 237–258.
33. CHANDAK, M.V., S.Y. LIN, W. JI & R.J. HIGGINS. 1997. Sorption and diffusion of VOCs in DAY zeolite and silicalite-filled PDMS membranes. J. Membr. Sci. **133**(2): 231–243.
34. VANKELECOM, F.J.I., E. MERCKX, M. LUTS & J. B. UYTTERHOEVEN. 1995. Incorporation of zeolites in polyimide membranes. J. Phys. Chem. **99**(35): 13187–13192.
35. DUVAL, J.M., A.J.B. KEMPERMAN, B. FOLKERS, et al. 1944. Preparation of zeolite filled glassy polymer membranes. J. Appl. Polym. Sci. **54**(4): 409–418.
36. SUKAPINTHA, W. 2000. Mixed Matrix Membrane for Olefin/Paraffin Separation. M.S. Thesis. The Petroleum and Petrochemical College, Chulalongkorn University. Bangkok, Thailand.

Nonlinear Parameter Estimation for Solution–Diffusion Models of Membrane Pervaporation

BING CAO[a] AND MICHAEL A. HENSON[b]

[a]*Department of Chemical Engineering, Louisiana State University, Baton Rouge, Louisiana, USA*

[b]*Department of Chemical Engineering, University of Massachusetts, Amherst, Massachusetts, USA*

> ABSTRACT: An optimization-based procedure for estimating unknown parameters in solution–diffusion models of membrane pervaporation is presented. Permeation of two components through a polymer membrane is described by distinct solution and diffusion models. The solution model is based on a modified form of Flory–Huggins theory that accounts for interactions between the two penetrants. The diffusion model is derived from Fick's law, where the diffusion coefficients are allowed to depend on the local concentration of each component in the membrane. A phenomenologic relation is used to account for the effect of temperature on the component fluxes. The solution and diffusion models, as well as the temperature–flux relation, contain parameters that are not directly measurable. It is shown that these parameters can be estimated effectively from sorption and flux data by the solution of suitably formulated nonlinear optimization problems. The separation of styrene and ethylbenzene with a polyurethane membrane is used to illustrate the parameter estimation procedure.
>
> KEYWORDS: pervaporation; solution–diffusion model; parameter estimation

INTRODUCTION

Pervaporation is a membrane-based process that has emerged as one of the most promising technologies for the separation of liquid mixtures.[1] A unique feature of pervaporation is that there is a phase change from a liquid on the feed side to a vapor on the permeate side. The separation is achieved by applying a vacuum on the permeate side, such that there is fugacity gradient across the membrane. Because mass transfer is largely independent of the vapor–liquid equilibrium properties, pervaporation is especially well suited for azeotropic and close boiling mixtures that are difficult to separate by distillation.[2] Typical applications include the removal of volatile organic compounds from water[3,4] and the separation of olefin/paraffin mixtures, such as ethylene/ethane or propylene/propane.[1,5]

Address for correspondence: Michael A. Henson, Department of Chemical Engineering, University of Massachusetts, Amherst, MA 01003-9303, USA. Voice: 413-545-3481; fax: 413-545-1647.

henson@ecs.umass.edu

Although other types of pervaporation membranes have been developed,[5] polymer membranes remain the most common material for the construction of pervaporation modules. The solution–diffusion model is the accepted mechanism for describing permeation in polymer membranes.[6,7] According to this mechanism, pervaporation involves the following three steps: (1) the liquid species are dissolved into the membrane surface, (2) the species diffuse through the membrane, and (3) the species desorb from the downstream membrane surface in the vapor phase. Quantitative prediction of pervaporation membrane performance requires the development of mathematical models for the sorption of the penetrants into the membrane and the transport of the penetrants through the membrane. This is commonly achieved by the development of separate solution and diffusion models. Flory–Huggins thermodynamics, the UNIQUAC model, and the penetrant solubility model have been used to describe the solution process.[1] Diffusion is typically modeled using free volume theory[1] or a phenomenologic approach.[8]

Regardless of the specific descriptions employed, solution–diffusion models invariably contain parameters that are not available in the literature and that cannot be measured directly. These unknown parameters provide degrees of freedom that allow the model to be fit to experimental data. Manual adjustment of parameters is inefficient and results in solution–diffusion models that are suboptimal with respect to their predictive capability. Despite its obvious importance, the development of automated parameter estimation techniques for solution–diffusion models has received little attention.

In this paper, a parameter estimation procedure based on nonlinear optimization is proposed for solution–diffusion pervaporation models. A specific estimation strategy is developed for a solution model based on Flory–Huggins theory, a diffusion model derived from Fick's law and a phenomenologic relation used to account for the effect of temperature on the component fluxes. The separation of styrene and ethylbenzene with a poly(hexamethylene sebacate) based polyurethane membrane[9,10] is used to illustrate the parameter estimation procedure. It is important to note that optimization-based parameter estimation is not restricted to the particular model forms used here. Similar estimation strategies could be developed for other types of solution–diffusion models.

ESTIMATION OF SOLUTION MODEL PARAMETERS

Theory

The component volume fractions sorbed into the membrane are predicted from thermodynamic properties of the liquid–polymer mixture. The solution model is based on Flory–Huggins theory[11] and utilizes the interaction parameter equations proposed by Mulder *et al.*[12] When a polymer film is exposed to a pure liquid, Flory–Huggins theory yields the following relation for the pure component activity a^P:

$$\ln a^P = \ln(1 - \phi_2) + \left(1 - \frac{V_1}{V_2}\right)\phi_2 + \chi \phi_2^2, \tag{1}$$

where ϕ_2 is the volume fraction of the polymer; V_1 and V_2 are the molar volumes of the liquid and polymer, respectively; and χ is the Flory–Huggins interaction parameter. The Gibbs free energy of mixing ΔG_{mix} of a ternary system comprising a binary liquid mixture and a polymer membrane is expressed by

$$\frac{\Delta G_{mix}}{RT} = n_1 \ln\phi_1 + n_2 \ln\phi_2 + n_3 \ln\phi_3 + \chi_{12} n_1 \phi_2 + \chi_{13} n_1 \phi_3 + \chi_{23} n_2 \phi_3, \quad (2)$$

where the subscripts 1 and 2 denote the liquid components and the subscript 3 denotes the polymer; n_i and ϕ_i are the mole fraction and volume fraction, respectively, of component i; χ_{ij} is the interaction parameter between components i and j; T is the temperature; and R is the gas constant. The volume fraction ϕ_i is defined by

$$\phi_i = \frac{n_i V_i}{\sum_{i=1}^{3} n_i V_i}. \quad (3)$$

The interaction parameter χ_{12} involving the two liquid components is calculated using excess functions. The following equation holds for binary mixtures:[13]

$$\frac{\Delta G_{mix}}{RT} = x_1 \ln x_1 + x_2 \ln x_2 + \frac{G_E}{RT}, \quad (4)$$

where G_E is the excess Gibbs free energy and x_i is the mole fraction of component i in the liquid phase. The free energy of mixing of the binary mixture is calculated using Flory–Huggins theory,[11]

$$\frac{\Delta G_{mix}}{RT} = x_1 \ln\psi_1 + x_2 \ln\psi_2 + \chi_{12} \psi_1 \psi_2 \left(x_1 + x_2 \frac{V_2}{V_1} \right), \quad (5)$$

where ψ_i is the volume fraction of component i in the liquid phase:

$$\psi_1 = \frac{V_1 x_1}{V_1 x_1 + V_2 x_2}, \quad \psi_2 = \frac{V_2 x_2}{V_1 x_1 + V_2 x_2}. \quad (6)$$

The excess Gibbs free energy is expressed by

$$\frac{G_E}{RT} = x_1 \ln\gamma_1^L + x_2 \ln\gamma_2^L, \quad (7)$$

where $\gamma_i^L = a_i^L / x_i$ is the activity coefficient of component i in the liquid phase and a_i^L is the activity of component i in the liquid phase. The liquid phase activities are estimated from vapor–liquid equilibrium data using the two-suffix Margules equation,[13]

$$RT \ln\gamma_1^L = A x_2^2, \quad RT \ln\gamma_2^L = A x_1^2. \quad (8)$$

The Margules constant A is determined from vapor–liquid equilibrium data using the following equation:

$$P^{sat} = \sum_{i=1}^{2} x_i \gamma_i^L P_i^{sat} = x_1 P_1^{sat} \exp\left(\frac{A}{RT} x_2^2\right) + x_2 P_2^{sat} \exp\left(\frac{A}{RT} x_1^2\right), \quad (9)$$

where P^{sat} is the vapor pressure of the binary mixture and P_i^{sat} is the temperature dependent vapor pressure of pure component i. The following equation for χ_{12} is readily derived by combining Equations (4)–(8):

$$\chi_{12} = \frac{1}{x_1 \psi_2} \left[x_1 \ln\left(\frac{x_1}{\psi_1}\right) + x_2 \ln\left(\frac{x_2}{\psi_2}\right) + \frac{A x_1 x_2}{RT} \right]. \quad (10)$$

The interaction parameters χ_{13} and χ_{23} between the liquid components and the polymer in the ternary mixture can be determined from solubilities of the pure liquids in the polymer. From (1) it follows that the activity of pure component i in the polymer is

$$\ln a_i^P = \ln(1 - \phi_3^{bi}) + \left(1 - \frac{V_i}{V_3}\right)\phi_3^{bi} + \chi_{i3}^b(\phi_3^{bi})^2, \qquad (11)$$

where the superscript b denotes a binary property of the ternary mixture and the superscript bi denotes a binary property between liquid component i and the polymer in the ternary mixture. Equilibrium between the polymer phase and the pure liquid phase requires that $a_i^P = a_i^L = 1$. The following equation for the binary interaction parameter χ_{i3}^b is obtained from Equation (11):

$$\chi_{i3}^b = -\frac{\ln(1 - \phi_3^{bi}) + (1 - V_i/V_3)\phi_3^{bi}}{(\phi_3^{bi})^2}. \qquad (12)$$

The polymer volume fraction ϕ_3^{bi} for a binary mixture of polymer and liquid component i is calculated as follows:

$$\phi_3^{bi} = \frac{1}{\dfrac{S_i^b}{\rho_i} + 1}, \qquad (13)$$

where S_i^b is the solubility of pure component i in the polymer and ρ_i is the density of component i. Typically the ternary interaction parameters χ_{i3} depend on the component concentrations in the polymer. They are calculated from the binary interaction parameter χ_{i3}^b using the relation proposed in Reference 12,

$$\chi_{13} = \chi_{13}^b + a_1 u_2^2 + a_2 u_2 + a_3(\phi_3 - \phi_3^{b1}), \qquad (14)$$

$$\chi_{23} = \chi_{23}^b + b_1 u_1^2 + b_2 u_1 + b_3(\phi_3 - \phi_3^{b2}), \qquad (15)$$

where u_i is the volume fraction of component i in the polymer on a polymer free basis, that is

$$u_1 = \frac{\phi_1}{\phi_1 + \phi_2}, \qquad u_2 = \frac{\phi_2}{\phi_1 + \phi_2}. \qquad (16)$$

Relation (1) can be extended to the ternary system that results when a polymer is exposed to a binary liquid mixture:

$$\begin{aligned}\ln a_1^P = {}& \ln\phi_1 + \phi_2 + \phi_3 - \phi_2\frac{V_1}{V_2} - \phi_3\frac{V_1}{V_3} + \chi_{12}\phi_2(\phi_2 + \phi_3) + \chi_{13}\phi_3(\phi_2 + \phi_3) \\ & - \chi_{23}\phi_2\phi_3\frac{V_1}{V_2} - u_1 u_2\phi_2\frac{\partial\chi_{12}}{\partial u_2} - u_1 u_2\phi_3\frac{\partial\chi_{13}}{\partial u_2} - \phi_1\phi_3^2\frac{\partial\chi_{13}}{\partial\phi_3} \\ & + \frac{V_1}{V_2}u_2^2\phi_3\frac{\partial\chi_{23}}{\partial u_1} - \frac{V_1}{V_2}\phi_2\phi_3^2\frac{\partial\chi_{23}}{\partial\phi_3} + \frac{V_1\rho_3}{M_c}\left(1 - \frac{2M_c}{M}\right)\left(\phi_3^{1/3} - \frac{1}{2}\phi_3\right)\end{aligned} \qquad (17)$$

$$\ln a_2^P = \ln\phi_2 + \phi_1 + \phi_3 - \phi_1\frac{V_2}{V_1} - \phi_3\frac{V_2}{V_3} + \chi_{12}\phi_1\frac{V_2}{V_1}(\phi_1 + \phi_3) + \chi_{23}\phi_3(\phi_1 + \phi_3)$$

$$- \chi_{13}\phi_1\phi_3\frac{V_2}{V_1} - \frac{V_2}{V_1}u_1^2\phi_2\frac{\partial\chi_{12}}{\partial u_2} + \frac{V_2}{V_1}u_1^2\phi_3\frac{\partial\chi_{13}}{\partial u_2} - \frac{V_2}{V_1}\phi_1\phi_3^2\frac{\partial\chi_{13}}{\partial\phi_3} \tag{18}$$

$$- u_1 u_2 \phi_3\frac{\partial\chi_{23}}{\partial u_1} - \phi_2\phi_3^2\frac{\partial\chi_{23}}{\partial\phi_3} + \frac{V_2\rho_3}{M_c}\left(1 - \frac{2M_c}{M_3}\right)\left(\phi_3^{1/3} - \frac{1}{2}\phi_3\right),$$

where ρ_3 is the density of the polymer, M_3 is the molecular weight of the polymer, and M_c is the molecular weight between two crosslinks of the polymer. Equilibrium between the two phase requires that

$$a_1^L = a_1^P, \qquad a_2^L = a_2^P. \tag{19}$$

Parameter Estimation Strategy

The objective is to solve the solution model for the component volume fractions sorbed into the polymer (ϕ_1, ϕ_2). To this end, the nonlinear algebraic Equations (17) and (18) are written to more clearly illustrate their functional dependencies:

$$f_1(a_1^P, \phi_1, \phi_2, \phi_3, \chi_{12}, \chi_{13}, \chi_{23}) = 0, \tag{20}$$

$$f_2(a_2^P, \phi_1, \phi_2, \phi_3, \chi_{12}, \chi_{13}, \chi_{23}) = 0. \tag{21}$$

Clearly, the three volumes fractions must sum to unity. Thus,

$$\phi_1 + \phi_2 + \phi_3 = 1 \Rightarrow f_3(\phi_1, \phi_2, \phi_3) = 0. \tag{22}$$

The activities a_i^P are determined explicitly from (8) and (19) as follows:

$$a_1^P = a_1^L = x_1\gamma_1^L = x_1\exp\left(\frac{Ax_2^2}{RT}\right), \quad a_2^P = a_2^L = x_2\gamma_2^L = x_2\exp\left(\frac{Ax_1^2}{RT}\right). \tag{23}$$

The interaction parameter χ_{12} is determined explicitly from (10). The other interaction parameters, χ_{13} and χ_{23}, are computed using (12)–(15). The five nonlinear algebraic equations, (14), (15), (17), (18), and (22), involving five unknowns ($\phi_1, \phi_2, \phi_3, \chi_{13}, \chi_{23}$) are solved simultaneously to yield the volume fractions ϕ_1 and ϕ_2. We use the MATLAB nonlinear equation solver fsolve for this purpose.

To use the computation procedure outlined above, it is necessary to generate estimates of the unknown solution model parameters. The unknown parameters are the Margules constant (A) and the six constants ($a_1, a_2, a_3, b_1, b_2, b_3$) associated with the ternary interaction parameters in (14) and (15). The Margules parameter is estimated from the nonlinear algebraic Equation (9) using vapor–liquid equilibrium data for the binary liquid mixture. This equation is rewritten

$$g(T, P, x_1, x_2, A) = 0. \tag{24}$$

Given vapor–liquid equilibrium data over a range of conditions, the constant A is determined by solving the following nonlinear optimization problem:

$$\min_{A} \sum_{j=1}^{N} [g(T_j, P_j, x_{1,j}, x_{2,j}, A)]^2 \tag{25}$$

$$\text{subject to: } 0 < A \leq A_u$$

where N is the number of data points; P_j, T_j, $x_{1,j}$, and $x_{2,j}$ are the pressure, temperature, component 1 mole fraction, and component 2 mole fraction, respectively, for the jth data point; and A_u is an upper bound on A. The problem is solved using the MATLAB constrained optimization routine fmincon.

Estimates of the six constants associated with the ternary interaction parameters are generated similarly. The nonlinear algebraic Equations (14) and (15) are rewritten

$$f_4(\chi_{13}, \chi_{13}^b, \phi_3^{b1}, u_2, \phi_3, a_1, a_2, a_3) = 0 \quad (26)$$

$$f_5(\chi_{23}, \chi_{23}^b, \phi_3^{b2}, u_1, \phi_3, b_1, b_2, b_3) = 0. \quad (27)$$

Assume the availability of solubility data in mass of component i sorbed per unit mass of polymer for the ternary mixture over a range of liquid compositions. Given the densities of the liquid components and the polymer, the binary (S_i^b) and the ternary (S_i) solubilities of component i in mass of component i sorbed per unit volume of polymer are readily computed. The binary polymer volume fractions ϕ_3^{bi} and binary interaction parameters χ_{i3}^b are calculated from the binary solubilities using (13) and (12), respectively. The ternary volume fractions ϕ_i are computed from the ternary solubilities using an equation analogous to (13).

The parameter estimation problem involves the least-squares minimization of the difference between the measured values of the ternary volume fractions (ϕ_i^m) and those predicted by the solution model (ϕ_i),

$$\min_{a_1, a_2, a_3, b_1, b_2, b_3} \sum_{j=1}^{M} \sum_{i=1}^{3} [(\phi_i^m)_j - (\phi_i)_j]^2, \quad (28)$$

where M is the number of ternary solubility data points and the subscript j denotes the data point. The quadratic objective function is minimized subject to equality constraints imposed by the solution model equations and inequality constraints corresponding to bounds on the parameter values (if such information is available). The set of nonlinear algebraic equations that comprise the equality constraints is obtained by combining (26) and (27) with (20), (21), and (22). These five equations have the following vector form representation, where only the unknown variables and parameters are shown explicitly:

$$f(\phi_1, \phi_2, \phi_3, \chi_{13}, \chi_{23}, a_1, a_2, a_3, b_1, b_2, b_3) = 0. \quad (29)$$

The resulting nonlinear optimization problem is solved using the MATLAB routine fmincon.

Application to Styrene/Ethylbenzene Pervaporation Membrane

The estimation procedure for solution model parameters is applied to a poly(hexamethylene sebacate) (PHS) based polyurethane membrane developed for styrene/ethylbenzene separations. FIGURE 1 shows the styrene and ethylbenzene uptake for a membrane of thickness 50μm as a function of the feed styrene concentration.[9] The styrene uptake increases with increasing feed styrene concentration, whereas the ethylbenzene uptake exhibits a maximum. The sorption mechanism favors styrene permeation. The necessary pure component physical property data are listed in TABLE 1. The Margules constant, A, was determined by solving the nonlinear optimization problem (25) using the vapor–liquid equilibrium data and the Antoine equations shown in TABLE 2.[14] The estimate obtained was $A = 163.9\,\text{Pa/mol K}$.

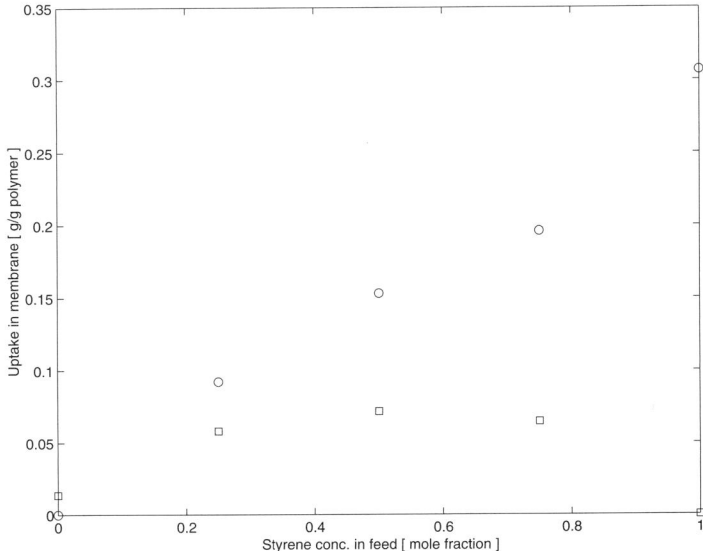

FIGURE 1. Effect of feed concentration on styrene (○) and ethylbenzene (□) uptake in a polyurethane membrane.[9]

The constants a_j and b_j were estimated from the data in FIGURE 1, expressed as mass of component i sorbed per unit mass of polymer. From these estimates the volume fractions ϕ_i of the ternary mixture were computed. In these calculations, the molecular weight between two crosslinks (M_c) was taken as the PHS molecular weight and the polymer molecular weight (M_3) was assumed to be very large compared to M_c. Because there are only six data points (styrene and ethylbenzene values at the three intermediate styrene feed concentrations) available to estimate the six unknown parameters, there are no degrees of freedom for optimization. In this case, the nonlinear optimization problem is reduced to solving Equation **(29)**. The resulting expressions for the interaction parameters are

TABLE 1. Pure component physical property data

Property	Styrene	Ethylbenzene	Polyurethane Membrane
Molar volume (cm^3/mol)	115.0	122.4	—
Density (g/cm^3)	0.9060	0.8670	0.96
Liquid viscosity (cP)	0.725	0.6428	—
Heat capacity (J/g K)	1.6907	1.752	—
Heat of vaporization (J/g)	421.7	335.0	—
Solubility (g/g polymer)	0.307	0.014	—

TABLE 2. Vapor–liquid equilibrium data for styrene/ethylbenzene mixtures at atmospheric pressure

Temperature (°C)	25.88	26.92	27.73	28.27	29.15	30.60	31.68	32.40
x_2	1.0	0.777	0.651	0.575	0.433	0.222	0.083	0.000
y_2	1.0	0.835	0.732	0.663	0.535	0.310	0.128	0.000

Vapor Pressure Equations

$$P_1^{sat} = 10\exp\left(7.2788 - \frac{1649.6}{230 + T}\right)$$

$$P_2^{sat} = 10\exp\left(6.95366 - \frac{1421.914}{212.931 + T}\right)$$

$$P_i^{sat} \text{ (mmHg)}, \quad T \text{ (°C)}$$

NOTE: 1, styrene; 2, ethylbenzene.

$$\chi_{13} = 1.142 - 8.951u_2^2 + 5.866u_2 - 5.652(\phi_3 - 0.754) \tag{30}$$

$$\chi_{23} = 3.297 + 13.537u_1^2 - 11.157u_1 + 8.694(\phi_3 - 0.984). \tag{31}$$

The predicted volume fractions obtained with the estimated parameters are shown in FIGURE 2. The model provides very accurate predictions of this, admittedly, limited data set. Note that the model captures the maximum in the ethylbenzene volume fraction. We have found that this effect cannot be captured with interaction parameter equations simpler than **(14)** and **(15)**.

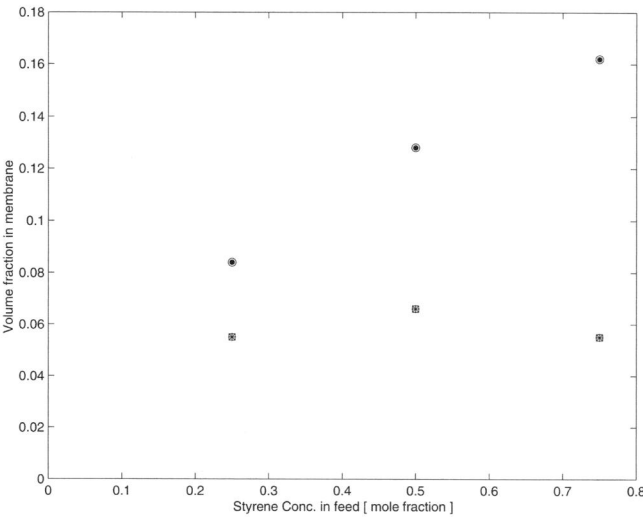

FIGURE 2. Comparison of experimental (● styrene and ★ ethylbenzene) and predicted (○ styrene and □ ethylbenzene) solubilities in a polyurethane membrane.

ESTIMATION OF DIFFUSION MODEL PARAMETERS

Theory

The component fluxes through the membrane are predicted with a diffusion model in which the component volume fractions on the feed-side surface of the membrane are obtained from the solution model. The diffusion model is based on Fick's law and uses the six parameter diffusion coefficient equations for binary liquid mixtures proposed by Brun et al.[8] Under the assumption of moderate membrane swelling, diffusion of component i through the polymer membrane is described by the following form of Fick's law:

$$J_i = -\frac{D_i}{1-\phi_i}\frac{dC_i}{dz}, \qquad (32)$$

where J is the mass flux, z is the flux direction, C is the mass concentration, and ϕ is the volume fraction.

The component diffusion coefficient D_i may depend on the concentration of each component in the liquid mixture. To account for this possibility, the identity $C_i = \rho_i \phi_i$ is used to rewrite (32) as follows:

$$J_i dz = -\frac{\rho_i D_i}{1-\phi_i} d\phi_i, \qquad (33)$$

where ρ_i is the density of component i. At $z = 0$ the component volume fraction ϕ_i is equal to the sorption value, which is denoted here by ϕ_i^0. The volume fraction on the permeate side of the membrane is approximately zero when the permeate pressure is maintained near vacuum. Under the assumption that D_i is constant, integration of (33) from $z = 0$ to $z = l$ yields

$$J_i^* l = \rho_i D_i \ln(1-\phi_i^0), \qquad (34)$$

where l is the membrane thickness and J_i^* denotes the component flux obtained under permeate vacuum.

Equation (34) suggests that a plot of $J_i^* l$ versus $\ln(1-\phi_i^0)$ should be linear. If the relationship is significantly nonlinear, then D_i is not constant but, rather, is a function of the component concentrations. In this case, the functional form of the concentration dependence can be deduced from the shape of the curve.[15] Although other nonlinear functions could be used, the diffusion coefficients are assumed to depend exponentially on the concentration of each component, as suggested by Brun et al.,[8]

$$D_1 = D_1^0 \exp(A_{11}C_1 + A_{12}C_2) \qquad (35)$$

$$D_2 = D_2^0 \exp(A_{21}C_1 + A_{22}C_2), \qquad (36)$$

where D_i^0 is the diffusion coefficient at infinite dilution for component i and the A_{ij} are constant parameters.

Parameter Estimation Strategy

The diffusion model is solved for the two component fluxes (J_1^* and J_2^*) by using the following procedure. The flux Equation (33) is combined with the diffusion coefficient relations (35) and (36) to yield

$$J_1 dz = -D_1^0 \rho_1 \frac{\exp(A_{11}\phi_1 + A_{12}\phi_2)}{1 - \phi_1} d\phi_1 \tag{37}$$

$$J_2 dz = -D_2^0 \rho_2 \frac{\exp(A_{21}\phi_1 + A_{22}\phi_2)}{1 - \phi_2} d\phi_2. \tag{38}$$

These equations are integrated from $z = 0$, where $\phi_i = \phi_i^0$, to $z = l$, where $\phi_i = 0$:

$$J_1^* = -\frac{D_1^0 \rho_1}{l} \int_{\phi_1^0}^{0} \frac{\exp(A_{11}\phi_1 + A_{12}\phi_2)}{1 - \phi_1} d\phi_1 \tag{39}$$

$$J_2^* = -\frac{D_2^0 \rho_2}{l} \int_{\phi_2^0}^{0} \frac{\exp(A_{21}\phi_1 + A_{22}\phi_2)}{1 - \phi_2} d\phi_2. \tag{40}$$

Approximating the integrals by Gaussian quadrature[16] yields

$$J_1^* \approx -\frac{D_1^0 \rho_1}{l} \sum_{k=1}^{P} \frac{\exp(A_{11}\phi_{1,k} + A_{12}\phi_{2,k})}{1 - \phi_{1,k}} w_k \tag{41}$$

$$J_2^* \approx -\frac{D_2^0 \rho_2}{l} \sum_{k=1}^{P} \frac{\exp(A_{21}\phi_{1,k} + A_{22}\phi_{2,k})}{1 - \phi_{2,k}} w_k, \tag{42}$$

where P is the number of quadrature points used, w_k is the quadrature weight at the quadrature point ξ_k (the quadrature point $\xi_k \in [0,1]$ is obtained as the root of the appropriate Jacobi polynomial), and

$$\phi_{1,k} = (1 - \xi_k)\phi_1^0, \qquad \phi_{2,k} = (1 - \xi_k)\phi_2^0. \tag{43}$$

Equations (41)–(43) allow the component fluxes, J_i^*, to be computed from the results of the solution model.

To solve the diffusion model, values must be specified for the six empirical constants in the diffusion coefficient Equations (35) and (36). The vector of six unknown parameters is denoted by $\theta = [D_1^0, D_2^0, A_{11}, A_{12}, A_{21}, A_{22}]^T$. Assume that N experiments are performed to obtain the sorption and flux data $\{J_{1,j}^*, J_{2,j}^*, \phi_{1,j}^0, \phi_{2,j}^0\}$, where j denotes the data point. The discretized flux equations, (41) and (42), are written for each data point to yield $2N$ nonlinear algebraic equations in the six unknown parameters θ. Parameter estimates are obtained by solving the following nonlinear least-squares estimation problem:

$$\min_{\theta} \sum_{j=1}^{N} [J_j^{*,m} - J_j^*]^T [J_j^{*,m} - J_j^*], \tag{44}$$

where $J_j^{*,m} = [J_{1,j}^{*,m} \; J_{2,j}^{*,m}]^T$ and $J_j^* = [J_{1,j}^* \; J_{2,j}^*]^T$ are the measured and predicted values, respectively, of the component flux. The minimization is performed subject to nonlinear equality constraints derived from the $2N$ component flux equations and inequality constraints on the estimated parameters (e.g., $D_i^0 > 0$).

Application to Styrene/Ethylbenzene Pervaporation Membrane

The estimation procedure for diffusion model parameters is applied to the polyurethane membrane considered previously. Component fluxes J_i^* are obtained from the flux and selectivity data[9] shown in FIGURE 3. Note that the styrene selectivity decreases rapidly as the feed styrene concentration is increased due to a loss of sorption selectivity. First, the assumed form of the diffusion coefficient Equations (35) and (36) are checked using (34), where $l = 50\mu m$ and values of ϕ_i^0 are obtained from the sorption data in FIGURE 1. The results in FIGURE 4 verify that the styrene diffusion coefficient has an exponential concentration dependence. An appropriate functional form for the concentration dependence of the ethylbenzene diffusion coefficient is less clear. Based on the styrene behavior, we also used the exponential function (36) for ethylbenzene. The diffusion model parameters were estimated from the sorption and flux data in FIGURES 1 and 3, respectively, by solving the constrained nonlinear optimization problem (44). There are a total of ten data points (styrene and ethylbenzene values for five styrene feed concentrations) available to estimate the six unknown parameters. The following relations were obtained:

$$D_1 = 8.78 \times 10^{-12} \exp(19.9 C_1 - 0.06 C_2) \tag{45}$$

$$D_2 = 8.13 \times 10^{-12} \exp(-13.7 C_1 + 58.5 C_2), \tag{46}$$

where styrene and ethylbenzene are designated components 1 and 2, respectively. Styrene has a slightly higher diffusion coefficient at infinite dilution than does ethylbenzene. On the other hand, the ethylbenzene diffusion coefficient increases more rapidly with increasing ethylbenzene concentration than does the styrene diffusion

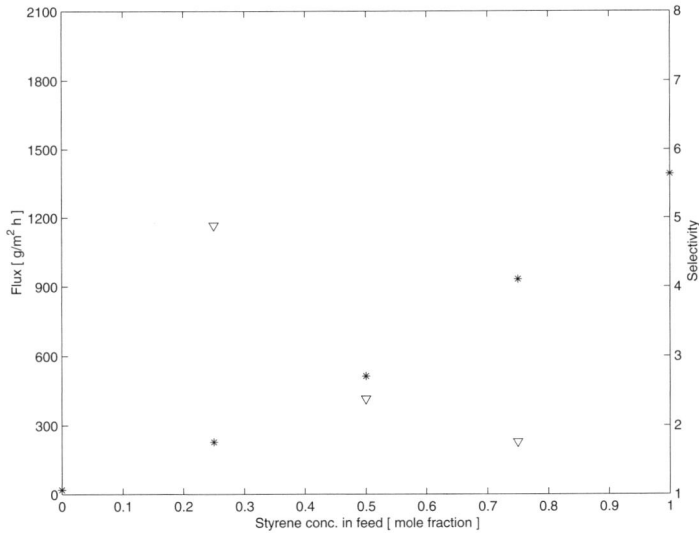

FIGURE 3. Effect of feed concentration on flux ($\triangledown$) and ($\star$) selectivity of a polyurethane membrane.[9]

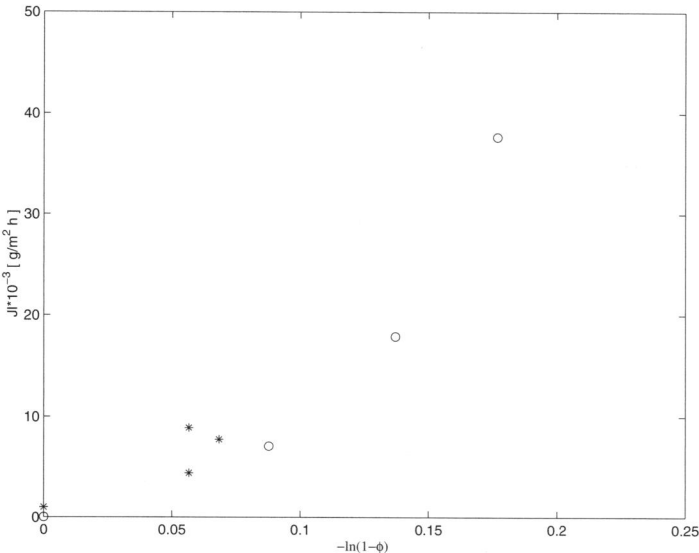

FIGURE 4. Concentration dependence of styrene (○) and ethylbenzene (★) diffusion coefficients.

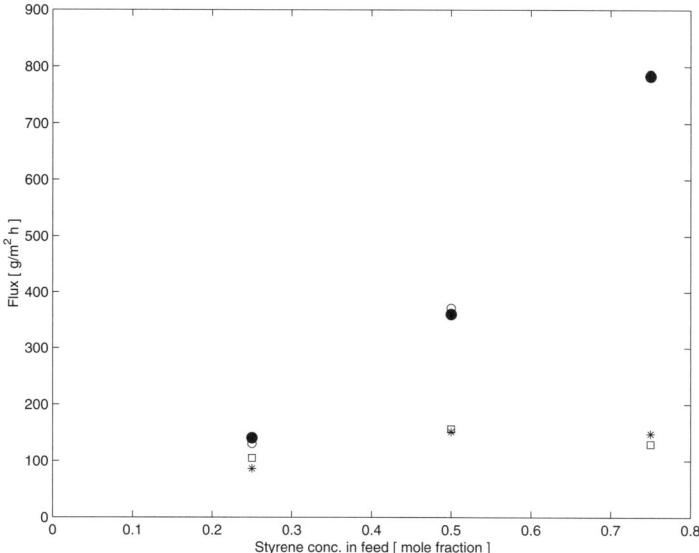

FIGURE 5. Comparison of experimental (○ styrene and □ ethylbenzene) and predicted (● styrene and ★ ethylbenzene) component fluxes in a polyurethane membrane.

coefficient with increasing styrene concentration. The ethylbenzene diffusion coefficient is significantly reduced by increasing styrene concentration, whereas the styrene diffusion coefficient is virtually unaffected by the ethylbenzene concentration. As a result of this behavior, sorption rather than diffusion is the primary mechanism that controls the styrene permselectivity of the polyurethane membrane.[9]

In FIGURE 5, component flux data are compared to calculated fluxes derived from the estimated diffusion coefficient Equations **(45)** and **(46)**. The diffusion model provides accurate predictions of both fluxes given the diffusion coefficient form used for ethylbenzene and the limited number of data points available. Note that the model is able to predict the maximum in the ethylbenzene flux. We have found that this effect cannot be captured if the coupling terms (A_{12} and A_{21}) are zero.

ESTIMATION OF TEMPERATURE DEPENDENT FLUX PARAMETERS

The potentially strong effect of temperature on the component fluxes[17] should be included in the solution–diffusion model to ensure accurate predictions of separation performance. Typically there is insufficient data to account for the temperature effect separately in the solution and diffusion models. A simpler and more direct alternative is to correct the fluxes derived from the diffusion model. Let T_0 denote the fixed reference temperature used for estimation of the solution and diffusion model parameters. The following phenomenologic relation[17] is used to account for temperature variations:

$$J_i^*(T) = J_i^*(T_0)\exp\left[-\frac{E_i}{R}\left(\frac{1}{T} - \frac{1}{T_0}\right)\right], \quad (47)$$

where $J_i^*(T_0)$ is the flux of component i generated by the solution–diffusion model, $J_i^*(T)$ is the temperature corrected flux of component i, E_i is an activation energy parameter for component i, and R is the gas constant.

The parameter estimation problem involves the determination of the unknown parameters E_i from temperature dependent flux data. Equation **(47)** shows that a plot of $\ln J_i^*(T)$ versus $T^{-1} - T_0^{-1}$ should be linear with slope $-E_i/R$. Therefore, nonlinear optimization is not required in this case. FIGURE 6 illustrates the procedure for the polyurethane membrane considered previously. The required temperature dependent flux data were obtained from Reference 9 for a feed styrene mole fraction $x_f = 0.5$ and reference temperature $T_0 = 25°C$. Both styrene and ethylbenzene exhibit an exponential dependence on temperature. The following activation energy parameters were derived: $E_1 = 1.343 \times 10^4$ J/mol and $E_2 = 2.986 \times 10^4$ J/mol, where styrene and ethylbenzene are designated components 1 and 2, respectively. The estimated activation energies demonstrate that the ethylbenzene flux is more strongly affected by temperature than is the styrene flux.

Membrane mass transfer coefficients can be computed directly from the temperature corrected component fluxes. The following calculation is valid if the permeate pressure is negligible, the liquid behavior is ideal, and boundary layer resistances are negligible. Then, the membrane mass transfer coefficient of component i (k_{mi}) is related to the component flux (J_i^*) as follows:

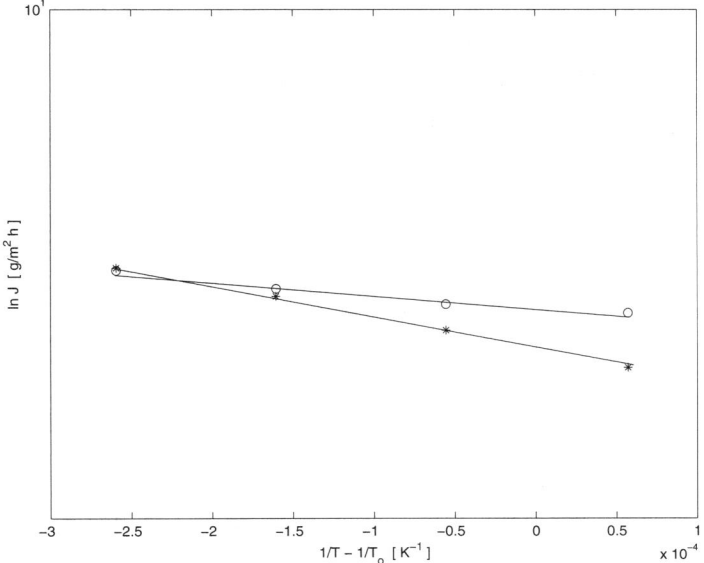

FIGURE 6. Temperature dependence of styrene (○) and ethylbenzene (★) flux.

$$J_i^* = k_{mi} x_i P_i^{\text{sat}}, \qquad (48)$$

where x_i is the liquid composition of component i and P_i^{sat} is the saturation pressure of component i at temperature T. This equation is rearranged to yield

$$k_{mi} = \frac{J_i^*}{x_i P_i^{\text{sat}}}. \qquad (49)$$

The mass transfer coefficients can be used directly to characterize a dense film membrane, or they can be combined with appropriate expressions for the liquid boundary layer resistance to determine the separation performance of a membrane module.[18]

SUMMARY AND CONCLUSIONS

An optimization-based procedure for estimating unknown parameters in solution–diffusion models of membrane pervaporation has been developed and evaluated. Although the general methodology is applicable to a wide variety of model types, we have presented a specific estimation strategy for a solution model based on Flory–Huggins theory, a diffusion model derived from Fick's law, and a phenomenologic temperature dependent flux relation. The parameter estimation problem is posed as nonlinear optimization problem in which the unknown parameters are the decision variables and the objective function is the least-squares difference between the measured data and the model predictions. The objective function is minimized subject to equality constraints imposed by nonlinear model equations and inequality constraints representing known bounds on the parameter values. The resulting nonlinear optimization problems are solved using the MATLAB routine fmincon, although other nonlinear programming codes also could be employed. Advantages of the optimization-based procedure as compared to manual parameter adjustment are illustrated by application to the separation of styrene and ethylbenzene with a polyurethane membrane.

ACKNOWLEDGMENTS

Financial support from the National Science Foundation (Grant CTS-9817298) is gratefully acknowledged.

REFERENCES

1. X. FENG & R.Y.M. HUANG. 1997. Liquid separation by membrane pervaporation: a review. Ind. Eng. Chem. Res. **36:** 1048–1066.
2. H.L. FLEMING & C.S. SLATER. 1989. Pervaporation Membrane Handbook. Van Nostrand and Reinhold, New York.
3. C. LIPSKI & P. COTE. 1990. The use of pervaporation for the removal of organic contaminants from water. Environ. Progr. **9:** 254–261.
4. R. PSAUME, P. APTEL, Y. AURELLE, et al. 1988. Pervaporation: importance of concentration polarization in the extraction of trace organics from water. J. Membr. Sci. **36:** 373.
5. J.C. DAVIS, R.J. VALUS, R. ESHRAGHI & A.E. VELIKOFF. 1993. Facilitated transport membrane systems for olefin purification. Sep. Sci. Technol. **28:** 463–476.
6. T. KATAOKA, T. TSURU, S. NAKAO & S. KIMURA. 1991. Permeation equations developed for prediction of membrane performance in pervaporation, vapor permeation and reverse osmosis based on the solution–diffusion model. J. Chem. Eng. Jpn. **24:** 334.
7. J.G. WIJMANS & R.W. BAKER. 1995. The solution-diffusion model: a review. J. Membr. Sci. **107:** 1.
8. J.P. BRUN, C. LARCHET, R. MELET & G. BULVESTRE. 1985. Modeling of the pervaporation of binary mixtures through moderately swelling non-reactive membranes. J. Membr. Sci. **23:** 257.
9. B. CAO, H. HINODE & T. KAJIUCHI. 1999. Permeation and separation of styrene/ethylbenzene mixtures through cross-linked poly(hexamethylene sebacate) membranes. J. Membr. Sci. **156:** 43.
10. B. CAO & T. KAJIUCHI. 1999. Pervaporation separation of styrene-ethylbenzene mixture using poly(hexamethylene sebacate)-based polyurethane membranes. J. Appl. Polym. Sci. **74:** 833.
11. P.J. FLORY. 1953. Principles of Polymer Chemistry. Cornell University Press, Ithaca.
12. M.H.V. MULDER, T. FRANKEN & C.A. SMOLDERS. 1985. Preferential sorption versus preferential permeability in pervaporation. J. Membr. Sci. **22:** 155.
13. J.M. PRAUSNITZ, R.N. LICHTENTHALER & E.G. AZEVEDO. 1986. Molecular Thermodynamics of Fluid-Phase Equilibria. Prentice-Hall, Englewood Cliffs.
14. P. CHAIYAVECN & M. VAN WINKLE. 1959. Styrene–ethylbenzene vapor–liquid equilibria at reduced pressures. J. Chem. Eng. Data **4:** 53–59.
15. J. CRANK. 1975. The Mathematics of Diffusion. Clarendon Press, Oxford.
16. B.A. FINLAYSON. 1980. Nonlinear Analysis in Chemical Engineering. McGraw-Hill, New York.
17. R. RAUTENBACH & R. ALBRECHT. 1985. The separation potential of pervaporation—Part 1. Discussion of transport equations and comparison with reverse osmosis. J. Membr. Sci. **25:** 1–23.

18. B. Cao & M.A. Henson. 2002. Modeling of spiral wound pervaporation modules with application to the separation of styrene/ethylbenzene mixtures. J. Membr. Sci. **197:** 117–146.

Poly(Vinyl Alcohol)–Based Polyelectrolyte Pervaporation Membranes

BENHUI SUN AND JIAN ZOU

Beijing University of Chemical Technology, Beijing, China

ABSTRACT: By modifying poly(vinyl alcohol) (PVA), phosphatic anionic PVA (P-PVA) and quaternary ammonium cationic PVA (C-PVA) with various degrees of substitution (D.S.) were synthesized. The effects of synthesis conditions on the degree of substitution were studied. With these two kinds of materials, polyelectrolyte complexes were formed and their solubilities in water were studied. Pervaporation composite membranes were prepared from P-PVA, C-PVA, and their polyelectrolyte complexes. Some of the membranes showed good separation performance. The polyelectrolyte complex membrane prepared by mixing P-PVA (D.S. 2.3%) and C-PVA (D.S. 2.9%) with weight ratio of 1/1, showed a permeation rate of $378\,g/m^2h$ and separation factor of 2,250 for dehydration of ethanol/water mixture (ethanol 95.4 wt%) at a feed temperature of 75°C. Some factors that influenced pervaporation performance included type of polyelectrolyte, degree of substitution, ratio of polyanion and polycation, feed temperature, feed concentration, and elapsed time. Solvent resistance of pervaporation composite membranes was also evaluated.

KEYWORDS: polyelectrolyte; poly(vinyl alcohol); pervaporation; dehydration; ethanol/water mixture

INTRODUCTION

Pervaporation is a novel membrane separation technique that exhibits many advantages over conventional separation processes. Because the separation property of pervaporation membranes is a key factor that affects equipment investment, it is very important to develop new membrane materials of high performance. Hydrophilic polymer materials used in organic solvent dehydration can be classified as non-ionic type or ionic type according to the hydrophilic group used. Poly(vinyl alcohol) is a representative pervaporation membrane material of non-ionic hydrophilic type that has good chemical resistance to organic solvents and long-term stability. In 1983, a German corporation, GFT, initialized the commercialization of ethanol/water separation using crosslinked PVA composite membrane.[1] Polymers with ionic hydrophilic groups, namely polyelectrolytes, include poly(acrylic acid) (PAA), chitosan (CS), alginic acid (AGA), and so on. Polyelectrolyte membranes often show good separation properties because of their high hydrophilicities, but often suffer a declination in separation performance because of the elution of low molecular weight counter ion. Karakane *et al.*[2] used several kinds of polycations to ionize poly(acrylic acid) to form polyelectrolyte complexes (PEC). Composite membranes with this kind of PEC separation layer not only have superior separation properties, but also long-term stability.

Based on poly(vinyl alcohol), phosphatic anionic PVA and quaternary ammonium cationic PVA were synthesized. Polyelectrolyte complexes between P-PVA and C-PVA were formed and their solubilities in water were studied. We prepared pervaporation dehydration composite membranes with P-PVA, C-PVA, and their complexes. Separation properties of these membranes were evaluated and some of the factors that influence pervaporation performance were studied.

EXPERIMENTAL

Experimental Reagent

PVA (Beijing Organic Chemical Plant) had a degree of hydrolysis equal to 99.5% and a degree of polymerization 1,740. The 3-chloro-2-hydroxypropyltrimetayl ammonium chloride (Tokyo Chemical) was used as a 60% aqueous solution. PAA (Fluka) had a degree of polymerization 2,100. The phosphoric acid was used as an 85% aqueous solution.

Material Synthesis

P-PVA

PVA was dissolved in hot distilled water to obtain a 3% aqueous solution; then the catalyst and phosphoric acid were added. After a period of stirring and heating, the reaction product was precipitated and dried in vacuum after washing away the residual phosphoric acid.

C-PVA

PVA was dissolved in hot distilled water to obtain a 3% aqueous solution, then the catalyst and 3-chloro-2-hydroxypropyltrimetayl ammonium chloride aqueous solution were added. After a period of stirring and heating, the reaction product was precipitated and dried in vacuum after washing away the residual 3-chloro-2-hydroxypropyltrimetayl ammonium chloride.

Degree of Substitution (D.S.)

The D.S. of P-PVA was calculated from the elemental content of phosphorus using a chemical analysis method proposed elsewhere.[3] If the elemental content of phosphorus is $x\%$, then the D.S. of P-PVA is equal to

$$\frac{x/31}{\frac{(100-124x/31)}{44}+\frac{x}{31}} \times 100. \tag{1}$$

The D.S. of C-PVA was calculated from the elemental content of nitrogen using an elemental analysis method with an elemental analyzer, Elementar vario EL. If the elemental content of nitrogen is $x\%$, then the D.S. of C-PVA is equal to

$$\frac{x/14}{\frac{(100-195.5x/14)}{44}+\frac{x}{14}} \times 100. \qquad (2)$$

Characterization of Chemical Structure of Materials

P-PVAs with various D.S. were cast into films with thickness 15–20 micron. The infrared spectra of these films were obtained by using a Perkin-Elmer FTIR 1650 spectroscope.

Phosphoric esterification can also be confirmed by potentiometric titration.[4] The detailed procedure was as follows: first, the substitute groups were converted into acid form. A sample (0.1–0.2 g) was added to 0.2 mol/L chlorhydric acid solution (70% methanol aqueous solution). After being soaked for 10 hours and stirred for three hours, the sample was washed with 70% methanol aqueous solution three times and then dissolved in 20 mL standard sodium hydrate solution (0.1 mol/L). Second, the solution was titrated with standard chlorhydric acid solution (0.1 mol/L). The pH value, measured with a pH meter, and the corresponding volume of buret were recorded.

Preparation of PEC

The P-PVA, C-PVA, and PAA were added to distilled water, heated, and stirred to obtain a 5% aqueous solution. PEC was formed by mixing solutions of opposite-charged polyelectrolyte in specified ratios.

Preparation of Composite Membrane

P-PVA and C-PVA were dissolved in water to obtain casting solutions. The P-PVA/C-PVA PEC casting solution was generated by mixing P-PVA and C-PVA solutions. All these solutions were purified by filtration and defoamed by applying vacuum. These solutions were then cast by using a tailor-made coating machine on a supporting poly(acrylonitrile) UF membrane that was reinforced with non-woven polyester cloth. The thickness of separation layer was about 2–3 microns.

Membrane Evaluation

The cross-section scanning electron micrographs of composite membrane were obtained on a scanning electron microscope (JEOL JSM6301F) with a magnification of 6,000.

Pervaporation was performed in a tailor-made apparatus. The flow was 143 L/h and the membrane area was 19.6 cm^2. A vacuum was used at the downstream side and the pressure was 500 Pa. Permeate was condensed and frozen by liquid nitrogen trap. The content of ethanol/water mixture was analyzed with a gas chromatography (Model SQ206, Beijing Analysis Apparatus Plant) equipped with 1 m-column packed with Porapak P. The separation property was characterized by the flux, J (g/m^{2}h), and separation factor, α, defined by

$$\alpha = \frac{Y_W/Y_A}{X_W/X_A}, \qquad (3)$$

where Y_W and X_W are the weight fractions of water in the permeate and in the feed, respectively. Y_A and X_A are the weight fractions of ethanol in the permeate and in the feed, respectively.

The solvent resistance of composite membranes was evaluated by soaking them in 95wt% ethanol aqueous solution, benzene, toluene/butanone mixture (weight ratio 40/60), respectively, at room temperature for a month. After these composite membranes were dried in air, they were used to study pervaporation.

RESULTS AND DISCUSSION

Effects of Reaction Conditions on D.S.

P-PVA was obtained by esterification of PVA with phosphoric acid in the presence of catalyst. The reaction products were white or pale yellow solids that were able to swell in water at ambient temperature and were easier to dissolve in water than unmodified PVA. The esterification reaction of PVA with phosphoric acid is depicted as follows:

$$-\text{[CH}_2-\text{CH]}_n- + m\,H_3PO_4 \xrightarrow{\text{catalyzer}} -\text{[CH}_2-\text{CH]}_m\text{[CH}_2-\text{CH]}_{n-m}- + m\,H_2O$$
$$\overset{|}{OH} \qquad\qquad\qquad\qquad\qquad \overset{|}{O}\qquad\quad\overset{|}{OH}$$
$$\qquad\qquad\qquad\qquad\qquad\qquad HO-P=O$$
$$\qquad\qquad\qquad\qquad\qquad\qquad \overset{|}{OH}$$

This reaction is a typical esterification of phosphoric acid and both monoester and diester are formed. Because of the presence of catalyst and the low concentration of PVA in the reaction solution, the reaction products were mainly monoester. The D.S. of P-PVA was affected by several factors, for example, the reaction temperature, the concentration of catalyst, and the concentration of phosphoric acid. Because D.S.

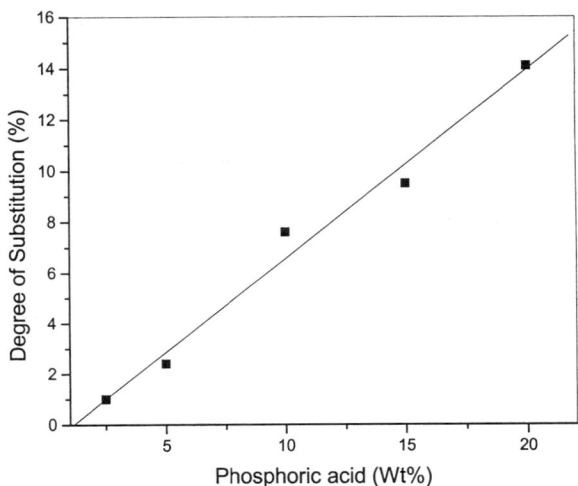

FIGURE 1. Effect of concentration of phosphoric acid on D.S.

TABLE 1. Effect of mole ratio of catalyst and etherification reagent on D.S.

Mole ratio of catalyst and etherification reagent	1/4	1/2	1/1
D.S. (%)	1.70	5.19	2.71

can be effectively controlled and, thus, desired materials can be prepared, we focus on the effect of phosphoric acid on D.S. of P-PVA in this paper.

FIGURE 1 suggests that D.S. of P-PVA is linearly related to the concentration of phosphoric acid. Thus, with other conditions remaining unchanged, P-PVA of the desired D.S. can be prepared by controlling the concentration of phosphoric acid. This D.S. can range from 1% to 20%.

C-PVA was obtained by etherification of PVA with the etherification reagent, 3-chloro-2-hydroxypropyltrimetayl ammonium chloride, in the presence of catalyst. The reaction products were white or pale yellow solids that were easier to dissolve in water than unmodified PVA. The etherification reaction of PVA is

$$\sim\!\!\!\!\underset{OH\ \ OH\ \ OH}{\wedge\wedge\wedge\wedge} + [Cl-CH_2CHCH_2N(CH_3)_3]^+Cl^-\ \underset{OH}{\xrightarrow{Catalyzer}} \sim\!\!\!\!\underset{\substack{O\ \ OH\ \ OH\\ CH_2\\ HO-CH\\ CH_2\\ H_3C-N-CH_3\\ \overset{+}{C}H_3\ Cl^-}}{\wedge\wedge\wedge\wedge}$$

The D.S. of C-PVA was affected by several key factors; the reaction temperature, the concentration of catalyst, and the concentration of etherification reagent, the latter two in particular. TABLE 1 shows that a high D.S. can be achieved with the mole ratio of 1/2. Mole ratios higher or lower than 1/2 result in a decreased D.S.

Characterization of D.S. and Chemical Structure

P-PVA and C-PVA with various D.S. were synthesized, two of each kind were selected for further experimental study, and their D.S. were analyzed (see TABLE 2). P-PVA, obtained by introducing phosphoric ester group into PVA, was dissolved in 20 mL standard sodium hydroxide solution (0.1 mol/L) and then titrated by standard chlorhydric acid solution. The first-order differential titration curve is shown in FIGURE 2. Two peaks can be seen, suggesting that P-PVA is a diprotic acid and this confirms the formation of phosphoric monoester. The infrared spectra are shown in FIGURE 3. An absorption peak occurs at 1,050 cm^{-1}, and the area of this peak increases with an increase in the D.S. This peak is the absorption peak of P–O–C single

TABLE 2. D.S. of materials

	Sample	D.S. (%)	Sample	D.S. (%)
P-PVA	A1	2.3	A2	5.0
C-PVA	C1	2.9	C2	5.2

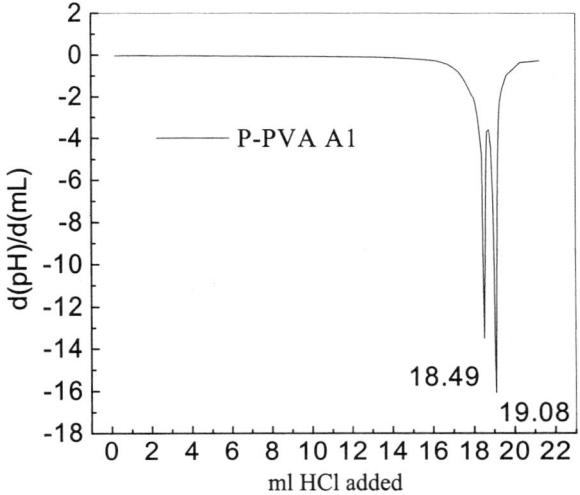

FIGURE 2. First-order differential of titration curve.

bonds[5] formed by phosphorus and alkyl carbon atoms. Thus, the infrared spectra also confirmed the esterification reaction.

C-PVA was synthesized by introducing quaternary ammonium groups into PVA, as confirmed by elemental analysis.

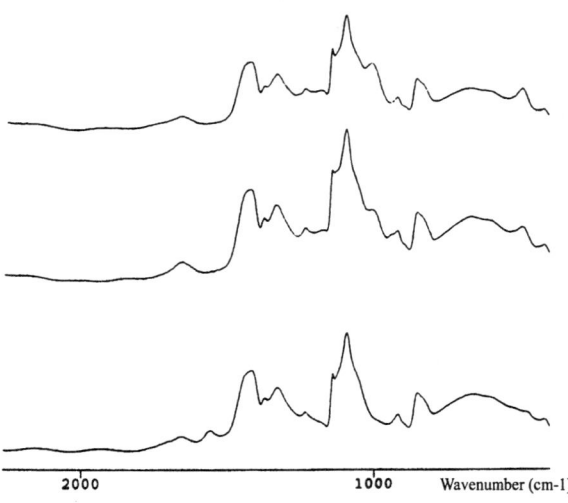

FIGURE 3. Infrared spectra of P-PVA: A2 (**top**), A1 (**middle**), PVA (**bottom**).

Formation of Polyelectrolyte Complex

In most circumstances, PEC precipitates form when solutions of oppositely charged polyelectrolytes are mixed. However, the formation of precipitates is affected by many factors; for example, molecular structure of the polyelectrolyte, the temperature of solution, the pH, and ionic strength. Most polyelectrolytes developed recently for pervaporation form PEC precipitates when mixed with polyelectrolyte of opposite charge. For example, PEC of PAA and various kinds of polycation prepared by Karakane et al.,[2] PEC of chitosan and PAA prepared by Lu et al.,[6] and PEC consisting of chitosan and alginic acid prepared by Zeng et al.[7] are all precipitates that are insoluble in water at room temperature. If precipitates form as soon as the solutions of oppositely charged polyelectrolytes are mixed, it is theoretically possible to prepare PEC PV membranes without an additional crosslinking process for separation of some organic/water mixtures. However, special coating methods, including the interfacial reaction method and the share-solvent method, are required because of the formation of PEC precipitates.

The interfacial reaction method, or two-layer complex method uses the polyelectrolyte solution coated on a support membrane. The solution of polyelectrolyte of opposite charge is coated on the first layer of the initial coated membrane dipped into the solution of polyelectrolyte of opposite charge. PEC forms *in situ* in the interfacial layer. The share-solvent method uses a specific solvent to dissolve the water-insoluble PEC and then to coat the solution onto a support membrane. The interfacial reaction method is more complicated, and more difficult to control that the ordinary coating process for composite membranes. In particular, when the second layer is cast, it is difficult to control the thickness and quality of separation layer and it is difficult to control the extent of complex formation, especially the ratio of polycation and polyanion. The problem of the share-solvent method is that it is not easy to find a suitable share solvent. The so called shielding solution is not feasible for the preparation of PV membrane in most circumstances. If the PEC is water-soluble, it is possible to use the ordinary coating process, which is more practical than the interfacial reaction method and the share-solvent method.

When the solutions of P-PVA and C-PVA obtained in this experiment were mixed, there was no precipitate. Experimental results are shown in TABLE 3, from which we can see that no precipitate is formed when solutions of P-PVA and C-PVA

TABLE 3. Water solubility of PEC

Polycation	Polyanion	Mixing Ratio (volume)	Precipitate
C1	A1	1/1	none
C1	PAA$^-$Na$^+$	1/1	white
C2	A2	1/3	none
C2	A2	1/1	none
C2	A2	3/1	none
C2	PAA$^-$Na$^+$	1/1	white

NOTE: The pH of the mixtures was between 5 and 8.

of various D.S. were mixed. Yet when the solutions of C-PVA and PAA were mixed, PEC precipitates. This may be explained by the fact that P-PVA and C-PVA were all obtained by modifying the same material, PVA, so the compatibility is very good. On the other hand, the D.S. was rather low, which means the Coulombic forces were weak. Thus, the hydrophilic intermolecular interaction was weak.

Because there is no PEC precipitate between P-PVA and C-PVA, it is feasible to use the ordinary coating process to manufacture PV composite membranes. This process is simple and it is easy to control the thickness of separation layer and the ratio of polycation and polyanion.

Composite Membrane Structure

A scanning electron micrograph of the PEC composite membrane (A1+C1 = 1/1) is shown in FIGURE 4 (the scale in FIG. 4 is 10 micron). As can be seen the separation layer is even with a thickness of about 3 micron. Beneath the separation layer lie the micropores of the supporting membrane. It can also be seen that the separation layer and the supporting membrane adhere tightly.

Effects of Polyelectrolyte Type and D.S. on Separation Properties

P-PVA and C-PVA of two different D.S., together with unmodified PVA, were used to prepare composite membranes. Pervaporation performances for dehydration of ethanol/water mixture at temperature of 75°C are listed in TABLE 4, from which it

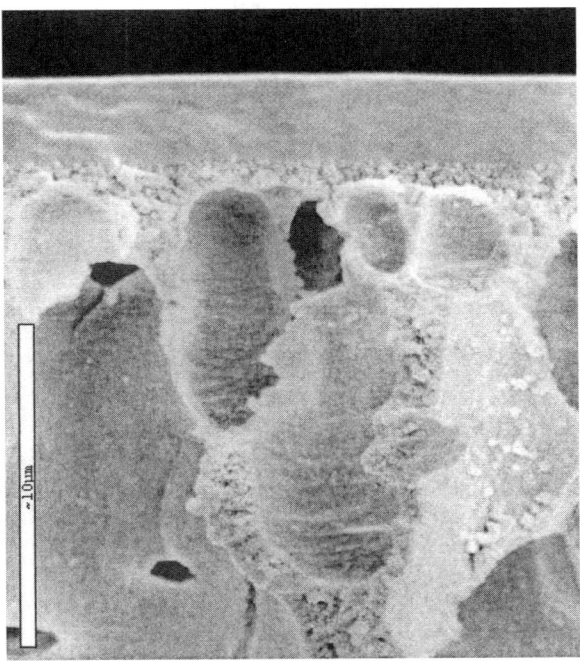

FIGURE 4. SEM of PEC composite membrane.

TABLE 4. Pervaporation performance of various materials

Material	D.S. (%)	Flux (g/m^{2}h)	Separation Factor
Unmodified PVA	0	247	721
P-PVA A1	2.3	398	637
P-PVA A2	5.0	72	407
C-PVA C1	2.9	216	424
C-PVA C2	5.2	249	1,540

NOTE: Feed: water 4.6 wt%, ethanol 95.4 wt%. $T = 75°C$.

can be seen that the flux of P-PVA with D.S. of 2.3% was significantly higher than that of unmodified PVA, whereas the flux of P-PVA with D.S. of 5.0% was much lower than that of unmodified PVA. The fluxes of C-PVA with D.S. of 2.9% and 5.2% were close to that of unmodified PVA.

These phenomena may be due to the chemical change and (or) structural change caused by the introduction of substitution groups. The chemical and physical properties and D.S. of substitution groups affects not only the hydrophilicity of separation layer, but also the second order and the third order structures of molecular chains resulting in changes in the separation properties of pervaporation membranes. The phosphoric acid group that was introduced into the PVA chains is a medium-strong acid whose hydrophilicity is stronger than that of hydroxyl group of PVA. This acid group interacted with counter ions through Coulombic force, which caused electric seduction to water molecules with strong polarity. The seduction resulted in permselectivity toward water molecules, and "salt out effect" toward organic molecules that only had weak polarity. This made the permselectivity to water even stronger. For P-PVA with D.S. of 2.3%, the interaction between phosphoric acid groups and its counterions increased permselectivity toward water when compared with unmodified PVA. Hence, the separation layer was able to swell reasonably even when the water concentration in feed was rather low, thus increasing the usable free-volume (pervaporation passage) and, thereby, increasing the flux. For P-PVA with D.S. of 5.0%, the hydrophilicity may be too strong. This, instead, caused P-PVA to swell excessively, decreasing the "pervaporation passage". In addition, the overly strong interaction between membrane and water molecules may subdue permeation of water, and this also decrease the flux. These two reasons together resulted in the flux of P-PVA with D.S. of 5.0% being lower than that of unmodified PVA.

In the case of C-PVA, because the introduction of quaternary ammonium group, whose hydrophilicity is not strong, was accompanied by introduction of hydrophobic groups, such as methyl and ether groups, the overall hydrophilicity of C-PVA was close to that of PVA. Thus, the permeability (flux) was also similar to that of unmodified PVA.

Effect of Polycation/Polyanion Ratio in PEC

For polyelectrolyte complex membranes, the ratio of cationic C-PVA to anionic P-PVA is also a significant factor that influences separation properties. The fluxes

TABLE 5. Separation properties of membranes with various P-PVA/C-PVA ratios

P-PVA/C-PVA Ratio	Flux (g/m^{2}h)	Separation Factor
A1	398	637
A1+C1 = 1/1	378	2,250
C1	216	424
A2	72	407
A2+C2 = 3/1	124	316
A2+C2 = 1/1	284	1,910
A2+C2 = 1/3	384	795
C2	249	1,540

NOTE: Feed: water 4.6 wt%, ethanol 95.4 wt%. $T = 75°C$.

and separation factors of membranes with various C-PVA/P-PVA ratios are listed in TABLE 5.

Among PEC membranes from P-PVA and C-PVA, the membrane with A1+C1 = 1/1 showed a high flux that was only a little lower than that of A1, and a high separation factor that was much higher than those of phosphatic PVA A1 and cationic PVA C1. For membranes formed by mixing A2 and C2, the flux of membrane A2+C2 = 1/3 was higher than those of membrane A2 and membrane C2, whereas the separation factor was higher than those of A2 and C2. These results further confirm the complexation of P-PVA and C-PVA and also suggest that by adjusting the ratio of cation and anion, the hydrophilicity and hydrophobicity can be well balanced and thereby optimal separation properties can be achieved. After the complexation of P-PVA and C-PVA, the flux and separation increased, which can be accounted for by two reasons. First, the hydrophilicity of polyelectrolyte complex is usually higher than that of any of its component polyelectrolytes, which means that PEC has a higher permselectivity towards water molecules and a stronger salt-out effect.[7] Second, that the introduced quaternary ammonium group and phosphoric ester group both exhibit steric hindrance, which causes the molecular chain arrangement to be looser and increases the usable free-volume, thereby being more favorable for small molecule penetration.

Effect of Feed Concentration on Separation Properties

FIGURE 5 shows the dependence of separation properties on feed concentration. As can be seen, the feed concentration affected flux and separation factor greatly. In FIGURE 5A, it can be seen that flux increased with the increase of water concentration in the feed. There may be two reasons for this. One reason is that with the increasing water concentration, the partial pressure of water on the upstream side is higher, thereby increasing the driving force for pervaporation of water molecules and contributing to the overall flux. Another reason is that, with increasing water content, the degree of swelling increased, which increases the usable free-volume and reduces resistance of components to permeate the membrane.

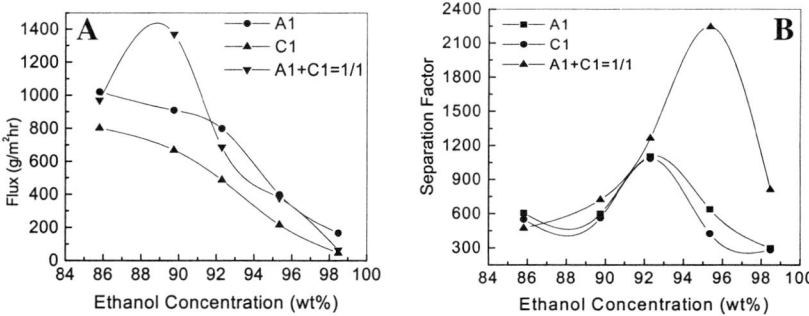

FIGURE 5. Feed concentration versus **(A)** flux and **(B)** separation factor at a feed temperature of 75°C.

In the case of PEC membrane A1+C1 = 1/1, however, the dependence of flux on feed concentration, which showed a maximum value of flux at an ethanol concentration of 89.7 wt%, was different from those of other membranes. Although this phenomenon may be special, there were similar circumstances for other PEC membranes. It is reported[8] that alginic sodium/chitosan PEC membrane reaches a maximum value of flux when water content is 70 wt%, for dehydration of water/alcohol mixtures. This kind of phenomenon may be due to macromolecular chain rearrangement within membrane caused by the presence of ethanol in the solvent, which causes the macromolecular chains to stretch. Thus, the membrane reaches its maximum swelling degree within the test feed concentration range, leading to a maximum value of flux.

In FIGURE 5B, it can be seen that all three membranes showed a maximum value of separation factor with an increase of water content in the feed. The reason for this may be that, with an increase of water concentration, the degree of swelling changes

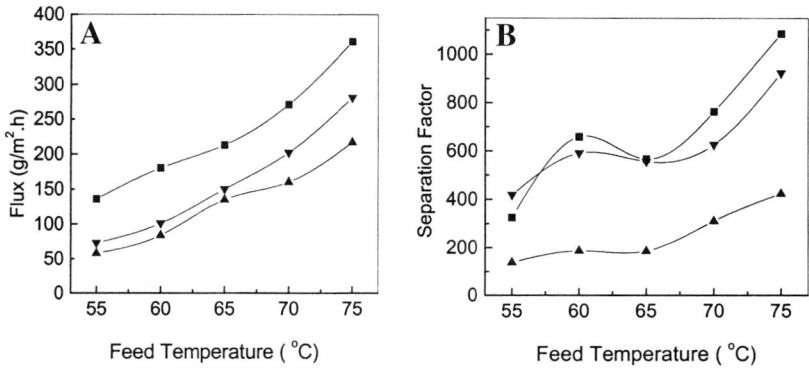

FIGURE 6. Feed temperature versus **(A)** flux and **(B)** separation factor with feed: water 4.6 wt%, ethanol 95.4 wt%. ■, A1+C1; ▲, C1; ▼, PDA.

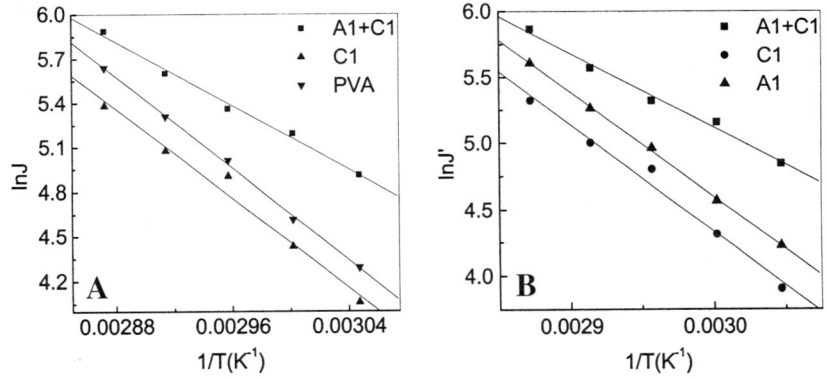

FIGURE 7. $1/T$ versus **(A)** $\ln J$ (overall flux) and **(B)** $\ln J'$ (water flux) with feed: water 4.6 wt%, ethanol 95.4 wt%.

and thereby the size of pervaporation passages change. Thus, at a certain feed concentration, the size of pervaporation passages reaches an optimal value for separating water from ethanol, that is, a maximum value for the separation factor.

Effect of Feed Temperature on Separation Properties

FIGURE 6 shows that flux and separation factor for all three membranes increased with increasing temperature, which may be explained as follows. As the temperature increases, the diffusion rate of penetrant increases, which leads to an increase in flux. At the same time, permselectivity toward water is enhanced and, thus, the separation factor increases.

FIGURE 7 suggests that the dependence of overall flux and water flux on temperature fits the Arrhenius equation. By means of linear regression, the overall permeation activation energy and the permeation activation energy for water molecule were obtained. (The permeation activation energy for membrane A1 was not available because of its separation property declination.) As can be seen in TABLE 6, the overall permeation activation energy and the permeation activation energy for water molecules in PEC membrane A1+C1 = 1/1 were much lower than those of unmodified PVA and C-PVA C1. This may be explained by the fact that the PEC A1+C1 = 1/1 had a stronger hydrophilicity and permeability to water than PVA. Therefore, PEC membrane showed a much lower energy barrier than that of PVA. However, the

TABLE 6. **Permeation activation energy**

Membrane Material	PVA	C-PVA C1	A1+C1 = 1/1
Overall permeation activation energy (KJ/mol)	64.39	62.41	44.83
Permeation activation energy for water (KJ/mol)	65.77	67.06	46.72

overall permeation activation energy and the permeation activation energy for water molecules in C-PVA C1 were close to those of unmodified PVA. This confirmed our previously mentioned contention that the overall hydrophilicity of C-PVA was close to that of PVA because the introduction of hydrophilic cationic groups into PVA was accompanied by the introduction of hydrophobic groups, such as methyl and ether groups.

Effect of Operation Time on Separation Properties

Some polyelectrolytes membranes showed high separation properties at the initial operating time, yet suffered a decline in separation performance as operation continued. For example, the PAA composite membrane showed excellent separation properties when PAA was converted into a salt form. However, the high separation properties decreased with time, attributed to the elution of alkali metal ion out of the membrane. Karakane *et al.*[2] used several kinds of polycations to ionize PAA to form PEC separation layer. This kind of PEC composite membrane, not only showed superior separation properties, but also long-term stability.

FIGURE 8 shows the operating time dependence of separation properties for P-PVA, C-PVA, and PEC membranes. From FIGURE 8, we can see that flux of P-PVA decreased sharply whereas fluxes of C-PVA and PEC changed little. The reason may be that there was some low molecular weight salt left in P-PVA membrane after membrane preparation. During operation, the salt form of phosphoric ester groups was converted into acid form. Because the salt form of phosphoric ester groups has stronger affinity for water and repellency towards organic molecules than that of acid form of phosphoric ester group, the separation properties of P-PVA decreased when the salt form was converted into acid form. However, cationic PVA was macromolecular salt itself and does not elute out of the membrane. Similarly, PEC membrane showed stable separation properties because the cationic PVA took the place of the easily eluted low molecular weight salt.

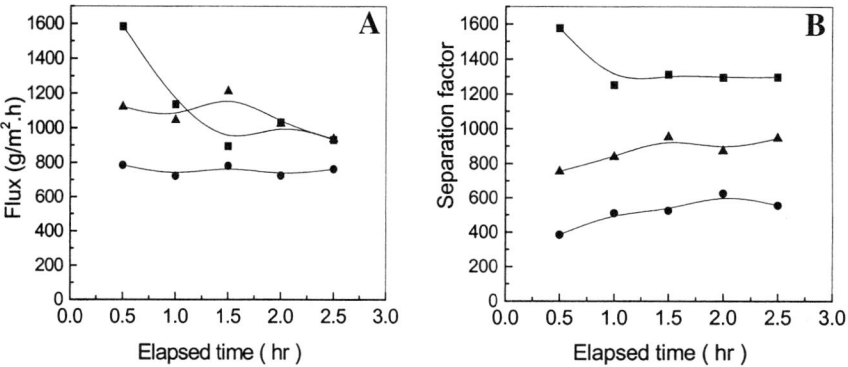

FIGURE 8. Elapsed time versus (**A**) flux and (**B**) separation factor with feed: water 4.6 wt%, ethanol 95.4 wt% at $T = 75°C$. ■, A1; ●, C1; ▲, A1+C1.

TABLE 7. Solvent resistance evaluation of PEC membrane

Separation Properties	Original Value	In Ethanol Aqueous Solution	In Benzene	In Toluene/Butanone Mixture
Flux (g/m^{2}h)	635	632	562	586
Separation Factor	1,050	1,510	1,350	1,940

NOTE: Membrane: A1+C1 = 1/1. Feed: water 7.7 wt%, ethanol 92.3 wt%. $T = 75°C$.

Solvent Resistance of PEC Membrane

TABLE 7 shows that the separation properties of PEC membrane soaked in three kinds of solvents for a month are similar to their original values. This suggests that this kind of PEC membrane has a degree of solvent resistance.

CONCLUSIONS

By chemically modifying PVA, P-PVA and C-PVA with various D.S. can be synthesized. Our study shows that within specified range the D.S. of P-PVA has linear relationship with the concentration of phosphoric acid. The D.S. of C-PVA is affected greatly by the mole ratio of catalyst and etherification reagent. C-PVA of high D.S. can be obtained with a mole ratio of 1/2. PECs formed by mixing the solutions of P-PVA and C-PVA are soluble in water, which means that it is possible to prepare PV composite membrane by means an ordinary coating process. This method is simple and it is easy to control the thickness of separation layer and the ratio of polycation and polyanion, compared with other methods for preparing PIC PV membranes. Using phosphatic PVA, cationic PVA and their polyelectrolyte complexes, we prepared pervaporation composite membranes. Phosphoric PVA membrane with D.S. of 2.3% showed high flux at the beginning of operation, although flux decreased sharply with elapsed time. A polyelectrolyte complex membrane, prepared by mixing P-PVA (D.S. 2.3%) and C-PVA (D.S. 2.9%) with weight ratio of 1/1, showed flux of 378 g/m^{2}h and separation factor of 2,250 for dehydration of ethanol/water mixture (ethanol 95.4 wt%) at a feed temperature of 75°C. The separation property of this membrane was stable and changed little after being soaked in 95 wt% ethanol aqueous solution, benzene, and toluene/butanone mixture, thus showing a certain degree of solvent resistance. It can be concluded that, by adjusting the ratio of cation and anion, the hydrophilicity and hydrophobicity can be well balanced and thereby optimal separation properties can be achieved.

REFERENCES

1. BRUESCHKE, H., Inventor; GFT. Ingenieurbuero fuer Industriealagenbau, Assignee. 1983. German Patent DE3220570. Date of application: Dec. 1.
2. KARAKANE, H., et al. 1991. Separation of water-ethanol by pervaporation through polyion complex composite membrane. J. Appl. Polym. Sci. **42:** 3229–3239.
3. HG2683-95. Standard of Chinese Chemical Engineering Ministry. Food additive: the analysis method of phytic acid.

4. ZHANG, J.W., *et al.* 1989. Analytic method of degree of substitution of carboxymethylated starch. Tianjing Chem. Eng. **4:** 53–57.
5. WANG, Z.M., *et al.* 1990. *In* Practical Infrared Spectrocopy, 2nd edit. 278–314. Beijing: Petroleum Industry Press.
6. LU, C.H., *et al.* 1996. Pervaporation separation of water-alcohol through chitosan/poly(sodium acrylate) composite membrane. Technol. Water Treat. **22**(2): 75–79.
7. ZENG, X., *et al.* 1998. Pervaporation characteristics of water/organic liquid mixtures through sodium alginate/chitosan polyelectrolyte composite membranes. J. Funct. Polym. **11**(3): 385–391.
8. OYAMA, H.T. & T. NAKAJIMA. 1984. Effects of complexation on the water vapor sorption of polymer alloys. J. Appl. Polym. Sci. **29:** 2143–2153.

Hydrophobic Pervaporation

Toward a Shortcut Method for the Pervaporation-Decanter System

ROBERT W. FIELD[a] AND VANESSA LOBO[b]

[a]*Department of Engineering Science, University of Oxford, Parks Road, Oxford, UK*

[b]*BP, Chertsey Road, Sunbury on Thames, Middlesex, UK*

ABSTRACT: Pervaporation is a relatively new technology. Although hydrophilic pervaporation has become established, hydrophobic pervaporation for recovery of organics from water has not been a commercial success. Technologic reasons for this are suggested. However, as the pressure to include waste minimization and to recycle, as well as pollution prevention, increases, there will be opportunity for the development of new wastewater treatment processes. This may lead to hybrid processes, including a coupling of pervaporation with conventional technology. The hybrid process examined herein is a simple pervaporation–decanter system that is applicable to organics with limited solubility in water. In this system, the PV unit produces permeate that after condensation gives two liquid phases. The organic phase is relative pure and concentrated (and available for reuse), whereas the aqueous phase can be recycled into the PV feed stream. For a given feed concentration and water purity target, there is a minimum membrane selectivity (α_{min}) that yields a two-phase condensate. If the membrane has a selectivity that is just greater than the minimum, the recycle rate of the aqueous phase relative to the feed rate will be very large and likewise the membrane area. Also, for a membrane with known organic permeability, one can define a theoretical A_{min}, the minimum membrane area required when the separation factor is so high that the aqueous phase is negligible. For membranes with $\alpha > \alpha_{min}$, values of the required membrane area, A, have been obtained for various representative duties, and a correlation between α/α_{min} and A/A_{min} has been obtained (compare with the Gilliland correlation used in distillation). An approximate relationship is $(A/A_{min} - 1)(\alpha/\alpha_{min} - 1) = 1$. Since α_{min} and A_{min} can be calculated readily, this relationship is a shortcut tool that permits estimation of A for any α and any duty. Finally, membrane areas for the above hybrid system and those for PV alone are compared.

KEYWORDS: hydrophobic pervaporation; hybrid pervaporation; recovery of organics; water treatment

---

Address for correspondence: Robert W. Field, Department of Engineering Science, University of Oxford, Parks Road, Oxford OX1 3PJ, UK. Voice: 44(0)1865 273814; fax: 44(0)1865 273905.
robert.field@eng.ox.ac.uk

INTRODUCTION

Pervaporation can be classified into the following types, as shown in FIGURE 1: hydrophilic pervaporation, organophilic (also known as hydrophobic) pervaporation, and target organic pervaporation. Hydrophilic pervaporation is an established industrial technology and its success has, in part, been attributed to the hybrid processes that have been developed around it.[1] The industrial applications of organophilic pervaporation have been few. The key areas of potential industrial application include:

- product recovery from dilute fermentation broths, where conventional technology is very expensive;
- aroma recovery in the fruit juice industry; to date, research has shown promising results;[2] and
- recovery of organics from wastewater streams with discharge and reuse of water.

Concerning the last area it should be noted that if the permeate shows phase separation, then the organic phase can be sent to a decanter, as shown in FIGURE 2. The organic phase is saturated with water, but the amount is typically very small. If this is acceptable, then the main advantage of this PV–decanter system over conventional technologies is that no further purification of the organic is required.[3] If polishing of the organic stream is required, then this can be accomplished with molecular sieve adsorption. The focus of this paper is to introduce, in a generic manner, a PV–decanter process that can achieve both pollution control and waste minimization.

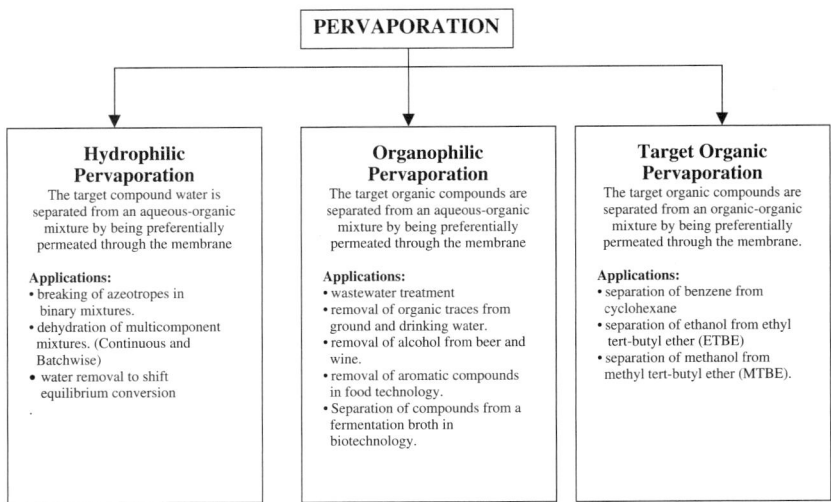

FIGURE 1. Pervaporation: its characterization and typical applications.

The technology can be used within a process to recover and reuse the organic and water, or as an end of pipe treatment.

There are two concepts for a pervaporation-based hybrid process:[1]

Type 1: An integrated combination achieving a binary split.

Type 2: A combination of consecutive separation processes achieving a split that (technically and economically) neither could achieve alone.

Type 1 is the only concept realized on an industrial scale and the PV-decanter is of this type. Although this certainly does not guarantee that the combination will be generally economic, it does suggest that it may be a candidate for niche applications particularly when waste minimization is emphasized. The decanter (as the integrated second stage) is more favorable than incineration or oxidation since it does not produce secondary pollutants and the organic recovered is an economic benefit. The outlet water from PV should meet the legislation for discharge or the specification for reuse within the process. The legislation depends on many factors, such as location, nature of organic, and environmental impact. Currently, in the U.K., and probably more generally as well, more attention is being given to the recovery of organics from discharge air streams than to the recovery of organics from aqueous effluent streams. It is, thus, worth noting that the shortcut approach developed below might well be adaptable for vapor permeation, which industrially is a growth area.

Organophilic pervaporation is preferred to hydrophilic pervaporation for organic recovery from dilute aqueous solutions since the membrane area is lower. One disadvantage of organophilic membranes is that they are case specific, since the permeating component varies with application, leading to higher investment costs and a separation that may not (without phase separation) be as clear cut as desired. Previous work has often dealt with high activity solutes that are relatively simple to separate by other means. For example, the separation of relatively pure water from an

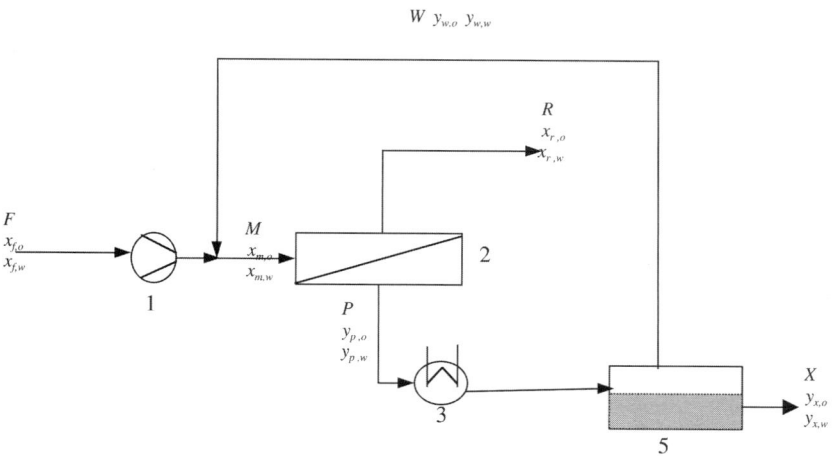

FIGURE 2. Schematic process diagram of a PV-decanter unit showing: 1, feed pump; 2, PV unit; 3, condenser; and 5, decanter. For clarity the vacuum system, recycle, and product pumps are not shown. X is the organic product and R is the purified water stream.

aqueous solution containing a chlorinated organic is generally more economic with conventional technology. The few sales in this area probably benefited from the apparent simplicity of pervaporation. It has been recommended[3] that the application of pervaporation is limited to the separation of dilute feed streams with low activity and this was confirmed in a more recent study.[4] In that study, pervaporation systems were classified into four types, with only two of the types having positive enhancement factors when compared to distillation. Furthermore, one of these two, labelled Type A, is adversely affected by increasing permeate pressure. One of the main examples of Type A is phenol, and the recovery of phenolics has been well researched.[5–8]

The performance, and hence, applicability of pervaporation depends, in part, on the quality of the membrane. Pervaporation in general offers the following advantages:

- separation can overcome organic–water azeotropes,
- organics can be concentrated by as much as 50 times the feed concentration,
- relatively low temperature operation is possible,
- low energy requirement, lowering operating costs, and
- systems are modular and compact, allowing easy scale-up and incorporation into existing processes.

Depending on the application, the disadvantages can include

- technology is unproven on an industrial scale,
- uncertain membrane lifetime,
- fouling leading to flux reductions, and
- high initial capital costs.

The aim of this paper is to explore a Type 1 organophilic pervaporation based hybrid process for wastewater treatment. It is taken as given that the aim is to obtain *two* streams, one principally water and the other principally organic, since this is probably the only outcome that is likely to prove to be industrial acceptable. The pervaporator–decanter process chosen allows the water purity target (for discharge or downstream biological treatment) to be met by appropriately sizing the pervaporator, whereas the recovery of organic can be achieved with the decanter. Therefore, this process accomplishes waste minimization and recycling, both of which are expected to be important in designing and implementing future technology. In particular a shortcut method is developed for sizing the PV unit. It is hoped that this will encourage more frequent screening of the pervaporation–decanter system.

DEVELOPMENT OF THE SHORTCUT METHOD

The driving force between the feed and permeate sides is the chemical potential, which leads to a concentration gradient across the membrane. The chemical potential gradient can be achieved in three ways:[9]

- vacuum—the total permeate pressure is low, thereby giving a partial pressure gradient across the membrane,

- sweep gas—the partial pressure of the components on the downstream side is lowered by adding a gaseous stream to the permeate side, and

- thermopervaporation—the condensation rate of the permeate stream on a chilled wall is fast enough to maintain a low pressure on the permeate side of the membrane, thus achieving a partial pressure gradient. This method is very rarely used.

The most common mode of operation is vacuum pervaporation and that is the only method considered here. The design of a membrane unit depends on the selectivity and flux of the membrane since these parameters influence the outlet water purity, the amount of organic recovered, recycle flows, and therefore, process economics. With a PV-decanter system the recycled water (W in FIG. 2) is saturated with organic and so a fraction of the organic passes through the membrane more than once. The organic flux depends upon the mass transfer coefficient of the boundary layer, membrane permeability, and driving force. The latter is influenced by the saturated vapor pressure of the components and the permeate pressure.

For a pervaporation–decanter system it is useful to define two parameters.

Required Minimum Selectivity, α_{min}

Consider a feed as shown in FIGURE 2 with an organic concentration of $x_{f,o}$, a desired retentate organic concentration of $x_{r,o}$, and an effective membrane selectivity, α. Then, irrespective of the overall *rate* of permeation, the permeate concentration ($y_{p,o}$) is fixed for a given mode of operation: counter-current, cocurrent, or crossflow. As shown elsewhere[10] the mode of operation is of secondary importance. In this work, the mode was crossflow and the retentate was assumed to be in plug flow. The mass transfer boundary layer above the membrane influences the selectivity and, therefore, reference has been made to "effective membrane selectivity" rather than just "membrane selectivity". As discussed by Ten and Field[4] the effective membrane selectivity α_{eff} relates the bulk concentration to the permeate concentration and at any point along the membrane:

$$\alpha_{eff} = \frac{y_o/y_w}{x_o/x_w},$$

where x and y are local values. For low permeate pressure the effective value of membrane selectivity can be approximately related to the ideal value, α^*, that would occur in the absence of a boundary layer resistance and as permeate pressure approaches zero:

$$\alpha_{eff} = \frac{\alpha^*}{\left(1 + \dfrac{R_{bl}}{R_m}\right)},$$

where R_{bl} and R_m are, respectively, the boundary layer and membrane resistances for the permeation of the organic. The latter is equal to l/P_o where l is the membrane thickness and P_o is the permeability coefficient of the organic through the membrane. Thus, for relatively thick membranes, α_{eff} approaches α^*.

From the above it follows that, for a given mode of operation and specified values of the three concentrations, $x_{f,o}$, $x_{r,o}$, and $y_{p,o}$, the effective selectivity that gives a permeate just rich enough in the organic for there to be phase separation after condensation can, for a given decanter temperature, be determined. Just above minimum effective selectivity, α_{min}, the membrane area is very large due to a very large W/F ratio. The parameter α_{min} is critical in determining which membranes are suitable.

Minimum Area, A_{min}

If the selectivity is sufficiently large for there to be no aqueous recycle (i.e., the permeate is so rich in the organic component that there is no separate aqueous phase in the decanter), then the membrane area is a minimum, A_{min}, for the given overall system specification (e.g., feed concentration and retentate concentration) and specified organic permeability.

In general, for selectivities greater than α_{min}, the recycle flows and membrane area can be determined using a standard approach.[10] In our work the operation was assumed to be isothermal and in crossflow. In practice the achievement of near isothermal operation implies a large number of reheaters. Since the emphasis is on the ratio A/A_{min}, thermal effects were not taken into account because the effects on numerator and denominator are similar. Clearly A/A_{min} can be computed for various values of α/α_{min}. For each membrane process application, there are various inlet conditions, outlet requirements, and organic properties. The effects of these conditions on selectivity and minimum selectivity were determined for a representative range of organics. As noted above, one of the aims of the present study was to determine if there is a general correlation between α/α_{min} and A/A_{min} so as to aid initial process calculations.

THEORETICAL STUDY

The organics with the greatest potential for treatment by the PV–decanter process were identified as including phenol, chloroform, and methyl-isobutylketone (MIBK). The physical properties of these components (e.g., solubility in water) set the limits of the various parameters. In examining the dependence of the recycle ratio W/F upon selectivity α, it was found that parameters, such as feed concentration, solubility limit of organic in water phase, and water purity target had little influence beyond a certain value of α. Typical curves are shown in FIGURES 3 and 4. Close to α_{min}, the feed rate and the amount of organic entering are small in comparison with the size of the recycle stream and the organic therein. Therefore, the feed concentration exerts a negligible impact on α_{min}. As expected, more demanding water purity targets (i.e., lower values of $x_{r,o}$) increase the value of α_{min} (see FIGURE 5).

To calculate membrane areas, the system was split into a number of sections. For both 100 and 500 ppm values of $x_{r,o}$, 25 sections gave accurate results. The effect of $P_{vac}/\gamma P_{sat}$ was investigated and the results were in line with those of others.[11] In this work the variation of the ratio A/A_{min} was of greater interest. A relationship between the ratio of A/A_{min} and α/α_{min} is analogous to the Gilliland relation in distillation, whereby area is similar to number of stages and α is similar to reflux ratio. The membrane area drops sharply for small increases in α/α_{min} and then almost plateaus for $\alpha/\alpha_{min} \geq 5$. The lower limit of the system is when α/α_{min} approaches unity, at which point the recycle flows are relatively very high, leading to infeasibly large

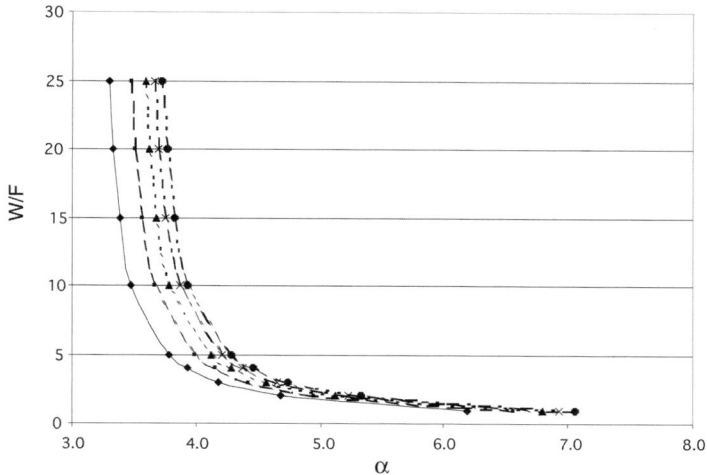

FIGURE 3. Effect of selectivity α and recycle ratio W/F for various values of organic solubility limit with $x_{f,o} = 0.01$, $x_{r,o} = 500$ ppm, and $y_{x,o} = 1.0$. $y_{w,o} = 1$ (♦), 1.2 (■), 1.35 (▲), 1.45 (×), and 1.55 (●) percent.

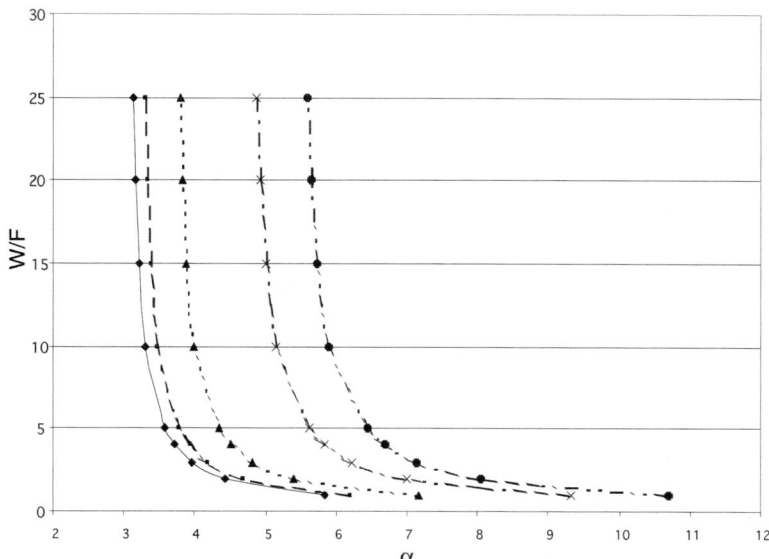

FIGURE 4. Effect of selectivity α on recycle ratio W/F for various values of retentate water purity with $x_{f,o} = 0.01$, $y_{w,o} = 0.01$, and $y_{x,o} = 1.0$. $x_{r,o} = 600$ (♦), 500 (■), 300 (▲), 100 (×), and 50 (●) ppm.

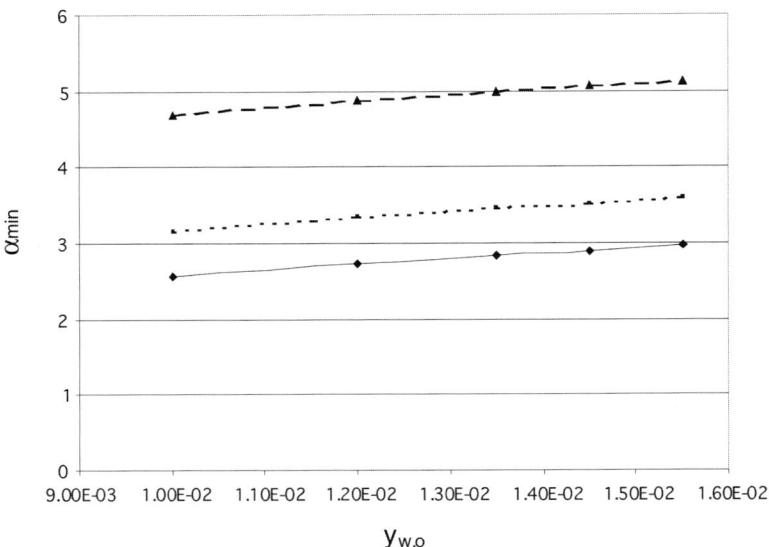

FIGURE 5. The effect of organic solubility limit $y_{w,o}$ and retentate water purity $x_{r,o}$ on α_{min}: $x_{r,o}$ = 1,000 (◆), 500 (■), and 100 (▲) ppm.

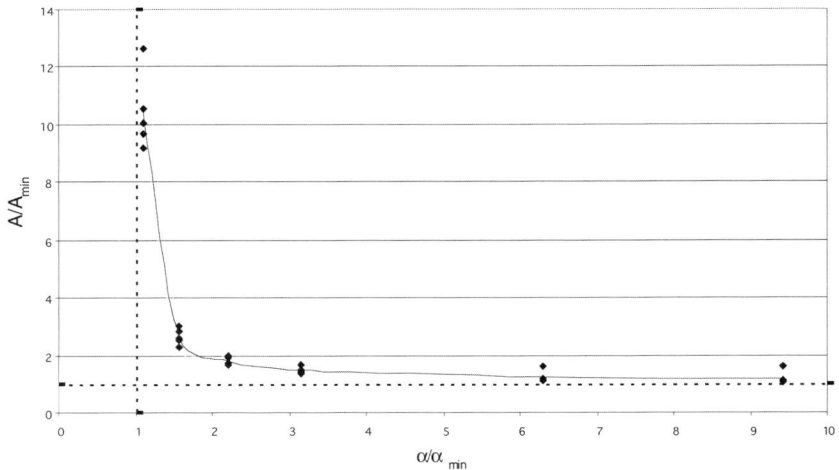

FIGURE 6. Overall correlation developed from examination of a range of solubility limits and performance targets.

membrane areas. The other limit is when $\alpha/\alpha_{min} \to \infty$, where there is no recycle and $A \to A_{min}$.

A graph was plotted using wide limits (of more than one order of magnitude) for the retentate concentration, $P_{vac}/\gamma P_{sat}^0$, fresh feed concentration, and recycle organic concentration in order to produce an overall correlation that can be used to estimate membrane area. The upper limits on feed concentration were close to saturation of water by organic. The outcome is presented in FIGURE 6. An approximate correlation is $(A/A_{min} - 1)(\alpha/\alpha_{min} - 1) = 1$. Further refinement is possible but is probably not justified at this stage. Since α_{min} and A_{min} can be calculated readily, this relationship is a shortcut tool that permits estimation of A for any α and any duty.

A single pervaporator–decanter system is applicable only when the single pervaporator is able to produce a permeate that exceeds the solubility limit of the organic in water otherwise there is no phase separation in the decanter. If this is not the case, a two-stage pervaporation system should be employed, as demonstrated elsewhere.[12] It was found that it is more economic to feed the recycle from the decanter into the second pervaporation unit. Finally, a comparison on the basis of membrane areas between the hybrid systems and PV alone showed that if $\alpha > 5\alpha_{min}$, then the membrane areas are similar.

In future developments, energy requirements could also be addressed. The recycled water will be cold, and heat exchange with the retentate might be beneficial. Thermal effects will probably be such that unless low grade waste heat is available the value of W/F will have to be less than unity. In the present study the question: "Does hydrophobic pervaporation have a future?" has not been answered, but a shortcut tool that may assist those with access to commercial data has been developed.

CONCLUSIONS

1. To achieve pollution control and waste minimization, pervaporation (when applied to sparingly soluble organics) is more viable when operated as a pervaporation–decanter hybrid process than in a stand-alone mode.

2. The overall correlation allows the empirical estimation of area, and the ideal operating range for α/α_{min} is probably between 1.5 to 4. This range gives low membrane area without investing in membranes with very high selectivity. The correlation can also be used for the qualification of the benefit of investing in higher selectivities in order to lower membrane area.

REFERENCES

1. LIPNIZKI, F., R.W. FIELD & P.K. TEN. 1999. Pervaporation-based hybrid processes: a review on process design, applications and economics. J. Membr. Sci. **153:** 183–210.
2. BORJESSON, J. 1996. Pervaporation of a model apple juice aroma solution: comparison of membrane performance. J. Membr. Sci. **119:** 229–239.
3. BENNETT, M. 1996. Pervaporation using Modified Polysiloxanes, Ph.D. Thesis, University of Bath, Bath, UK.
4. TEN, P.K. & R.W. FIELD. 2000. Organic pervaporation: an engineering analysis of component transport and the classification of behaviour with reference to the effect of permeate pressure. Chem. Eng. Sci. **55:** 1425–1445.

5. BENNETT, M., B.J. BRISDON, R. ENGLAND & R.W. FIELD. 1997. Performance of PDMS and organofunctionalised PDMS membranes for the pervaporative recovery of organics from aqueous streams. J. Membr. Sci. **137:** 63–88.
6. BODDEKER, K.W., H. PINGEL & K. DEDE. 1992. Continuous pervaporation of aqueous phenol on a pilot plant scale. Proceedings 6th International Conference on Pervaporation Processes in the Chemical Industry, Ottawa, Canada. 514–519. Bakish, New Jersey.
7. ASADA, T. 1992. Future of pervaporation. *In* Proceedings 6th International Conference on Pervaporation Processes in the Chemical Industry. Ottawa, Canada. 554–558. Bakish, New Jersey.
8. WU, P., R.W. FIELD, B.J. BRISDON & R. ENGLAND. 2001. A fundamental study of organofunctionalised PDMS membranes for the pervaporative recovery of phenolic compounds from aqueous streams. J. Membr. Sci. **190:** 147–157.
9. NEEL, J. 1991. Introduction to pervaporation. *In* Pervaporation Membrane Separation Processes. R.Y.M. Huang, Ed.: 1–110. Elsevier, Amsterdam.
10. LIPNIZKI, F. & R.W. FIELD. 1999. Simulation and process design of pervaporation plate-and-frame modules to recover organic compounds from waste water. Trans. IChemE **77**(Part A): 231–240.
11. WIJMANS, J.G. & R.W. BAKER. 1993. A simple predictive treatment of the permeation process in pervaporation. J. Membr. Sci. **79:** 101–113.
12. LOBO, V. 2001. BEng research project report. University of Bath, Bath, UK.

Membrane Bioreactors for Treating Waste Streams

J.A. HOWELL, T.C. ARNOT, AND W. LIU

University of Bath, Bath, UK

ABSTRACT: Membrane bioreactors (MBRs) have a number of advantages for treating wastewater containing large quantities of BOD. This paper reviews the inherent advantages of an MBR, which include high potential biomass loadings, lower sludge yields, and retention of specialized organisms that may not settle well in clarifiers. A major problem in effluent treatment occurs when mixed inorganic and organic wastes occur with high concentrations of pollutants. Inorganics that might cause extremes of pH and/or salinity will inhibit microbial growth and only specialized organisms can survive under these conditions. Refractory organics are only biodegraded with difficulty by specialized organisms, which usually do not resist the extreme inorganic environments. The use of membrane bioreactors to help separate the micro-organisms from the inorganic compounds, yet permit the organics to permeate, has been developed in two different designs that are outlined in this paper. The use of membrane contactors in a multimembrane stripping system to treat acidic chlorinated wastes is proposed and discussed.

KEYWORDS: membrane bioreactor; waste stream; organic–inorganic waste

INTRODUCTION

Membrane bioreactors are used to provide a combination of biologically or biochemically catalyzed reaction and separation or mass transfer control. The functions that a membrane bioreactor can provide include:

- selective/controlled feed of reactants,
- selective removal of products,
- retention of catalyst,
- retention of unreacted products,
- control of mass transfer to/from the catalyst, and
- separating two liquid phases with reaction at or near the interface.

These functions can be used in waste treatment to advantage in a number of ways. In general wastes that are biologically oxidized to smaller molecules, ideally CO_2 and water, also result in the production of fresh microorganisms that can be separated

Address for correspondence: J.A. Howell, Department of Chemical Engineering, University of Bath, Claverton Down, Bath BA2 7AY, UK.
  j.a.howell@bath.ac.uk

from the smaller reaction products and water by a membrane. In some cases the reactant oxygen can be supplied efficiently through a membrane at higher mass transfer rates and with higher oxygen utilization efficiency than is achieved by bubbling air. In other cases, reactants can be supplied to feed the microorganisms by being extracted through a membrane from a complex mixture that might otherwise inhibit the biological reaction.

The environment can affect the living organism in many ways. Organisms respond to the environment by controlling their metabolism. We are able to affect the behavior of an organism and ultimately its catalytic activity by exercising control over the signals reaching the organism. These effects can be beneficial or deleterious to the activity we desire and, thus, it is important that we understand the sort of behavior that can be changed.

CONFIGURATIONS OF MBRs IN WASTEWATER TREATMENT

Membranes are not the only separation devices to be coupled to bioreactors but they do possess distinct advantages. In conventional activated sludge treatment the biomass must be separated from the product (clarified effluent) and recycled back to the bioreactor (activated sludge tank) in order to maintain a supply of organisms that grow at a slow rate. The reciprocal of the growth rate of the organisms is much larger than the liquid residence time in the treatment tank. If the biomass were not settled and recycled it would need to be fed continuously to the bioreactor, which would be expensive. The conventional settling tank is large but is cheaper than a membrane of the same separation capacity. Why then might a membrane have advantages?

Potential advantages of MBR include:

- total solids retention at all biomass concentrations,
- high intensity of biochemical oxygen demand (BOD) removal,
- lower sludge yields,
- sludge is less fouling,
- effluent is disinfected and almost virus free from UF, and
- easier to treat difficult wastes.

The membrane gives complete control over how much biomass is retained and, hence, the sludge age or solids residence time, while retaining completely independent control over the hydraulic residence time (assuming an appropriate design providing sufficient membrane area).

For the activated sludge process there are two possibilities for failure due to inadequate separation. The biomass might not settle completely and so tend to leak slowly from the combined reactor settler. This adds to the BOD in the effluent stream, lowering the concentration of permitted BOD in a dissolved form. Because BOD is the substrate for the microorganisms, their growth rate is lowered and the necessary volume of the aerator increases. Furthermore, the concentration of sludge that can be recycled from the settler is limited, thus limiting the maximum sludge concentration in the bioreactor.

Use of a membrane can avoid these two problems. There is a much higher limit on the biomass that can be successfully maintained in the bioreactor by a membrane system (in excess of 25,000 mg/L) compared to the settler (6,000 mg/L). The retention is near 100% of biomass and consequently a higher dissolved BOD in the effluent leads to a higher growth rate and BOD removal rate per unit of biomass. With both factors in our favor the membrane bioreactor can be economic. The reactor volume can be reduced, but a higher intensity of aeration may be required. This does create the risk of producing more sludge at the higher growth rate but this may be offset in two possible ways. A sludge digester can be used in a two-stage system, where the second stage is fairly compact. The second approach is to increase the sludge residence time since this is also controlled independently and thereby reduce the sludge yield. Operational experience with conventional activated sludge has shown that at very high growth rates and at very high sludge retention times the settleability of the biomass deteriorates. In the case of the membrane separator these factors are not serious. The cells continue to grow until they reach very high cell densities in the membrane bioreactor (MBR). These densities are unsustainable in an activated sludge plant.[1]

In time, the very high cell density starts to affect the metabolism of the cells. More of the cells become degraded by endogenous metabolism and, as is shown in FIGURE 1, the observed yield decreases progressively as the sludge age increases. Eventually the biomass in the MBR approaches an asymptotic value. This is close to the value observed in MBRs operated by Kubota for the treatment of waste water. No activated sludge plant could operate in this fashion. It can be seen that, under the assumption of no inert solids in the feed, the net sludge yield or observed yield is close to zero. This is an immense advantage in that sludge disposal costs are reduced. On the other hand increased aeration intensity is required to carry out full sludge oxidation and, thus, complete mineralization of the incoming solids.[2]

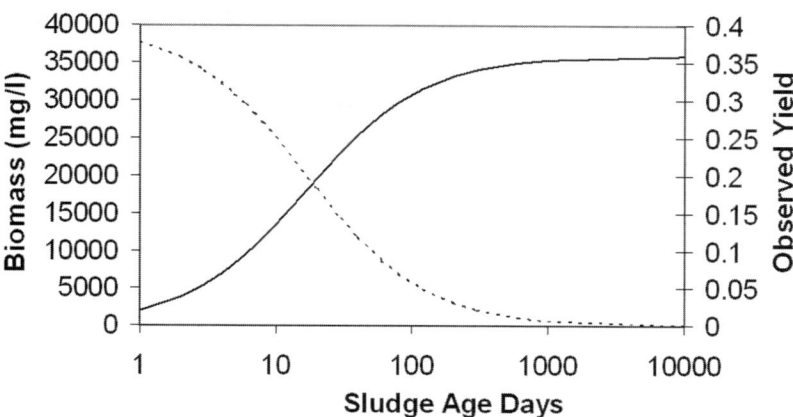

FIGURE 1. Net observed yield --- and biomass concentration —— as a function of sludge age in a membrane bioreactor. HRT = 2.7 h, $Y = 0.4$, $k = 0.07 \, d^{-1}$, $k_d = 0.06 \, d^{-1}$.

Since the reactors can also be more compact with shorter hydraulic retention times and since the high solids loading makes the liquor more viscous, pressurized oxygen or high intensity aeration has been used.

Recent work at Compiegne[3] has shown that, for the same solids loading, the sludge from an MBR has about a quarter of the viscosity of that from a standard aeration tank (although it is slightly less shear thinning). Fortunately, it is also considerably less fouling, and fouling that was observed appeared to be almost totally reversible. Interestingly the fouling resistances measured for the individual components were not additive; more that 65% of the resistance was ascribed to suspended solids and 30% to colloids.

Other work from Seoul investigated membrane fouling with activated sludge and an MBR. Chang and Lee[4] found that when a foaming sludge was filtered the fouling was much greater than when a non-foaming sludge was filtered. Although foaming can occur at high solids residence time (SRT) values, the filtration of sludges with higher SRT showed much lower fouling potential. The degree of extracellular polysaccharide (EPS) was found to be the most significant factor in fouling. Polyhydroxybutyrate (PHB), one of these EPS materials, is known to accumulate under nitrogen or oxygen limitation[5] in *Alcaligenes*, and it can be degraded under balanced growth. Although essential in the operation of an activated sludge plant in order to achieve settling, EPS is not required in an MBR and is deleterious to flux. This factor more than any other suggests that MBRs need to be operated in rather different modes than used in the conventional activated sludge for wastewater treatment (WWT).

The situation is actually more complex. It is desirable to operate WWT plants in a low sludge production mode. Whereas the previous references indicate that nitrogen limitation can lead to PHB and other EPS production, other results suggest that control of the nitrogen supply may well be beneficial in suppressing biomass formation. Holubar *et al.*[6] found that under nitrogen limitation cell yield was only 5% compared to the more normal 40% when degrading toluene and *n*-heptane in a biofilm reactor. A similar result was found by Scott *et al.*,[7] who found very low sludge yields when treating ice-cream wastewater under nitrogen limitation. Further work by Liu demonstrated that nitrogen supplementation in this case increased sludge yields but had no effect of chemical oxygen demand (COD) degradation rates.[8]

It appears that careful manipulation of the nitrogen environment and the use of high SRT values can result in relatively low fouling sludges with low yields. There is a risk, presumably only with some wastes, of EPS production with a deleterious effect on fluxes but the conditions for this to occur in an MBR have not been determined.

Membrane bioreactors are available in a number of configurations. The classical approach is to have a membrane in a side loop adjacent to a bioreaction vessel. Broth is pumped around the loop, through the membrane module, and past the membrane surfaces at an elevated Reynolds number to reduce the fouling potential. These have arisen as an outgrowth of the more conventional membrane filtration of wastewaters where cross flow is normal.

More recently commercial systems that have been more successful use very low power consumption. Owen *et al.*[9] calculated the optimum power consumption for a membrane system treating a secondary sewage effluent. They concluded that the

TABLE 1. Typical performance data for an MBR treating municipal sewage

	Waste Water Feed Average	Waste Water Feed Range	Final Effluent Average	Percent Removal
Suspended solids ($mg\,L^{-1}$)	230	<30–800	<1	>99.5
BOD ($mg\,O_2\,L^{-1}$)	224	<30–650	<0.4	>98.1
Fæcal coliforms ($10^6/100\,mL$)	8.7	0.9–26	<0.000022	99.9998 > log 5.6

optimum crossflow velocity is about 1 m/sec, although higher fluxes are obtained at higher cross-flows. The conclusion was reached on the basis of a computer model and involved a trade-off between membrane costs and energy costs. Since 1995 membrane costs have dropped by a factor of about eight, with longer lifetimes and lower unit replacement costs. Energy costs have fluctuated about more or less the same value. The consequence is that very low energy costs are now sought for the largest scale systems. Immersed membrane bioreactors are now normal. One immersed system is the Kubota flat sheet system that uses flat sheets of membrane suspended in frames placed over aeration spargers and distributed in a rectangular aeration tank. Tank depths are a few meters and treated permeate effluent is removed by gravity at controlled low pressure, or by permeate pumping at a controlled rate of withdrawal.

In the case of industrial wastes it may be useful to operate under nitrogen limited conditions. If nitrogen is supplied at a sufficient rate to allow enzyme regeneration but limit the formation of new cellular material, a very low yield coefficient is observed. There may be some drawbacks since authors report that nitrogen promotes the formation of extracellular polysaccharides, including PHB. The conditions for EPS formation in MBRs are, however, not determined and other authors have found that the sludge from an MBR is less fouling to membranes than that from a conventional activated sludge system.

The major contribution to waste treatment from MBRs that has been reported is the reduced sludge yields that are obtained. In practice yield values of about 0.3 are reported. In theory, endogenous respiration of the microorganisms can be encouraged by very high sludge ages and high sludge concentrations. Conventional kinetic data suggests that one might have a sludge yield of zero at a sludge age of just over 100 days. Under these circumstances one would expect a biomass concentration exceeding 30,000 mg/L MLVSS which turns out to be rather too high to maintain good mixing and aeration rates.

The degree of cell lysis that occurs in the aerobic digestion process observed during normal aeration may be autocatalytically affected by the lysis itself. Another aspect of changing sludge yield is to control how the lysis of cells proceeds. In *Penicillium chrysogenum* it has been shown that autolysis starts under carbon or nitrogen limitation by the action of serine and aspartyl proteases. These peaked as soon as endogenous energy reserves started to be mobilized under the starvation conditions.

Later the formation of metalloproteinases was discovered and at this point wholesale disintegration of the culture structures occurred and respiration ceased.[10] A parallel mechanism for the lysis of bacterial cells may occur. This suggests that encouraging cell lysis in the main aerobic reaction vessel may not be optimal in terms of overall treatment of the incoming load.

TABLE 1 shows typical operating performances for an MBR system using the Wessex Water Kubota plant at Porlock as an example. The data from Churchouse[11] show that the effluent BOD remains consistently acceptable under large fluctuations in the incoming load. Furthermore, disinfection of the effluent is easily achieved with greater than 5 log reduction in fæcal coliforms.

FOULING PREVENTION

Membrane bioreactors are prone to serious fouling by the nature of the material being processed. The tendency to foul can be mostly controlled by simple fouling reduction strategies that increase or maintain good shear rates across the membrane surface. Backwashing with permeate, air, or hypochlorite solutions is used regularly to remove cake formed on the surface. It turns out that using immersed membranes and having relatively low transmembrane pressures also limits fouling to a level that is easily controlled by the simple techniques mentioned above. The need to have serious chemical cleaning of the membranes occurs only every few months in a well-run system.

As well as the standard immersed membrane aeration MBR, two versions of extractive MBR systems have been proposed, one of which is now in commercial operation. The demand occurs when a stream of organics contains salts and acid or alkali to such a level that no bacterium is known to be capable of degrading the

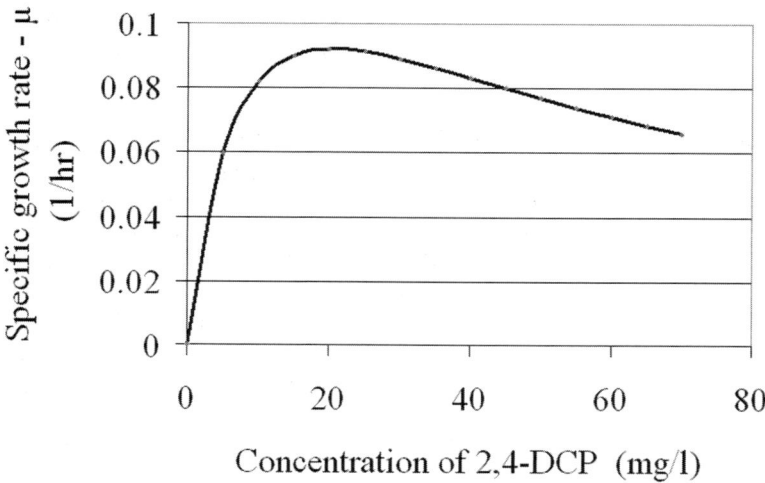

FIGURE 2. Maximum specific growth rate as a function of chlorinated organic concentration.

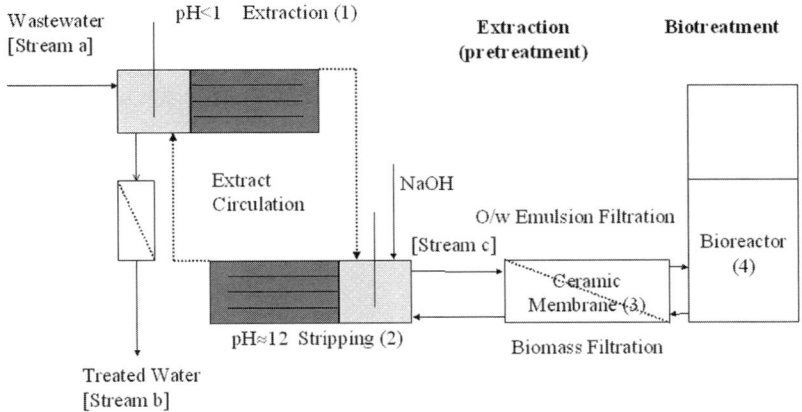

FIGURE 3. A schematic showing a possible stripping system with two mixer settler units operating. The stripping solution contains 20 mg/L chlorinated organic and can extract from an effluent stream at 0.1 mg/L using a pH gradient as the driving force.

organic in the salt/acid environment. This happens when the organic is of a particularly difficult nature. Under these circumstances, one often finds organisms that may exist in the salt acid environment, and organisms that will degrade the difficult organic in a neutral low salt environment, but no organism or consortium is capable of surviving in the combined environment.

Livingston and his colleagues[12–15] have used a PDMS (polydimethyl siloxane) hollow fiber membrane to extract organics from a waste stream into a bioreactor. The bioreactor operates with the bundle of hollow fibers immersed in the ærated vessel.

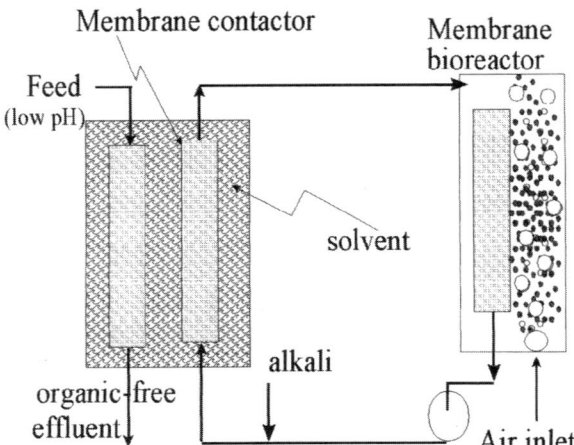

FIGURE 4. Membrane contactors could be used as an alternative to mixer settlers.

A high degree of stripping is obtained, but as the passage of the organic through the membrane is by diffusion alone, the membrane acquires a biofilm that can accelerate the rate of removal of the organic by consuming it. A disadvantage of direct extraction is the need to remove the organic to very low levels in the effluent stream. As FIGURE 2 shows, the rate of removal of a chlorinated organic by a bacterium is inhibited at high concentration, but at very low concentrations (less that 0.1 mg/L) it is also low. The best operating condition to minimize aerator volume is at about 20 mg/L.

The problem is how to extract, using a stripping solution at 20 mg/L, from an effluent stream at 0.1 mg/L.

FIGURE 3 shows a possible stripping system with two mixer settler units operating between a feed pH of less that 1 to a stripping pH of greater that 10. Under these circumstances a ratio of about 200 is achieved. The stripping solution receives alkali to allow the neutralization of the acid formed by the degradation of the chlorinated organic in the bioreactor. The membrane bioreactor retains the sludge and can, therefore, use bacteria that may not easily flocculate. The pH adjustment is made on the permeate, which recycles through the mixer settler to remove the organic from the solvent and deliver it to the bacterial mass for disintegration.

The membrane in the bioreactor can operate in two directions to allow frequent backflushing effects if two parallel membrane modules are used. The membrane does have to operate in a sidestream to take advantage of the high pH in the stripping solution.

As an alternative to the use of mixer settlers, membrane contactors could be used for the extraction process, as shown in FIGURE 4. These are less costly and the system very compact. Some recirculation of liquid is still required, but the membrane can be immersed in the bioreactor.

Experiments on isolated chlorinated organic[8] have shown that it can be degraded in a membrane bioreactor at a rate about 250 times higher than achieved in conventional waste treatment systems that demand the effluent from the bioreactor meets the discharge consent condition.

CONCLUSIONS

Membrane bioreactors are effective in treating wastewaters. They offer advantages of compactness over conventional technology, as well as producing a very high quality disinfected effluent. Sludge production can be minimized by careful control of the operational parameters. If difficult wastes containing inorganics and refractory organics are to be treated, extractive membrane bioreactors are able to separate the inorganics from the organics and remove the organics down to a few parts per billion. Where solvent extraction is used, low transmembrane pressure and high frequency backflushing allow a membrane on the final effluent to be maintained virtually free of solvent. Use of membrane contactors as extraction devices provides high mass transfer area, clean separations and economic treatment. If an acidic or basic organic is being treated pH differentials across the system may enhance extraction and final effluent quality.

REFERENCES

1. VAN DIJK, L. & G.C.G. RONCKEN. 1997. Membrane bioreactors for wastewater treatment: the state of the art and new developments. Water Sci. Technol. **35:** 35–41.
2. EIKELBOOM, D.H., et al. 1993. Workshop on integrated water management. Miyzaki, Japan.
3. DEFRANCE, L., M. JAFFRIN, B. GUPTA, et al. 2000. Contribution of various constituents of activated sludge to membrane bioreactor fouling. Biores. Technol. **73:** 105–112.
4. CHAND, I. & C.H. LEE. 1998. Membrane filtration characteristics in membrane coupled activated sludge system—the effect of physiological states of activated sludge on membrane fouling. Desalination **120:** 221–233.
5. DU, G.C., J.A. CHEN, H.J. GAO, et al. 2000. Effects of environmental conditions on cell growth and polyhydroxybutyrate accumulation in *Alcaligenes eutrophus*. World J. Microbiol. Biotechnol. **16:** 9–13.
6. HOLUBAR, P., C. ANDORFER & R. BRAUN. 1999. Effects of nitrogen limitation on biofilm formation in a hydrocarbon-degrading trickle-bed filter. Appl. Microbiol. Biotechnol. **51:** 536–540.
7. SCOTT, J.A., J.A. HOWELL, T.C. ARNOT, et al. 1996. Enhanced system $k_l a$ and permeate flux with a ceramic membrane bioreactor. Biotechnol. Tech. **10:** 287–290.
8. LIU, W. 2000. A Novel Extractive Membrane Bioreactor to Treat Chlorinated Wastes in the Presence of Salt. Ph.D. Thesis, University of Bath.
9. OWEN, G., M. BANDI, J.A. HOWELL & S.J. CHURCHOUSE. 1995. Economic assessment of membrane processes for water and waste applications. J. Membr. Sci. **102:** 77–91.
10. MCINTYRE, M., D.R. BERRY & B. MCNEIL. 2000. Role of proteases in autolysis of *Penicillium chrysogenum* chemostat cultures in response to nutrient depletion. Appl. Microbiol. Biotechnol. **53:** 235–242.
11. CHURCHOUSE, S. 2001. Problems to clear solutions—experience of full scale membrane bioreactor effluent treatment applications in Europe. NAMS 2001 Conference, 15–20 May, Kentucky, USA.
12. LIVINGSTON, A.G. 1993. A novel membrane bioreactor for detoxifying industrial wastewater: I. biodegradation of phenol in a synthetically concocted wastewater. Biotechnol. Bioeng. **41:** 915–926.
13. LIVINGSTON, A.G. 1993. A novel membrane bioreactor for detoxifying industrial wastewater: II. biodegradation of 3-chloronitrobenzene in an industrially produced wastewater. Biotechnol. Bioeng. **41:** 927–936
14. PAVASANT, P., L.M.F. DOS SANTOS, E.N. PISTIKOPOULOS & A.G. LIVINGSTON. 1996. Prediction of optimal biofilm thickness for membrane-attached biofilms growing in an extractive membrane bioreactor. Biotechnol. Bioeng. **52:** 373–386.
15. STRACHAN, L.F. & A.G. LIVINGSTON. 1997. The effect of membrane module configuration on extraction efficiency in an extractive membrane bioreactor. J. Membr. Sci. **128:** 231–242.

Why and How Membrane Bioreactors with Unsteady Filtration Conditions Can Improve the Efficiency of Biological Processes

ISABELLE DAUBERT, MURIEL MERCIER-BONIN, CLAUDE MARANGES, GÉRARD GOMA, CHRISTIAN FONADE, AND CHRISTINE LAFFORGUE

Laboratoire Biotechnologie–Bioprocédés, Institut National des Sciences Appliquées, UMR INSA CNRS 5504, UMR INSA INRA 792, Toulouse, France

> ABSTRACT: A membrane bioreactor (MBR), an association of a bioreactor with a crossflow filtration unit, enables continuous processes with total cell retention within the reactor to be realized. Provided that high dilution rates can be applied and that inhibition processes are avoided, very high biomass concentrations can be reached, thereby improving the volumetric productivities. These membrane bioreactors have been successfully applied to various microbial bioconversion, such as alcoholic fermentation, solvents, organic acid production, starters, and wastewater treatment. On the basis of the biological reaction characteristics and bibliographic results, the potentialities and bottlenecks of this methodology are discussed. Depending on the application, it is shown how the performance of the membrane bioreactor can be enhanced by acting either on the biological reaction achievement, by controlling the balance between cell growth and death, or on the dilution rate, by increasing the permeate flux through the filtration unit. This discussion is based on results obtained in specific biological treatments applied to polluted liquid and gas.
>
> KEYWORDS: membrane bioreactors; environment; liquid and gaseous effluents; unsteady flows

INTRODUCTION

In all biotechnology processes, the reactor itself must be associated with a separation unit in order to recover either the biomass, the products that are in solution in the broth, or the purified liquid when environment problems are involved. Many separation methodologies are available: for example, sedimentation (decantation or flocculation), centrifugation, or dead-end filtration. Such methods lead to a sequential operation, the various steps of the entire process being achieved successively. To overcome this inconvenience and as a result of progress in membrane technology, the association of a biological reactor and a cross-flow filtration (membrane bioreactor, MBR) was proposed in the 1980s. This methodology was first used for research purposes, because the accumulation of the biomass and the possibilities to efficiently control the parameters inside of the reactor allow the productivity of the reaction to be, *a priori*, strongly increased. It was subsequently applied in industrial

---

Address for correspondence: Isabelle Daubert, Laboratoire Biotechnologie–Bioprocédés, Institut National des Sciences Appliquées, UMR INSA CNRS 5504, UMR INSA INRA 792, 135 avenue de Rangueil, 31077 TOULOUSE cedex 4, France.

situations, essentially because it is a continuously operating process and that it improves the unit compactness. For this reason, many studies were devoted to crossflow filtration, and especially to the characteristics and production technology of the membranes. In fact, the reaction and filtration operating conditions give the properties of the broth (biomass concentration and viscosity), but conversely the fluid properties modify the characteristics of the process (viability and activity of the cells, permeability and selectivity of the membranes). The performance of the MBR then results from these mutual interactions. The purpose of the present paper is first to develop a critical analysis of such interactions (advantages and bottlenecks) and then to deduce some process modifications that can enhance this performance (on the basis of applications essentially developed in our laboratory). Even if in most cases biomass concentration governs the relative importance of biological and physical disturbances, with an increase in physicomechanical limitations at high cell densities, a specific analysis of the effective bottleneck must be performed for each application in order to determine the direction in which improvement can be achieved.

MEMBRANE BIOREACTORS PRINCIPLE, PERFORMANCES, AND BOTTLENECKS

How a Membrane Bioreactor Structure Can Improve the Performance of Microbial Bioprocesses

Microbial process performance can be considered based on various criteria depending on the aim of the project; such criteria include product concentration, substrate residual concentration, conversion yield of substrate into product, or production rate. In most cases, when a compound is produced by a microorganism, the main criterion involved in the evaluation of process performance, according to economic considerations, is the volumetric productivity P, defined as follows:

$$P = \nu_p X,$$

where ν_p is the specific activity of the microorganism and X is the cell concentration. Higher productivity can be obtained by increasing cell concentration provided that the specific activities are maintained. The activities are controlled by the physicochemical microorganism environment, the optimal level of which depends on the metabolic behavior of the cells. For both biomass concentration and microbial activity, high levels can be realized by operating in a continuous culture mode (optimizing the operating conditions for improved control of cell activity) associated with a cell recycle process (enhancement of cell concentration). This is the structure of a membrane bioreactor (MBR) and in such systems the volumetric productivity is

$$P = \nu_p X = CD,$$

where C is the product concentration and D is the dilution rate. D is defined as the ratio of the feed liquid flowrate Q to the broth volume V. Because in a MBR the flowrate Q is equal to the permeate flowrate, the greater the permeate flowrate, the greater the dilution rate.

Studies have shown that the cell activity increases with dilution rate (see FIGURE 1) because a high value of D avoids the nutritional limitations that can occur with cell

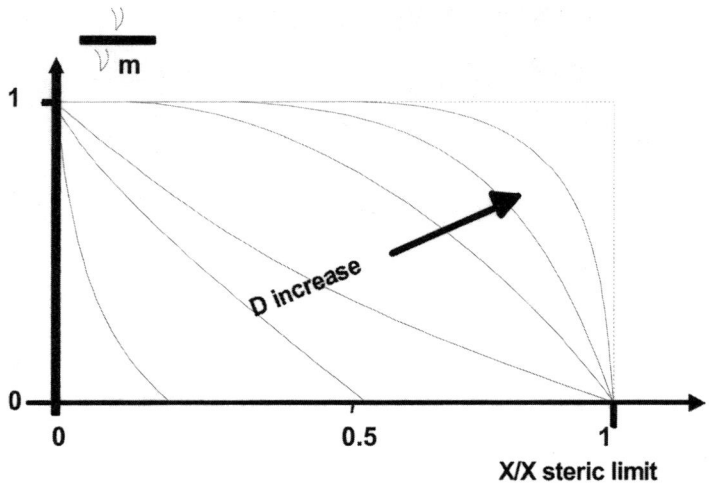

FIGURE 1. Influence of relative total cell concentration ($X/X_{\text{steric limit}}$) and dilution rate ($D$) on relative specific activity (v/v_m).

concentration. Thus, with respect to productivity, the best operating conditions for a fermentation process are high product concentrations and dilution rates.

Numerous processes have been developed to simultaneously increase these two parameters (e.g., centrifugation, flocculation, and decantation), but MBR has shown the best results in terms of biomass concentration and productivity. According to the problem to be treated, the filtration unit can be inside or outside of the reactor[1] (see FIGURE 2). In the usual laboratory setup, a side stream filtration unit is associated

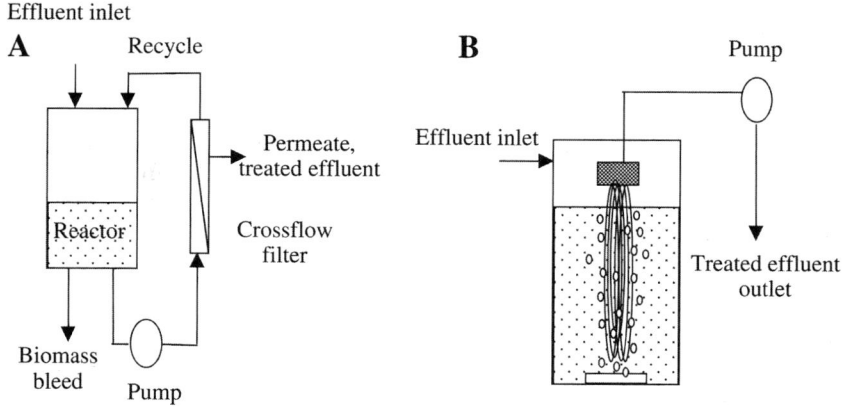

FIGURE 2. Flow chart for a membrane bioreactor: (**A**) with an external crossflow unit and (**B**) with immersed hollow fibers.

TABLE 1. Specimen results obtained in membrane bioreactors

Product	Microorganism	Membrane	$X_{dry\ mass}$ g L^{-1}	P g L^{-1}	D h^{-1}	D_x h^{-1}	P g L^{-1} h^{-1}	Duration, d	Reference
Ethanol	S.c	SFEC M14 0.14 μm	330	65	0.5	0	33	6.5	3
	S.c	HF 0.2 μm	120	70	1	0.06	70	8	4
	S.c	HF 50,000 Da	100	34			170		5
	S.c	HF 0.1 μm	85	46.5	0.58	0	27		6
	K.m	Craflo 0.45 μm	16	19.5	0.44		8.6	1	7
	C.t	HF 50,000 Da	3.1	8.6	0.24	0.12	2.2	3	8
ABE	C.a		20	13	0.5		6.5		9
	C.a	Carbosep M1	125	13.6	0.33		4.5	20	10
Exo-β-amylase	C.t	Ceraflo 0.2 μm	14	220 U/ml			48 U/ml/h	5.5	11
Butyrate	C.t	Techsep M6	35	29.7	0.32	0.032	9.5	12	12
	P.a	Techsep M6	100	15	0.95		14.3		13
Propionate	P.a	SFEC M14	112	8.5	0.25		2.14	12	14
	P.a	Techsep M9	50				1.2	23	15
	P.t	Techsep M6	20	10	0.1		1	10	16
Acetate	C.t	HF 500,000 Da	5	16	0.25		4	54	17
Lactate	L.r	Techsep 0.08 μm	88	88	0.4		35.2	4	18
	L.b	HF 30,000 Da	40	89	0.25		22.5	12	19
	L.a	FC 30,000 or 500,000 Da		50	0.5		25	10	20
	L.c	Techsep M6	88	49	1.05		51.5	2	21

S.c, *Saccharomyces cerevisia*; K.f, *Kluyveromyces fragilis*; K.m, *Kluyveromyces marxianus*; C.t, *Clostridium thermosaccharolyticum*; S.p, *Schizosaccharomyces pombe*; C.a, *Clostridium acetobutylicum*; C.t, *Clostridium thermosulfurogenes*; C.t, *Clostridium tyrobutyricum*; P.a., *Propionibacterium acidipropionici*; P.t, *Propionibacterium throenii*; C.t, *Clostridium thermoaceticum*; L.r, *Lactobacillus rhamnosus*; L.b, *Lactobacillus bulgaricus*; L.a, *Lactobacillus amylovorus*; L.c, *Lactobacillus cremoris*.

HF, hollow fibers; d, day; h, hours; ABE, acetone–butanol–ethanol.

with the reactor. In all cases, and because a high velocity is needed in the external loop, the reactor can be expected to show uniform or well-mixed behavior:[2] any oxygen limitation is avoided during solid/liquid separation. This results in æration of the broth during the entire process and in continuous removal of any toxic or inhibitory product that might be produced during the fermentation.

Thus, in comparison with other techniques, the membrane bioreactor structure insures the total retention of the cells and also allows the limitation phenomenon to be solved by easier control of the microenvironment of the cells. Moreover, this continuous operating mode can help solve the limitation phenomena that reduce cell performance by a possible increase in substrate flow rate at higher dilution rates. It also permits the accumulation of toxic or inhibitory products (for microorganism activity) to be reduced.

Examples of Performance

Significant results obtained with membrane bioreactors are listed in TABLE 1. It can be seen that, in most cases, very high values were reached for both cell concentration X (up to 300 g/L dry mass for yeast and more than 100 g/L dry mass for bacteria) and dilution rate. Thus, high dilution rate combined with high cell concentration allow high volumetric productivity to be achieved.

Bottlenecks in the Process

The bottlenecks in membrane bioreactor processes essentially come from the high cell concentrations that, when in excess of particular critical values, exert significant negative influences on physical properties of the fluid (mass transfers and membrane fouling) and on the biological reactivity (physiology and death).

Physical Limits

Membrane Fouling. Membrane fouling is due to suspended solids, soluble compounds, complex substrates, or products. As shown in FIGURE 3, three phenomena contribute to fouling: the adsorption on the membrane or pore surfaces, pore clogging, and cake formation over the membrane. These modify the two performance indicators of the filtration unit, permeate flux and selectivity.

Molecular adsorption depends on the medium composition and on its evolution throughout the culture. For example, a 15-min contact period of a fresh yeast culture

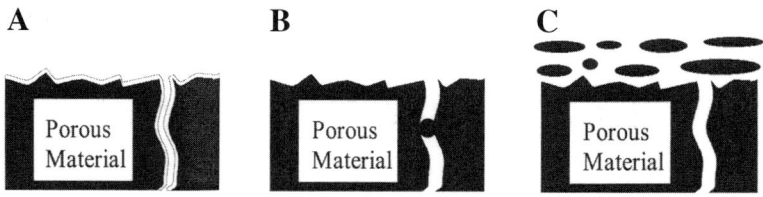

FIGURE 3. Schematic representation of the main phenomena involved in membrane fouling during microbial bioconversion in membrane bioreactors: **A**, adsorption; **B**, pore clogging; and **C**, particle deposition.

medium with a microfiltration membranes (Carbosep M14, Techsep France), without filtration, reduces the water permeability of 40%.[22] The pore clogging intensity is due, in part, to the size distribution of the pores, and also, importantly, to the cell fragments (lysis and mechanical rupture). Cake formation (deposit of particles) creates a high concentration boundary layer that generates an additive resistance to mass transfer.

Membrane fouling has two major consequences. The first, linked with a modification of selectivity, is the retention of toxic compounds or of the expected product in the reactor. The second is flux reduction. The decrease in permeate flowrate depends on both the biomass concentration and the culture duration and results in a lower capacity to maintain an adequate hydraulic dilution rate, thereby reducing the volumetric productivity.[23] During a culture of *Saccharomyces cerevisiæ*, the permeate flowrate decreases quickly as biomass concentration increases until a stable level is reached (generally five to ten times lower than the initial permeate flow). The flux decline has two main consequences. First, the microorganisms specific activity decreases, essentially because of nutritional limitations when D is reduced. The second effect is a lower volumetric productivity ($P = CD$). Hence, in order to maintain high dilution rates, it is necessary either to limit biomass accumulation by bleeding a part of the cells (but lower productivity) or to develop new concepts to reduce fouling of the crossflow filters.

Other Physical Limitations. The rheologic behavior of the broth depends very much on the cell concentration X:[23] when X increases, the fluid passes from Newtonian behavior to a non Newtonian one (close to a Bingham model). Mixing is more difficult to achieve and mass transfer performance is affected: sometimes a maximal value of cell concentration is required.[18]

For ærobic processes, and when high cell concentrations and activity are concerned, the oxygen transfer capacity of the usual æration systems is insufficient to insure the biologic reaction demand.[24,25] This ultimately limits the capacity of the microbial population to support fully ærobic growth. Moreover, the viscosity increase contributes to poor gas transfer for both consumed gases and produced gases. Thus, in alcoholic fermentation, the gas hold up can reach 30% of the total reactor volume.[2]

Biologic Limitations

Decrease in Cell Specific Activity. Microbial specific activity decreases as cell concentration increases.[2,18,21] Thus, volumetric productivities are not enhanced as expected. Above a critical cell concentration, the growth rate is apparently zero.[26] Nutritive limitations could partly explain this result, but in most cases an increase in dilution rate that could solve this problem cannot be realized because of membrane fouling.[12] To justify the decrease in the microbial specific activity, studies have also pointed out the possible retention by the filter of toxic compounds,[27] but in most cases a suitable choice of the membrane can avoid this problem. For each biotransformation, there is a special microbial concentration for which the gain in productivity is balanced by the problems due to biomass accumulation, thus fixing a limit on the MBR operation.

Modifications of Cell Integrity. Because of the accumulation of biomass in the membrane bioreactor, the apparent mean age of the cell is increased. Moreover, the impact of mechanical stresses on cell integrity (cell lysis and cell death) has been noted.[6,11,21,28] In consequence, the natural phenomenon of cell death is enhanced. Some authors have studied the relationship between the nutriment supply and the viability of the population, and have defined a maximal admissible value for biomass concentration related to the substrate supply.[3,29]

The accumulation in the reactor of dead biomass,[4,11,30] cell lysates, and molecular compounds produced by the cells themselves strongly contributes to physical perturbations (an increase in viscosity, mass transfer modifications, etc.).

Stacking Bound. As well as the physical and biological limitations, it is impossible to overlook the stacking bound that corresponds to a maximal pile up of the microbial cells. Obviously, a maximal cell concentration can be defined for each microorganism and this depends on the microbial environment and on the physiologic state of the cells. In membrane bioreactors, unusual cell concentrations can be obtained; for example, a 300 g/L dry yeast mass concentration[30] or a 100 g/L dry bacteria mass concentration.[10] In the case of yeast, noticeable morphologic modifications were observed with an important fall in intracellular water content from four grams of water per gram of dry mass[4] to one gram of water per gram of dry mass.

POSSIBILITIES FOR ENHANCING MBR PERFORMANCE

Action on the Characteristics of the Biological Reaction: Application to Wastewater Treatment

In activated sludge processes, the dissolved organic pollutants are transformed into water, carbon dioxide, and biomass. This results in an excess production of sludge, the elimination of which is one of the main problems (partly because of increasingly stringent regulations). Biomass growth depends on various factors: mass loading, pollutant characteristics (biodegradability), predator activity, cell lysis, endogenous respiration of microbial cells, and growth on intracellular products. Because the residence times of the biomass and the liquid are independent parameters in an MBR structure, such processes are very efficient to set up under the best operating conditions for any given organic load. Higher volumetric throughput can then be obtained than for classical continuous culture setups—up to 20 kg/m^3/day with a purification yield of about 95% COD and 100% solid matter.[31] The so-obtained critical conditions correspond to a too low conversion factor, to a fall in biomass activity and viability, and/or to a too low permeate flux value (viscosity effect).

To reduce the biomass production, two main strategies can be considered. The first consists in reducing part of consumed substrate used for growth, mainly by increasing maintenance through control of the process parameters and, thereby, microbial activity. The second strategy corresponds to an intensification of the natural microbial death phenomenon to induce cryptic growth by physical and/or chemical treatment. In the first method, enhancement of the maintenance phenomenon can be achieved by adequate control of the biomass and liquid residence times. This maintenance concept, introduced by Pirt,[32] describes the reduction of substrate/biomass conversion yield under limiting substrate conditions. Taking into account the maintenance coefficient m, and the specific activity μ, of the cell

$$\frac{\mu}{R} = \frac{\mu}{Y_{max}} + m,$$

where R is the observed growth yield and Y_{max} is the limit growth yield on the substrate. Hence, and for an MBR steady state operation, the net sludge yield, R, is given by the relation

$$R = \frac{D_p}{m + D_p/Y_{max}},$$

where D_p is the bleed rate.

Several studies have demonstrated the effective reduction of biomass yield under substrate limitation conditions in a membrane bioreactor (see FIGURE 4) and these indicate the importance of biomass state (active, viable, resting, dead, etc.) in clearly understanding the results obtained.[31,32–34] Observed yields result from a balance between the growing activity of viable cells, their death, and the lysis of viable and/or resting and dead cells. Obviously, especially when a substrate limitation occurs, viable cells die naturally and their lysis products are used as neosubstrates for cryptic growth. However, the kinetics of these phenomena are very slow and the net biomass production yield remains at about 0.3 kg/kg DCO.[31,33–35]

The second possibility to minimize biomass production is to increase the natural cell death and lysis by using external processes. The net yield can be expressed as the product of the yields of consecutive reactions (see FIGURE 5). Assuming that the yield of the lysis step is equal to 1, the net yield of biomass production in this multiple-step operation is, in each case, less than the yield of classical operation without biomass denaturation.

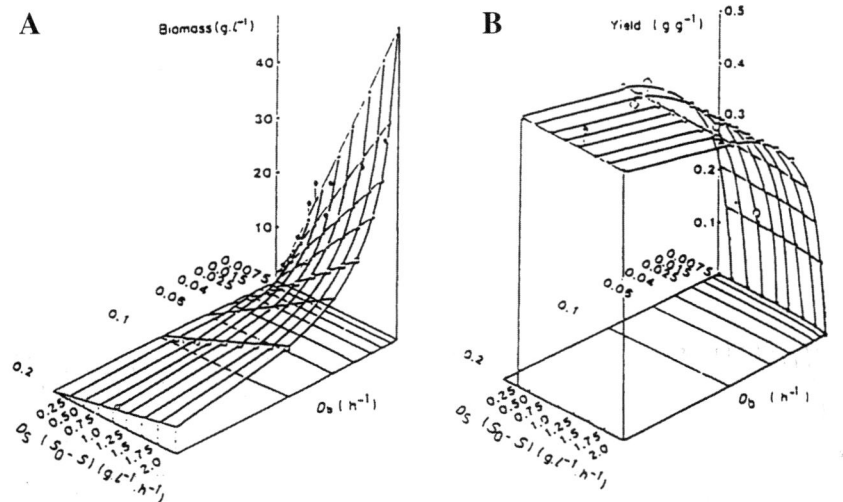

FIGURE 4. Representation of (**A**) biomass concentration and (**B**) experimental yield at the apparent steady state versus the substrate consumption rate D_s and the biomass bleeding rate D_p according to Bouillot.[33]

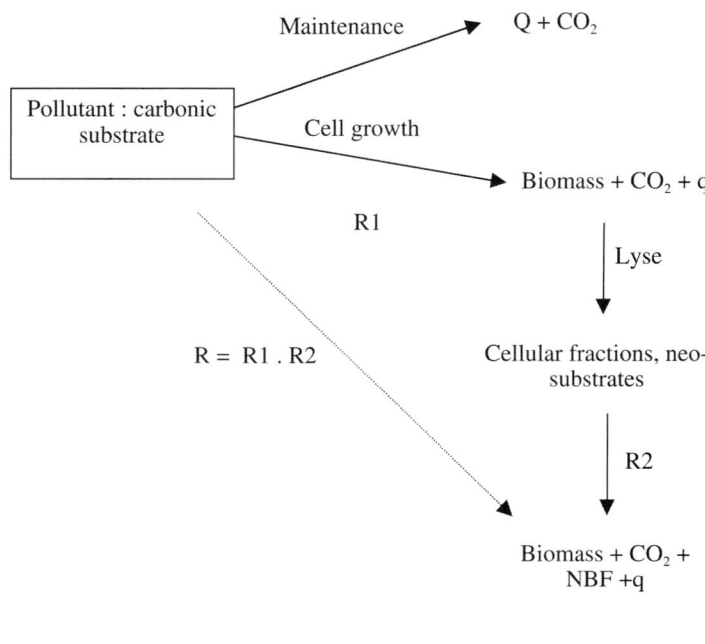

FIGURE 5. Effect of consecutive reactions of cell growth, lysis, and use of neosubstrates on the net growth yield.

Cell denaturation and cell breakage can be realized by chemical and/or physical treatments, such as acid or basic treatments, thermal inactivation (from 60°C to 100°C) or ultrasound (see FIGURE 6). The optimal operating conditions are generally deduced from biodegradability tests applied to the treated sludge. For example, in the case of *Ralstonia eutropha*, the best results are obtained at pH 10 after 20 min of incubation at 60°C.[36]

Action on the Performance of the Filtration System: Application to the Treatment of a Phenol Polluted Effluent

As discussed above, through the influence of dilution rate and, therefore, the NS value of the permeate flux, the reduction in membrane fouling is a truly interesting economic objective for membrane bioreactors, whatever the application (wastewater treatment[1] or production of valuable molecules[4,7,19]). A large part of the fouling is due to deposition of suspended cells on the membrane, and the mechanical stability of this deposit depends on a balance between the self-adhesive forces of cells over the membrane and the hydrodynamic forces (wall shear stresses). Numerous methods have been studied to increase these hydrodynamic forces: turbulence generation, natural (Taylor or Dean vortices), or forced unsteadinesses. Among these, the use of two-phase gas/liquid flow has been shown to be efficient in both

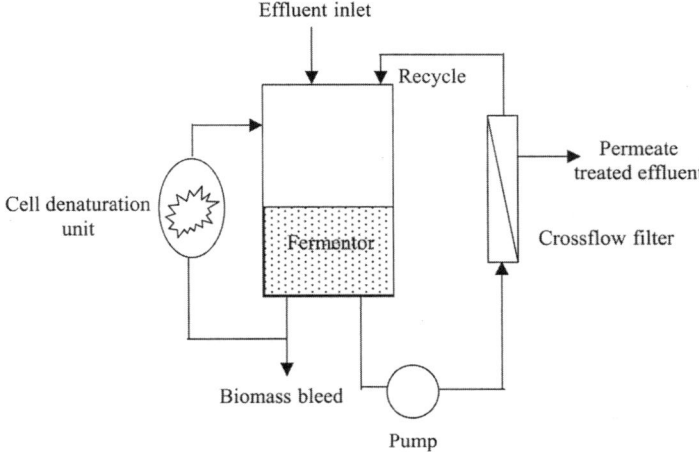

FIGURE 6. Schematic flow sheet for a combined wastewater treatment membrane bioreactor with a side sludge reduction unit.

ultrafiltration and in microfiltration, irrespective of the membrane configuration (tubular membranes[37,38] or flat sheet membranes[39]). Two-phase flows are obtained by a continuous injection at a gas flow rate Q_G cocurrently with the main liquid flow at rate Q_L. The structure of the flow obtained depends on the ratio Q_G/Q_L (bubbly flow for small values and slug flow for $Q_G/Q_L > 0.5$). All these studies deal with clarification of a baker's yeast suspension, and the permeate flux enhancements were always significant, ranging from 50 to 250%, according to the membrane type, the yeast concentration, and the two-phase flow pattern. These results can be attributed to enhancement of the wall shear stresses acting to erode the cake[40] or to the role of bubble-induced secondary flows promoting local mixing in the bubble wake.[41,42]

Using this type of unsteady flow, maximal enhancement factors of 2 (ultrafiltration) and 1.6 (microfiltration) were obtained during an alcoholic fermentation achieved with a MBR.[43] The enhancement increased with the cell concentration as particle fouling become more severe. Using another method, it was shown that unsteady flow must be applied at the very beginning of the filtration operation, since cells and extracellular organic compounds form an adhesive cake that cannot be subsequently removed. However, these very interesting enhancements of the permeate flux are countered by a negative influence of an increase in the wall shear stresses on the selectivity characteristic of the membrane. For example, when filtering an enzyme/yeast mixture (unwashed yeast cells suspended in physiologic serum) with and without air sparging, it was shown that gas/liquid slug flow strongly decreases the transmission of the enzyme (70% decrease).[44] However, if selectivity is an important parameter in the problem to be solved, the application of a bubbly flow is more interesting: even when the enhancement factor of the permeate flux was only 1.5, the transmission remained at a high value, so that the enzyme mass flux was 25% higher.

The following example shows the very positive role of enhancing the permeate flux when an inhibition by the substrate limits the performance of the biological reaction. It deals with the treatment of industrial effluents polluted by phenolic compounds. Such substrates are degraded by *Ralstonia eutropha,* bacteria that use the phenol as source of carbon and energy.[45,46] Study of the degradative pathways[47] and preliminary experiments achieved in a classical bioreactor suggested that the activity of the bacteria is affected by the concentration of phenol in the broth, above 50 mg/L,[48] and that this was the cause of a poor performances of the process. In such problems, high biomass concentration and dilution rate are required, but a large amount of oxygen is also needed because oxygen is involved in the catabolic pathways, in addition to the respiratory chain requirements.

The use of a gas/liquid two-phase flow in the MBR with full cell recycle enabled a twofold gain in permeate flow to be obtained and increased the oxygen transfer capacity by a factor of 1.5.[49] Moreover, a comparative evaluation of the economic cost of such a process demonstrated a significant energy saving (about 50%) for the two-phase flow crossflow filtration process. The maximal cell concentration was 60 g/L, the dilution rate $0.5\,h^{-1}$, and the phenol feed concentration 8 g/L. With these characteristics, the MBR was able to degrade a maximal phenol load of 96 kg/m^3/day with a constant 99% efficiency throughout the entire cell culture (see FIGURE 7).

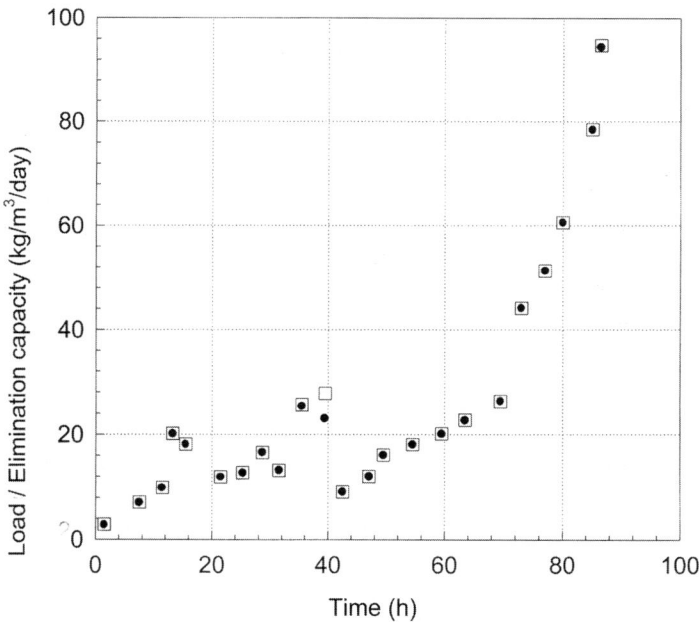

FIGURE 7. Variation with time of the applied load (□) and the elimination capacity (●) during phenol biodegradation by *Ralstonia eutropha* in an MBR with full cell recycle.

Association of an MBR with a Gas–Liquid Contactor: Application to the Treatment of Volatile Organic Compounds

Membrane processes can also be applied as a part of gaseous effluent treatment process. In this case, an association with a gas/liquid contactor is necessary to allow the gaseous pollutants to be solubilized into the liquid phase. Pollutant elimination from the liquid phase can then be achieved in a membrane bioreactor in order to benefit from high productivity, leading to a precise control of biomass activity and reduction in plant size. The feasibility of this combination has been shown[50] by associating an aero-ejector (gas/liquid contactor) and a membrane bioreactor (see FIGURE 8). The aero-ejector was placed just before the filtration module and could be used in a dual role, combining the functions of an aeration system with filtration enhancement (limitation of fouling phenomenon). The yeast *Candida utilis* was used to ensure ethanol degradation. The high cell concentration (80 g/L) maintained with a biomass bleeding enabled total degradation of the ethanol contained in the gaseous effluent (1 m^3/h of a gaseous effluent polluted with 7 g/m^3 of ethanol). The high treatment efficiency was maintained over an extended period (more than 350 hours) despite modifications in operating conditions, demonstrating process stability and reliability. In this application, the use of a membrane bioreactor allowed high elimination capacities, close to 140 kg of ethanol/m^3/day, to be achieved (see FIGURE 9).

The treatment efficiency of this combined process was also applied to the treatment of a more complex gaseous effluent, polluted with three different VOC (ethanol, methyl ethyl ketone, and butyl acetate). In this application, a mixed culture was

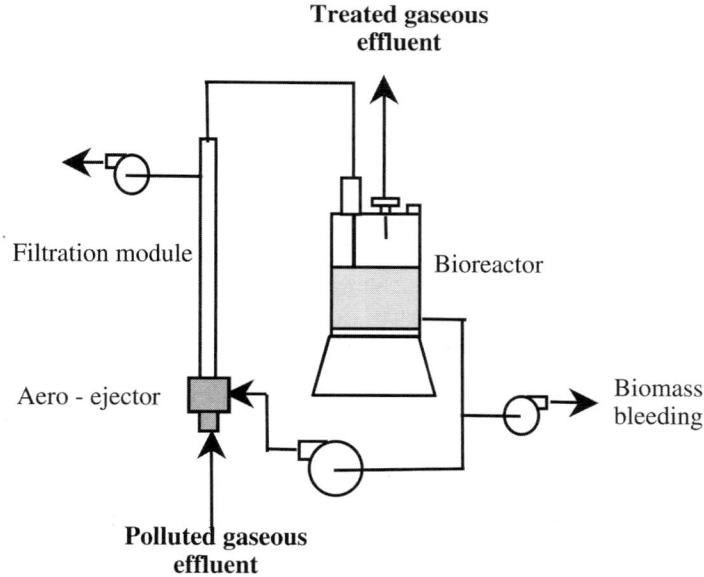

FIGURE 8. Experimental setup for a gaseous effluent treatment unit associating an aero-ejector and a membrane bioreactor.

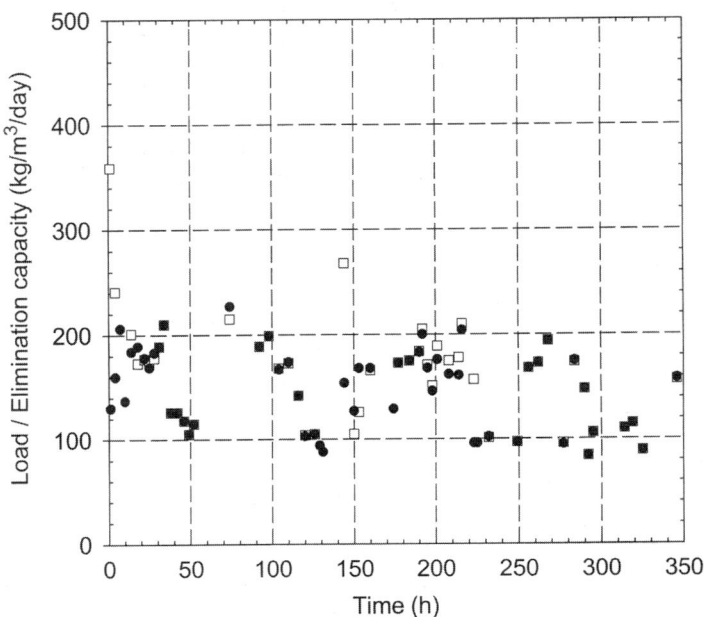

FIGURE 9. Variation with time of the applied load (□) and the elimination capacity (●) during ethanol biodegradation by *Candida utilis* in an MBR.

used to ensure pollutant degradation in the liquid phase. In this case, the possibility of a high dilution rate overcame a biological problem due to the transitory presence of an hindering metabolite that was consequently washed out.[51]

CONCLUSIONS

The membrane bioreactor principle shows *a priori* very interesting characteristics because, on one hand, it is a continuously operating system, and on the other hand, because it allows very high productivities to be reached due to distinct residence times for solid and liquid phases. However, biochemical and physical limitations force a choice of optimal structure and operating conditions for the particular problem to be solved, taking into account the economic factors. If intensive processes are examined, such as production of high added value compounds, high productivities are required economically, but the complexity of the process can be increased; at the other extreme, for extensive problems, such as are encountered in environmental applications, the simplicity and robustness of the unit are very important parameters and a different balance between basic performance and process complexity is required. The field of environmental applications with effluent treatment processes in MBR is certainly worth wide investigation in future studies. First, new designs involving MBR could be proposed toward strategies for development of autolytic wastewater treatment processes; second, mimicking the various steps occurring during efficient natural ruminant digestion, the design of optimal bioreactor technology including multistage operations and integrated extractive bioreactions should be studied more deeply.

Whatever the objective, it is obvious that any work achieved in order to enhance performance must simultaneously take into account the biological reaction and the physical properties, and their mutual interaction. For this reason, future research work must apply, on one hand, to understanding microorganism behavior when they suffer from extreme microenvironment conditions, especially at high cell concentrations (cryptic growth, forced lysis, etc.); and on the other hand, to the possibility of better control of this microenvironment (critical times, mixing and heat, and mass transfer in fluids showing complex rheology properties, membrane fouling, etc.). In this way, the structure and operating conditions of an MBR will be enlarged and this will allow a more judicious choice for the solution that is best adapted to the problem to be solved.

ACKNOWLEDGMENT

The authors thank Christophe Ellero for his technical support.

REFERENCES

1. STEPHENSON, T., S. JUDD, B. JEFFERSON & K. BRINDLE. 2000. Membrane Bioreactors for Wastewater Treatment. IWA Publishing, London.
2. MOTA, M., C. LAFFORGUE, P. STREHAIANO & G. GOMA. 1987. Fermentation coupled with microfiltration: kinetics of ethanol fermentation with cell recycle. Bioproc. Eng. **2:** 65–68.
3. LAFFORGUE-DELORME, C., P. DELORME & G. GOMA. 1994. Continuous alcoholic fermentation with *Saccharomyces cervisiæ* recycle by tangential filtration: key points for process modelling. Biotechnol. Lett. **16**(7): 741–746.
4. WARREN, R.K., G.A. HILL & D.G. MACDONALD. 1994. Continuous cell recycle fermentation to produce ethanol. Trans. I. Chem. E. **72**(C): 149–157.
5. MEHAIA, M.A. & M. CHERYAN. 1984. Hollow fibre bioreactor for lactose by *Kluyveromyces fragilis*. Enz. Microbiol. Technol. **6:** 117–120.
6. NISHIZAWA, Y., Y. MITANI, M. TAMAI & S. NAGAI. 1983. Ethanol production by cell recycling with hollow fibers. J. Ferment. Technol. **61**(6): 599–605.
7. TIN, C.S.F. & A.J. MAWSON. 1993. Ethanol production from whey in a membrane recycle bioreactor. Proc. Biochem. **28:** 217–221.
8. MISTRY, F.R. & C.L. COONEY. 1989. Production of ethanol by *Clostridium thermosaccharolyticum*: I effect of cell recycle and environmental parameters. Biotechnol. Bioeng. **34:** 1295–1304.
9. PIERROT, P., M. FICK & J.M. ENGASSER. 1986. Continuous acetone-butanol fermentation with high productivity by cell ultrafiltration and recycling. Biotechnol. Lett. **8:** 253–256.
10. FERRAS, E., M. MINIER & G. GOMA. 1986. Acetonobutylic fermentation: improvement of performances by coupling continuous fermentation and ultrafiltration. Biotechnol. Bioeng. **28:** 523–533.
11. NIPKOW, A., J.G. ZEIKUS & P. GERHARDT. 1989. Microfiltration cell-recycle pilot system for continuous thermoanaerobic production of exo-α-amylase. Biotechnol. Bioeng. **34:** 1075–1084.

12. MICHEL-SAVIN, D., R. MARCHAL & J.P. VANDECASTEELE. 1990. Butyric fermentation: metabolic behaviour and production performances of *Clostridium tyrobutyricum* in a continuous culture with cell recycle. Appl. Microbiol. Biotechnol. **34:** 72–177.
13. BOYAVAL, P., C. CORRE & S. TERRE. 1987. Continuous lactic acid fermentation with concentrated product recovery by ultrafiltration and electrodialysis. Biotechnol. Lett. **9**(3): 207–212.
14. BLANC, P. & G. GOMA. 1987. Kinetics of inhibition in propionic acid fermentation. Bioproc. Eng. **2:** 175–179.
15. COLOMBAN, A., L. ROGER & P. BOYAVAL. 1993. Production of propionic acid from whey permeate by sequential fermentation, ultrafiltration, and cell recycling. Biotechnol. Bioeng. **42:** 1091–1098.
16. BOYAVAL, P., M.N. MADEC & C. CORRE. 1992. Concentrated starter productions in a two stage cell-recycle bioreactor plant. Biotechnol. Lett. **14**(7): 589–592.
17. PAREKH, S.R. & M. CHERYAN. 1994. Continuous production of acetate by *Clostridium thermoaceticum* in a cell recycle membrane bioreactor. Enzyme Microb. Technol. **16:** 104–109.
18. XAVIER, A.M.R.B., L.M.D. GONÇALVES, J.L. MOREIRA & M.J.T. CARRONDO. 1995. Operational patterns affecting lactic acid production in ultrafiltration cell recycle bioreactor. Biotechnol. Bioeng. **45:** 320–327.
19. TEJAYADI, S. & M. CHERYAN. 1995. Lactic acid from cheese whey permeate. Productivity and economics of a continuous membrane bioreactor. Appl. Microbiol. Biotechnol. **43:** 242–248.
20. ZHANG, D.X. & M.CHERYAN. 1994. Starch to lactic acid in a continuous membrane bioreactor. Proc. Biochem. **29:** 145–150.
21. BIBAL, B., Y. VAYSSIER, G. GOMA & A. PAREILLEUX. 1991. High concentration cultivation of *Lactococcus cremoris* in a cell-recycle reactor. Biotechnol. Bioeng. **37:** 746–754.
22. MERCIER, M., O. PICHOT, C. FONADE, *et al.* 1997. Membrane bioreactors in fermentation process: two-phase flow may be a solution to enhance crossflow filtration flux. *In* Bioreactors and Bioprocess Fluid Dynamics. A.W. Nienow, Ed.: 331–347. BHR Group Conf. Series. Publication No.25.
23. MALINOWSKI, J., C. LAFFORGUE & G. GOMA. 1987. Rheological behaviour of high density continuous cultures of *Saccharomyces cerevisiae*. J. Ferment. Technol. **65**(3): 319–323.
24. FONADE, C., G. GOMA & C. LAFFORGUE. 1988. A new approach of momentum and mass transfer in high cell concentration biological processes. Second International Conference in Bioreactor Fluid Dynamics, Cambridge. 309–318.
25. CHANG, H.N., I.K. YOO & B.S. KIM. 1994. High density cell culture by membrane based cell recycle. Biotech. Adv. **12:** 467–487.
26. SAN-MARTIN, R., D. BUSHEL, D.J. LEAK & B.S. HARTLEY. 1994. Cultivation of an L-lactate dehydrogenase mutant of *Bacillus stearothermophilus in* continuous culture with cell recycle. Biotechnol. Bioeng. **44:** 21–28.
27. DAMIANO, D., C.S. SHIN, N.H. JU & S.S. WANG. 1985. Performances, kinetics and substrate utilization in a continuous yeast fermentation with cell recycle by ultrafiltration membranes. Appl. Microbiol. Biotechnol. **21:** 69–77.
28. NARODOSLAWSKY, M., H. MITTMANNSGRUBER, W. NAGL & A. MOSER. 1988. Modelling of alcohol fermentation in a tubular reactor with high biomass recycle. Bioproc. Eng. **3:** 135–140.
29. BLANC, P. & G. GOMA. 1987. Propionic acid fermentation: improvement of performances by coupling continuous fermentation and ultrafiltration. Bioproc. Eng. **2:** 137–139.
30. LAFFORGUE, C., J. MALINOWSKI & G. GOMA. 1987. High yeast concentration in continuous fermentation with cell recycle obtained by tangential microfiltration. Biotechnol. Lett. **9**(5): 347–352.
31. CANALES, A., A. PAREILLEUX, J.L. ROLS, *et al.* 1994. Decreased sludge production strategy for domestic wastewater treatment. Water Sci. Technol. **30:** 97–106.
32. PIRT, S.J. 1965. The maintenance energy of bacteria in growing cultures. Proc. Roy. Soc. Lond. **163**(B): 224–231.

33. BOUILLOT, P. 1988. Bioréacteurs à recyclage cellules par procédés membranaires: application à la dépollution des eaux en aérobiose. Ph.D. Thesis, Institut National des Sciences Appliquées, Toulouse, France.
34. BOUILLOT, P., A. CANALES, A. PAREILLEUX, et al. 1990. Membrane bioreactors for evaluation of maintenance phenomena in wastewater treatment. J. Ferment. Bioeng. **69:** 178–183.
35. CANALES, A. 1991. Croissance cryptique en bioréacteur à membrane: Application au traitement d'eaux résiduaires urbaines. Ph.D. Thesis, Institut National des Sciences Appliquées, Toulouse, France.
36. ROCHER, M., G. GOMA, A. PILAS BEGUE, et al. 1999. Towards a reduction in excess sludge production in activated sludge process: biomass physicochemical treatment and biodegradation. Appl. Microbiol. Biotechnol. **51:** 883–890.
37. MERCIER, M., C. FONADE & C. LAFFORGUE-DELORME. 1997. How slug flow can enhance the ultrafiltration flux in mineral tubular membranes. J. Membr. Sci. **128:** 103–113.
38. MERCIER, M., C. MARANGES, C. FONADE & C. LAFFORGUE-DELORME. 1998. Yeast suspension filtration: flux enhancement using an upward gas/liquid slug flow—application to continuous alcoholic fermentation with cell recycle. Biotechnol. Bioeng. **58:** 47–57.
39. MERCIER-BONIN, M., C. LAGANE & C. FONADE. 2000. Influence of a gas/liquid two-phase flow on the ultrafiltration and microfiltration performances: case of a ceramic flat sheet membrane. J. Membr. Sci. **180:** 93–102.
40. MERCIER-BONIN, M., C. MARANGES, A. LINÉ, et al. 2000. Hydrodynamics of slug flow applied to crossflow filtration in narrow tubes. AIChE J. **46:** 476–488.
41. CUI, Z.F. & K.I.T. WRIGHT. 1996. Flux enhancements with gas sparging in downwards crossflow ultrafiltration: performance and mechanism. J. Membr. Sci. **117:** 109–116.
42. CABASSUD, C., S. LABORIE, L. DURAND-BOURLIER & J.M. LAINE. 2001. Air sparging in ultrafiltration hollow fibers: relationship between flux enhancement, cake characteristics and hydrodynamic parameters. J. Membr. Sci. **181:** 57–69.
43. MERCIER-BONIN, M., I. DAUBERT, D. LÉONARD, et al. 2001. How unsteady filtration conditions can improve the process efficiency during cell cultures in membrane bioreactors. Sep. Purif. Technol. **22–23:** 601–615.
44. MERCIER-BONIN, M. & C. FONADE. 2001. Enzyme transmission during crossflow filtration of yeast suspensions using gas/liquid two-phase flows. Presented at the Advanced Membrane Technology Meeting, Barga, Italy, October 14–19.
45. JOHNSON, B.F. & R.Y. STAINER. 1971. Dissimilation of aromatic compounds by *Alcaligenes eutrophus*. J. Bacteriol. **107:** 468–475.
46. HUGHES, E.J.L. & R.C. BAYLY. 1983. Control of catechol meta-cleavage in *Alcaligenes eutrophus*. J. Bacteriol. **154:** 1363–1370.
47. LEONARD, D. & N.D. LINDLEY. 1998. Carbon and energy flux constraints in continuous cultures of *Alcaligenes eutrophus* grown on phenol. Microbiology **144:** 241–248.
48. HILL, G.A. & C.W. ROBINSON. 1975. Substrate inhibition kinetics: phenol degradation by *Pseudomonas putida*. Biotechnol. Bioeng. **17:** 1599–1615.
49. LEONARD, D., M. MERCIER-BONIN, N.D. LINDLEY & C. LAFFORGUE. 1998. A novel membrane bioreactor with gas/liquid two-phase flow for high performance degradation of phenol. Biotech. Prog. **14:** 680–688.
50. DAUBERT, I., C. LAFFORGUE, C. FONADE & C. MARANGES. 2001. Feasibility study of a compact process for biological treatment of highly soluble VOCs polluted gaseous effluent. Biotech. Prog. **17**(6): 1084–1092.
51. DAUBERT, I. 2001. Etude de l'association d'un contacteur gaz/liquide et d'un réacteur biologique pour le traitement d'effluents gazeux industriels. Ph.D. Thesis, Institut National des Sciences Appliquées, Toulouse, France.

Effect of Immobilization Site and Membrane Materials on Multiphasic Enantiocatalytic Enzyme Membrane Reactors

NA LI,[a] LIDIETTA GIORNO, AND ENRICO DRIOLI

Research Institute on Membrane Technology, ITM-CNR, University of Calabria, Rende (CS), Italy

ABSTRACT: In the experimental work reported here the optical resolution of racemic naproxen methyl ester with crude lipase immobilized in a membrane reactor was studied. The multiphasic enantiocatalytic enzyme membrane reactor consisted of an organic phase that dissolved the naproxen methyl ester, a lipase-loaded membrane, and an aqueous phase that extracted the reaction product. Lipase preferentially converted the (S)-naproxen methyl ester to (S)-naproxen acid that was simultaneously separated by a membrane. The effect of the immobilization site and membrane type on the performance of the enzyme membrane reactor was studied. Capillary polyamide membrane with 10 kDa nominal molecular weight cutoff (NMWCO) and polysulfone membrane with 30 kDa NMWCO were applied, with lipase loaded in the sponge layer or on a membrane thin layer. With various immobilization sites and membrane types, the enzyme membrane reactors showed different productivity and enantioselectivity resulting from the varying amounts of immobilized enzyme and varying microenvironment of the hydrolysis reaction. Higher amounts of immobilized enzyme led to increased productivity and generally higher enantioslectivity of the membrane reactor. It seems that the location of organic/aqueous interface on membrane, which plays an important role in a multiphasic enzyme membrane reactor, was influenced by the immobilization site and membrane type and this affected the productivity and enantioselectivity. As much as 90% enantioexcess was obtained with a polyamide membrane, and lower values with a polysulfone membrane. A sponge layer of polyamide membrane is the preferred immobilization site for its higher productivity than a thin membrane layer, and for its higher enantioselectivity than a polysulfone membrane. In comparison with the unstable hydrolysis activity of free lipase in stirred tank reactor, a stable lipase activity can be obtained with the lipase-immobilized membrane reactor, irrespective of whether a polyamide or polysulfone membrane is used.

KEYWORDS: multiphasic membrane reactor; immobilized lipase; kinetic resolution; (S)-naproxen; enantioselectivity

Address for correspondence: Lidietta Giorno, Research Institute on Membrane Technology, ITM-CNR, c/o University of Calabria, Via P. Bucci 17/C, 87030 Rende (CS), Italy. Voice: (39) 0984 492040; fax: (39) 0984 402103.
l.giorno@itm.cnr.it.
[a]Present address: Department of Environmental and Chemical Engineering, Xi'an Jiao Tong University, Xi'an 710049, China.

INTRODUCTION

Among the new methods of producing pure enantiomers, the use of biocatalysis combined with membrane operations in so-called biocatalytic membrane reactors seems very promising.[1–9] This kind of biocatalytic membrane reactor takes advantage of enantiospecific enzymes to carry out the kinetic resolution of a racemic mixture and an appropriate membrane system to simultaneously separate the converted isomer. The porous structure and specific physicochemical characteristics of membranes make them good media for the reaction and simultaneous downstream separation. Enzyme immobilization usually allows a more stable enzyme activity over time.[10,11]

Lipases (triacylglycerol acyl hydrolases, EC 3.1.1.3), one of the most studied type of biocatalyst, are able to discriminate between enantiomers and are frequently used in kinetic resolution of racemic mixtures.[12–14] Furthermore, because of their ability to act at the organic/water interface with various substrates and their resistance to organic solvents, lipases are used in multiphase reactions.[15–22] In these systems the interfacial area affects the reaction rate more than the substrate concentration in the bulk. Lipases are more active on water insoluble substrates compared to those on water soluble ester substrates. Catalytic reactions are accelerated by adsorption of the enzyme to an interface containing the substrate.[7,23] Because of these peculiar characteristics, the lipases are feasible for use by immobilization at the interface of biphasic organic/aqueous membrane reactors.

Enzyme membrane reaction systems for the resolution of racemic mixtures have been studied in our laboratory; lipase, in particular, was used in membrane reactors for the resolution of (R,S)-naproxen methyl ester mixture to produce (S)-naproxen. It was known that lipase preferentially hydrolyzes the (S)-naproxen methyl ester into (S)-(+)-2-(6-methoxy-2-naphthyl)-propionic acid (naproxen) that is 28 times more active than its (R)-naproxen isomer when functioning as non-steroidal anti-inflammatory drug.

The multiphase membrane reaction systems for the resolution of (R,S)-naproxen methyl ester mixtures consisted of the following components: an organic phase containing (R,S)-naproxen methyl ester as substrate and isooctane as organic solvent, lipase-loaded membrane, and an aqueous buffer solution to extract the converted isomers. (R,S)-naproxen methyl ester was brought into contact with lipase molecules and (S)-naproxen ester was preferentially hydrolyzed into (S)-naproxen, which then diffused through the membrane and was extracted into aqueous phase. In previous work, the influence of operation conditions (including pH, temperature, substrate concentration, and lipase concentration) on the catalytic activity and enantioselectivity of lipase using polyamide membrane with 50kDa NMWCO was studied. One of the key results showed that the enantioselectivity of immobilized lipase was affected by the organic/aqueous interface.

Since the organic/aqueous interface plays an important role on the reaction rate and enantioselectivity of lipase hydrolysis reaction, the conditions favoring a stable organic/enzyme/aqueous configuration can improve the reactor performance. In addition, it was reported that the activity of enzyme changes after immobilization onto polymeric porous membranes.[24–26] This phenomenon may be related to the amount of immobilized enzyme, the effect of various kinds of membrane matrix and immobilization sites, denaturation of the enzyme due to interaction between enzyme and support materials, and the block of the active site on the enzyme. However, most of reported studies have mainly focused on the activity of enzyme and only a few have interpreted the effect of the membrane support property, both on the activity and the enantioselectivity of enzyme.[24,26] Various membrane materials and immobilization sites may differently affect the performance of immobilized enzyme in membrane reactors, especially when the enzyme is used for sensitive resolution process. The related research should provide a helpful guide for choosing suitable immobilization sites and membrane materials for optical resolution with an enzyme membrane reactor.

The aim of the experimental study reported here was to investigate the effect of immobilization site and membrane type on the performance of a multiphase enantiocatalytic enzyme membrane reactor. To achieve this, polyamide membrane of 10 kDa NMWCO and polysulfone membrane of 30 kDa NMWCO were applied and lipase was immobilized into the sponge layer or onto the thin layer of membranes. The catalytic behavior of immobilized lipase in membrane reactors is discussed and compared in terms of productivity, enantioselectivity, enantioexcess, and conversion, The data are also compared with that from the previous work using polyamide membrane of 50 kDa NMWCO. The results show that both membrane type and immobilization site have effect on performance of lipase-immobilized membrane reactors.

MATERIALS AND METHODS

Chemicals

Crude lipase powder from *Candida rugosa* (Sigma), containing 12% protein, was used without further purification. The racemic naproxen ester mixture in powder form (provided by Dr. F. Trotta, University of Torino) was the substrate, with isooctane as solvent. Isooctane from Sigma-Aldrich was reagent grade and this was used without further purification. Standard pure (S)-naproxen ((S)-6-methoxy-α-methyl-2-naphthaleneacetic acid) isomer was obtained from Sigma-Aldrich. The (R)-naproxen acid was obtained by reaction with the (S)-naproxen acid and standardized with HPLC. (The retention time of (S)-naproxen acid was taken as the commercial standard and the retention time of (R)-naproxen acid was calculated from the difference; its concentration was determined by the percentage area method. Standard solutions for calibration curves were then prepared from these solutions.)

Sodium dihydrogen phosphate (NaH_2PO_4) and disodium hydrogen phosphate (Na_2HPO_4) were purchased from Fluka (Buchs, Germany) and Sigma (St. Louis, MO). Isopropanol, HPLC grade, was obtained from Ashland Italia S.p.A. Acetic Acid, analysis grade, was obtained from Merck KgaA, Germany. *n*-Hexane for HPLC was obtained from Prolabo (CE).

Equipment

The membranes used in this study included polyamide membrane of 10 kDa NMWCO and 1.5/2.2 mm inner/outer diameter (kindly provided by Forschungsinstitut Berghof, Germany; called PA 10 kDa here), and polysulfone membrane of 30 kDa

NMWCO and 0.83/1.74 mm inner/outer diameter (from Romicon, Germany; called PS 30 kDa here). The laboratory-made membrane modules were prepared by assembling four capillary membranes inside a Pyrex glass cylinder with 1.2 cm I.D. and 20 cm length. All modules had an internal dense layer on the lumen side and external sponge layer on the shell side. The module with PA 10 kDa had an internal membrane area of $35.8\,cm^2$ and external membrane area of $53.2\,cm^2$. The module with PS 30 kDa had an internal membrane area of $18.76\,cm^2$ and external membrane area of $39.34\,cm^2$. The module with PA 50 kDa membrane has internal membrane area of $28.6\,cm^2$ and external membrane area of $57.38\,cm^2$.

The multiphasic enzyme-catalyzed membrane reaction system was run using two flowing methods, differing in having the enzyme immobilized on the shell side or on the lumen side of the membrane module. FIGURE 1 shows a scheme for the hydrolytic reaction and separation process of the multiphasic membrane reactor, with lipase immobilized in the sponge layer. In this case, the organic phase was fed into the shell side and the aqueous phase was separately fed into the lumen side by two gear pumps. In case of lipase immobilized in the lumen side, the organic phase was circulated in the lumen side and aqueous phase circulated in the shell side. Two control panels equipped with valves, flowmeters, and pressure gauges were used to separately control the fluid dynamics conditions of the two circuits. The membrane module

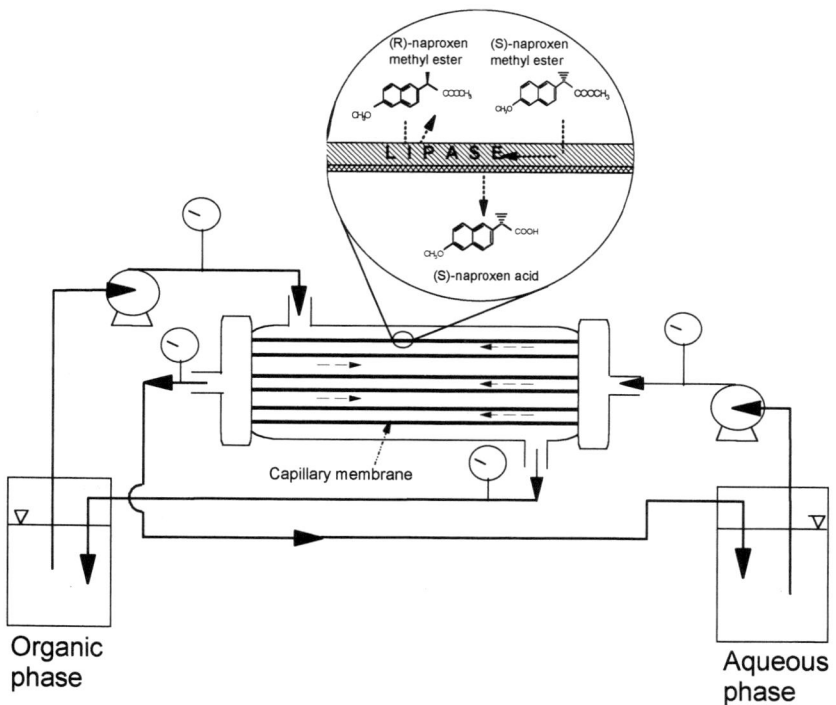

FIGURE 1. Schematic diagram of reaction and separation process in multiphasic enzyme membrane reactor with sponge layer of membrane as immobilization site.

was dipped into a thermostated bath to maintain a constant reaction temperature of 30°C. Additionly, stirred tank reactors (STR) were used to investigate the performance of native free lipase in emulsion.

Enzyme Activity Test

The native hydrolytic activity of enzyme in solution was tested using triacylglycerol acyl in the olive oil as substrate. This method was used to measure the hydrolytic activity of enzyme solutions so as to calculate the amount of immobilized lipase: 23 mL of 50 mM phosphate buffer at pH 7.0 were mixed with 1 mL of olive oil in a 50-mL bottle to form an emulsion, and then 2 mL of lipase solution was added. The mixture was stirred with magnetic stirrer and maintained at constant temperature of 30°C. The reaction was monitored by automatic titration method using a Mettler DL21 Titrator.

Lipase Immobilization

The lipase was immobilized into the sponge layer or onto the thin layer of membranes by crossflow filtration of 2 g/L or 3 g/L of lipase in buffer solution (pH 7.00) along the shell side or along the lumen side of the module, respectively, under a pressure of 50 kPa or 90 kPa, at room temperature. The immobilization process lasted a few hours until the permeation flux was constant.

Immobilized enzyme activity in the membrane was obtained by an activity-balance calculation of enzyme solutions before and after immobilization. This included the original lipase solution, the retentate lipase solution, the permeate lipase solution, and the washings from the lipase immobilization process. The enzyme activity was tested with triacylgylcerol acyl in olive oil as substrate by the titration method. Enzyme activity was proportional to its concentration in buffer solution in the of enzyme and substrate concentration range experimentally investigated (see RESULTS AND DISCUSSION). Since the immobilized enzyme activity in membrane was obtained by an activity-balance calculation, the mass amount of crude lipase immobilized in the membranes could be obtained from the linear relationship between enzyme activity and enzyme concentration in the buffer solution.

Operations in Enzyme Membrane Reactor and Stirred Tank Reactor

In the membrane bioreactor, 200 mL of 5 mM naproxen methyl ester/isooctane solution (organic phase) and 250 mL of phosphate buffer solution (aqueous phase, pH = 7.00) were separately circulated in the two circuits of the system at an axial flow rate of 330 mL/min. The organic phase pressure was maintained at 60 kPa and the aqueous phase pressure was maintained at 35 kPa; therefore, a transmembrane pressure of 25 kPa was applied from the organic phase to the aqueous phase, to prevent the aqueous phase passing through the membrane into the organic phase. The reaction temperature was 30°C. Three-milliliter samples were taken from the aqueous phase at appropriate time intervals for analysis of the concentration of (S)- and (R)-naproxen by HPLC. An equivalent volume (3 mL) of fresh buffer was added to the aqueous phase after each time sample was removed so as to maintain a constant volume of the aqueous phase. The mass accumulation of isomers in the aqueous phase; that is, the production of isomers as function of time, was calculated at each time taking into account the sample volume.

In the stirred tank reactor, 23 mL of 50 mM phosphate buffer (pH = 7.0), 3 mL of 2 g/L (or 3 g/L) crude lipase buffer solution, and 20 mL of 5 mM naproxen ester in isooctane solution were mixed with a magnetic stirrer to form an emulsion. The reaction temperature was 30°C. The converted isomers were dissolved into the aqueous phase and the increase in concentration of (S) and (R) isomers inside the aqueous phase was analyzed by HPLC. The volume of the sample was replaced with an equivalent volume of fresh buffer and was taken into account when calculating production in the aqueous phase as function of time.

Productivity, Enantioselectivity, Enantioexcess and Conversion Measurement

The concentration of (S)-naproxen and its isomer (R)-naproxen, both in the membrane bioreactor and in the stirred tank reactor, were measured by HPLC (Merck Hitachi) coupled with an UV detector (Merck Hitachi L-7400) at 270 nm using the external standard method. The column was a Chiracel OD (Daicel Chemical Ind., Ltd.), 25×0.46 cm; the mobile phase was a mixture of hexane/isopropanol/acetic acid (97:3:1 v/v) and the analysis were performed at a flow rate of 1.0 mL/min. The operation temperature and pressure were 30°C and 20 bar, respectively. Before analysis the naproxen was extracted from aqueous samples into toluene and the toluene phase was injected into HPLC. The error in the concentration measurements was estimated to be ±0.2%.

Production data were calculated according to the concentration of naproxen in the aqueous phase and the volume of the aqueous phase as follows:

$$P = V_t \times C_t + \sum_n V_n \times C_n, \quad (1)$$

where V_t is the total volume of aqueous phase, C_t is the naproxen concentration of the sample, V_n is the volume of the nth previous sample and C_n is the naproxen concentration of the nth previous sample. Productivity was obtained by dividing the production by the reaction time.

When the substrate conversion is very low, enantioselectivity can be expressed as follows:[27]

$$E = \frac{P_S}{P_R}, \quad (2)$$

where P_S and P_R are the production of (S)- and (R)-naproxen, respectively.

Enantioexcess of (S)-isomer in the product, ee_p, was calculated from

$$ee_p = \frac{P_S - P_R}{P_S + P_R} \times 100\%. \quad (3)$$

Conversion of substrate was calculated from

$$\text{Conversion}\% = \frac{P_S + P_R}{M_s} \times 100, \quad (4)$$

where M_s is the mass of the substrate.

For the various series of experiments, the total production was calculated by summing over many samples. Therefore, the error associated with each concentration measurement was taken into the account for the total production. To account for other experimental errors, an additional 3% was assumed on the basis of the deviation calculated for experiments performed under the same operating conditions. Therefore, a maximum error of 5% is associated with the production, enantioexcess, and selectivity values.

RESULTS AND DISCUSSION

The Relationship between Lipase Activity and Its Concentration in the Buffer Solution

The immobilized amount of crude lipase was calculated from the activity of immobilized lipase in the membrane and the linear relationship between the amount of crude lipase in buffer solution and the hydrolytic activity of lipase with the olive oil as the substrate. The activity immobilized in the membrane was calculated from the enzyme activity balance. FIGURE 2 shows the enzyme activity as a function of lipase concentration in aqueous buffer solution. The concentration of crude lipase varied from 1 g/L to 3 g/L in pH 7.00 buffer solution. The results show that the enzyme activity increased linearly with the increase in crude lipase concentration for the tested substrate concentrations. Therefore, activity immobilized in the membrane was obtained by means of an activity-balance calculation, and the immobilized amount of crude lipase in the membrane was calculated from the coefficient derived from FIGURE 2 (namely, 1.1469).

Effect of Immobilization Process on Enzyme Activity

The long recycling-flow time of lipase solution along membrane module during the immobilization process may cause some denaturation of lipase and this can result

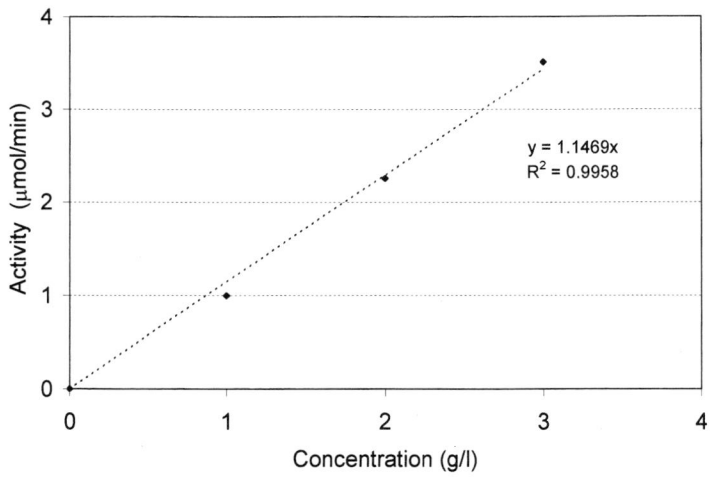

FIGURE 2. Enzyme activity as function of lipase concentration tested by the titration method.

TABLE 1. Effect of recycling flow during immobilization process

Flowing Circuit	Running Time (hours)	Flowing Velocity (mL/min)	Enzyme Activity Lost (%)
Along the shell side	2	330	13.7%
Along the lumen side	2	330	12.4%

in the loss of lipase activity. To investigate the effect of the immobilization process on lipase activity, 350 mL of 2 g/L lipase buffer solution was circulated along the shell side or along the lumen side of PS 30kDa membrane for two hours at a flow rate of 330 mL/min, without applying any pressure and at room temperature. This operation condition is the same as that during the immobilization process, except for the pressure. After two hours, the enzyme activity on olive oil was tested and was compared with that of the initial lipase activity. The decrease in enzyme activity is shown in TABLE 1. The results show that at most 14% of the original activity was lost during immobilization process. This loss might have resulted from adsorption of proteins by the membrane and/or denaturation of lipase due to the effect of shear stress. Moreover, the decrease in lipase hydrolytic activity is equivalent for lipase solution recycled along the shell side or along the lumen side. This means that the effect of the immobilization process on lipase activity with lipase recycling in the shell side of membrane is the same as that for lipase recycling in the lumen side of membrane. Therefore, it is assumed that any different performance of enzyme membrane reactor does not result from the immobilization process in the two different recycling circuits of the reactor system.

Effect of Immobilization Site on Enzyme Membrane Reactor Performance

The various enzyme immobilization sites included the sponge layer and the thin layer of PA 10kDa membrane, and those of PS 30kDa membrane. FIGURE 3 shows the production of (S)-naproxen and its (R)-isomer catalyzed by lipase immobilized at the various sites of PA 10kDa membrane as a function of reaction time. With lipase in the sponge layer, the production of (S)-naproxen was higher than that with lipase on the thin layer of membrane. FIGURE 4 shows the enantioexcess and conversion corresponding to the two immobilization sites on the PA 10kDa membrane. The enantioexcess of lipase immobilized in the sponge layer was higher than that of lipase on the thin layer.

FIGURES 5 and 6 illustrate the production, enantioexcess, and conversion of lipase immobilized at various sites of the PS 30kDa membrane as a function of reaction time.

It is evident that for both type of enzyme membrane reactors the production and enantioexcess were higher with lipase immobilized in the sponge layer than that with lipase on the thin layer. The results are mainly related to the loading capacity of the two sides of the membranes. The lumen side has a limited capacity to host large amounts of enzyme due to the low molecular weight cutoff. In fact, even using higher concentration of enzyme solution with respect to immobilization in the shell side, it was not possible to increase the amount of enzyme immobilized on the thin layer.

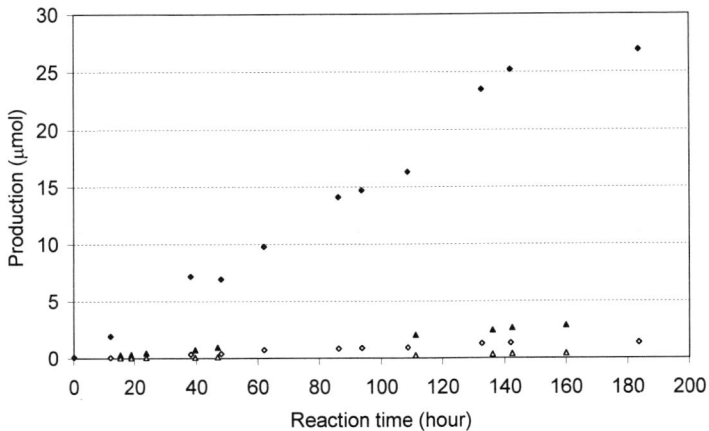

FIGURE 3. Production of (S)-naproxen and (R)-naproxen by lipase immobilized in the sponge layer (◆ and ◇, respectively) or on the thin layer (▲ and △, respectively) of PA 10 kDa membrane as function of reaction time.

The hydrolytic activity of enzyme-loaded membrane is related to the amount of lipase immobilized on the membrane. The amounts of immobilized lipase on PA 10 kDa membrane and PS 30 kDa membrane and related results are listed in TABLE 2. In this table experimental results with PA 50 kDa membrane from a previous study are also listed for comparison.[9] It is evident that a larger amount of lipase was immobilized into the sponge layer of membrane than onto the thin layer. This resulted in more lipase participating in the reaction and, thus, higher productivity

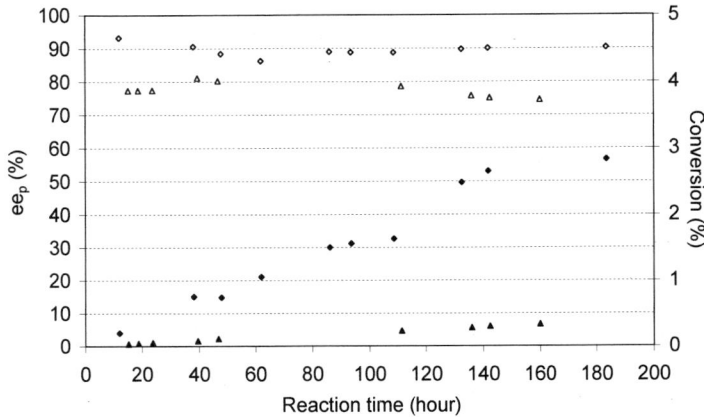

FIGURE 4. Enantioexcess and conversion by lipase immobilized in the sponge layer (◇, ee_p; ◆, conversion) or on the thin layer (△, ee_p; ▲, conversion) of PA 10 kDa membrane as function of reaction time.

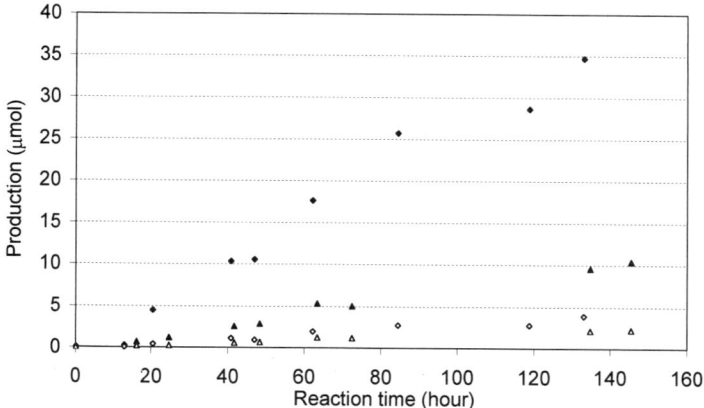

FIGURE 5. Production of (S)-naproxen and (R)-naproxen by lipase immobilized on the sponge layer (◆, S isomer; ◇, R isomer) or on the thin layer (▲, S isomer; △, R isomer) of PS 30 kDa membrane as function of reaction time.

was obtained with the sponge layer of membranes as immobilization site. This phenomenon is related to the different structure of the sponge layer from that of the thin layer of the membranes. The sponge layer can provide more immobilization space than the thin layer surface. During the immobilization process of lipase onto the thin layer, most of lipase molecules were rejected by the membrane and gelled on the membrane top surface as a cake layer because of the small pore size of the membrane with respect to lipase molecular weight (67 kDa). Furthermore, a washing step was used after the cross-filtration immobilization step in order to remove the loose lipase

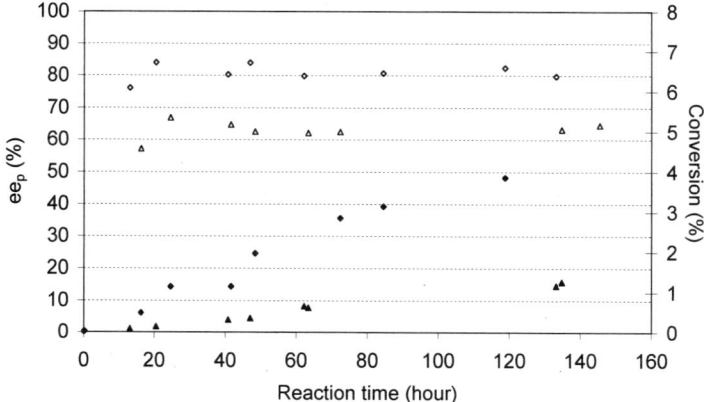

FIGURE 6. Enantioexcess and conversion by lipase immobilized in the sponge layer (◇, ee_p; ◆, conversion) or on the thin layer (△, ee_p; ▲, conversion) of PS 30 kDa membrane as function of reaction time.

TABLE 2. Experimental results of enzyme membrane reactors

Membrane Type	MWCO (kDa)	Immobilization Site	Lipase loading (mg/cm^2)	(S)-Naproxen Productivity (μmol/h)$\times 10^2$	Productivity per Immobilized Mass (μmol/h mg)$\times 10^2$	Enantioexcess[a] (ee_p)%	Enantioselectivity[a] (E)	Conversion[b] (%)
PA	10	sponge layer	0.3	16	1.35	90	21	2.7
PA	10	thin layer	0.06	1.9	1.68	77	7.8	0.3
PS	30	sponge layer	1	26	0.66	80	10	3.9
PS	30	thin layer	0.17	7.1	2.22	63	4.2	1.4
PA	50	sponge layer	0.61	22.7	0.95	91	21	3.0
PA	50	thin layer	0.19	2.3	0.65	90	21	0.3

[a] Average value for various reaction times.
[b] Value after 150 hour reaction time.

that would otherwise desorb during reaction. The lipase on membrane top surface is exposed to washing solution and part of lipase could be removed by flow under shear stress, whereas the lipase inside the sponge layer was less affected by the washing step because the lipase "hides" inside the membrane sponge layer. Therefore, immobilization from the shell side/sponge layer allowed a more stable enzyme-immobilized membrane to be obtained.

FIGURE 7 shows the productivity as a function of the amount of immobilized lipase in the sponge layer of membranes. Evidently productivity increased with an increase in immobilized lipase, but not always linearly. This means that over a certain range not all lipase participated in the reaction, since lipase is dissolved in the aqueous phase and hydrophobic naproxen ester exists in the organic phase and the organic/aqueous biphasic interface on the membrane, where lipase hydrolyzes the substrate, is rather limited. Thus, when the amount of lipase on the membrane increases to the point where all interfacial area is saturated, productivity will not increase proportionally any further. An increase in interfacial area is of great importance in increasing the number of substrate molecules available to the enzyme.

The experiment results also show that the enantioexcess, for PA 10 kDa membrane and PS 30 kDa membrane, was generally higher with lipase in the sponge layer of membranes than on the thin layer of membranes. One possible reason for this is related to the different amounts of lipase between the sponge layer and the thin layer of membranes. A higher amount of lipase can create a microenvironment that favors higher enantioselectivity. The same effect of the amount of immobilized lipase on the enantioselectivity was also reported in a previous study with PA 50 kDa membrane. That is, the enantioselectivity was improved from 5.5 to 21 when the amount of lipase in the sponge layer membrane was increased from 8.1 mg to 35 mg. It is notable that the lumen side of PA 50 kDa membrane led to a good enantioexcess compared to PA 10 kDa membrane. This may be related to different amounts of lipase on the membrane, but the productivity is still low compared to that when

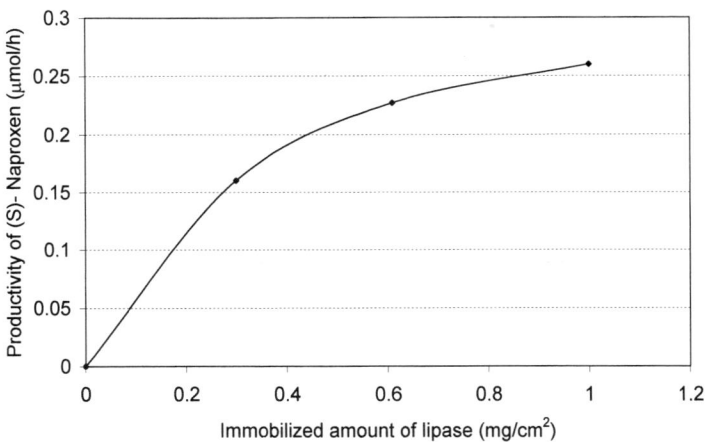

FIGURE 7. Productivity of (S)-naproxen by lipase immobilized in the sponge layer of membranes as function of amount of lipase immobilized.

lipase is immobilized in the sponge layer. TABLE 2 also indicates that increased lipase immobilization in the sponge layer of membrane leads to higher conversion.

Effect of Membrane Materials on Enzyme Membrane Reactor Performance

According to the experimental data we obtained by using various membrane materials, including polyamide and polysulfone, the membrane material does affect the productivity and enantioselectivity of an enzyme membrane reactor. From TABLE 2 it can be seen that, with the lumen side of membrane as the immobilization site, PS 30 kDa membrane has a higher load of lipase than the PA 10 kDa membrane, leading to higher production; however, with a lower enantioexcess. With the sponge layer of membrane as immobilization site, the same results were obtained, that is, the enzyme-immobilized PS 30 kDa membrane showed higher production but lower enantioexcess than enzyme-immobilized PA 10 kDa membrane. This is related to the properties of membrane materials that affect the amount of adsorption and configuration of the enzyme on the membrane surface. The initial lipase concentration of 3 g/L and the operation pressure of 90 kPa were applied to immobilize lipase on the thin layer of a PA 10 kDa membrane whereas a lower initial lipase concentration of 2 g/L and a lower pressure of 50 kPa were applied to immobilize lipase on the thin layer of PS 30 kDa membrane. This means that, even with higher initial enzyme concentration and higher operation pressure, a smaller amount of enzyme was immobilized on the PA 10 kDa membrane than on the PS 30 kDa membrane. Consequently, lower productivity was obtained with the PA 10 kDa membrane. The tendency of lipase to be adsorbed on hydrophobic surfaces has been reported,[28–30] which conforms with the present study.

Different enantioselectivities were obtained for lipase-immobilized PA 10 kDa membrane and lipase-immobilized PS 30 kDa membrane. This was also related to the different material properties of polysulfone and polyamide membranes, such as hydrophilicity/hydrophobicity. Lipase immobilized on the thin layer of a PA 10 kDa membrane showed higher enantioselectivity than that on a PS 30 kDa membrane. Furthermore, the thin layer of the PS 30 kDa membrane immobilized a similar amount of lipase than the thin layer of a PA 50 kDa membrane. However, the former showed a higher productivity and lower enantioselectivity. This suggests that polysulfone membrane materials may affect the configuration of lipase molecules on the membrane surface and, thus, the resolution capability between isomers, leading to lower enantioselectivity, but this does not dramatically affect their hydrolytic activity. It is estimated that, in a multiphase enantiocatalytic enzyme membrane reactor, membrane materials can affect the enantioselectivity in the following two ways. First, by affecting the configuration of immobilized enzyme on the membrane, allowing different parts of the enzyme molecules to be exposed to the substrate and participate in the hydrolysis, thus leading to different enantioselectivities. Second, by affecting the location site of organic/aqueous interface on membrane. The experiment to date show the overall effect of the various materials on the lipase-loaded multiphase membrane reactor system. The specific mechanism responsible for the observed behavior needs further investigation. In addition, the fact that the lipase on the thin layer of a PA 50 kDa membrane showed lower productivity than that on a PS 30 kDa membrane could mean that only a limited part of the lipase on PA 50 kDa membrane came into contact with organic phase and participated in the hydrolysis of substrate at the biphasic interface. This may be also related to the difference in hydrophilic/hydrophobic character between polyamide and polysulfone membranes. V.M. Balcào et al. described that, in the case of hydrophobic carriers, the immobilized enzyme is soaked in the organic liquid phase, whereas in hydrophilic carriers the immobilized enzymes is soaked in water.[31] It is suggested that more lipase was soaked in the aqueous phase than in the organic phase because of the hydrophilicity of polyamide membrane, consequently, reducing the contact between lipase and organic phase and leading to lower productivity of the membrane reactor.

Reaction Stability of Membrane Reactors Compared to Stirred Tank Reactors

FIGURE 8 illustrates the production and enantioexcess of free lipase in emulsion stirred tank reactor. By comparison with results from an enzyme-loaded membrane, it is evident that the stability of lipase on the membrane reactor was superior to free lipase in a stirred tank reactor. No matter whether the polyamide membrane or polysulfone membrane was used or whether lipase was immobilized inside membrane pores or on membrane lumen surface, lipase always showed stable activity during the 200 hours operation. The production of (S)-isomer increased linearly with time and no decrease of lipase activity was observed, which means that the productivity is constant. However, in case of a stirred tank reactor, lipase activity decreased significantly with time and its halflife time was just 15 h. As a result, it is evident that stable lipase activity and its repeated use can be achieved with a lipase-immobilized membrane reactor. The membrane materials and structure did not show different effects on the stability of lipase activity. This suggests that different materials affected the initial activity and selectivity of an enzyme-loaded membrane, but after the lipase experienced initial deactivation, its stability remained constant during the reaction period.

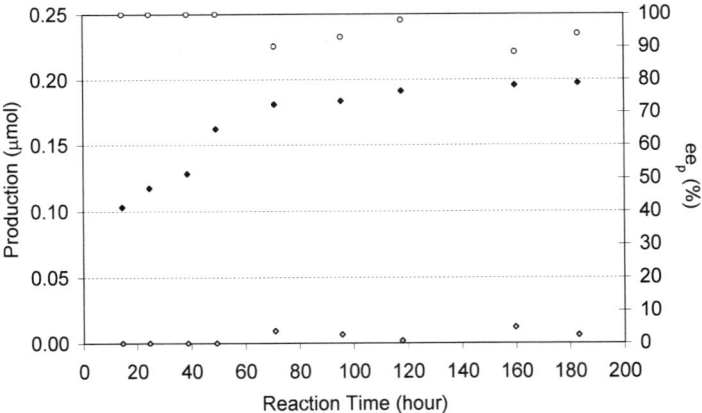

FIGURE 8. Production of (S)-naproxen and enantioexcess by free lipase in an emulsion stirred tank reactor: ◆, S isomer; ◇, R isomer; ○, ee_p.

CONCLUSIONS

In this paper, the effect of immobilization site and membrane material on the optical resolution of racemic naproxen methyl ester applying lipase-immobilized membrane reactor was studied. After recycling enzyme solution along the shell side or along the lumen side of membrane modules, the initial enzyme solution lost about 14% of activity due to adsorption and/or shear stress. Both immobilization site and membrane materials affected the performance of lipase-immobilized membrane reactor.

The higher the amount of immobilized enzyme, the higher the productivity the membrane reactor. Under the same operation conditions, the sponge layer of membranes immobilized larger amounts of lipase than the thin layer of membranes, thus leading to higher productivity and higher conversion. The polysulfone membrane loaded more lipase than polyamide membrane. It was also shown that the productivity of immobilized enzyme did not always increase linearly with an increase in immobilized amount.

Concerning enantioselectivity, in general, higher enantioselectivity was obtained with the polyamide membranes; lower with polysulfone membranes. With the sponge layer of membrane as the immobilization site, 90% of (S)-naproxen enantioexcess was obtained with 10kDa polyamide membrane, which was similar to that for 50kDa polyamide membrane. Using the thin layer of membrane as immobilization site, the enantioexcess was lower by 20–30% for polyamide 10kDa membrane than that for 50kDa polyamide membrane (90%), resulting from the different amounts of lipase immobilized on the membrane.

In general, compared to the sponge layer of membrane, the lumen side has a low loading capability that results in low productivity and enantioselectivity in the membrane reactor; the polysulfone membrane materials affect the catalytic properties of the immobilized lipase compared to polyamide membrane. We conclude that the sponge layer of polyamide membrane is the preferable immobilization site for its higher productivity than the thin layer of membrane, and its higher enantioselectivity than polysulfone membrane.

In all cases, the lipase-immobilized membrane reactor showed much higher stability as a function of time than the free enzyme in stirred tank reactor, no matter whether polyamide membrane or polysulfone membrane was used.

ACKNOWLEDGMENTS

This work was supported by the National Research Council (CNR) in Italy.

REFERENCES

1. WU, D.-R., S.M. CRAMER & G. BELFORT. 1993. Kinetic resolution of racemic glycidyl butyrate using a multiphase membrane enzyme reactor: experiments and model verification. Biotechnol. Bioeng. **41:** 979.
2. ZHANG, X.-M. & I.W. WAINER. 1993. On-line determination of lipase activity and enantioselectivity using an immobilized ezyme reactor coupled to a chiral stationary phase. Tetrahedron Lett. **34:** 4731.

3. INDLEKOFER, M., et al. 1995. Continuous enantioselectivity transesterification in organic solvents. Use of suspended lipase preparations in a microfiltration membrane reactor. Biotechonol. Progr. **11:** 436.
4. GIORNO, L., et al. 1997. Hydrolysis and regioselectivity transesterification catalyzed by immobilized lipases in membrane bioreactors. J. Membr. Sci. **125:** 177.
5. LOPEZ, J.L. & S.L. MATSON. 1997. A multiphase/extractive enzyme membrane reactor for production of diltiazem chiral intermediate. J. Membr. Sci. **125:** 189.
6. TASAI, S.W. & C.-M. HUANG. 1999. Enantioselective synthesis of (s) soprofen ester prodrugs by lipase in cyclohexane. Enzyme Microb. Technol. **25:** 682–688.
7. THEIL, F. 2000. Lipases as a tool for the synthesis of chiral intermediates. Chem. Today **18:** 61–64.
8. GIORNO, L. & E. DRIOLI. 2000. Biocatalytic membrane reactors applications and perspectives, Trends Biotech. **18:** 339–348.
9. SAKAKI, K., L. GIORNO & E. DRIOLI. 2001. Lipase-catalyzed optical resolution of racemic naproxen in biphasic enzyme membrane reactors. J. Membr. Sci. **184:** 27–38.
10. KAMORI, M., et al. 2000. Immobilization of lipase on a new inorganic ceramic support, toyonite, and the reactivity and enantioselectivity of the immobilized lipase. J. Molec. Catal. B: Enzymatic **9:** 269–274.
11. ARICA, M.Y., et al. 2001. Reversible immobilization of lipase on phenylalanine containing hydrogel membranes. Proc. Biochem. **36:** 847–854.
12. VULFSON, E.N. 1994. Industrial applications of lipases. In Lipase, Their Structure Biochemistry and Application. P. Wooley & S.B. Peterson, Eds.: 271–288. Cambridge University Press, Cambridge.
13. RIVA, S. 1996. Regioselectivity of hydrolases in organic media. In Enzymatic Reactions in Organic Media. A.M.P. Koskinen & A.M. Klibanov, Eds.: 140–169. Chapman & Hall, Glasgow.
14. BORNSCHEUER, U.T. & R.J. KAZLAUSKAS. 1999. Hydrolases in Organic Synthesis–Regio- and Stereoselective Biotransformations. VCH-Wiley, Weinheim.
15. KLIBANOV, A.M. 1990. Asymmetric transformation catalyzed by enzymes in organic solvents. Acc. Chem. Res. **23:** 114.
16. MIYAKE, Y., M. OHKUBO & M. TERAMOTO. 1991. Lipase-catalyzed hydrolysis of 2-naphtyl esters in biphasic systems. Biotechnol. Bioeng. **38:** 30–36.
17. PRONK, W. & K.V. RIET. 1991. The interfacial behavior of lipase in free form and immobilized in a hydrophilic membrane reactor. Biotech. Appl. Biochem. **14:** 146.
18. GUIT, R.P.M., et al. 1994. Lipase kinetics: hydrolysis of triacetin by lipase from *Candida cylindracea* in a hollow-fiber membrane reactor. Biotech. Bioeng. **38:** 727.
19. PRONK, W.P., et al. 1991. A hybrid membrane-emulsion reactor for the enzyme hydrolysis of lipids. J. Am. Oil Chem. Soc. **68:** 852.
20. MOLINARI, R., M.E. SANTORO & E. DRIOLI. 1994. Study and comparison of two enzyme membrane reactors for fatty acids and glycerol reproduction. Ind. Eng. Chem. Res. **33:** 2591.
21. GIORNO, L., et al. 1995. Performance of a biphasic organic/aqueous hollow fiber reactor using immobilized lipase. J. Chem. Tech. Biotech. **64:** 345.
22. SHIOMORI, K., et al. 1995. Hydrolysis rates of olive oil by lipase in a monodispersed o/w emulsion system using membrane emulsification. J. Ferment. Bioeng. **6:** 552–558.
23. DEREWENDA, U., et al. 1992. Catalysis at the interface: the anatomy of a conformational change in a triglyceride lipase. Biochemistry **31:** 1532–1541.
24. BUTTERFIELD, D.A., et al. 2001. Catalytic biofunctional membranes containing site-specifically immobilized enzyme arrays: a review. J. Membr. Sci. **181:** 29–37.
25. GONZÁLEZ-SÁIZ, J.M. & C. PIZARRO. 2001. Polyacrylamide gels as support for enzyme immobilization by entrapment. Effect of polyelectrolyte carrier, pH and temperature on enzyme action and kinetics parameters. Euro. Polym. J. **37:** 435–444.
26. FERNÁNDEZ-LORENTE, G., et al. 2001. Modulation of lipase properties in macro-aqueous systems by controlled enzyme immobilization: enantioselective hydrolysis of a chiral ester by immobilized *Pseudomonas lipase*. Enzyme Microbial Technol. **28:** 389–396.
27. CHEN, C.S. & C.J. SIH. 1989. General aspects and optimization of enantioselective biocatalysis in organic solvents: the use of lipases. Angew. Chem. Int. Ed. Engl. **28:** 695.

28. SUGIRA, M. & M. ISOBE. 1976. Studies on the mechanism of lipase reaction. IV. Adsorption of *Chromobacterium lipase* on hydrophobic glass beads. Chem. Pharm. Bull. **24:** 1822–1828.
29. BASTIDA, A. *et al.* 1998. A single step purification, immobilization and hyperactivation of lipases via interfacial adsorption on strongly hydrophobic supports. Biotechol. Bioeng. **58:** 486–493.
30. FERNÁNDEZ-LAFUENTE, R., *et al.* 1998. Immobilization of lipases by selective adsorption on hydrophobic supports. Chem. Phys. Lipids **93:** 185–197.
31. BALCÀO, V.M., A.L. PAVIA & F.X. MALCATA. 1996. Bioreactors with immobilized lipases: state of the art. Enzyme Microb. Technol. **18:** 392–416.

CO_2 Capture by Means of an Enzyme-Based Reactor

R.M. COWAN,[a] J.-J. GE,[a] Y.-J. QIN,[b] M.L. McGREGOR,[c] AND M.C. TRACHTENBERG[a,b]

[a]*Sapient's Institute, New Brunswick, New Jersey, USA*

[b]*Carbozyme, Inc., Bordentown, New Jersey, USA*

[c]*McGregor & Associates, Spring, Texas, USA*

ABSTRACT: We report a means for efficient and selective extraction of carbon dioxide (CO_2) at low to medium concentration from mixed gas streams. CO_2 capture was accomplished by use of a novel enzyme-based, facilitated transport contained liquid membrane (EBCLM) reactor. The parametric studies we report explore both structural and operational parameters of this design. The structural parameters include carbonic anhydrase (CA) concentration, buffer concentration and pH, and liquid membrane thickness. The operational parameters are temperature, humidity of the inlet gas stream, and CO_2 concentration in the feed stream. The data show that this system effectively captures CO_2 over the range 400 ppm to at least 100,000 ppm, at or around ambient temperature and pressure. In a single pass across this homogeneous catalyst design, given a feed of 0.1% CO_2, the selectivity of CO_2 versus N_2 is 1,090:1 and CO_2 versus O_2 is 790:1. CO_2 permeance is 4.71×10^{-8} mol m^{-2} Pa^{-1} sec^{-1}. The CLM design results in a system that is very stable even in the presence of dry feed and sweep gases.

KEYWORDS: carbon dioxide; carbonic anhydrase; CO_2 separation; contained liquid membrane; gas separation

INTRODUCTION

CO_2 separation technologies are becoming ever more important in fields such as life support for space exploration, agricultural greenhouses, reduction of atmospheric greenhouse gas levels, and enrichment (sweetening) of natural gas. Many different technologies have been developed to separate CO_2 from gas mixtures.[1-6]

At low to moderate pCO_2, chemical and physical absorption in liquid media and physical adsorption to solids are the most effective approaches.[6,7] Chemical absorption relies on a reactive absorbent, that is, a promoter, facilitator, or catalyst that reacts with the CO_2, preferably at the gas–liquid interface. This reaction increases either or both the rate and magnitude of CO_2 solubility or the subsequent evolution of the gas from the liquid. Beginning with the Girdler patent of 1931, there has been extensive study of bulk fluid, homogeneous reactant, alkanolamine-based chemical

Address for correspondence: M.C. Trachtenberg, Sapient's Institute, Cook College, Rutgers University, 20 Ag Extension Way, New Brunswick, NJ 08901-8500, USA. Voice: 732-932-8875; fax: 732-932-7931.

miket@aesop.rutgers.edu

absorption, and facilitated transport.[6] However, cost, maintenance, safety, and environmental issues that surround the use of amines in bulk fluid systems present significant impediments to their application outside of industrial acid gas scrubbing. Therefore, there still exists a substantial need for the development of alternative approaches.

There are two approaches to take in addressing the limitations of the bulk fluid alkanolamine based CO_2 separation systems. One is to change the chemistry of the system by using alternative promoters, facilitators, and/or catalysts. Ideally the alternative chemistry will yield a system that is more efficient, cheaper, faster, safer, and more environmentally friendly. None of the alternative chemistries identified previously (alkali carbonates,[8] glycinates,[9] arsenites,[8] novel amines,[10,11] or crown ethers[12]) can meet the desired requirements. However, the biological catalyst (enzyme) carbonic anhydrase (CA–EC 4.2.1.1)[13–19] shows great promise. CA is the most efficient catalyst for CO_2 hydration and dehydration (absorption and desorption),[13–15] particularly for application at low pCO_2 and near ambient temperature. The second approach is to move from a bulk fluid design to a membrane design. The benefits are higher efficiency, lower energy costs, and greater separation per unit volume of reactor. Liquid membrane designs are attractive in comparison with nonporous polymer membranes of equal thickness because they decrease transport resistance so that high permeance is achieved together with high selectivity. Use of this concept for CO_2 separation was first explored in the 1960s.[8,18] We demonstrate in this work that the design parameters of a liquid membrane can be carefully tailored to the desired product and operating conditions. This is a distinct benefit over other membranes.

The mammalian isozyme, human CA type 2 (HCAII), found in erythrocytes has a maximum turnover number of 1.6×10^6 mol $[CO_2]$/mol [CA] sec.[19] The maximum turnover number for the bovine analog (BCAII) is about 33% lower. The catalytic site is a zinc atom covalently bound to three histidine molecules.[20] CA catalyzes the reversible hydration of CO_2 by a two-step process.[21] In the first step CA hydrolyzes water to capture a hydroxyl releasing a proton (Rxn 3). In the second the hydroxyl attacks the carbonyl bond of CO_2 to yield bicarbonate and the unreacted enzyme (Rxn 2). This is the hydration step and it occurs at the CO_2-rich side. At the CO_2-lean side the reverse reactions (Rxn 2, Rxn 3, Rxn 1, respectively) occur, liberating CO_2, which can then evolve into the permeate stream. As shown later in FIGURE 3B, CA changes the rate at which the hydration and dehydration reactions occur at a given pH. This allows for increased transport of CO_2 as the dissolved ionic species, HCO_3^- and CO_3^{2-}, and therefore increased separation performance through increases in both CO_2 permeance and selectivity of CO_2 versus other gases.

From the viewpoint of solution and diffusion mechanisms, the enzyme functions as a facilitator. More correctly, it is a catalyst and, unlike amines and other chemical combining reactants, does not itself participate in the transport process as a chaperone. Only small ionized species, principally bicarbonate and carbonate, diffuse across the membrane.

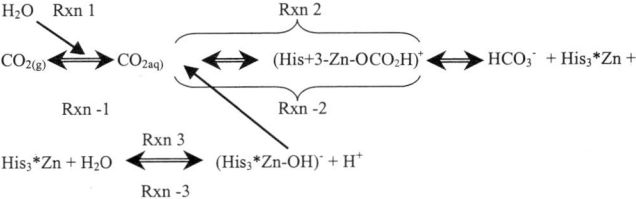

The earliest studies related to CA-based membrane systems typically used supported liquid membrane (SLM) designs, with CA dissolved in an aqueous buffer solution. The solution was trapped in the pores of a filter membrane or in a proteic matrix.[8,14,18,22] These studies successfully demonstrated the ability of these designs to selectively extract CO_2 from mixed gas streams. The majority of these early studies used a feed CO_2 concentration of about 5% at 1 atm. However, feed CO_2 concentrations as high as 100% have also been studied.[23] The most common approach taken in the early work was to not add any external buffer, but instead to allow pH to be controlled by the carbonic acid–bicarbonate–carbonate solution that formed spontaneously;[13–15] others used dilute phosphate buffers.[8,17,18,22] The one notable exception to the use of dilute buffer solutions was the work done by Ward.[8] He used CA in a high concentration (2 M) $KHCO_3$ solution to facilitate CO_2 uptake. However, Ward did not attempt to optimize the high strength bicarbonate–CA system and it is likely his results were far from optimal. The use of bicarbonate–carbonate buffers has been shown to decrease the rate at which CA catalyzes CO_2 hydration[14,18,24,25] with the 50% inhibitory concentration for bicarbonate generally reported to be near 60 mM.[13–15]

Alper and Deckwer studied the effect of buffer systems on the kinetics of absorption of CO_2 from air into a well-mixed solution, that is, hydration rate.[26] Using a feed CO_2 concentration of 100% at 1 atm they found little difference between the performance of a 0.5 M carbonate buffer at pH 9.6 and that of 0.5 M phosphate buffer at pH 11.1. They also studied interactions between the buffer strength and CA concentration, finding that there needs to be a balance between these (i.e., higher enzyme concentrations require higher buffer concentrations) or the adsorption rate can become limited through depletion of OH^- (i.e., decreased local pH) at the gas liquid interface. This is important because, as is shown later in FIGURE 2A, the rate of CA catalysis of CO_2 hydration falls off quickly as pH decreases below pH 7.5 to 7.0.[17,27] A similar concern with respect to pH, buffer strength, and CA concentration occurs at the permeate side of the membrane where the dehydration reaction occurs. In this case the rate of the CA catalyzed reaction decreases rapidly as pH is increased above 6.5 to 7.0.[17,27]

In contrast to a 5% feed CO_2, the concentration of CO_2 in air is about 0.035% and in closed life support systems, for example, a space station, the uppermost value is limited to no more than 0.5% at 1 atm. Plants grown in greenhouses prefer CO_2 levels of 0.05–0.15% at 1 atm. Operation at higher levels of CO_2 (e.g., greater than 5%) may be useful for some applications (e.g., removal of CO_2 from stack gas) and it is convenient for an initial feasibility study (so as to simplify the analytical requirements). If the ultimate use of the technology involves the separation of CO_2 from air and/or respiratory gases, it is important to perform experiments within the relevant CO_2 concentration range and with the relevant gas mixture.

Using a CA catalyzed SLM, Trachtenberg and his colleagues showed that CO_2 concentrations as low as 0.1% and as high as 10% could be extracted from air at 1 atm.[24] Later studies by Trachtenberg[28] showed that CO_2 could be extracted from air having CO_2 concentrations as low as 0.04%. However, despite rigorous control of the feed and the sweep side humidity it was impossible to maintain SLM designs operative for periods in excess of one day. This is as expected since a known disadvantage of SLM designs is their instability due to evaporative water loss.[29] Typically solvent loss leads to catalyst loss or inactivation. This was not the case here because of the large size of CA (30 kD) and the fact that it can be repeatedly dried and rehydrated without loss of activity.[24] Furthermore, in contrast to the limited lifetime of the SLM, we have successfully operated the flat sheet EBCLM for eight days without failure and a more recent hollow fiber EBCLM for 40 days.[30] Finally, crosslinked CA has been shown to remain active for as long as three months while in use within a contained gel membrane.[22]

Many strategies have been investigated for improving the stability of supported liquid membranes, however, the most reliable approach that can be used to ensure stability is a contained liquid membrane, CLM, instead of an SLM. CLMs are made by sandwiching the liquid between two polymer membranes and stabilizing against failure due to solvent loss through use of a reservoir of liquid membrane solvent to continuously resupply the core liquid.[31–34] (For a thorough review of methods for stabilizing SLMs see Kemperman et al., 1996,[35] subsequent work by Kemperman and colleagues,[36–38] and the work of others of others.[39])

In this paper, we report on an examination of the influences that both structural and operational parameters of a CA based contained liquid membrane exert on separation performance (CO_2 permeance and selectivity of CO_2 versus N_2 and O_2). The structural parameters studied include: carbonic anhydrase (CA) concentration, buffer concentration and pH, and liquid membrane thickness. The operational parameters studied are temperature, humidity of inlet gas stream, and CO_2 concentration in feed stream. The results of these studies provide a first level optimization of CA-based contained liquid membrane (CLM) system and prove its utility as a means for separating CO_2 from low concentration CO_2 sources, such as air and respiratory gas.

EXPERIMENTAL

Materials

The enzyme catalyst used was bovine carbonic anhydrase type 2 (activity 3,240 W-A units, BCAII, Sigma Chemical Co.). The buffer was compounded from monobasic anhydrous potassium phosphate (99.0% purity, Sigma) and dibasic anhydrous sodium phosphate (99.5% purity, Fisher Scientific). Microporous polypropylene (Celgard PP-2400) was used to contain the liquid membrane. PP-2400 is 25.4 μm thick, with 41 nm × 120 nm pores, 30% porosity, and has a tortuosity factor of 5.[29] The thickness of the liquid membrane was controlled by the use of woven polyester mesh spacers (Small Parts Inc., CMY-200) constructed as an annulus around the active liquid membrane area. Air (Matheson) was ultra zero, argon (Matheson) was ultra high purity, and CO_2 (Matheson) was high purity.

Reactor Configuration and Test Stand Setup

The liquid membrane (22 mm diameter, effective area 380 mm^2) was constructed by sandwiching a CA containing phosphate buffered solution between two polypropylene membranes (see FIGURE 1A). The thickness of the aqueous phase was varied from 70 μm to 670 μm by the use of annular spacers. The liquid membrane fluid volume was maintained by hydrostatic fluid addition from a reservoir. This slight positive pressure within the liquid phase also served to ensure a constant liquid

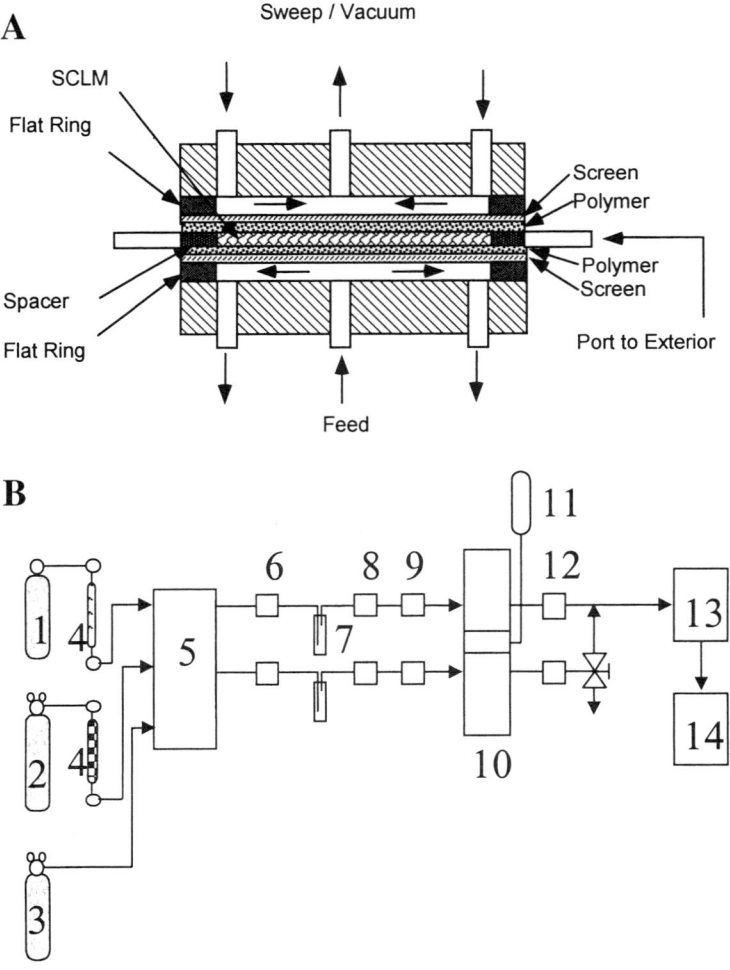

FIGURE 1. Schematic illustration of (**A**) test cell and (**B**) test stand. KEY: 1, argon; 2, air; 3, CO_2; 4, CO_2 purified column; 5, mass flow controller system; 6, humidifier; 7, trap; 8, pressure gauge; 9, temperature/humidity probe; 10, test cell; 11, reservoir; 12, mass flow meter; 13, mass spectrometer; and 14, computer.

membrane thickness and prevent separation between the polymer membrane and the metal support. The test stand setup is schematically illustrated in FIGURE 1B. The feed gas stream consisted of a mixture of CO_2 in air. Argon, treated to remove any residual CO_2 (OMI-4, Supelco), was used as the sweep gas stream. Delivery of all gases was regulated by an Environics mass flow controller (Series 2020, NIST-traceable calibration). The feed and sweep gases could be humidified to the extent desired by means of polysulfone humidifiers operated in reverse. A humidity monitor (Vasalia) was present in the sweep gas between the reactor and the mass spectrometer. After passing through the test cell the permeate (sweep) was delivered to a residual gas analyzing mass spectrometer (Questor IV, ABB Extrel) where concentrations of N_2, O_2, argon, CO_2, and H_2O were measured and recorded to electronic data files.

Separation performance was calculated based on the flow rate and the composition of the feed and sweep gases. The separation performance measures we used were permeance (Q) and selectivity (α). $Q_i = J_i/\Delta P_i$, J is the gas flux through the membrane, ΔP is the partial pressure difference across the membrane,

$$\alpha_{CO_2:O_2} = \frac{Q_{CO_2}}{Q_{O_2}}, \text{ and } \alpha_{CO_2:N_2} = \frac{Q_{CO_2}}{Q_{N_2}}.$$

Q is expressed in units of $mol\,m^{-2}\,Pa^{-1}\,sec^{-1}$ and α is a dimensionless ratio. All permeance and selectivity values given are plotted as means with error bars (plus and minus one standard deviation of the mean) to represent the measurement error.

Unless otherwise noted, the feed gas contained 0.1% CO_2 in air; argon was used as the sweep gas and evaluations were carried out at 30°C. CO_2-free gas was used to initiate each test. CA was dissolved in phosphate buffer as a homogeneous catalyst.

RESULTS

CA Concentration

Factors inherent to the enzyme include the isozyme, the spatial distribution in the liquid (homogeneous versus heterogeneous) and the concentration. Spatial distribution of CA is important because the catalytic reaction is limited to the gas–liquid interface. With a homogeneously distributed enzyme catalysis, capacity at the interface is linearly related to the CA concentration. Historically, CA concentrations used in homogenous mode were less than 4 mg/mL (133.33 µM),[14,18,24,25] although one example of a concentration in excess of 30 mg/mL is known.[23] We tested four enzyme concentrations—1, 2, 3, and 5 mg/mL (33, 67, 100, and 166 µM).

The effect of CA concentration on selectivity and CO_2 permeance is shown in FIGURE 2A. Both the selectivity of CO_2 versus N_2 or O_2 and CO_2 permeance increase as the CA concentration increases from 33 µM (1 mg/mL) to 100 µM (3 mg/mL). There is a leveling for each at 166 µM (the slight decline seen for the selectivity of $CO_2:N_2$ was not significant).

Buffer Concentration

Experiments were performed using four concentrations of the phosphate buffer: 20, 50, 75, and 100 mM. FIGURE 2B illustrates the separation performance as a function of buffer concentration. As the buffer concentration increased from 20 mM to 75 mM, the selectivity increased, to be followed by a slight decline at 100 mM. CO_2

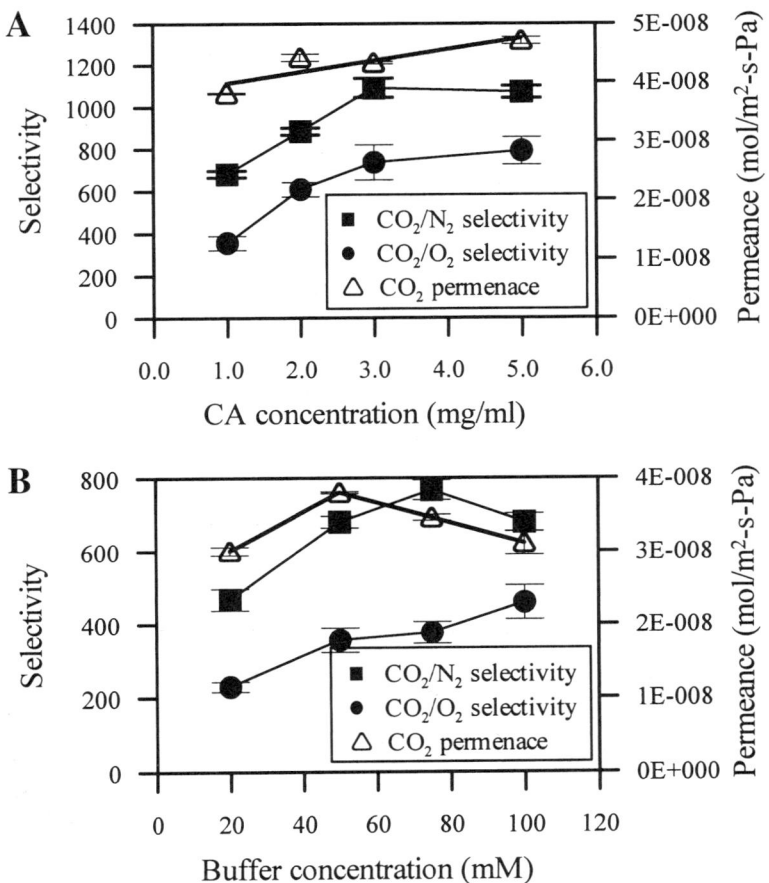

FIGURE 2. Effect of (**A**) CA concentration and (**B**) phosphate buffer concentration on separation performance. Test conditions: (**A**) buffer, 50 mM phosphate; pH = 8.0; thickness, 330 μm; CO_2, 0.1% in air; CA, variable; T = 30°C. (**B**) buffer, variable; pH = 8.0; thickness, 330 μm; CO_2, 0.1% in air; CA, 33.3 μM (1 mg/mL); T = 30°C.

permeance increased with the change from 20–50 mM but decreased as the buffer concentration was elevated to 75 and 100 mM. The permeance values observed all fell in a narrow range between 2.5×10^{-8} and 3.0×10^{-8} mol/m² sec Pa but the selectivity change was much larger, with maxima for both CO_2 versus N_2 and CO_2 versus O_2 at a buffer strength of 75 mM. This suggests that the best phosphate buffer concentration to use for a feed of 0.1% CO_2 is in the range of 50–75 mM.

pH

As shown in FIGURE 3A, pH greatly impacts both CA activity and the uncatalyzed rates of CO_2 hydration and dehydration. At pH of 6.4 the rates of the CA catalyzed

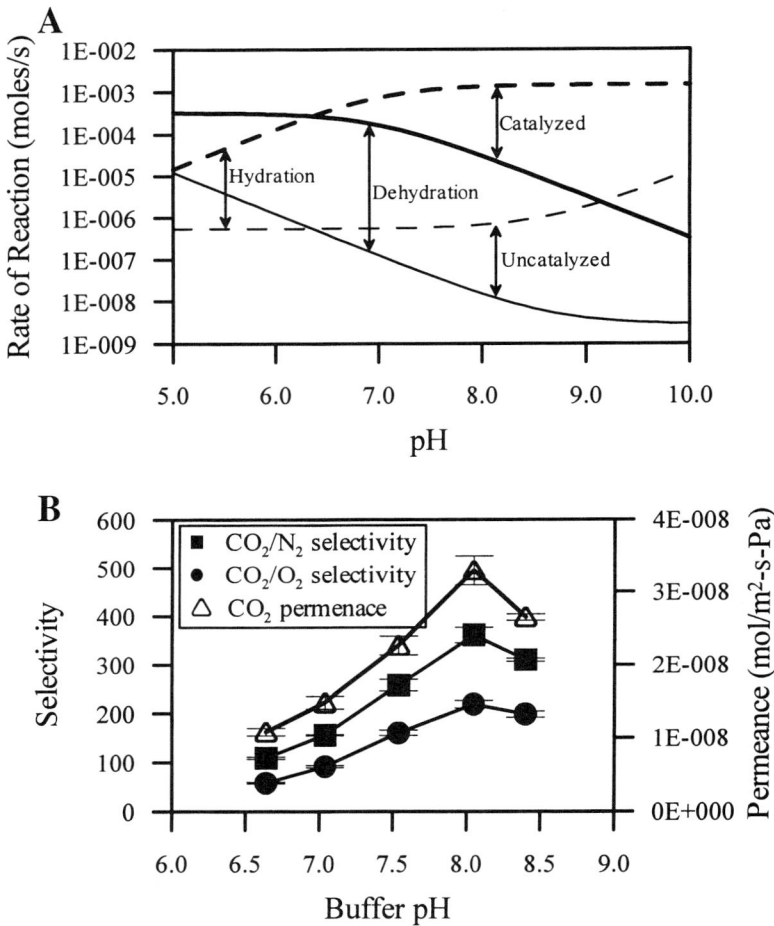

FIGURE 3. (**A**) Effect of pH on rates of CO_2 hydration and dehydration with and without CA catalysis, and (**B**) effect of pH on liquid membrane performance. Test conditions: buffer, 20 mM phosphate; pH, variable; thickness, 180 μm; CO_2, 0.1% in air; CA, 33.3 μM (1 mg/mL); $T = 30°C$.

hydration and dehydration reactions are equal.[17,27] As pH increases, the rate of CO_2 hydration by CA increases at much lower pH values than does the uncatalyzed rate. At the same time, the rate of bicarbonate dehydration increases with decreasing pH at much higher pH values than for the uncatalyzed reaction.[17,27] In our system, the best separation performance occurs at whatever pH allows for the highest balanced rate of hydration and dehydration. Although initially this might be expected to be pH 6.1, the addition of CO_2 to the phosphate buffered solution tends to decrease the pH (the exact amount depends on the balance between the pCO_2 and the buffer strength).

As shown in FIGURE 3B, we evaluated separation performance at five pH values: 6.80, 7.04, 7.5, 8.0, and 8.5. The data shown in FIGURE 3B clearly illustrate the optimum performance for the conditions studied (20 mM phosphate buffer, 0.1% CO_2 in 1 atm air) to occur at pH 8. The reasons for this are given in the discussion.

Thickness of the Liquid Phase

Most membrane separation methods force a tradeoff between permeance and selectivity on the basis of membrane thickness. Whereas selectivity increases with membrane thickness, thus providing greater separation, permeance decreases due to the increase in diffusion path length. The advantage in selectivity occurs because

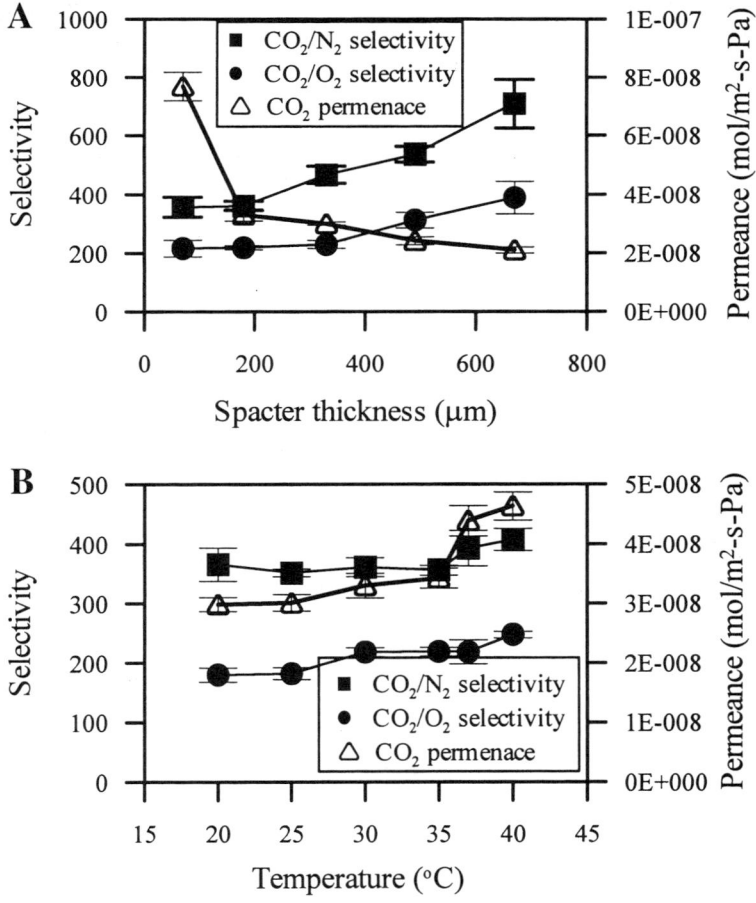

FIGURE 4. Effect of **(A)** liquid membrane thickness and **(B)** temperature on separation performance. Test conditions: **(A)** buffer, 20 mM buffer; pH = 8.0; thickness, variable; CO_2, 0.1% in air; CA, 33.3 mM (1 mg/mL); T = 30°C. **(B)** buffer, 20 mM phosphate; pH = 8.0; thickness, 180 μm; CO_2, 0.1% in air; CA, 33.3 mM (1 mg/mL); T, variable.

the rate of decrease in CO_2 permeance is lower than the rate in decrease of O_2 and N_2 permeance. Prior work tended to use liquid impregnated filters (SLM) with the membrane thickness (typically 60µm to 160µm) given by the swollen thickness of the filter.[16,24]

Unlike the SLM, in which maximum thickness is limited by physical parameters, such a capillary force, the CLM design allows us to freely vary thickness without such limitations. We examined liquid membranes of five different thicknesses—70, 180, 330, 490, and 670µm. FIGURE 4A illustrates the separation performance versus liquid phase thickness. As expected, we observed that selectivity increased with increasing thickness, whereas CO_2 permeance decreases. The permeance values of CO_2, O_2, and N_2 were all very high when the liquid membrane thickness was only 60µm (low selectivities show that O_2 and N_2 permeances were also high). Second, for a liquid membrane of 180µm thickness and above the rate at which CO_2 permeance decreases with increasing thickness is relatively constant (first order). Third, the selectivity of CO_2 versus O_2 and CO_2 versus N_2 increase relatively linearly with thickness. FIGURE 4A illustrates the sharp decrease in permeance as thickness increased from 75µm to 200µm. This decrease in permeance continued, although more slowly, as the liquid membrane thickness was increased to 670µm. In contrast to the varying rate of change seen for CO_2 permeance, the rate of change in selectivity was relatively constant. It was seen to be almost linear over the entire range of thickness studied. The slope of the selectivity versus thickness curve for $CO_2:N_2$ is far steeper than that for $CO_2:O_2$.

Temperature

Temperature is very important to this system as it impacts both CA activity and the solubility and diffusivity of all the dissolved species (O_2, N_2, CO_2, and the CO_2 equivalents HCO_3^- and CO_3^{2-}). BCAII starts to denature at about 40°C.[40] To detail the effect of temperature on membrane transport performance we examined the following temperatures: 20, 25, 30, 37, and 40°C. FIGURE 4B illustrates the effect of temperature on separation performance.

The data show a positive relationship between both selectivity and permeance with temperature. Up to 35°C selectivity and permeance changed modestly. However, further increases in temperature to 37°C and 40°C showed substantial increases in CO_2 permeance, but only moderate increases in selectivities. These results suggest that substantial increases in performance are possible with only moderate increases in temperature.

Humidity

Unlike an SLM, which is subject to catastrophic failure from liquid loss due to evaporation, when a CLM design is used there is little risk of failure. Furthermore, provided effective management of the liquid membrane chemistry (i.e., buffer and CA concentration) is provided, the CLM based system should show little or no sensitivity to humidity in the feed or sweep gases. In the experiments conducted here we tested four humidity values: 4, 27, 53, and 74%.

FIGURE 5A shows that the EBCLM performed largely independent of changes in humidity in the inlet stream over the test range of humidity, although there was a slight selectivity increase as humidity increased. As expected, the system proved to provide stable operation even with very low humidity feed gas (operation for up to 24 hours). This has been further supported in more recent studies (data not shown), where dry feed gas (0% humidity) was used for a period of eight days without loss of system performance.

CO_2 Concentration

The design requirements for this reactor are directly influenced by the feed stream CO_2 concentration (pCO_2). In a space station (e.g., International Space Station Freedom) or a transport vehicle (e.g., Space Shuttle) the CO_2 concentration should be

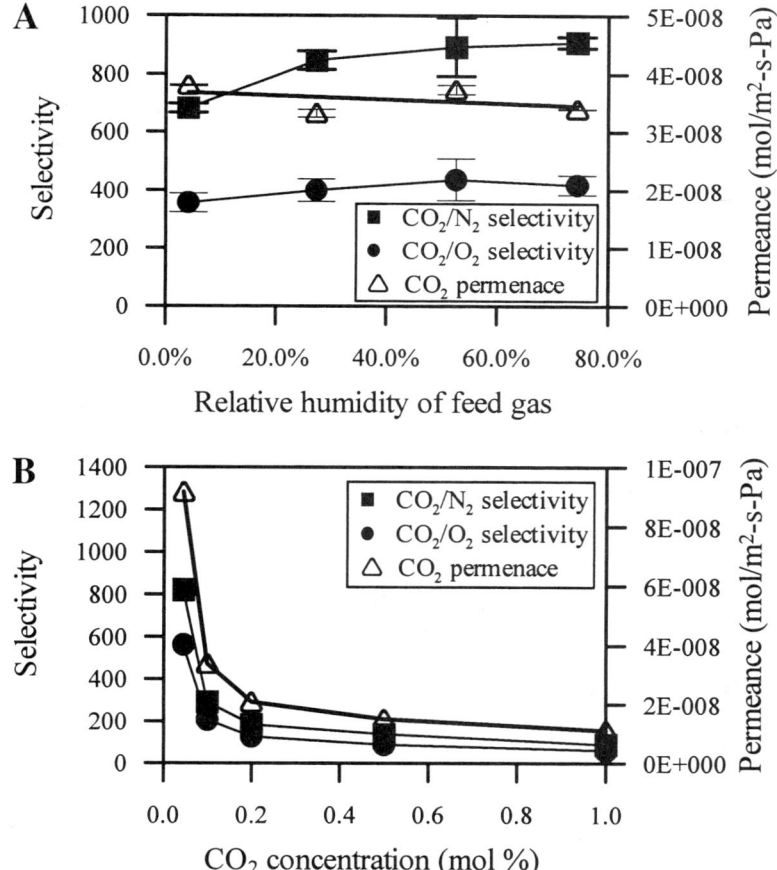

FIGURE 5. (A) Effect of feed stream humidity on separation performance, and (B) separation performance as a function of CO_2 concentration in the feed stream. Test conditions: (A) buffer, 50 mM phosphate; pH = 8.0; thickness, 330 μm; CO_2, 0.1% in air; CA, 33.3 μM (1 mg/mL); T = 30°C. (B) buffer, 20 mM phosphate; pH = 7.85; thickness, 330 μm; CO_2, variable; CA, 166.67 μM (5 mg/mL); T = 30°C.

controlled to values lower than 0.1% ($pCO_2 = 101.30$ Pa). In contrast, in an extravehicular mobility unit (EMU–spacesuit) the CO_2 concentration should be no more than 0.5% at a total pressure of 29.65 kPa ($pCO_2 = 148.24$ Pa). In these experiments we investigated CO_2 concentrations over a range from 0.04% to 1.00% ($pCO_2 =$ 40.52 to 1,013.00 Pa) in air at 1 atm.

FIGURE 5 B illustrates permeance and selectivity as a function of the CO_2 concentration in the feed. There is a rapid drop in both CO_2 permeance and selectivity as CO_2 concentration is increased from 0.04 to 0.1%. This rate of decline decreases markedly as CO_2 concentration is increased further, with only slight decreases in permeance and selectivity occurring as the CO_2 concentration is increased from 0.2 to 1.0%.

DISCUSSION

We previously performed a detailed analysis of the various parameters that can influence EBCLM performance.[28] This analysis revealed a high level of interdependence among the various parameters. The optimum value of a given parameter depends not only on the separation requirements (i.e., feed stream CO_2 concentration), but also on the value of other adjustable parameters. For example, the optimum buffer pH depends not only on the feed stream CO_2 concentration, but also on the buffer strength and perhaps even the CA concentration (and *vice versa*). Therefore, to obtain a true global optimum for even a single separation condition it is necessary to perform studies that are more extensive than those reported here. In this work we report initial efforts toward this end by separately varying several structural and operational characteristics of this complex system, such that a first level optimization might be achieved. Future work will involve more extensive studies needed to more closely approach a global optimum. In general, for a given pCO_2, selectivity is the goal most easily controlled accepting whatever tradeoff in permeance is needed.

CA Concentration

Transport of CO_2 in this facilitated reactor design includes two scenarios. One is the uncatalyzed diffusion of CO_2 directly across the membrane. The other is that CO_2 is catalytically hydrated to bicarbonate ions on contact with the CA and then the bicarbonate ions diffuse through the membrane to the permeate surface. At the permeate side, the bicarbonate ions are dehydrated and CO_2 released from the membrane. Because solubility of CO_2 in water (0.053 wt%, 0.012 M, at 25°C) is much less than that of bicarbonate (with Na^+ as a counter ion, bicarbonate solubility is 9.32 wt%, 2.1 M, at 25°C; i.e., 175 times greater),[15] the transport capacity for CO_2 is much greater in HCO_3^- form than for $CO_{2(aq)}$ even though the diffusivity for $CO_{2(aq)}$ is greater than that for HCO_3^-. Furthermore, although the hydration and dissociation of CO_2 to form bicarbonate (and the reverse dehydration) does occur naturally without catalysis by CA, the catalyzed rates (depending on pH) can easily be hundreds to thousands of times higher. Thus, the results suggest that for the conditions used (pH, CO_2 in feed, etc.) when the CA concentration was low there was less CA present than that required to carry out conversion of all the CO_2 at a maximum rate (i.e., the available CA was likely saturated). Therefore, increasing the amount of CA available increased CO_2 permeance. Increasing selectivities were observed because neither O_2 nor N_2 react with CA. Therefore, the permeance of each of these gases is not directly effected by the presence of more CA and continues to diffuse across as dissolved gases (N_2 solubility is 0.0134 STP mL/mL H_2O and is 0.0261 STP mL/mL H_2O for O_2 at 30°C). The leveling off of the increase in selectivity provides the most obvious evidence of the decreasing benefit of adding more CA. Note that when feed streams with higher CO_2 concentrations are used there will be a need for greater rates of hydration and dehydration of CO_2 and, therefore, higher CA concentrations are likely to be needed to achieve maximum selectivity and CO_2 permeance. This is also true if changes in other parameters increase the required hydration and dehydration rates.

Buffer Concentration and pH

A robust buffer is needed to moderate the pH changes that can accompany the reactions catalyzed by CA. As stated above (see RESULTS) the reactions that occur at the feed side produce hydrogen ions forcing the local environment toward decreased pH values. This decreases the activity of the CA for catalyzing CO_2 hydration. The opposite trend occurs at the sweep side of the liquid membrane, where consumption of hydrogen ions tends to force the pH to increase. This also decreases the activity of CA for catalysis of the dehydration reaction.

From the data presented in FIGURE 3A it appears the optimum pH should be close to 6.4 because that is the pH where the maximum rate of hydration and dehydration occur if we require them to be equal and to be carried out at the same pH. This is what would be expected for a perfect buffer. However, with the use of weaker buffers we know the feed side pH will be lower than the bulk pH (pH will tend toward the pH of the phosphate buffer in equilibrium with the feed stream CO_2 concentration). We also know that the sweep side pH should tend toward the value in equilibrium with the sweep stream CO_2 concentration (for the phosphate buffer this is the make-up concentration—i.e., that reported in FIG. 3B). Therefore, it appears, at the pH optimum of 8 that we observed (for this 0.1% CO_2 feed, 20 mM phosphate buffer, 33 µM CA, 180 µm-thick liquid membrane), this occurred as a result of substantial acidification of the system. Looking at FIGURE 3A, assuming: (1) the rate of hydration equals the rate of dehydration; (2) the uncatalyzed reactions were insignificant; and (3) the sweep side local pH was 8.0, it appears the feed side pH was close to 5.4 thus indicating the buffer strength was too low. This is in good agreement with the results presented in FIGURE 2B that indicated improved performance at increased buffer strengths. However, the true optimum can only be found through a combined search of the pH and the buffer concentration (and this optimum is a function of pCO_2). From our analysis it seems clear that, in the presence of a stronger buffer, the optimum pH will be lower and will approach pH 6.4 forcing this curve to shift to the left for the same CO_2 load.

CLM Thickness

Because CO_2 permeance decreases and selectivity increases with increased liquid membrane thickness there will exist some optimum thickness at which the best overall separation performance is achieved. However, this optimum point is not only a function of the feed composition and the need for removal of CO_2 from the feed, but also the requirements of the desired composition of the permeate (sweep). As is well known,[41] if the membrane is particularly thin, permeance increases dramatically. This is the case at 60µm; at the same time, the selectivity is low because of the relatively high N_2 and O_2 permeances due to bulk movement of these gases across this thin water film. To effect adequate selectivity it is necessary to have sufficient thickness to serve as a barrier resistance to the flux of unwanted species, thereby emphasizing the benefits of the chemical facilitation.

Temperature

Enzyme activity is typically a linear function of temperature with a Q_{10} of about 2. For CA, the upper temperature bound, using ordinary mammalian isozymes, is about 40°C. At the same time, water is evaporating from the membrane cooling the reactor and the higher the operating temperature the greater the evaporative cooling. We can calculate a first approximation to the surface temperature. A third contributor is that the water vapor flow may disturb any boundary gas layers. Thus, while both selectivity and permeance increase with temperature at present, due to the actions of at least three processes, the non-linear response cannot be adequately explained.

In future studies we have the option of using other isozymes, including those known to operate at higher temperature, even to 70°C.[42] By means of such isozymes it may be possible to attain greater selectivity and permeance by operating at a higher temperature.

Humidity

The stability of the CLM bioreactor design is particularly important since changes in solvent level and loss of solvent have, historically, been a principal reason underlying the failure of SLM type liquid membranes. Early efforts to reduce evaporation of water from the core solution of SLMs relied on increasing the humidity of the inlet gas stream. This approach was not satisfactory.[33] The current design, modeled after that of Sirkar and colleagues,[32,33] exhibits the level of stability needed for long-term use under field conditions.

CO_2 Concentration

The ability of an enzyme-based CLM (EBCLM), using a biological catalyst, to capture CO_2 is principally a function of the CO_2 concentration, CA concentration, buffer concentration, membrane area, and the feed and sweep gas flow rates. Proper management of these parameters can achieve as high a permeance as is required. The data in FIGURE 5B reflect the performance of this particular reactor, given the operating conditions we selected. As demonstrated in the RESULTS section the ability of this liquid membrane to maintain the desired pH at the feed and permeate sides is crucial to successful operation.

CONCLUSIONS

The CA membrane bioreactor presented here performed extremely well. The best operating conditions demonstrated here for the 0.1% feed, phosphate buffered design were pH 8.0, buffer strength 50 mM, membrane thickness of 330 µm, and temperature 30°C. When these conditions were applied together we found separation performance corresponding to a CO_2 permeance of $4.71 \times 10^{-8}\,\text{mol}\,\text{m}^{-2}\,\text{Pa}^{-1}\,\text{sec}^{-1}$ ($1.41 \times 10^{-4}\,\text{cm}^3\,\text{cm}^{-2}\,\text{cmHg}^{-1}\,\text{sec}^{-1}$), and selectivities of CO_2 versus N_2 and O_2 of 1,090:1 and 790:1, respectively.

The data show that the preferred concentrations of CA and of buffer are a function of the feed CO_2 concentration. Preferred membrane thickness is determined principally by the desired selectivity and secondarily by desired permeance. The preferred temperature is a function of CA activity and of water evaporation and cooling. The optimal buffer pH is a function of the CO_2 concentration and of the buffer strength. Work is ongoing in our group toward improving the response properties of this design and to further understand this reactor design.

ACKNOWLEDGMENTS

This work was supported by NASA Grants NAGW-9-1021 and NAGW-9-1923. We appreciate the help of Mary Ellen Opper, Marisol Toro, and Dr. Harry Janes.

REFERENCES

1. HAO, J.-H. 2000. Polymeric membrane materials for CO_2-selective separation (I). Chem. J. Internet. **2**(8).
2. HAO, J.-H. & S.C. WANG. 1998. Influence of quench medium on the structure and gas permeation properties of cellulose acetate membranes. J. Appl. Polym. Sci. **68**: 1269–1276.
3. KIM, J.H., S.Y. HA, S.Y. NAM, et al. 2001. Selective permeation of CO_2 through pore-filled polyacrylonitrile membrane with poly(ethylene glycol). J. Membr. Sci. **186**: 97–107.
4. OSADA, K., T. OHNISHI, Y. SHIN, et al. 1999. Development of inorganic membranes by sol–gel method for CO_2 separation. Greenhouse Gas Control Technology, Proc. 4th Int. Conf. 43–45.
5. MATSUYAMA, H., A. TERADA, T. NAKAGAWARA, et al. 1999. Facilitated transport of CO_2 through polyethylenimine/poly(vinyl alcohol) blend membrane. J. Membr. Sci. **163**: 221–227.
6. ASTARITA, G., D.W. SAVAGE & A. BISO. 1989. Gas Treating with Chemical Solvents. John Wiley & Sons, New York.
7. HALMANN, M.J. & M. STEINBERG. 1999 Greenhouse Gas Carbon Dioxide Mitigation. Lewis Publ., Boca Raton.
8. WARD, W.J., III & W.L. ROBB. 1967. Carbon dioxide–oxygen separation: facilitated transport of carbon dioxide across a liquid film. Science **156**: 1481–1484.
9. GUHA, A.K., S. MAJUMDAR & K.K. SIRKAR. 1990. Facilitated transport of CO_2 through an immobilized liquid membrane of aqueous diethanolamine. Ind. Eng. Chem. Res. **29**: 2093–2100.
10. TERAMOTO, M., K. NAKAI, K. OHNISHI, et al. 1996. Facilitated transport of carbon dioxide through supported liquid membranes of aqueous amine solutions. Ind. Eng. Chem. Res. **35**: 538–545.

11. KUMAZAWA, H. 2000. Absorption and desorption of CO_2 by aqueous solutions of sterically hindered 2-amino-2-methyl-1-propanol in hydrophobic microporous hollow fiber contained contactors. Chem. Eng. Commun. **182:** 163–179.
12. NAKABAYASHI, M., K. OKABE, E. FUJISAWA, et al. 1995. Carbon dioxide separation through water-swollen-gel membrane. Energy Convers. Mgms. **36:** 419–422.
13. DONALDSON, T.L. & J.A. QUINN. 1974. Kinetic constants determined from membrane transport measurements: carbonic anhydrase at high concentrations. Proc. Natl. Acad. Sci. USA **71:** 4995–4999.
14. DONALDSON, T.L. & J.A. QUINN. 1975. Carbon dioxide transport through enzymatically active synthetic membranes. J. Chem. Eng. Sci. **30:** 103–115.
15. QUINN, J.A. & T.L. DONALDSON. 1980. Facilitated diffusion of CO_2 and carbonic anhydrase activity. In Biophysics and Physiology of Carbon Dioxide. C. Bauer, G. Gros & H. Bartels, Eds.: 23–25.
16. SCHULTZ, J.S. 1980. Facilitation of carbon dioxide through layers with a spatial distribution of carbonic anhydrase. In Biophysics and Physiology of Carbon Dioxide. C. Bauer, G. Gros & H. Bartels, Eds.: 15–22.
17. SUCHDEO, S.R. & J.S. SCHULTZ. 1974. Mass transfer of CO_2 across membranes: facilitation in the presence of bicarbonate ion and the enzyme carbonic anhydrase. Biochem. Biophys. **352:** 412–440.
18. ENNS, T. 1967. Facilitation by carbonic anhydrase of carbon dioxide transport. Science **155:** 44–47.
19. SILVERMAN, D.N. & S. LINDSKOG. 1988. The catalytic mechanism of carbonic anhydrase: implications of a rate-limiting protolysis of water. Acc. Chem. Res. **21:** 30–36.
20. TU, C.K., D.N. SILVERMAN & C. FORSMAN. 1989. Role of histidine 64 in the catalytic mechanism of human carbonic anhydrase II studied with a site-specific mutant. Biochemistry **28:** 7913–7918.
21. POKER, Y. & D.W. BJORKQUIST. 1977. Stopped-flow studies of carbon dioxide hydration and bicarbonate dehydration in water and water-d2. Acid–base and metal ion catalysis. J.A.C.S. **16**(26): 5698–5707.
22. THOMAS, D. & G.B. BROUN. 1976. Artificial enzyme membranes. Meth. Enzymol. (Immobilized Enzymes) **44:** 901–929.
23. TSAO, G.T. 1972. The effect of carbonic anhydrase on carbon dioxide absorption, Chem. Engin. Sci. **27:** 1593–1600.
24. TRACHTENBERG, M.C., C.K. TU, R.A. LANDERS, et al. 1999. Carbon dioxide transport by proteic and facilitated transport membranes. Life Supp. Biosphere Sci. **6:** 293–302.
25. POCKER, Y. & C.H. MIAO. 1987. Molecular basis of ionic strength effects: interaction of enzyme and sulfate ion in CO_2 hydration and HCO_3^- dehydration reactions catalyzed by carbonic anhydrase II. Biochemistry **26:** 8481–8486.
26. ALPER, E. & W.D. DECKWER. 1980. Kinetics of absorption of CO_2 into buffer solutions containing carbonic anhydrase. J. Chem. Eng. Sci. **35:** 549–557.
27. BAIRD, T.T., JR., A. WAHEED, T. OKUYAMA, et al. 1997. Catalysis and inhibition of human carbonic anhydrase IV. Biochemistry **36:** 2669–2678.
28. TRACHTENBERG, M.C., R.M. COWAN, J.J. GE & M.L. MCGREGOR. 2000. Separation of CO_2 from air by a carbonic anhydrase(II) bioreactor, Intl. Conf. Life Support & Biospheric Sci., Aug. 6–9, 2000.
29. SUDIPTO, M., A.K. GUHA & K.K. SIRKAR. 1988. New liquid membrane technique for gas separation. AIChE J. **34:** 1135–1145.
30. GE, J.-J., Y.-J. QIN, R.M. COWAN & M.C. TRACHTENBERG. 2002. High efficiency capture of CO_2 from mixed gas streams. Presentation and Abstracts of Papers, 223rd ACS National Meeting, Orlando, FL, USA, April 7–11.
31. MAJUMDAR, S., A.K. GUHA, Y.T. LEE & K.K. SIRKAR. 1989. A two-dimensional analysis of membrane thickness in a hollow-fiber-contained liquid membrane permeator. J. Membr. Sci. **43:** 259–276.
32. CHEN, H., A.S. KOVVALI, S. MAJUMDAR & K.K. SIRKAR. 1999. Selective CO_2 separation from CO_2–N_2 mixtures by immobilized carbonate-glycerol membrane. Ind. Eng. Chem. Res. **38:** 3489–3498.
33. SENGUPTA, A., R. BASU & K.K. SIRKAR. 1988. Separation of solutes from aqueous solutions by contained liquid membranes. AIChE J. **34:** 1698–1708.

34. GE, J.-J., M.C. TRACHTENBERG, M.L. MCGREGOR & R.M. COWAN. 2003. Enzyme based CO_2 capture for advanced life support. Life Supp. Biosphere Sci. In press.
35. KEMPERMAN, A.J.B., D. BARGEMAN, TH. VAN DEN BOOMGAARD & H. STRATHMANN. 1996. Stability of supported liquid membranes: state of the art. Sep. Sci. Technol. **31:** 2733–2762.
36. KEMPERMAN, A.J.B., H.H.M. ROLEVINK, TH. VAN DEN BOOMGAARD & H. STRATHMANN. 1997. Hollow-fiber-supported liquid membranes with improved stability for nitrate removal. Sep. Purif. Technol. **12:** 119–134.
37. KEMPERMAN, A.J.B., B. DAMINK, TH. VAN DEN BOOMGAARD & H. STRATHMANN. 1997. Stabilization of supported liquid membranes by gelation with PVC. J. Appl. Polym. Sci. **65:** 1205–1216.
38. KEMPERMAN, A.J.B., H.H.M. ROLEVINK, D. BARGEMAN, *et al.* 1998. Stabilization of supported liquid membranes by interfacial polymerization top layers. J. Membr. Sci. **138:** 43–55.
39. YANG, X.J., A.G. FANE, J. BI & H.J. GRIESSER. 2000. Stabilization of supported liquid membranes by plasma polymerization surface coating. J. Membr. Sci. **168:** 29–37.
40. PERSSON, M., U. CARLSSON & N.C.H. BERGENHEM. 1996. GroEL reversibly binds to, and causes rapid inactivitation of human carbonic anhydrase II at high temperature. Biochim. Biophys. Acta **1298:** 191–198.
41. OTTO, N.C. & J.A. QUINN. 1971. The facilitated transport of carbon dioxide through bicarbonate solutions. Chem. Eng. Sci. **26:** 949–961.
42. ALBER, B.E. & J.G. FERRY. 1996. Characterization of heterologously produced carbonic anhydrase from *Methanosarcina thermophila*. J. Bacteriol. **178:** 3270–3274.

Affinity Membranes as a Tool for Life Science Applications

HEIKE BORCHERDING,[a] HANS-GEORG HICKE,[b] DIERK JORCKE,[a] AND MATHIAS ULBRICHT[a]

[a]*ELIPSA GmbH, Berlin, Germany*

[b]*GKSS Forschungszentrum Geesthacht GmbH, Institut für Chemie, Teltow, Germany*

> ABSTRACT: As a matrix for affinity membrane technology we chose a flat-sheet microfiltration membrane based on polypropylene. Using photopolymerization to graft epoxy groups onto the pore surface, we worked with glycidylmethacrylate as a monomer. We developed optimized, efficient, and mild UV irradiation conditions for the two-step photografting process practically preserving the given pore structure of the base membrane. A grafting degree of up to $1.2\,mg/cm^2$ per surface area of the membrane was obtained. The polypropylene membrane surface became significantly more hydrophilic. Introduction of epoxy groups allowed a stable covalent immobilization of the protein streptavidin serving as receptor for affinity ligand binding. A relatively high streptavidin immobilization capacity of about $65\,\mu g/cm^2$ per surface area of the membrane was obtained. Apparently, only about two of the binding sites of the immobilized streptavidin were available for biotin recognition. We also found that the oriented immobilization of biotinylated alkaline phosphatase onto the surface via a streptavidin bridge increased the specific enzymatic activity about sixfold compared with random immobilization of this enzyme.
>
> KEYWORDS: membrane; polypropylene; photografting; affinity; streptavidin; biotin

INTRODUCTION

Using proteins for affinity or sensor technology and bioanalytical assays, a large number of immobilization methods onto various polymeric surfaces were reported.[1,2] Immobilization methods range broadly from passive adsorption to more advanced covalent attachment procedures using chemically activated polymers, covalent binding, or crosslinking. As well as porous particles, plane surfaces (e.g., wells of microtiter plates) are favored for immobilization and coating of proteins. However, flat-sheet or hollow-fiber membranes are often used in life science applications and are, therefore, of particular interest for protein immobilization.

The choice of a suitable membrane matrix is a key problem. Membranes that are used nowadays for protein immobilization[3,4] are based on modified cellulose or modified polysulfone, polyvinylidene fluoride, and polyamide (nylon). For further improvements there are requirements that extend this list of membranes. Polypropy-

---

Address for correspondence: Dr. Heike Borcherding, ELIPSA GmbH, Köpenicker Str. 325, D-12555 Berlin, Germany. Voice: 0049-30-65762915; fax: 0049-30-65762941.
borcherding@elipsa.de

lene (PP) membranes are one of the favored candidates. To date, hydrophobic PP supports were mainly applied in membrane technology as contactors, in chemistry for SPOT synthesis[5] and for molecularly imprinted polymer (MIP) technology.[6] However, for the microfiltration of biological solutions PP membranes were mainly used after additional hydrophilization (see, e.g., Ref. 7). Several techniques, such as plasma-, radiation-, and photoinduced graft polymerization, are known to functionalize polymeric supports. We made use of photoinduced grafting, which is a particularly useful technique for the functionalization of PP membranes due to facile control of the chemistry.[8]

Affinity technology enables the recognition of a molecule on the basis of its individual chemical structure or biological function. The streptavidin–biotin system is one of the popular toolboxes for bioanalytical and biomedical research. Streptavidin is well-characterized[9] as a tetrameric protein (60kDa) with extraordinarily high biological affinity for d-biotin ($K_a = 2.5 \times 10^{13} M^{-1}$) and, therefore, for biotinylated ligands. The bond formation is rapid and unaffected by wide extremes of pH, temperature, organic solvents, detergents, or denaturating agents. In homogeneous solution, streptavidin has four binding sites for d-biotin with cooperative binding behavior.

We developed a flat-sheet affinity membrane to capture d-biotin and biotinylated molecules. The membrane can be used in life science (e.g., molecular biology and biomedicine) and can be fitted into filtration housings or filter plates. The major advantages of using flat-sheet membranes as a support for immobilized receptors are the high receptor capacity, the advanced hydrodynamic properties and high flow rates, the low consumption of reagents, and especially the small elution volume.

The streptavidin membrane was characterized with respect to the free binding sites of streptavidin using fluorescent biotin. We found that the specific activity of biotin-anchored alkaline phosphatase is increased about sixfold compared to the randomly immobilized enzyme. Low unspecific protein binding and high biotin binding capacities make this membrane well suitable for capturing biotinylated ligands, including oligonucleotides and hybridization probes for PCR-ELISA, antibodies, drugs, or biotin-tagged recombinant proteins.

EXPERIMENTAL

Materials

The flat-sheet microfiltration membrane, Accurel PP 2E HF (R/P), based on polypropylene (PP) is commercially available from Membrana GmbH (Germany). The characteristics of the membrane (Charge 96146, No. 6001) were specified by the manufacturer as a mean thickness of $160 \pm 20 \mu m$ and a mean pore size $d_p \approx 0.4 \mu m$.

Glycidylmethacrylate (GMA) from Aldrich and benzophenone (BP, >99% purity) from Merck were used as received. HEPES and CHES buffer substance were purchased from Biomol. Purified streptavidin (*Streptomyces avidinii*) and biocytin Alexa 594 (as biotinylated fluorescent dye) were received from Molecular Probes. Surfact Amps 20 (purified non-ionic detergent Tween 20) and immunopure biotinylated calf intestinal alkaline phosphatase (rBAP) (Lot. AC39286) were purchased from Pierce. Alkaline phosphatase (AP) from bovine intestinal mucosa (E.C.

3.1.3.1.), the substrate p-nitrophenylphosphate (p-NPP), ovalbumin, and the Ponceau S solution (P7170) were from Sigma. Deionized water with 18.2 MΩ and less than 2 ppb TOC content was used for all experiments.

Methods

Photoinitiated Graft Polymerization

Polypropylene membranes were functionalized to obtain PPgGMA membranes using a two-step photografting process already described in detail elsewhere.[8] Here a F300S UV illumination system from Fusion UV Systems (Alton, UK) (glass filter with $\lambda > 300$ nm, 80 mW/cm^2) was used. In brief, membranes (area 50 cm^2) were soaked in 100 mM BP in methanol for 18 h. Membranes still wet with the BP solution, were immersed in 30 mL of the monomer mixture, containing 8.2 mg/mL GMA, 0.07 mM BP, 1 mM NaIO$_4$, and 10% (v/v) methanol in water. One side of the membrane was exposed to UV light for 2–14 min. Removed after five minutes, the membranes were washed with methanol and extracted with acetone for four hours in a soxhlet extractor. The degree of grafted monomer (DG value) was determined gravimetrically according to[8] $DG = \{(m_{gr} - m_0)/m_0\}m_{spec}$, where m_0 is the initial membrane sample weight, m_{gr} is the weight after modification, and m_{spec} is the specific weight per surface area of the membrane (see TABLE 1).

TABLE 1. Characteristic data of the initial PP and a selected PPgGMA membrane

Membrane	Unmodified PP	Selected PPgGMA
Degree of grafting per surface area of membrane, DG (mg/cm^2)	—	1.11 ± 0.08
BP coating relative to surface area of membrane (μmol/cm^2)	0.85 ± 0.14	—
Dry membrane thickness, d (μm)	165.0 ± 4	174.0 ± 5
Weight per surface area of membrane, m (mg/cm^2)	3.41 ± 0.04	4.52 ± 0.04
Weight per membrane volume, m (g/cm^3)	0.209	0.260
Advancing contact angle, in captive-bubble mode, CA (°)	166.4 ± 3	49.7 ± 3
BET surface area (m^2/g)	23.1 ± 1	15.6 ± 1
Porosity in water-swollen state (vol%)	76.3 ± 0.2	74.7 ± 0.8
Pure water permeability, J_p (L/h m^2 bar)	16,130.0 ± 730	15,440.0 ± 45
Streptavidin capacity (only buffer washer membrane) per surface area of membrane (μg/cm^2)	113.9	58.2
Streptavidin covalently immobilized per surface area of membrane (μg/cm^2)	12.3 ± 9[a]	65.6 ± 8

[a]Remaining residue of non-covalently adsorbed fraction.

Characterization of the Functionalized Membrane

FT-IR spectra in transmission and ATR mode were obtained with a Nicolet-Magna-IR 530 FT-IR spectrometer. For *contact angle* measurements the goniometer equipment from Zeiss was used. In captive-bubble mode, samples were pretreated with ethanol followed by a careful exchange to water. Air bubbles were applied to the surface of the water filled membrane. In sessile-drop mode, water droplets were placed onto the dry membrane. *BET surface area* measurements based on a nitrogen isotherm were performed using an ASAP 2000 from Micromeritics. Membranes were predried at 50°C. *Membrane porosity* was retrieved from gravimetric data for dry and water swollen state (after methanol pretreatment). *Water permeability* was determined using stirred Berghof cells (active filtration area $37.39\,cm^2$). The membranes were pretreated with methanol. After pressurizing the membranes with water to 0.1 bar for five minutes, a water flux J_w was measured under constant pressure conditions at $p = 0.01$ bar.

Protein Immobilization

PPgGMA and initial PP membranes were conditioned with methanol. Thereafter, water and incubation buffer (100 mM HEPES, pH 7.0) were filtered through the membrane (surface area of the membrane, $1.13\,cm^2$), which had been placed in a syringe filter housing.

To prepare the streptavidin membrane, 0.6 mL of 0.2 mg/mL streptavidin in incubation buffer was filtered through the membrane by using a syringe pump at a continuous filtration rate of 50 mL/h. Membranes were then incubated with 0.4 mL fresh protein solution for 18 h. After rinsing with incubation buffer, the washing solution (incubation buffer supplemented with 0.1%w/w, Surfact Amps 20, 150 mM NaCl) was applied to remove non-covalently bound streptavidin fraction. Membranes were shaken in or perfused three times with 1 mL washing solution for at least one hour. Finally, membranes were adapted to the buffer used for application (e.g., 100 mM HEPES, pH 7.0).

Alkaline phosphatase (AP) was coupled to PPgGMA membrane (DG = 0.5 mg/cm^2, surface area $1.13\,cm^2$) using 14 µg AP in incubation buffer. After 18 h incubation, the membrane was treated with washing solution (see above) and equilibrated in 100 mM CHES buffer (pH 9, 1 mM MgCl$_2$).

Quantitative Protein Assay

Ponceau S reversible staining method was used to quantify the amount of protein (adsorptive or covalently) bound to the membrane. Manufacturers protocol was adapted to our membranes. In brief, membranes (surface area of the membrane $1.13\,cm^2$) were immersed in the dye solution and then thoroughly washed with water and acetic acid. Destaining was performed by agitating the membranes in 0.84 mL 0.1 M NaOH for four hours. After removing the membranes, the dye solution was neutralized with 14 µL 6 M HCl and measured at $\lambda = 515$ nm against the reference (blank membranes with very low background).

Protein amounts were calculated based on a calibration that had been performed by applying known quantities of protein (20 to 100 µg) completely adsorbing onto unmodified PP membranes.

Receptor–Ligand System

The streptavidin membrane was prepared as mentioned above. The membrane was then incubated for three hours with 0.8 mL of biocytin Alexa 594 (10 μM in 100 mM HEPES, pH 7.0). The bound biotinylated dye was estimated as the difference in fluorescence intensity of the supernatant before and after applying the biotin solution to the membrane. Standards were obtained from biocytin Alexa 594 fluorescence (EX 594 nm and EM 606 nm).

The streptavidin membrane was also incubated with a mixture of 0.2 mg/mL biotinylated alkaline phosphatase (rBAP) and 2 mg/mL ovalbumin for 18 h in in 100 mM HEPES, pH 7.0. After washing the membrane (see above) the cumulative protein content was determined. The exclusion of ovalbumin binding was electrophoretically determined using standard protocols for SDS-PAGE.

For measurements of enzyme activity, 10 μg rBAP in binding buffer (100 mM CHES buffer, pH 9, 1 mM $MgCl_2$) was incubated with the streptavidin membrane (from PPgGMA with $DG = 0.5$ mg/cm^2). The receptor to ligand ratio was set to 10:1. After binding, the membrane was treated with washing solution (binding buffer, supplemented with 0.1%w/w Surfact Amps 20 and 150 mM NaCl). The activity assay was performed with 3 μmol p-NPP at 25°C in 2 mL 100 mM CHES buffer (pH 9, supplemented with 1 mM $MgCl_2$). The formation of *p*-nitrophenole was measured as a function of time at $\lambda = 402$ nm. Differences in the extinction values were used to calculate the enzymatic activity (1 U = 1 μmol/min).

The specific enzymatic activities of the enzymes used, rBAP and AP, were the same, with about 420 U/mg per soluble protein.

RESULTS AND DISCUSSION

Functionalization of the Matrix Microfiltration Membrane

Photografting of polymers onto various base membranes is extensively described elsewhere.[8,10,11] It was found that the grafting reaction is strongly promoted by BP coated onto the pore surface. To functionalize PP membranes with respect to the capability of covalent attachment of biomolecules, we chose GMA to place reactive epoxy groups onto the surface. Due to the limited solubility of GMA in water, the photografting process was performed in an organic–aqueous mixture (e.g., methanol–water). Whereas membranes (e.g., from nylon) are often sensitive to solvent dependent swelling,[8] PP is more rigid.

Only short UV reaction times, up to 14 min, were used for fine adjustment of the degree of graft polymerization. The DG value linearly increased with irradiation time up to 1.2 mg/cm^2 per surface area of the membrane with good reproducibility (variation less than 10%). Homopolymer precipitation of poly-GMA in the aqueous reaction mixture occurred at longer reaction times, but poly-GMA was exhaustively washed out of the pore structure by acetone.

With FT-IR spectroscopy, the structure of the modified membrane was unambiguously identified as PPgGMA. The characteristic epoxy v_{C-O} peak at 908 cm^{-1} and the ester $v_{C=O}$ peak at 1,732 cm^{-1} could be detected (spectra not shown). From FTIR-ATR spectroscopic data it appears that both sides of the membrane were grafted, but as expected, the non UV-exposed side grafted to a lower extent. This gradient distribution of graft epoxy polymer over the membrane thickness was supported by data from energy dispersive X-ray spectroscopy (not shown), similar to the data of Ulbricht[8] for polyacrylic acid grafted onto a PP membrane.

The initial PP membrane is characterized by a high advancing contact angle of about 160° considering the data from both modes of the contact angle measurements (166.4 ± 2.8° captive-bubble and 154.6 ± 2.3° sessile-drop mode). The receding contact angle was 65.6 ± 2.3° in the captive-bubble mode and 113.5 ± 3.6° in the sessile-drop mode. The discrepancy of the receding contact angle between the two techniques gives an indication of the high roughness of the facial surface of the membrane,[12] which might also explain the high advancing contact angle. The advancing contact angle decreased drastically for the UV-exposed side of the membrane, but as expected, the other side also became more hydrophilic (see FIGURE 1). Receding contact angles for modified membranes are lower than for the unmodified PP and remained constant at about 25°. Reflecting the dependence of the advancing and receding contact angles on the DG values, it can be assumed[13] that both sides of the membrane are covered with reactive epoxy groups.

Measurements of BET revealed a considerable loss in the specific membrane area with higher DG value (compare TABLE 1). This finding could be interpreted by a pore size reduction and by a decrease in roughness of the pore surface. However, the porosity in water-swollen state is only slightly diminished in value (TABLE 1), indicating that the grafted layer has only little effect on pore size reduction. SEM investigations (not shown) revealed no significant change in the membrane pore structure. Additionally, water permeability of PPgGMA membranes with moderate DG values (up to 1.2 mg/cm^2) was decreased by only 10% to about 15,000 L/h m^2 bar (compare TABLE 1). However, interpreting permeability measurements is difficult considering the opposite effects of pore size reduction on the one hand and the introduction of a hydrophilic pore surface on the other hand.

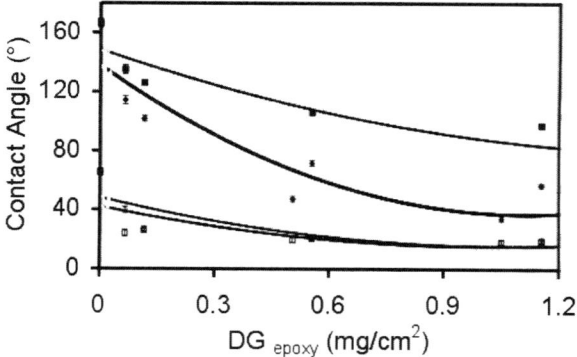

FIGURE 1. Contact angles in captive-bubble mode as a measure of change in membrane surface hydrophobicity after photografting with the GMA monomer. Advancing contact angle of the membrane surface: ■ back side, ◆ UV-exposed side; receding contact angle: □ back side, ◇ UV-exposed side.

In conclusion, we propose that moderately epoxy-functionalized membranes are suitable for the application as affinity filtration membranes in syringe filter, spin-columns, and filter plates.

Streptavidin Immobilization

The epoxy functionalization of the PP membrane was performed with the aim of enabling covalent immobilization of streptavidin. Due to their hydrophobic nature, the initial PP membranes showed high protein adsorption out of a low concentration streptavidin solution with about 110 μg/cm^2 per surface area of the membrane. Based on the high specific surface area of the membranes used (TABLE 1) and assuming a spherical protein shape (8 nm diameter) a loading capacity of about 130 μg/cm^2 per surface area of the membrane was estimated. We attempted to remove surface adsorbed streptavidin by simple washing steps. However, washing with Tween 20 (see METHODS) was still far from optimal desorption conditions (see TABLE 1). Nevertheless, about 90% of the protein initially adsorbed onto PP was removed. The washing step is an important tool permitting one to differentiate between simple attached and covalently bound protein.

Streptavidin coupling to PPgGMA membranes with high DG values seems to be of exclusively covalent nature since no difference in protein capacity was determined between buffer- and detergent-washed samples (see FIGURE 2). At low degrees of epoxy-grafting, a supplementary amount of adsorbed protein was found, which could be removed by detergent. Considerably high streptavidin immobilization capacities were obtained even for low DG values (FIG. 2).

For PPgGMA membranes with DG values of 1.0–1.2 mg/cm^2 the streptavidin immobilization capacity was about 65 μg/cm^2 per surface area of the membrane (see TABLE 1 and FIG. 2). As mentioned above, taking into account the specific membrane surface area (TABLE 1), immobilized streptavidin (about 8 nm sphere diameter) covered the available pore surface of the membrane to about 60%. To drastically

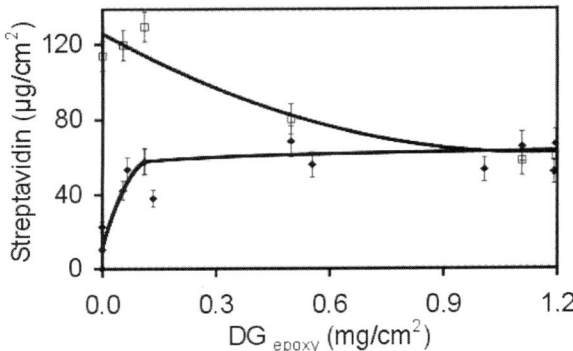

FIGURE 2. Bound streptavidin amounts on PP and PPgGMA membranes depending on the degree of grafted epoxy polymer: □ only buffer washed, that is, with a residue of physically adsorbed fraction; ◆ detergent treated, that is, immobilized protein.

increase the receptor capacity of the membrane a looser three-dimensional architecture of the grafted polymer chains[11] needs to be realized on the pore surface.

Streptavidin–Biotin System

Our streptavidin membrane could bind up to $3\,nmol/cm^2$ biotin per surface area of the membrane, which corresponds to about $700\,pmol/mg$ per membrane weight. The binding capacity is in the same range as described for streptavidin coupled particles (Merck), but is more than 12-fold higher than for this protein coupled onto a plane solid surface (e.g., well of a microtiter plate[14]). Biotin binding was measured using the biotin labeled photostable fluorescent dye Alexa 594. For most experiments using grafted membranes in the optimal range of DG values we found a molar ratio of $1:2.5$ for streptavidin to biotin (see FIGURE 3). Similar results were obtained using biotin-fluorescein (data not shown). We conclude that the multipoint coupling of streptavidin through $-NH_2$ group residues onto the epoxy surface blocks at least one of the four possible binding sites for biotin. The assumption, that a site directed streptavidin immobilization might improve the accessibility for biotin ligand and, therefore, the binding efficiency, needs to be proved by further experiments.

Using the streptavidin membrane we linked biotin-conjugated alkaline phosphatase (rBAP). The amount of bound rBAP was in the same range (about $2.5\,nmol/cm^2$ per surface area of the membrane) as measured for the dye conjugate. Out of a mixture with ovalbumin only the binding of rBAP took place (not shown). Thus, our streptavidin membrane showed a high specificity towards biotinylated molecules.

Under the linking condition, as stated in EXPERIMENTAL specific enzymatic activity of bound rBAP was found to be about $18\,U/mg$. As a control, non-labeled alkaline phosphatase (AP) was immobilized onto a PPgGMA membrane. Immobilized AP showed a specific enzymatic activity of about $3\,U/mg$. As a result, we determined a sixfold increase of the specific activity in favor of the directed enzyme coupling. Vishwanath and coworkers[15] already noted, for an alkaline phosphatase with an

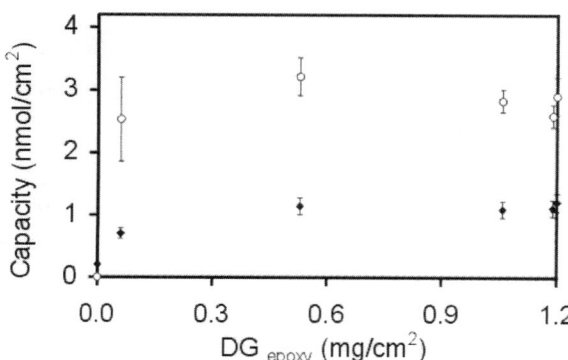

FIGURE 3. Biotin binding to immobilized streptavidin depending on the degree of grafted epoxy polymer: ○ biotin-Alexa 594 binding capacity and, for comparison, ◆ streptavidin immobilization capacity.

octapeptide tag (FLAG tag), that the ordered immobilization to its receptor on a macroporous membrane led to an increase in enzyme activity (minimum 10%) as compared with random immobilization.

CONCLUSIONS

With photoinitiated graft polymerization we introduced epoxy groups onto the surface of a membrane based on polypropylene. The grafting process was performed reproducibly under mild UV irradiation conditions. Using the monomer glycidylmethacrylate we worked in aqueous solution containing 10% methanol without considerable swelling of the matrix. The degree of grafting was time-controlled and was found to be optimal in the range $0.5–1.2\,mg/cm^2$ per surface area of the membrane. The morphology of the grafted membrane remained practically constant. The high water permeability of about $15,000\,L/h\,m^2\,bar$ favored this membrane, especially for fitting into filter housings applicable in life science. The introduced epoxy groups were used to couple streptavidin. A receptor capacity of about $65\,\mu g/cm^2$ per surface area of the membrane may be sufficient for various applications. About two biotin binding sites of the immobilized streptavidin are available in the receptor–ligand system.

ACKNOWLEDGMENTS

This work was kindly supported by the German "Bundesministerium für Bildung und Forschung" (BMBF 01SF9965).

REFERENCES

1. HERMANSON, G.T. 1992. Immobilized Affinity Ligand Techniques. Academic Press, Inc., London.
2. TISCHER, W. & F. WEDEKIND. 1999. Immobilized enzymes: methods and applications. Biocatal. Topics Curr. Chem. **200:** 95–126.
3. CHARCOSSET, C. 1998. Purification of proteins by membrane chromatography. J. Chem. Technol. Biotechnol. **71:** 95–110.
4. MORAIS, S., A. MAQUIEIRA & R. PUCHADES. 1999. Selection and characterisation of membranes by means of an immunofiltration assay. Application to the rapid and sensitive determination of the insecticide carbaryl. J. Immunol. Meth. **224:** 101–109.
5. WENSCHUH, H., et al. 2000. Coherent membrane supports for parallel microsynthesis and screening of bioactive peptides. Biopolymers **55:** 188–206.
6. PILETSKY, S.A., et al. 2000. Surface functionalization of porous polypropylene membranes with molecularly imprinted polymers by photograft copolymerization in water. Macromolecules **33:** 3092–3098.
7. MA, H.M., C.N. BOWMAN & R.H. DAVIS. 2000. Membrane fouling reduction by backpulsing and surface modification. J. Membr. Sci. **173:** 191–200.
8. ULBRICHT, M. 1996. Photograft-polymer modified microporous membranes with environment-sensitive permeabilities. React. Funct. Polym. **31:** 165–177.
9. WILCHEK, M. & E.A. BAYER. 1999. Foreword and introduction to the book (strept) avidin–biotin system. Biomol. Eng. **16:** 1–4.

10. ULBRICHT, M., et al. 1996. Photo-induced graft polymerization surface modifications for the preparation of hydrophilic and low-protein-adsorbing ultrafiltration membranes. J. Membr. Sci. **115:** 31–47.
11. ULBRICHT, M. & M. RIEDEL. 1998. Ultrafiltration membrane surfaces with grafted polymer "tentacles": preparation, characterization and application for covalent protein binding. Biomaterials **19:** 1229–1237.
12. DRELICH, J., J.D. MILLER & R.J. GOOD. 1996. The effect of drop (bubble) size on advancing and receding contact angles for heterogeneous and rough solid surfaces as observed with sessile-drop and captive-bubble techniques. J. Coll. Interface Sci. **179:** 37–50.
13. HOLLÄNDER, A., J. BEHNISCH & H. ZIMMERMANN. 1994. Chemical derivatization as a mean to improve contact angle goniometry of chemically heterogenous surfaces. J. Polym. Sci. Part A. **32:** 699–709.
14. KOCH, T., et al. 2000. Photochemical immobilization of antraquinone conjugated oligonucleotides and PCR amplicons on solid surfaces. Bioconjug. Chem. **11:** 474–483.
15. VISHWANATH, S.K., et al. 1997. Kinetics studies of site-specifically and randomly immobilized alkaline phosphatase on functionalized membranes. J. Chem. Technol. Biotechnol. **68:** 294–302.

Enzyme Transmission during Crossflow Filtration of Yeast Suspensions Using Gas/Liquid Two-Phase Flows

MURIEL MERCIER-BONIN AND CHRISTIAN FONADE

Centre de Bioingénierie Gilbert Durand, UMR INSA-CNRS 5504, UMR INSA-INRA 792, Toulouse, France

ABSTRACT: The optimal conditions for recovery of an enzyme were determined using gas/liquid two-phase flows. When filtering the enzyme-only solution under single-phase flow conditions, severe fouling occurred. This fouling was manifest as a decline in flux to less than 2% of the initial water flux and a decline in protein concentration in the permeate to 30% of its initial value, during a five-hour filtration period. When yeast cells were added under the same experimental conditions, enzyme transmission was maintained at 100% for the five-hour period and the enzyme mass flux was twofold higher. During gas-sparged microfiltration of the enzyme/yeast mixture in a permeate-recycling mode at the same liquid flow rate, gas/liquid slug flow strongly decreased the transmission of the enzyme (70% decrease), even though the permeate flux was improved (140% improvement). As a result, the mass flux of the enzyme was significantly reduced. However, with a bubble flow pattern, the permeate flux was 1.5 times higher and the transmission was maintained at a high level. The enzyme mass flux was then 25% higher when compared to single-phase flow filtration conditions. During diafiltration experiments with a bubble flow pattern, a 13% higher enzyme recovery was achieved.

KEYWORDS: two-phase flow; enzyme recovery; yeast suspension; flux; transmission

INTRODUCTION

Microfiltration is a membrane process that uses size-based separation to achieve downstream purification as well as separation. Although it is already used in the biotechnology and pharmaceutical industries to recover valuable soluble proteins from fermentation broths and from cell homogenate suspensions, it is still limited by membrane fouling, which results in a decrease in both the permeate flux and protein transmission. One approach to reducing the deposition of suspended cells or particles on the membrane is to use improved fluid hydrodynamics. In this respect, several studies have pointed out the value of gas/liquid two-phase flows to enhance the flux in ultrafiltration and microfiltration for various applications (drinking water production,[1] biological treatment,[2–4] and macromolecule separation[5–8]). In all these

Address for correspondence: Muriel Mercier-Bonin, Centre de Bioingénierie Gilbert Durand, UMR INSA-CNRS 5504, UMR INSA-INRA 792, 135, Avenue de Rangueil, 31077 Toulouse cedex 4, France. Voice: +33 (0) 5 61 55 94 19; fax: +33 (0) 5 61 55 94 00.
mercierb@insa-tlse.fr

studies, the permeate flux enhancements were always significant, ranging from 20 to 320%, according to the application, the membrane type and the two-phase flow pattern. With slug flow conditions inside tubular and hollow fiber membranes, the flux improvement was attributed to variations in the wall shear stress, which acted on the erosion of the cake[9] and bubble-induced secondary flows promoting local mixing in the bubble wake.[10,11] However, less work has been devoted to the analysis of the variation in selectivity due to the gas/liquid two-phase flow inside the membrane, especially when a microfiltration membrane is used for the recovery of a target molecule, such as an enzyme. The goal of this work was, therefore, to determine whether gas/liquid two-phase flows could achieve greater enzyme transport through a microfiltration membrane than that with standard crossflow operation. The possibility of yeast cells aiding in enzyme transmission and recovery through the formation of the dynamic cake layer of yeast was particularly explored. Changes in protein transmission and permeate flux with time for microfiltration of enzyme solutions alone were first studied using a ceramic tubular membrane. These changes were compared with those observed for microfiltration of enzyme/yeast mixtures under various experimental conditions (gas sparging/no gas sparging, liquid, and gas flow rates) and operating modes (permeate recycling, diafiltration). Optimal conditions for enzyme recovery were then determined.

MATERIALS AND METHODS

Feed Suspensions

Invertase/yeast mixtures and invertase-only solutions were prepared by mixing commercially available baker's yeast *Saccharomyces cerevisiae* (Lesaffre, Marcq-en-Baroeul, France) and invertase (β-D-fructofuranosidase EC 3.2.1.26) as a purified industrial powder preparation (Maxinvert 200000 MG), generously provided by DSM Food Sp. (Seclin, France). Invertase is a large enzyme (molecular weight 270,000) consisting of two subunits and containing up to 50% of its mass as carbohydrate in the form of nine high-mannose oligosaccharide chains.[12] It was used without further purification. In all experiments, the invertase was dissolved at 0.5 g/L in a 0.1 M sodium acetate buffer pH = 4.5, freshly prepared for each experiment in distilled water. The yeast suspension was prepared at a concentration of 20 g/L. The determination of the dry cell weight (filtration on 0.45 µm pore size filter and drying to constant weight under partial vacuum 24 h, 200 mmHg, 60°C) showed that only 72% of the commercially available yeast was present in the form of whole cells. The remaining 28% was cell debris and soluble compounds like proteins.

Filtration Module

The tubular membrane used in this work was manufactured by SCT (Bazet, France). It was a ceramic monotubular membrane, vertically mounted, with a pore size of 0.2 µm, an internal diameter of 15 mm and a length of 750 mm, giving a membrane area of 0.0353 m². It was chemically cleaned after each run using Alkaline P-3 Ultrasil 13 (Henkel-Ecolab, SNC, Issy-les-Moulineaux, France) (1.5%w/v, 80°C, 10 min without permeation and 10 min with permeate flux) and HNO_3 (Carlo Erba

Reagenti, Val de Reuil, France) purity of 65% (2%w/v, 80°C, 10min without permeation and 10min with permeate flux). Finally, the system was rinsed completely with tap water filtered on serial cartridges of 0.45 and 0.22µm. Following cleaning, the pH of the solution was checked to confirm that the cleaning solutions had been removed. To assess the cleanliness of the membrane, the new permeability was measured with distilled water. The water flux decreased during the first five minutes to a steady value of $1,250 \pm 250 L/h/m^2$ at a transmembrane pressure of 1 bar and a temperature of 20°C.

Experimental Setup and Operating Procedures

The experimental setup is shown schematically in FIGURE 1. The filtration pilot rig operated at controlled transmembrane pressure and, unless otherwise specified, all runs were performed during 300min under various operating conditions (feed suspensions, liquid and gas flow rates) in a batch retentate-recycling mode. This protocol was developed to give time for steady-state conditions to be established. A few experiments (diafiltration) were extended to eight hours. The feed suspension was driven from the two-liter feed tank by a volumetric pump (Netzsch Nemo NL20A,

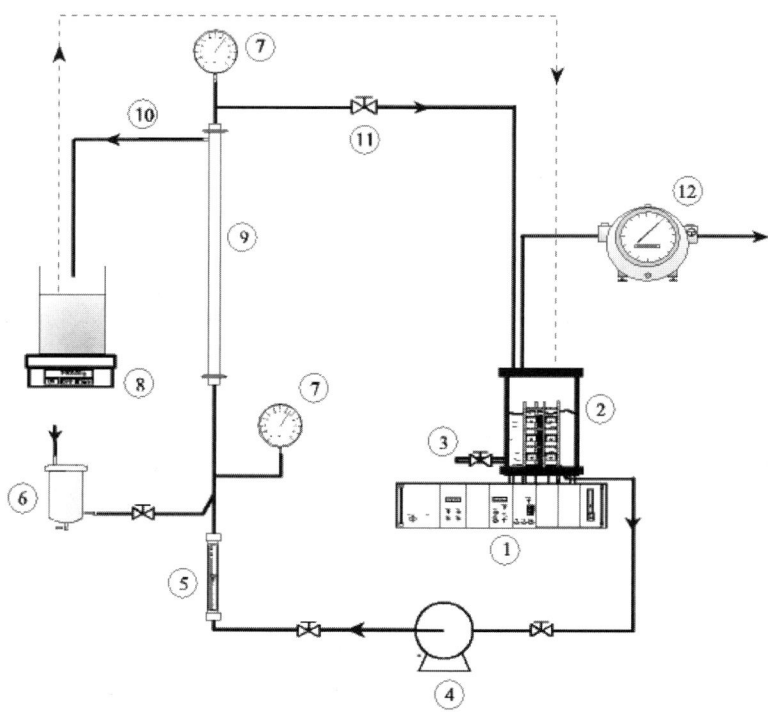

FIGURE 1. Experimental setup: **1**, controller (temperature); **2**, 2–L feed tank; **3**, retentate sampling; **4**, volumetric pump; **5**, liquid flowmeter; **6**, air sparging system; **7**, manometer; **8**, balance; **9**, tubular membrane; **10**, permeate outlet; **11**, retentate recycling; **12**, gas flowmeter.

Pontault Combault, France) and circulated through the membrane with a gas/liquid cocurrent flow. The flow rate of the pump could be adjusted within the range 0–1,000 L/h and measured with a rotameter. The permeate flux was deduced from weight recordings using an electronic balance (Sartorius, Goettingen, Germany) linked to a personal computer. Permeate was then returned to the feed tank in order to maintain a constant feed concentration. During two-phase flow experiments, compressed air was continuously supplied to the inlet of the module through a Y-tubular piece. The feed tank acted as a gas/liquid separator. The gas flow rate was measured at the feed tank exit and corrected according to the operating pressure to obtain the gas flow rate Q_G inside the membrane. Pressure was measured at the inlet and outlet of the membrane with Bourdon-type manometers. A mean transmembrane pressure (TMP) of 0.25 bar was applied by adjusting the valve downstream from the filtration module. Experiments were carried out at 20°C to guarantee the enzyme stability. Under single-phase flow experiments, the liquid flow rate was in the range 240–1,020 L/h (U_{LS} = 0.38–1.60 m/sec) corresponding to a Reynolds number Re_L in the range 5,700–24,000 (turbulent conditions). For two-phase flow experiments, the liquid flow rate was kept constant (240 L/h) and the gas flow rate was varied in the range 14–670 L/h (U_{GS} = 0.02–1.05 m/sec) corresponding to a Reynolds number between 6,030 and 21,450.

The superficial velocities U_{LS} (liquid phase) and U_{GS} (gas phase) and the Reynolds number associated with the liquid phase Re_L are defined as follows:

$$U_{LS} = \frac{Q_L}{S} \qquad U_{GS} = \frac{Q_G}{S} \qquad (1)$$

$$Re_L = \frac{\rho_L U_{LS} D_H}{\mu_L} \quad \text{(single-phase flow)}$$

$$Re_L = \frac{\rho_L (U_{LS} + U_{GS}) D_H}{\mu_L} \quad \text{(two-phase flow)}, \qquad (2)$$

where S is the membrane cross-section (m^2), μ_L is the dynamic viscosity of the liquid phase (Pa.sec), ρ_L is the density of the liquid phase (kg/m^3), and D_H is the hydraulic diameter of the membrane (m).

At the beginning of each run, the permeate valve was firstly closed to adjust the flow rates and pressures and then rapidly opened, and the permeate flux was recorded.

The feed suspension and permeate were assayed for the invertase enzyme. Samples of permeate (permeate outlet) and retentate (feed tank) were taken at various intervals (t = 0, 30, 60, 120, 210, and 300 min). For experiments with invertase/yeast mixtures, the retentate samples were spun in a microcentrifuge (Hettich Mikro 12-24, Tuttlingen, Germany) at 15,000 RPM for five minutes and the assay sample was taken from the supernatant. The invertase activity was then measured as follows: the reaction was started by the addition of the diluted sample (0.1 mL) to a 0.4 M sucrose (Merck, Darmstadt, Germany) solution in a 0.1 M sodium acetate buffer pH = 4.5 (5 mL). The variation in concentration of reducing sugars resulting from invertase activity at 40°C was measured by the 2,4-dinitrosalicylic acid method[13] as a function of time to determine the initial reaction rate. Standardization was obtained with various concentrations of an equimolar mixture of D-glucose and D-fructose in acetate buffer. One unit of invertase activity corresponds to the amount of enzyme that catalyzes the hydrolysis of 1 μmol sucrose/min under the conditions of assay. Assays were carried out repeatedly and the reported results proved to be reproducible (error less than 5%). During preliminary experiments, it was verified that the enzyme activity was directly proportional to its concentration:

$$Ac = 440C, \qquad (3)$$

where Ac is the enzyme activity (U/mL) and C is the enzyme concentration (g/L). Enzyme transmission was then calculated by taking the ratio of the permeate activity to the feed activity.

For the diafiltration experiments (invertase 0.5 g/L + yeast 20 g/L), the protocol was as follows:

1. A first period allowed the permeate flux to remain quasi constant (about 2 h). During this period, the TMP and flow rates were adjusted to the chosen operating conditions (no gas sparging: Q_L = 240 L/h, TMP = 0.25 bar; gas sparging: Q_L = 240 L/h, Q_G/Q_L = 5% and 150%, TMP = 0.25 bar) and the permeate was recycled to the feed tank.

2. The diafiltration period was then started by collecting the permeate and adding 0.1 M acetate buffer pH = 4.5 to the feed tank at the same permeate flow rate, keeping the feed volume constant at 3 L. During this phase, samples were taken from the permeate and the retentate streams and assayed for the invertase.

RESULTS AND DISCUSSION

Effect of Added Yeast on Enzyme Transmission and Flux

FIGURE 2 presents permeate flux versus time for the single-phase flow experiment done using a 0.5 g/L enzyme solution at a liquid flow rate of 240 L/h and a TMP of 0.25 bar. Although microfiltration membranes are expected to allow proteins to freely pass through their large pores, it was shown that the flux decreased initially with time to reach a long term value of 20 L/h/m^2, which represented less than 2% of the clean membrane flux. According to several papers dealing with the microfiltration of protein solutions with or without added yeast,[14,15] the initial decline is due to the internal fouling of the membrane caused by protein aggregates (that represent a very small fraction of the total protein) depositing inside the pores or near the pore entrances. Once internal fouling is substantial (so that the pores are highly blocked or constricted), the rejected protein begins to form a layer on the external membrane surface. This hypothesis of the formation of an external protein fouling was confirmed here since, during a two-phase flow experiment achieved with a Q_G/Q_L ratio of 150%, the final permeate flux was increased twofold (result not shown). However, in our case, it was not possible to easily quantify the part of enzyme aggregates formed during the initial stage of filtration. FIGURE 2 also shows permeate flux versus time for the single-phase flow experiment achieved with the enzyme/yeast mixture under the same operating conditions (initial enzyme concentration, liquid flow rate, TMP). The flux declined to a lower long term value of 10 L/h/m^2 (representing less than 1% of the clean membrane flux), due to the additional resistance caused by the yeast cake. However, it should be noticed that the initial flux decline was slower for the enzyme/yeast mixture than for the enzyme-only solution.

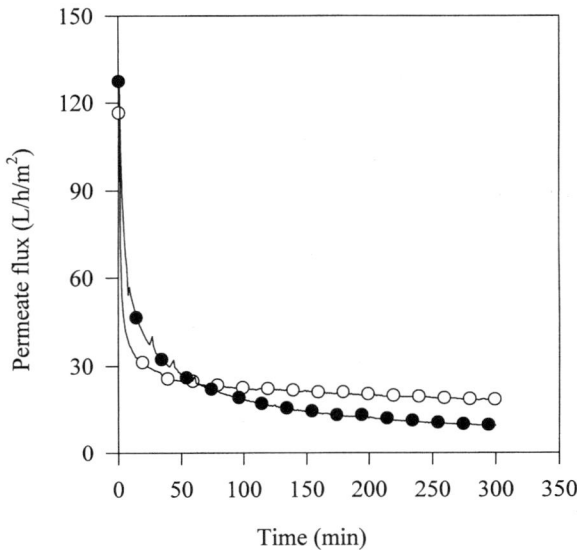

FIGURE 2. Variation with time of the permeate flux without air sparging for the enzyme-only solution (○) and the enzyme/yeast mixture (●). Conditions: Q_L = 240 L/h, TMP = 0.25 bar, [enzyme] = 0.5 g/L, [yeast] = 20 g/L.

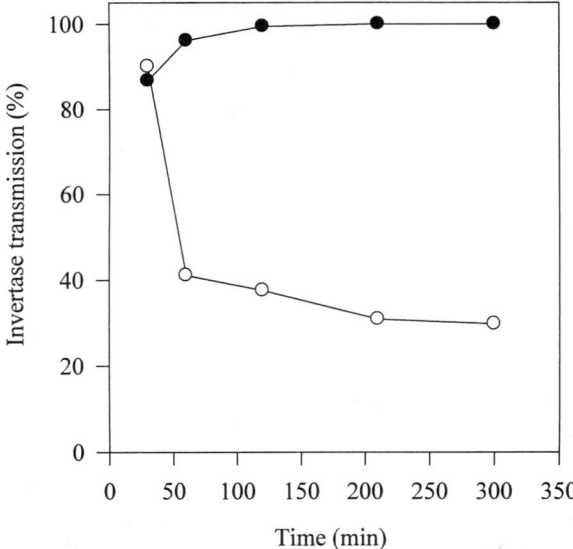

FIGURE 3. Variation with time of the invertase transmission without air sparging for the enzyme-only solution (○) and the enzyme/yeast mixture (●). Conditions: Q_L = 240 L/h, TMP = 0.25 bar, [enzyme] = 0.5 g/L, [yeast] = 20 g/L.

FIGURE 3 presents the transmission versus time for the experiments shown in FIGURE 2. For the enzyme-only solution, the transmission showed a fast decrease from nearly 100% to 40% during the first 60 min of filtration. As seen previously, the formation of external fouling could occur during this period. This lower enzyme transmission was thus probably due to the external fouling layer, which had a very low permeability for invertase molecules. The transmission then gradually declined to 30% over the rest of the experiment, due to the consolidation of the fouling layer. With the enzyme/yeast mixture, the transmission curve was very different than that for the enzyme-only solution: the transmission remained constant at 100% throughout the experiment. As explained already, the primary reason for the decrease in invertase transmission was the presence of an external enzyme layer. However, the yeast cake prevented or retarded its formation, resulting in a higher long-term transmission. It should be noted that, under these conditions, the porosity and permeability of the yeast cake were probably high enough to allow the free passage of the invertase molecules through the membrane. As a matter of fact, the final enzyme mass flux with the enzyme/yeast mixture was nearly twice that obtained with the enzyme-only solution. In conclusion, the cells in the enzyme/yeast mixture produced a dynamic membrane on the top of the original membrane that was capable of entrapping at least some of the enzyme aggregates. This enhanced the enzyme transmission and mass flux by slowing the formation of the protein fouling layer on the surface of the primary membrane.

Effect of Gas/Liquid Two-Phase Flow on Permeate Flux and Enzyme Transmission with Enzyme/Yeast Mixtures

The effect of gas sparging on the permeate flux as a function of time for the enzyme/yeast mixture is shown in FIGURE 4 for single-phase flow conditions and two-phase flow conditions (Q_G/Q_L ranging from 5 to 280%) with operating conditions Q_L = 240 L/h and *TMP* = 0.25 bar. With air sparging, the pseudo-steady-state fluxes, obtained after 300 min of filtration, were systematically higher than the flux obtained without air sparging. For small values of the Q_G/Q_L ratio (5%) corresponding to a bubble flow pattern, the flux was slightly improved (+50%), whereas for slug flow conditions (Q_G/Q_L ranging from 25 to 280%), the enhancement effect depended on the flow rate ratio up to a threshold value of 100%. As much as a 140% increase in permeate flux was then achieved. This permeate flux behavior was in good agreement with previous studies of microfiltration with only yeast in the feed.[4]

The invertase transmission is plotted versus time in FIGURE 5 for the experiments shown in FIGURE 4. As previously observed, without air sparging, the enzyme transmission remained constant at 100% throughout the experiment. With air sparging, two different behaviors were obtained, according to the two-phase flow pattern. For bubble flow conditions, the transmission profile was close to that achieved under single-phase flow conditions with a nearly constant level reaching 90%. For slug flow conditions, whatever the flow rate ratio, the transmission firstly showed a fast decrease (the higher the Q_G/Q_L ratio, the faster the decrease) and it was not quite stabilized after five hours even though the permeate flux had already reached a pseudo-steady state (FIG. 4). The final value was about 30%. About the possible effect of gas sparging on protein denaturation, it was verified that no significant loss of enzyme activity in the retentate samples occurred throughout all these two-phase flow experiments, even with high Q_G/Q_L ratios.

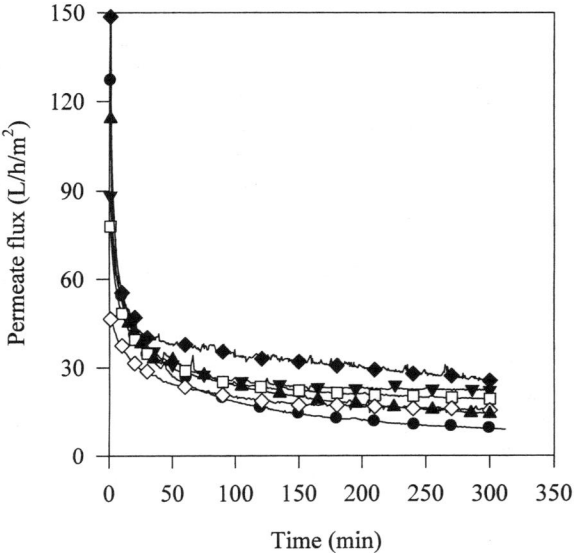

FIGURE 4. Variation with time of the permeate flux for the enzyme/yeast mixture without air sparging (●) and with air sparging $Q_G/Q_L = 5\%$ (▲), $Q_G/Q_L = 25\%$ (◇), $Q_G/Q_L = 40\%$ (□), $Q_G/Q_L = 150\%$ (▼), $Q_G/Q_L = 280\%$ (◆). Conditions: $Q_L = 240\,L/h$, $TMP = 0.25\,bar$, [enzyme] = 0.5 g/L, [yeast] = 20 g/L.

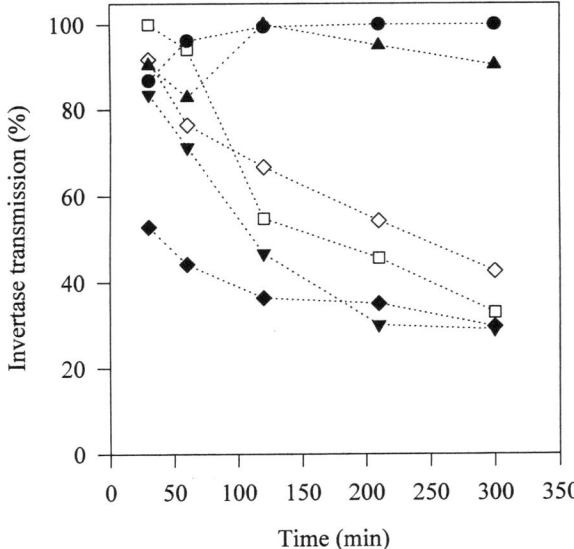

FIGURE 5. Variation with time of the invertase transmission for the enzyme/yeast mixture without air sparging (●) and with air sparging $Q_G/Q_L = 5\%$ (▲), $Q_G/Q_L = 25\%$ (◇), $Q_G/Q_L = 40\%$ (□), $Q_G/Q_L = 150\%$ (▼), $Q_G/Q_L = 280\%$ (◆). Conditions: $Q_L = 240\,L/h$, $TMP = 0.25\,bar$, [enzyme] = 0.5 g/L, [yeast] = 20 g/L.

The enzyme-extraction mass flux through the membrane was then determined. The maximal final mass flux (5 g/h/m^2) was obtained with the bubble flow conditions with an improvement of 25% compared to the reference value without air sparging (3.9 g/h/m^2). This could be explained by the combined effects of high enzyme transmission and enhanced permeate flux. However, final mass fluxes corresponding to a slug flow pattern were systematically lower (between 2.6 and 3.3 g/h/m^2 according to the flow rate ratio). In fact, the gain in flux was not sufficient to counterbalance the reduction in enzyme transmission.

To explain this apparent contradiction of slug flow hydrodynamics producing a large increase in flux and a drastic decrease in enzyme transmission, it was supposed that, under these conditions, the cake layer of cells was thinner but more tightly packed than that formed in single-phase flow filtration with low liquid flow rate. During crossflow microfiltration of a yeast suspension, it was reported the preferential deposit of smaller cells,[16] leading to a thinner and less porous cake with higher velocities.[17,18] In the same way, during microfiltration of skimmed milk for the separation of casein micelles from whey proteins, it was observed that an increase in the wall shear stress decreased the deposit thickness and porosity by preferential removal of larger micelles away from the membrane, thereby producing a sharp decrease in whey protein transmission.[19] This hypothesis of the formation of a deposit with lower porosity and thickness by increasing wall shear stress was verified under single-phase flow filtration experiments. Indeed, it was shown that increasing liquid flow rate from 240 L/h to 1,020 L/h led to lower enzyme transmissions (reduction by a factor greater than 3) even though the permeate flux was greatly improved (+120%) (see FIGURE 6). No loss of enzyme activity was observed in the feed stream during single-phase flow experiments, whatever the liquid flowrate used. It should be noted that similar flux

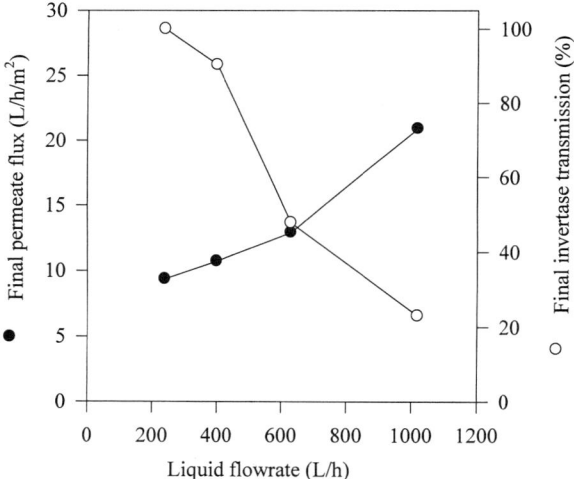

FIGURE 6. Variation of the final permeate flux (●) and the final invertase transmission (○) with the Q_L flowrate for the enzyme/yeast mixture under single-phase flow conditions. Conditions: TMP = 0.25 bar, [enzyme] = 0.5 g/L, [yeast] = 20 g/L.

values achieved with gas sparging (Q_L = 240 L/h, Q_G/Q_L = 150%) and without gas sparging (Q_L = 1,020 L/h), here about 20 L/h/m², corresponded to equivalent levels of transmission (29% and 23%, respectively), therefore reinforcing the hypothesis that flux and transmission were linked through the physicomechanical characteristics of the cake deposit (porosity and thickness). FIGURE 7 presents the schematic of the yeast cake structure obtained with low Q_G/Q_L (or Q_L) and high Q_G/Q_L (or Q_L) and the subsequent permeate flux and enzyme transmission.

Diafiltration for Enzyme Recovery

After a two-hour period with permeate-recycling to provide a pseudo-steady state flux, the diafiltration protocol was applied for six hours at a liquid flow rate of 240 L/h and a transmembrane pressure of 0.25 bar under single-phase flow conditions, bubble flow conditions (Q_G/Q_L = 5%) and slug flow conditions (Q_G/Q_L = 150%). The main conclusions drawn on the basis of experiments with permeate recycling were confirmed here, in terms of the final flux and transmission values obtained after eight hours of filtration. With bubble flow conditions, a slight increase in flux was achieved (+30%) and the enzyme transmission was maintained at 100%. With slug flow conditions, the flux improvement was significant (+200%) but the transmission was reduced by a factor of three (results not shown). The decrease in enzyme activity with time as permeate passed through the membrane was also clearly put in evidence for the three experiments and, considering the initial and final enzyme activities, the recovery percentage at the end of the diafiltration period was evaluated according to

$$\text{Recovery} = 1 - \frac{Ac_{(t=8\text{h})}}{Ac_0}. \tag{4}$$

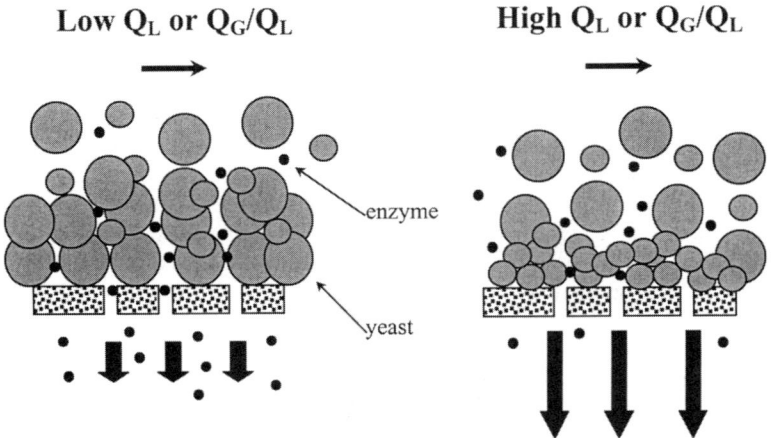

FIGURE 7. Schematic of the yeast cake structure obtained with low Q_G/Q_L (or Q_L) and high Q_G/Q_L (or Q_L) and the subsequent permeate flux and enzyme transmission.

Without gas sparging, the recovery reached 60%. For bubble flow and slug flow conditions, it was 68% and 45%, respectively. The maximal enzyme recovery was thus achieved with bubble flow conditions (13% gain in recovery) and, for slug flow conditions, this recovery was strongly reduced due to the drastic reduction in enzyme transmission outweighing the flux improvement. These results demonstrate that the choice of the two-phase flow pattern during filtration of valuable molecules depends on whether permeate or retentate streams are the principle concern. For applications where a maximal protein mass flux through the membrane is needed, enzyme recovery being one example, bubble flow conditions should be preferred (slight flux improvement and high protein transmission). On the other hand, slug flow conditions are required for applications dealing with protein concentration (high permeate flux and low protein transmission).

CONCLUSIONS

In this study, air sparging inside a tubular module was used in order to obtain quantitative information about the transport of a target enzyme (invertase) through a microfiltration membrane. Although the individual enzymes were much smaller than the microfiltration membrane pores, severe fouling occurred when filtering the enzyme-only solution under single-phase flow conditions. This fouling was manifest as a decline in flux to less than 2% of the initial water flux and a decline in protein concentration in the permeate to 30% of its initial value during a five-hour filtration period. When yeast cells were added under the same experimental conditions, simulating the industrial situation in which the protein has to be purified from a fermentation broth containing cells, the enzyme transmission was maintained at 100% for the five-hour period and the enzyme mass flux was twofold higher. It was supposed that the yeast cake on top of the primary membrane acted as a secondary membrane that retained eventual enzyme aggregates, thereby reducing protein fouling of the primary membrane. During gas-sparged microfiltration of the enzyme/yeast mixture in a permeate-recycling mode at the same liquid flow rate, it was shown that gas/liquid slug flow strongly decreased the transmission of the enzyme (70% decrease), even though the permeate flux was improved (140% improvement). To explain this apparent contradiction of high flux and low transmission, it was supposed that an increase in wall shear stress (high Q_G/Q_L or Q_L) decreased the cake thickness and porosity by preferential removal of larger cells away from the membrane. As a result, the mass flux of the enzyme was significantly reduced. However, with a bubble flow pattern, the permeate flux was 1.5 times higher and the transmission was maintained at a high level. The enzyme mass flux was then 25% higher, compared to single-phase flow filtration conditions. All these observations were confirmed during diafiltration experiments and it was notably found that a bubble flow pattern led to a 13% higher enzyme recovery. In conclusion, this study has enabled the optimal operating conditions for the recovery of a target enzyme to be determined when a gas/liquid two-phase flow is used. Bubble flow is more suitable for the recovery of valuable molecules in the permeate stream, whereas slug flow is more beneficial for the concentration of suspensions and/or solutions. Further experimental study on real biological mixtures arising from cell cultures is now required.

ACKNOWLEDGMENTS

The authors wish to thank Christelle Lagane, Juan Rios, and Christophe Ellero for their technical help. We are grateful to Monique Sudérie for the helpful discussions about invertase.

REFERENCES

1. CABASSUD, C., S. LABORIE & J.M. LAINE. 1997. How slug flow can improve ultrafiltration flux in organic hollow fibres. J. Membr. Sci. **128:** 93–101.
2. LEE, C.K., W.G. CHANG & Y.H. JU. 1993. Air slugs entrapped crossflow filtration of bacterial suspensions. Biotechnol. Bioeng. **41:** 525–530.
3. LÉONARD, D., M. MERCIER-BONIN, N.D. LINDLEY & C. LAFFORGUE. 1998. A novel membrane bioreactor with gas/liquid two-phase flow for high performance degradation of phenol. Biotechnol. Prog. **14:** 680–688.
4. MERCIER, M., C. MARANGES, C. FONADE & C. LAFFORGUE-DELORME. 1998. Yeast suspension filtration: flux enhancement using an upward gas/liquid slug flow—application to continuous alcoholic fermentation with cell recycle. Biotechnol. Bioeng. **58:** 47–57.
5. BELLARA, S.R., Z.F. CUI & D.S. PEPPER. 1996. Gas sparging to enhance permeate flux in ultrafiltration using hollow fibre membranes. J. Membr. Sci. **121:** 175–184.
6. CUI, Z.F. & K.I.T. WRIGHT. 1994. Gas–liquid two-phase crossflow ultrafiltration of BSA and dextran solutions. J. Membr. Sci. **90:** 183–189.
7. LI, Q.Y., Z.F. CUI & D.S. PEPPER. 1997. Fractionation of HSA and IgG by gas sparged ultrafiltration. J. Membr. Sci. **136:** 181–190.
8. LI, Q.Y., R. GHOSH, S.R. BELLARA, *et al.* 1998. Enhancement of ultrafiltration by gas sparging with flat sheet membrane modules. Sep. Purif. Technol. **14:** 79–83.
9. MERCIER-BONIN, M., C. MARANGES, A. LINÉ, *et al.* 2000. Hydrodynamics of slug flow applied to crossflow filtration in narrow tubes. AIChE J. **46:** 476–488.
10. CUI, Z.F. & K.I.T. WRIGHT. 1996. Flux enhancements with gas sparging in downwards crossflow ultrafiltration: performance and mechanism. J. Membr. Sci. **117:** 109–116.
11. CABASSUD, C., S. LABORIE, L. DURAND-BOURLIER & J.M. LAINE. 2001. Air sparging in ultrafiltration hollow fibers: relationship between flux enhancement, cake characteristics and hydrodynamic parameters. J. Membr. Sci. **181:** 57–69.
12. CHU, F.K., W.W. WATOREK & F. MALEY. 1983. Factors affecting the oligomeric structure of yeast external invertase. Arch. Biochem. Biophys. **223:** 543–555.
13. SUMNER, J.B. & S.F. HOWELL. 1935. A method for the determination of invertase activity. J. Biol. Chem. **108:** 51–54.
14. GÜELL, C., P. CZEKAJ & R.H. DAVIS. 1999. Microfiltration of protein mixtures and the effects of yeast on membrane fouling. J. Membr. Sci. **155:** 113–122.
15. KUBERKAR, V.T. & R.H. DAVIS. 1999. Effects of added yeast on protein transmission and flux in crossflow membrane microfiltration. Biotechnol. Prog. **15:** 472–479.
16. FOLEY, G., P.F. MACLOUGHLIN & D.M. MALONE. 1992. Preferential deposition of smaller cells during crossflow microfiltration of a yeast suspension. Biotechnol. Tech. **6:** 115–120.
17. RIESMEIER, B., K.H. KRONER & M.R. KULA. 1987. Studies on secondary layer formation and its characterization during crossflow filtration of microbial cells. J. Membr. Sci. **34:** 245–266.
18. RIESMEIER, B., K.H. KRONER & M.R. KULA. 1989. Tangential filtration of microbial suspensions: filtration resistances and model development. J. Biotechnol. **12:** 153–172.
19. LE BERRE, O. & G. DAUFIN. 1996. Skimmilk crossflow microfiltration performance versus permeation flux to wall shear stress ratio. J. Membr. Sci. **117:** 261–270.

Turnup Turndown of Membrane Operation of Membrane Bioreactors

J.A. HOWELL, T.C. ARNOT, AND H.C. CHUA

Department of Chemical Engineering, University of Bath, Claverton Down, Bath, United Kingdom

ABSTRACT: Membrane bioreactors can be operated with intermittent permeation and continuous aeration. Aeration close to the surface of a submerged membrane helps to maintain a membrane surface that is free from fouling. The conditions under which this occurs depend on the interaction between flux and aeration rate. Increased flux is possible without severe fouling if the aeration rate is increased. Results of performing membrane operation under the dual intermittency of aeration rate and permeation rate, with permeation also interrupted on a regular cycle, are presented. The results show that membrane plants designed for optimal operation at moderate flux can survive effectively with higher flux operation for restricted periods. Designing to account for such effects could reduce overall plant costs.

KEYWORDS: dynamic membrane operation; membrane bioreactors; interrupted permeation

INTRODUCTION

It is conventional in waste water treatment plants to design a system to cope with three times the average minimum flow rate or flow received by the plant in dry weather. This allowance is regulated in the United Kingdom and in other countries may be slightly less. FIGURE 1 shows the frequency of different flow rates received by a plant over a two-year period. Although the highest flow rates do indeed exceed the modal flow by a factor of about three, they occur for short periods and at low frequencies. The use of membrane bioreactors for municipal waste water treatment is increasing and is found at larger and larger scales. The largest plant currently in operation is at Swanage in the South of England, on the coast, to the South of Bath. Many smaller plants have been constructed. By designing a membrane plant to be at optimal operation at three times dry weather flow, as suggested for economic efficiency by Owen *et al.*,[1] an uneconomic design may be produced when compared to competing technologies. Davies *et al.*[2] showed that significant savings result from lowering the overdesign factor. Howell[3] postulated that it might be possible to obtain a more economic design using 1–1.5 times dry weather flow (DWF) so long as operation at 3×DWF can be achieved by increasing the energy input to the plant. It was

---

Address for correspondence: J.A. Howell, Department of Chemical Engineering, University of Bath, Claverton Down, Bath BA2 7AY, UK.
 j.a.howell@bath.ac.uk

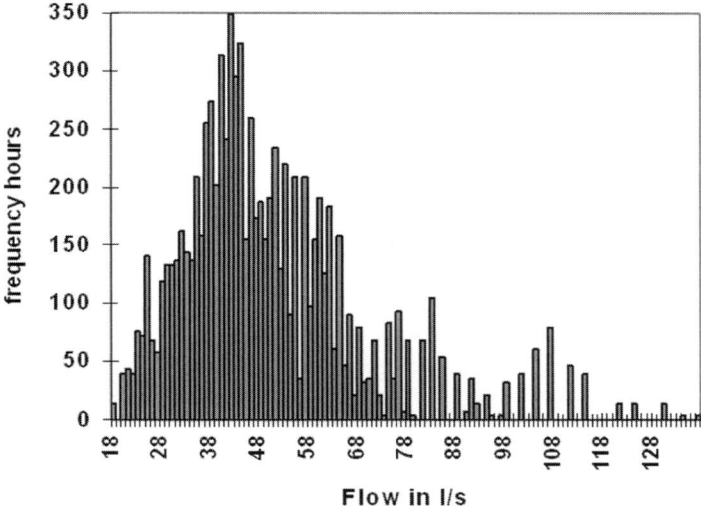

FIGURE 1. Example data: frequency of various flow rates into a sewage treatment works during a period of two years.

assumed in that exercise that increasing energy input by increasing flow over the membrane surface would increase the flux. In steady state operation higher energy inputs are known to lead to higher fluxes. It was not known whether this would happen with changing flow rates. The concept might have potential value since the most economic MBR designs use relatively low fluxes and low energy inputs. These designs have been made possible by the submerged membrane systems of Zenon, Kubota, and Mitsubishi Rayon. These systems use the energy of the æration bubbles to help reduce the fouling on the membrane surfaces. Coupled with a reduced membrane cost it is now economic to use larger membrane areas and lower crossflow energy than was previously considered optimal.

It has been shown by many workers, e.g., Owen et al.,[1] that increased cross flow velocities can lead to higher steady state fluxes. It was postulated by Howell[4] that operating at fluxes close to the critical flux can result in relatively little fouling. It has also been shown that the critical flux for microfiltration is a function of the shear rate at the membrane surface (Wu et al.[5]). The hypothesis behind the concept that one can increase and decrease flux at will by increasing and decreasing the energy input to the system is that any fouling resulting from a high flux operation will be removed during low flux operation. At the beginning of this work such a hypothesis had not been validated.

Subcritical flux operation with no fouling is possible in a laboratory with a simple defined suspension that is being filtered but difficult with a complex mixture. An alternative is to operate with controlled fluxes that slightly exceed the critical flux and thus allow fouling. Fouling may then removed by reducing the driving pressure force holding the cake to the membrane surface and to allow self-cleaning.

TABLE 1. Composition of simulated sewage

Component	Concentration ($g\,L^{-1}$)
Peptone	0.2
Meat extract	0.14
Urea	0.01
$CaCl_2 \cdot 2H_2O$	0.004
$MgSO_4 \cdot 7H_2O$	0.002
K_2HPO_4	0.011
NaCl	0.007

MATERIALS AND METHODS

A 3.5 L laboratory scale air-lift MBR was constructed using acrylic sheet. A Kubota flat sheet membrane (A4 size, area $0.106\,m^2$, pore size $0.4\,\mu m$) was immersed in the riser. The design incorporated acrylic sheets around the membrane so that the gap between sheets and membrane duplicated that in a full scale plant. Aeration was by a sparger at the bottom of the reaction vessel. A membrane free section comprised half the depth of the vessel. Permeate flow was controlled using a flowmeter (Cole-Parmer) and a computer controlled pump (Flowgen) using the Geniedac software. The transmembrane pressure (TMP) was recorded using a pressure transducer (Druck). Simulated sewage (see TABLE 1) was used. The operating conditions of the MBR are shown in TABLE 2. The reaction vessel is shown in FIGURE 2.

Initial experiments ran the system with a controlled flux and various æration velocities. Flux levels once set were held for a minimum of 90 minutes. In the next set the æration was continued, but the flux was interrupted periodically with eight minutes of permeation followed by two minutes without permeation. In each case, transmembrane pressures were recorded throughout.

RESULTS

FIGURE 3 shows the transmembrane pressure (TMP) recorded as a function of time as the æration rate was increased from 2 to 6 and then $10\,L\,min^{-1}$ before being reduced again in steps, all while the flux was controlled at $10\,L\,m^{-2}h^{-1}$ ($2.8\,\mu m.sec^{-1}$). It can be seen that, at flows below $10\,L\,m^{-2}h^{-1}$ the membrane TMP

TABLE 2. Operating conditions of MBR

HRT	6 h
SRT	infinite
Temperature	20°C
MLVSS	$6.78–8.72\,g\,L^{-1}$

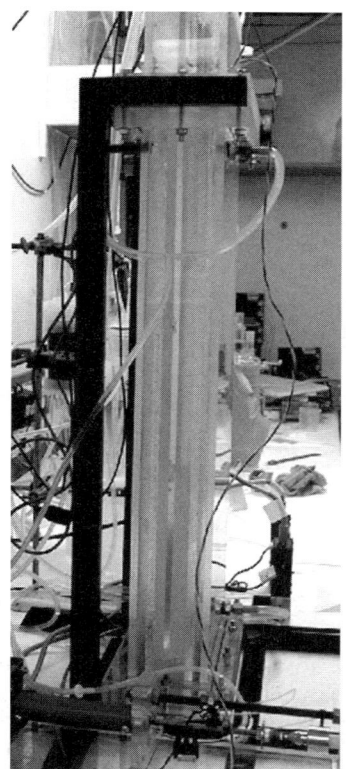
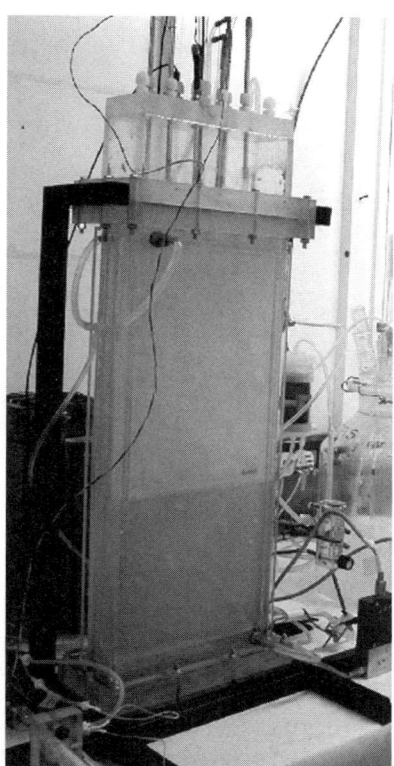

FIGURE 2. Photographs of a laboratory scale MBR based around a scaled down design of the Kubota submerged air lift reactors.

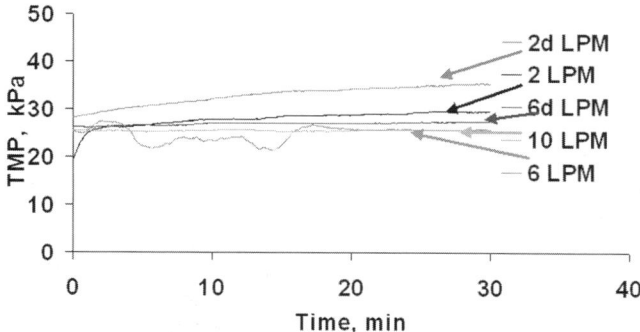

FIGURE 3. TMP as a function of time for increasing (2, 6, and 10) and decreasing (6 d and 2 d) air flow rate, for a constant flux of $10\,L\,m^{-2}\,h^{-1}$.

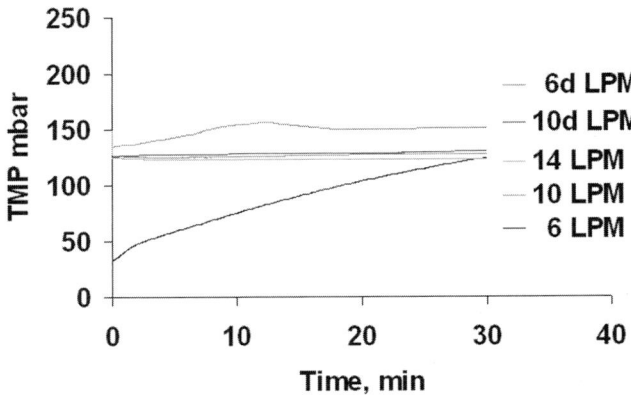

FIGURE 4. TMP as a function of time for increasing (6, 10, and 14) and decreasing (10d and 6d) air flow rate, for a constant flux of $20 L m^{-2} h^{-1}$.

increases demonstrating that fouling was occurring. At the lowest air flow rate the fouling was fastest. By contrast at $10 L min^{-1}$ fouling did not apparently occur and the TMP remained steady.

FIGURE 4 shows similar experiments at the higher flux of $20 L min^{-1}$. In these experiments the fouling is more severe at an airflow rate of $6 L min^{-1}$ and can be observed at $10 L min^{-1}$. Only when the air flow rate was increased to $14 L min^{-1}$ did fouling stop and a steady flux was observed. Other experiments at even higher fluxes (not shown) resulted in fouling at even higher air flow rates and a difficulty in finding any experimentally attainable air flow rate that would eliminate fouling completely. Kubota commercial membrane systems will operate with minimal fouling at this higher flux and we presume that the different scales of the full scale and laboratory reactors contribute to this. The vertical height of our reactors was less than 1 meter. The commercial designs are at least 2 and up to 3 meters liquid depth. These higher depths can generate much higher bubble velocities and, hence, a higher cleaning affect across the membrane.

Despite the lower attainable fluxes, we believe that the qualitative observations made in the laboratory have validity. From the experiments summarized in FIGURES 3 and 4 we observe that a high air flow rate can lead to a steady non-fouling flux across the membrane with a suspension of activated sludge organisms and simulated settled sewage. We also observe that the higher the air flow rate the higher the steady, non-fouling flux that can be sustained. Following Wu et al.[5] the flux at which fouling is just avoided is the critical flux. The critical flux is increased by increasing air flow across the surface of the membrane. This flow has been shown by Ghosh and Cui[6] to influence the shear rate at the membrane surface, which is critical to removing particles from the surface of the membrane. Known mechanisms for particle migration from the surface include shear induced diffusion or inertial lift and electrostatic forces if the boundary layer thickness is considered to be a key parameter controlling the critical stability of the particle membrane interaction, as suggested by Bacchin et al.[7]

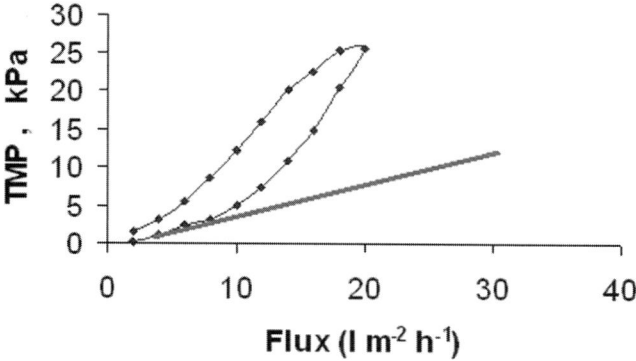

FIGURE 5. TMP plotted against flux to show hysteresis for increasing/decreasing controlled flux and a fixed air flow rate of $2\,\mathrm{L\,min^{-1}}$ (●). The *straight line* shows the same results, but for a clean water system.

Instead of changing the air flow rate at a constant flux, the flux may be changed at a constant air flow rate. FIGURE 5 shows the effect of first increasing the flux and then decreasing it: hysteresis is obvious. The lower part of the curve shows the effect as the flux increases, and the upper part the effect of decreasing flux. The TMP initially rises linearly with flux. The straight line is the TMP observed when the flux is applied to a clean water stream. It can be seen that initially the MBR has the same TMP/flux relationship, but as flux increases surface fouling increases the TMP. When flux reduces, initially the cake formed is not removed and the TMP is higher than for the rising curve. At the lowest flux there is some residual fouling since the TMP has not returned to its initial value.

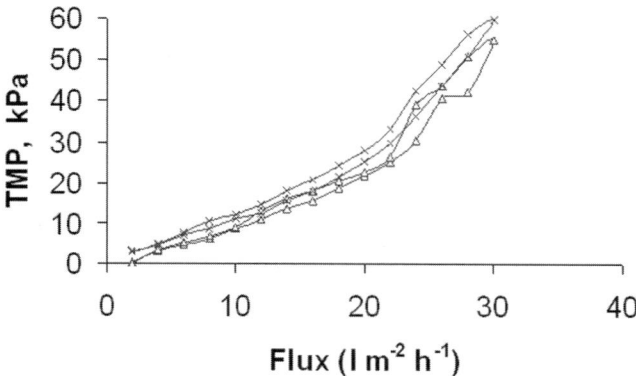

FIGURE 6. TMP plotted against flux to show hysteresis for increasing/decreasing controlled flux and fixed air flow rates of 6 (△) and 8 (×) $\mathrm{L\,min^{-1}}$.

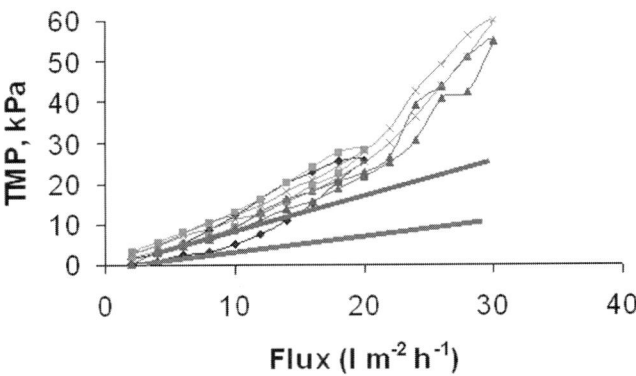

FIGURE 7. Results from FIGURES 5 and 6 combined: ◆, 2; ■, 4; ▲, 6; and ×, 8 LPM.

At a higher air flow rate the flux can be increased further, as is shown in FIGURE 6. In this case there is much less hysteresis and, furthermore, the net increase in TMP is reduced. The two curves shown in FIGURE 6 follow on from lower air flow rate experiments and there is no additional increase in the minimum TMP observed.

FIGURE 7 puts all of these experiments together and shows that once the first experimental cycle was completed the initial rise of TMP with flux remained linear, but with a greater slope than with the clean water line. This suggests that there is a residual layer of fouling that remains at the end of each cycle. This layer is then not increased until a minimum flux is exceeded and fouling deposits start to build up once more.

To reduce the air flows necessary to maintain a clean membrane it has been suggested that cleaning of the membrane can occur when the direct flux is stopped, thus removing the transmembrane pressure and the forces tending to hold a particle within the pores of the membrane once it has stuck there. These forces were analyzed by Kuiper et al.[8] who refer to the *critical pressure* as that required to maintain the

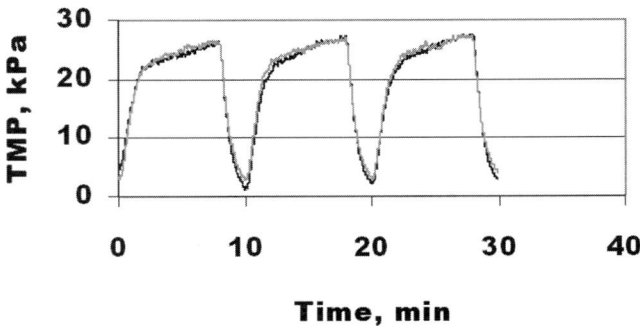

FIGURE 8. TMP plotted against time for a constant flux of $10 \, L \, m^{-2} h^{-1}$. In this experiment filtration took place for eight minutes, followed by a two-minute break, during which TMP was reduced to zero: ——, 2 LPM; ——, 3 LPM.

particle at the surface once it has deposited. This is also a function of the membrane shear rate. FIGURE 8 shows experimental results for TMP versus time measurements at constant flux once more. It demonstrates how effective stopping the flux for two minutes every 10 minutes can be in allowing a dynamic yet steady flux time pattern where, even with low air flow rates of 2 or $3 L\,min^{-1}$, the long-term TMP does not increase with time. The traces of TMP versus time show that the fouling occurs quite quickly and continues to increase during the operation with permeate flow. After a two-minute quiescent period the flux is once again restored. During the quiescent period air flow continues ærating the microorganisms and also cleaning the surface of the membrane. Once the permeation is restored, the TMP is once again low, but initially rising sufficiently quickly to suggest that the cake layer has not been completely removed, but that the situation is relatively stable.

FIGURE 9 shows a series of intermittent cycles carried out at various air flow rates for a flux of $20 L\,m^{-2}h^{-1}$. In contrast to the steady state experiments, which could not achieve stability below $14 L\,min^{-1}$, the intermittent experiments only require $6 L\,min^{-1}$ of air flow to achieve a stable dynamic system. FIGURE 10 shows a similar series of experiments where the flux was set at $30 L\,m^{-2}h^{-1}$ and stable dynamic operation was possible with an air flow rate of $10 L\,min^{-1}$.

With these experiments as a background, it was decided to test the hypothesis that sustained operation at high and then lower fluxes could be achieved repeatedly by simultaneously changing the air flow rate. In these experiments a higher MLVSS or biomass concentration was present in the aeration vessels and a higher air flow rate was needed to maintain the stable dynamic operation. FIGURE 11 shows the result of maintaining dynamic operation for three hours at a flux of $30 L\,m^{-2}h^{-1}$ with an air flow rate of $45 L\,min^{-1}$ followed by a period of three hours at a flux of $10 L\,m^{-2}h^{-1}$ at an air flow rate of $15 L\,min^{-1}$. The two cycles were then repeated and overlaid on the plot as the second cycle. It can be seen that the TMP rises over the initial period of the first phase, but that when the flux was reduced it proceeded to fall. The second cycle showed slightly lower TMP values, suggesting that there will be no serious problem in sustained cyclic operation. Turnup and turndown of membrane plants of

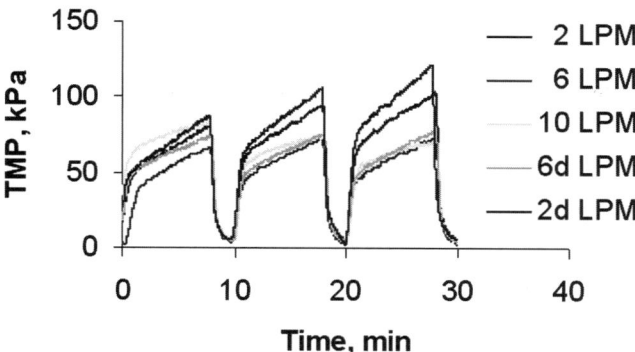

FIGURE 9. TMP plotted against time for a constant flux of $20 L\,m^{-2}h^{-1}$. In this experiment, filtration was again cycled: eight minutes of filtration followed by a two-minute break, during which TMP was reduced to zero.

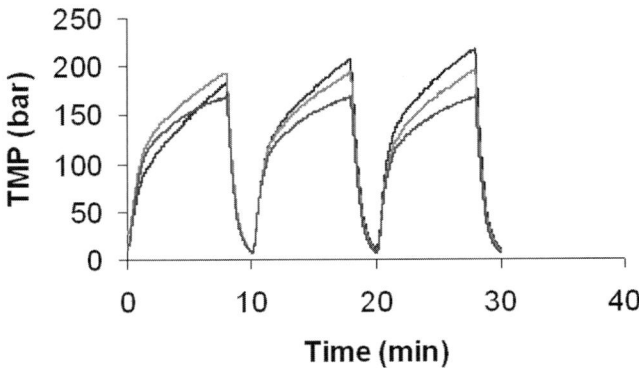

FIGURE 10. TMP plotted against time for a constant flux of $30\,L\,m^{-2}\,h^{-1}$. Filtration was again cycled: eight minutes of filtration followed by a two-minute break, during which TMP was reduced to zero: ——, 3 LPM; ——, 6 LPM; ——, 10 LPM.

the submerged bioreactor type seems to be eminently feasible. Additional experiments on a longer time scale are obviously required to establish the limits of this type of operation. Nevertheless, this work suggests that more flexible design should be possible giving significant economies of design for large-scale plants.

CONCLUSIONS

During constant permeation a steady flux was maintained with higher æration rates and low fluxes in an airlift MBR treating synthetic settled sewage. At lower æration rates, fouling occurred and TMP increased. At higher fluxes, higher æration

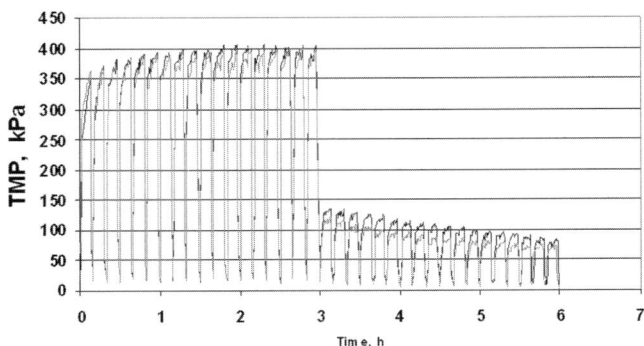

FIGURE 11. The effect of maintaining dynamic operation for three hours for a flux of $30\,L\,m^{-2}\,h^{-1}$ and an air flow rate of $45\,L\,min^{-1}$, and then another three hours for a flux of $10\,L\,m^{-2}\,h^{-1}$ and an air flow rate of $15\,L\,min^{-1}$. The cycles were then repeated and superimposed on each other: —, first cycle; —, second cycle.

rates are required to maintain a steady flux. If permeation was intermittently stopped while æration was maintained, fouling occurred during permeation but the cake was reproducibly removed to a base level during the non-permeating period. A sustainable dynamic operation was possible at lower airflow rates for a given flux than with steady permeation.

If a very high flux was maintained, even a high airflow rate resulted in progressively increased fouling during intermittent operation. However, if the flux was reduced after several hours at high flux, then intermittent operation once more steadily reduced the remaining cake back to its base level. This was repeatable. These experiments suggest that plants can be operated under fluctuating loads efficiently when input energies are adjusted to reflect the throughput demanded only when required.

REFERENCES

1. OWEN, G., M. BANDI, J.A. HOWELL & S.J. CHURCHOUSE. 1995. Economic assessment of membrane processes for water and waste applications. J. Membr. Sci. **102:** 77–91.
2. DAVIES, W.J., M.S. LE & C.R. HEATH. 1998. Intensified activated sludge process with submerged membrane microfiltration. Water Sci. Technol. **38**(4–5): 21–27.
3. HOWELL, J.A. 1995. Membrane Processes for the Treatment of Activated Sludge. Euromembrane 95. W.R. Bowen, R.W. Field & J.A. Howell, Eds.: 462–468. European Membrane Society.
4. HOWELL, J.A. 1995. Sub-critical flux operation. J. Membr. Sci. **107:** 165–171.
5. DENGXI, WU., J.A. HOWELL & R.W. FIELD. 1999. Critical flux measurement for model colloids, J. Membr. Sci. **152:** 89.
6. GHOSH, R. & Z.F. CUI. 1999. Mass transfer in gas-sparged ultrafiltration: upward slug-flow in tubular membranes. J. Membr. Sci. **162:** 92–102.
7. BACCHIN, P., P. AIMAR & V. SANCHEZ. 1996. Influence of surface interaction on colloid transfer during colloid ultrafiltration. J. Membr. Sci. **115:** 49–63.
8. KUIPER, S., C.J.M. VAN RIJN, W. NIJDAM, et al. 2000. Determination of particle release conditions in microfiltration: a single particle model tested on a model membrane. J. Membr. Sci. **180:** 15–28.

Designing Blood Oxygenators

S.R. WICKRAMASINGHE, A.R. GOERKE, J.D. GARCIA, AND BINBING HAN

Department of Chemical Engineering, Colorado State University, Fort Collins, Colorado, USA

ABSTRACT: Extracorporeal blood oxygenators are used to provide cardiopulmonary support during open heart surgery. In the study reported here, mass transfer correlations were determined for commercially available blood oxygenators. Two configurations used commercially, flow outside and across bundles of hollow fibers and flow in thin channels between parallel flat sheet membranes, were investigated. Water and glycerol/water mixtures were used as a substitute for blood. Diffusion of oxygen into and out of these solutions was studied. For flow across bundles of hollow fibers, the mass transfer correlations derived here are in agreement with analogous correlations for crossflow heat exchangers. However, for flow in thin channels, the rate of mass transfer is often less than predicted from theory. This compromised mass transfer can be explained by considering slight variations in the thickness of the blood flow channels. The mass transfer correlations developed here could be used to design better blood oxygenators.

KEYWORDS: blood oxygenators; mass transfer correlations; microporous membranes; hollow fibers

NOMENCLATURE:

A	membrane surface area
B	half the base of triangular channel (see FIG. 2)
C	outlet oxygen concentration
C_0	inlet oxygen concentration
C^*	equilibrium oxygen concentration
ΔC	concentration difference
d_e	equivalent diameter
D	diffusion coefficient
D_i	outside diameter of inner core (see FIG. 1)
D_0	inside diameter of module casing (see FIG. 1)
K	overall mass transfer coefficient
L	length of hollow fiber, length of rectangular channel
L_0	length of module (see FIG. 1)
N	total molar flux
Q	flow rate
u	velocity
W	width of triangular channel (see FIG. 2)

Greek Symbols

ν	kinematic viscosity

Dimensionless Numbers

Gr	Graetz number, $d_e^2 u/DL$
Re	Reynolds number, $u d_e/\nu$
Sc	Schmidt number, ν/D
Sh	Sherwood number, $K d_e/D$

Address for correspondence: S.R. Wickramasinghe, Department of Chemical Engineering, Colorado State University, Fort Collins, CO 80523-1370, USA. Voice: 970-491-5276; fax: 970-491-7369.
wickram@engr.colostate.edu

INTRODUCTION

In 1953 Gibbon[1,2] performed the earliest successful open-heart operation using a film oxygenator. Although the blood surface area in contact with oxygen in these devices was large, the gas transfer efficiency was often compromised by channeling the blood flow.[3,4] The next generation of blood oxygenators (BOs) was the bubble oxygenators. In these devices the gas exchange efficiency was increased by dispersing bubbles of oxygen in the blood resulting in significantly reduced priming volumes compared to film oxygenators. Since bubbling gas through blood leads to foam formation, silicone compounds were used as defoaming agents.[5] Although increasing the gas flow rate (i.e., the number of gas bubbles) leads to an increase in the gas transfer efficiency, it also leads to greater blood damage and microemboli formation.

In film and bubble oxygenators, direct contact between the blood and gas phases occurs. To improve the hemocompatibility of BOs, a nonporous membrane was placed between the blood and gas phases giving rise to membrane BOs.[6,7] The greatest limitation of early membrane BOs was a lack of membrane materials that were sufficiently permeable to respiratory gases. Furthermore, these oxygenators were designed with little regard to fluid mechanics. Thus, large concentration boundary layers developed on the blood side significantly compromising the gas transfer efficiency.[8] Advances in membrane technology in the 1960s led to the introduction of silicone rubber membranes that were much more permeable to respiratory gases, giving rise to the first commercial membrane BOs.[9,10] In the 1960s the first hollow fiber BOs were also developed using silicone rubber tubing.[11] Blood flow inside and outside the fibers was studied.[11–16] Although blood flow outside the fibers leads to higher rates of gas transfer,[17] early commercial designs were restricted to blood flow inside the fibers due to complexities in designing BOs with blood flow outside the fibers.

The next major advance in the development of membrane BOs came with the introduction of microporous hydrophobic membranes. Since the membrane pores are gas filled, the membrane resistance to gas transfer is negligible because the respiratory gases pass through the membrane pores rather than the membrane material. Furthermore, problems with carbon dioxide removal as a result of the permeability of the membrane to carbon dioxide were eliminated.[18–20] Early microporous membrane BOs included the Travenol Modulung-Teflo and Travenol TMO.

In 1980 membrane BOs accounted for only 20% of all BOs sold in the United States.[8] The benefits of membrane versus bubble oxygenators were often debated.[21,22] The Travenol TMO was considered complex to use.[23] Furthermore, hollow fiber BOs (blood flow inside the fibers) required surface areas as high as $5.5\,m^2$. During the early 1980s the designers of membrane BOs started to focus on passive mixing of the blood in order to reduce the blood side resistance to gas transfer. For example, the Cobe CML contained a microporous flat sheet membrane with a screen in the blood channels to induce mixing.[24] The membrane surface was reduced to $2.5\,m^2$. The Johnson and Johnson Extracorporeal Maxima used cross-wound hollow fibers where the blood flowed outside the fibers.[25] By increasing the gas transfer efficiency of these devices, the membrane surface area and priming volume were reduced. In addition, these BOs were easier to use. As a consequence of these improvements, by 1986, membrane BOs accounted for more than 60% of all BOs sold in the United States.

Today more than 99% of BOs sold in the United States contain microporous membranes. Additional increases in the rate of gas transfer have been achieved by carefully designing the blood channels to increase mixing. Devices with very low membrane surface areas (e.g., Sarns Turbo and Cobe Optima have surface areas of 1.9 and 1.7 m^2, respectively) have been built using carefully spaced mats of woven hollow fibers.[26,27] These woven hollow fibers provide uniform flow channels that minimize channeling of the blood. Since the membranes used are hydrophobic (e.g., polypropylene and Teflon) the pores are gas filled, resulting in a negligible membrane mass transfer resistance. The major resistance to oxygen transfer from the gas phase to the blood (and carbon dioxide transfer in the opposite direction) is the blood side concentration boundary layer. During the past 50 years tremendous advances have been made in the design of BOs. In the next 50 years, designing improved BOs will be far more complex.

In the past, the body temperature of a patient was lowered by cooling the blood during cardiopulmonary bypass (CPB), thus reducing the oxygen requirement. In current practice there is a trend towards normothermic coronary perfusion, which increases the oxygen requirement.[28] In addition, the need to minimize the transfusion of donated blood due to possible transmission of pathogens drives module design toward lower priming volumes.[8]

During CPB surgery, hematological and immune responses are observed as a result of contact between the blood and oxygenator surfaces. An increased gas transfer efficiency would allow a reduction in membrane surface area and, hence, a reduction in the average contact time between the blood and foreign surfaces. Since the membrane is often the most expensive component in a BO, minimizing the membrane surface area will reduce manufacturing costs.

Although disrupting the blood side mass transfer boundary layer leads to higher gas transfer efficiencies, it also leads to increased shear stresses on the blood cells. Damage to blood cells depends on both the applied shear stress and the time for which the shear stress is applied.[29] Thus, although mixing on the blood side is desirable, it is essential to ensure that the cells are not damaged.

Designing improved BOs will be complex, given the interdependence of the important design variables, such as the rate of gas transfer, the membrane area, priming volume, and blood damage. In the work reported here, gas transfer studies were conducted using flat sheet and hollow fiber BOs, since these are the two geometries used commercially. Water and glycerol/water solutions were used as a substitute for blood. Results for the oxygenation and deoxygenation of these solutions have been obtained using either pure oxygen or nitrogen as the gas phase. Mass transfer correlations were developed. These correlations are compared with analogous correlations for heat exchangers.

THEORY

The transfer of oxygen to water and glycerol/water solutions may be described by the following equation:[30]

$$N = K\Delta C, \quad (1)$$

where N is the total molar flux, ΔC is the overall concentration difference, and K is the overall mass transfer coefficient. K may be determined experimentally using the equation,

$$K = \frac{Q}{A}\ln\left(\frac{C_0 - C^*}{C - C^*}\right), \quad (2)$$

where Q is the liquid flow rate, A is the membrane surface area, and C_0, C, and C^* are, respectively, the inlet and outlet oxygen concentration in the liquid and the concentration of oxygen in the liquid if it were in equilibrium with the gas phase.[31,32] The overall mass transfer coefficient results from three contributions due to three individual mass transfer coefficients that describe the transfer of oxygen from the bulk gas to the membrane surface (gas transfer across the gas side concentration boundary layer), through the membrane, and from the membrane surface to the bulk liquid (gas transfer across the liquid side concentration boundary layer).[33,34]

Since the gas streams were either pure oxygen or nitrogen, the concentration boundary layer on the gas side is negligible. Furthermore, since the membranes are hydrophobic the membrane pores were gas filled. Therefore, the membrane resistance to gas transfer is also expected to be negligible.[35–38] Thus, the major resistance to mass transfer was on the liquid side and the overall mass transfer coefficient could be approximated by the liquid side mass transfer coefficient. Experimentally, this means that the overall mass transfer coefficient was independent of the gas flow rate.

Mass transfer coefficients are usually presented in terms of mass transfer correlations. For flow outside and across mats of hollow fibers, the mass transfer correlation has the form $Sh = aRe^b Sc^c$, where Sh, Re, and Sc are the Sherwood, Reynolds, and Schmidt numbers, respectively. In flat sheet BOs, the liquid stream flows in thin channels. Here the mass transfer correlation is usually of the form $Sh = aGr^b$, where Gr is the Graetz number.[17]

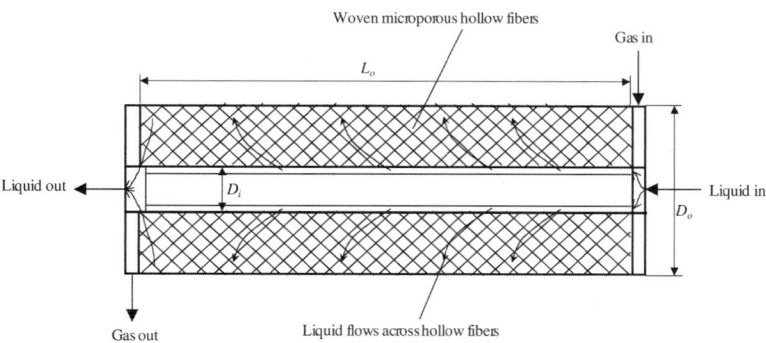

FIGURE 1. Schematic representation of a hollow-fiber membrane BO.

EXPERIMENTAL

FIGURES 1 and 2 are schematic diagrams of the hollow fiber and flat sheet BOs tested. TABLE 1 provides additional details of the BOs studied. All BOs were provided by Cobe Cardiovascular, Arvada, CO. The experimental setup is shown in FIGURE 3. The liquid stream consisted of deionized water and glycerol/water mixtures. TABLE 1 also gives the liquid streams used with the various BOs. The liquid stream was pumped from the feed reservoir using a Masterflex peristaltic pump (Cole-Parmer, Vernon Hills, IL). Having passed through a rotameter, the liquid stream was introduced into the first BO. The gas sweep to the first BO consisted of pure oxygen. Before flowing into the BO, the oxygen was saturated by bubbling it through a sample of the liquid stream. The liquid stream leaving the first BO was then fed to the second BO. In this device the sweep gas consisted of nitrogen saturated by the liquid stream. Thus, the liquid stream was oxygenated in the first BO and deoxygenated in the second BO. The outlet from the second BO was returned to the feed reservoir. Two gas flow rates, 0.81 and 2.1 L min^{-1}, were tested. Liquid flow rates ranged from 0.45 to 11 L min^{-1}.

The oxygen concentration of the liquid stream flowing into and out of the first BO and out of the second BO was measured using MI-730 oxygen electrodes

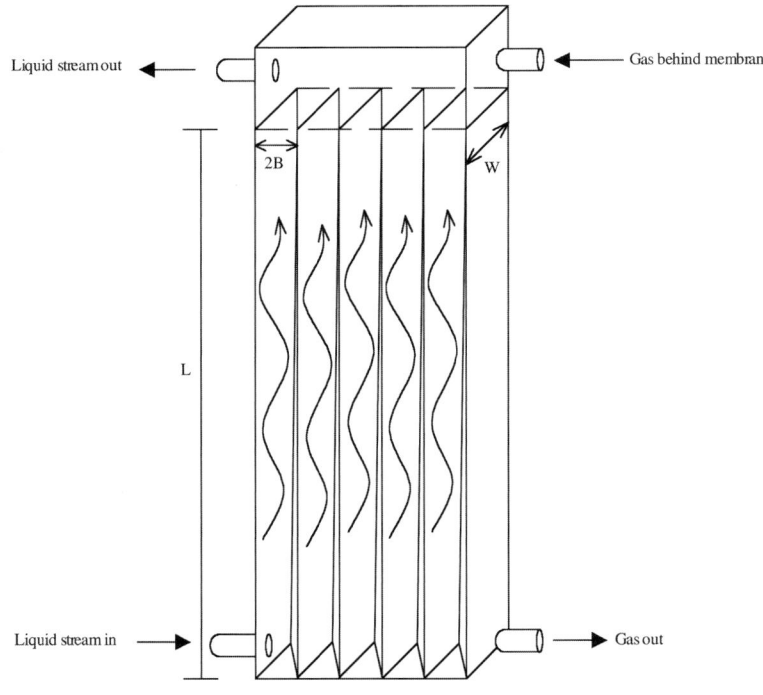

FIGURE 2. Schematic representation of a flat-sheet membrane BO.

(Microelectrodes Co., Bedford, NH). The oxygen concentration in the gas streams leaving both BOs was also measured using oxygen electrodes. The oxygen concentration in the inlet gases was either 100% (oxygen) or 0% (nitrogen).

RESULTS

Experiments were conducted in order to determine the rate of oxygen transfer to and from the various liquid streams. The viscosity of water and various glycerol/water mixtures was determined from tabulated data.[39] The diffusivity of oxygen in the various glycerol/water mixtures was estimated from the Wilke–Chang equation.[40]

TABLE 1. Oxygenators tested

Oxygenator	Membrane	Comments
Cobe Optima XP	There are 14,500 microporous polypropylene hollow fibers ($200\pm30\,\mu m$ inside diameter, $300\pm30\,\mu m$ outside diameter). Total surface area, equivalent diameter, and void fraction are $1.9\,m^2$, $480\,\mu m$, and 0.615, respectively.	Blood flows outside the fibers whereas gas flows in the fiber lumen. Water:glycerol ratio in liquid streams tested: 100:0, 95:5, 90:10, 80:20, 70:30, 60:40, and 50:50.
Cobe Optimin	There are 7,000 microporous polypropylene hollow fibers ($200\pm30\,\mu m$ inside diameter, $300\pm30\,\mu m$ outside diameter). Total surface area, equivalent diameter, and void fraction are $1.0\,m^2$, $516\,\mu m$, and 0.633, respectively.	Blood flows outside the fibers whereas gas flows in the fiber lumen. Water:glycerol ratio in liquid streams tested: 100:0, 80:20, and 60:40.
Cobe VPCML Plus	A flat sheet microporous polypropylene membrane (thickness $50\pm5\,\mu m$) is crimped to form blood channels $230\,\mu m$ thick.	This device consists of two compartments, one with a membrane surface area of $0.4\,m^2$, the other $0.85\,m^2$. In these experiments each compartment was tested separately. Liquid stream consisted of deionized water.
Cobe CML Duo	A flat sheet microporous polypropylene membrane (thickness $50\pm5\,\mu m$) is crimped to form blood channels $160\,\mu m$ thick.	Two oxygenation compartments (each $1.3\,m^2$) are linked in series. Experiments were conducted using membrane surface areas of 1.3 and $2.6\,m^2$. Water:glycerol ratio in liquid streams tested: 100:0, 80:20, 60:40.

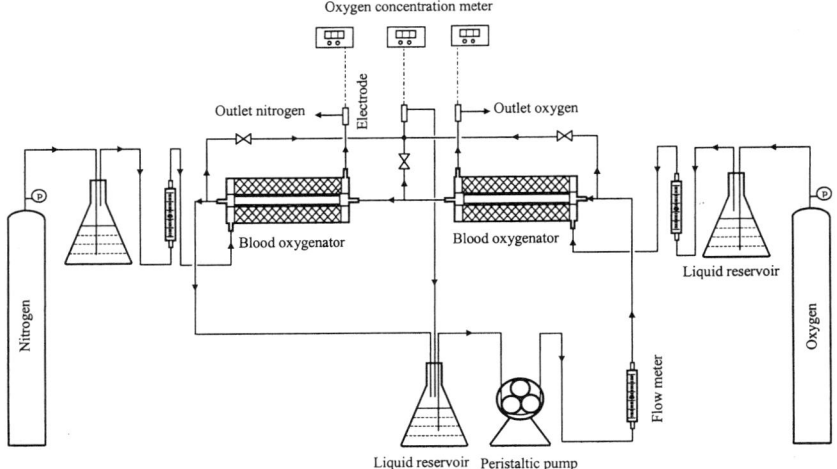

FIGURE 3. Experimental setup.

Flow outside Hollow Fibers

FIGURE 4 shows the variation of Sherwood number with Reynolds number for Schmidt numbers ranging from 470 to 11,880. The Sherwood, Reynolds, and Schmidt numbers are defined as Kd_e/D, ud_e/ν, and ν/D, respectively, where d_e, D, ν, and u are the equivalent diameter, diffusion coefficient of oxygen in the liquid stream, the kinematic viscosity of the liquid stream, and the liquid velocity,

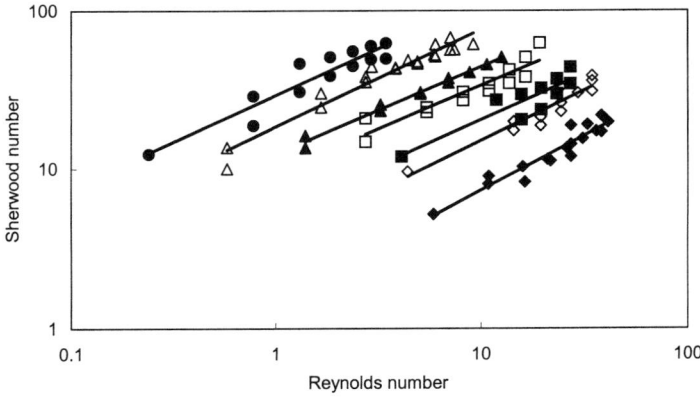

FIGURE 4. Variation of Sherwood number with Reynolds number for flow across hollow fibers. Data for oxygenation and deoxygenation at two gas flow rates have been combined. Schmidt numbers used were 470 (◆), 570 (◇), 730 (■), 1,230 (□), 2,300 (▲), 4,710 (△), and 11,880 (●).

respectively. The equivalent diameter is defined as (4 × [volume of liquid]/[wetted surface area]). As can be seen, the data for a given Schmidt number fall on a straight line. The average slope of the lines representing the various Schmidt numbers is 0.59. Data for oxygenation and deoxygenation for both BOs at both gas flows are indistinguishable.

In FIGURE 5 the Sherwood number divided by the Schmidt number raised to the one-third power is plotted against the Reynolds number. As can be seen, all the mass transfer data may be collapsed on to the same line given by the equation,

$$Sh = 0.8Re^{0.59}Sc^{0.33}. \tag{3}$$

The data in FIGURE 4 show some scatter. However, when these results are compared to those of previous studies[17,31,33] a similar level of scatter is observed.

Flow in Thin Channels

Mass transfer results for the flat sheet BO configurations are given in FIGURE 6, where the Sherwood number is plotted against the Graetz number. The Sherwood number is defined as $(K4B/D)$. The Graetz number is defined as $([4B]^2 u/DL)$. In these dimensionless numbers, B and L are the average half-thickness of the rectangular flow channel and the length of the channel (see FIG. 2). Here, different symbols are used for different liquid streams. The results for oxygenation and deoxygenation for the different BOs, with various blood channel thicknesses, at both gas flow rates, fall on the same curve. Furthermore, the results for the various liquid streams fall on the same curve. Although there is some scatter, the level of scatter is similar to that observed in a previous study.[17]

Shah and London[41] have modified the Lévêque solution for flow in rectangular ducts. The analogous mass transfer correlation is

$$Sh = 6.4Gr^{0.33}. \tag{4}$$

This equation is shown by a solid line in FIGURE 6. The Lévêque solution assumes fully developed laminar flow and a developing concentration boundary layer. Under

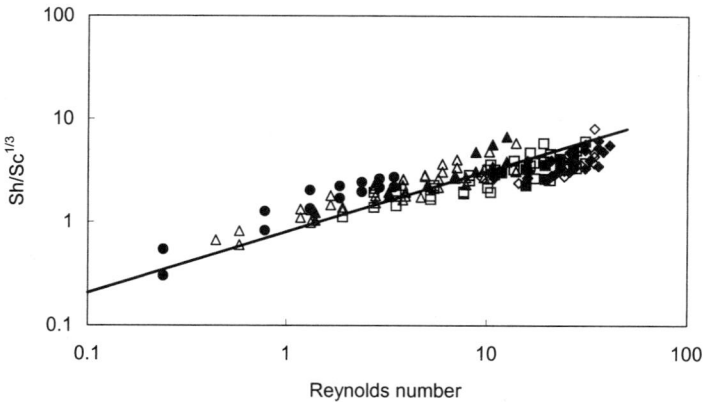

FIGURE 5. Variation of $Sh/Sc^{0.33}$ with Reynolds number at Schmidt numbers 470 (◆), 570 (◇), 730 (■), 1,230 (□), 2,300 (▲), 4,710 (△), and 11,880 (●).

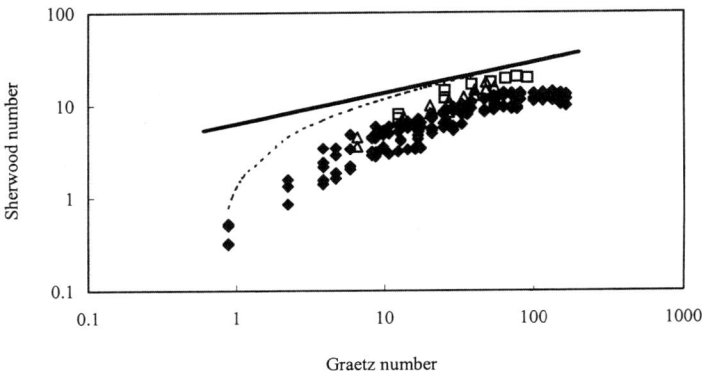

FIGURE 6. Variation of Sherwood number with Graetz number for flow in thin channels. Data are shown for liquid streams consisting of water (◆), 80:20 (△), and 60:40 (□) water glycerol mixtures. Data for oxygenation and deoxygenation in blood channels of various thicknesses and for two different gas flow rates have been combined. The equation $Sh = 6.4\,Gr^{0.33}$ is represented by the *solid line*. The *dotted curve* gives the prediction obtained when polydispersity in the blood channel thickness is considered.

these conditions, it is found that the Sherwood number does not depend upon the Schmidt number.[17] Thus, as observed in FIGURE 6, the experimental results for various Schmidt numbers fall on the same curve. For Graetz numbers above 10 the experimental data lie below but parallel to the theoretical correlation. However, at Graetz numbers below 10, the experimental and theoretical results deviate.

The experimental results are summarized in TABLE 2. The first column gives the flow geometry and the second column gives the Reynolds and Schmidt number ranges investigated. The third column gives the mass transfer correlation that was derived

TABLE 2. Experimentally derived correlations

Geometry	Flow Range Investigated	Inferred Correlation	Literature Correlation	Comments
Flow across fibers	$0.1 < Re < 100$ $470 < Sc < 11{,}880$	$Sh = 0.8Re^{0.59}Sc^{0.33}$	$Sh = 0.39Re^{0.59}Sc^{0.33a}$ $Sh = 0.8Re^{0.47}Sc^{0.33b}$	Good agreement between heat transfer and mass transfer
Flow in thin channels	$0.5 < Gr < 500$ $470 < Sc < 4{,}710$	$Sh = 0.5Gr$ $(0.5 < Gr < 10)$ $Sh = 3.0Gr^{0.33}$ $(10 < Gr < 500)$	$Sh = 6.4Gr^{0.33c}$	Rate of mass transfer less than predicted from theory

[a]Correlations derived from analogous heat transfer results.[43]
[b]Correlations derived from analogous mass transfer correlations.[31]
[c]Correlations derived from theory.[41]

from the experimental data. The fourth column gives analogous mass and heat transfer correlations that were developed by others. These results are discussed in detail in the next session.

DISCUSSION

Mass transfer to and from woven hollow fibers for liquid flowing across the fibers may be compared to flow across banks of smooth tubes in compact cross flow heat exchangers.[42] The tubes in a cross flow heat exchanger are very carefully spaced. Inline and staggered arrangements exist. Variations of the heat transfer coefficient with the arrangement and pitch of the tubes have been tabulated.[42]

The hollow fibers used in BOs are woven together in a mat. These mats are then wound around a central core (FIG. 1) forming several layers. Each layer contains carefully spaced parallel fibers (transverse pitch). However, since the layers are wound around the central core, they are less carefully spaced (longitudinal pitch). Thus, the three-dimensional arrangement of the fibers is far less regular than the tubes in a heat exchanger.

TABLE 2 shows that the mass transfer correlation developed here is very similar to two previously described correlations. The mass transfer correlation derived from the analogous heat transfer correlation[43] for closely spaced tube banks in a cross flow heat exchanger predicts that

$$Sh = 0.39 Re^{0.47} Sc^{0.33}. \tag{5}$$

Wickramasinghe et al.[31] determined mass transfer correlations for a handmade hollow fiber module consisting of hollow fiber fabric wound around a central porous tube. In these modules the liquid entered the central porous tube and was forced to flow back and forth across the fibers by a series of plugs and O rings. Wickramasinghe et al. found that, for liquid flow outside the fibers, the liquid side mass transfer coefficient could be represented by the correlation

$$Sh = 0.8 Re^{0.47} Sc^{0.33}. \tag{6}$$

In Equations (5) and (6) the dependence of the Sherwood number on the Schmidt number was not verified. By varying the kinematic viscosity of the liquid stream and the diffusion coefficient of oxygen in liquid, the dependence of the Sherwood number on the Schmidt number has been determined experimentally in this study. FIGURE 5 shows that in fact the Sherwood number does vary with the Schmidt number raised to the one-third power, as has been assumed in the past.[17,31,33,38,44]

The flat sheet BOs studied contain a screen in the blood channel to prevent the membrane layers from collapsing on top of each other. Areas of contact between the screens and the membrane surface are not available for mass transfer, resulting in the effective membrane surface area for mass transfer being less than the actual membrane surface area present. Thus, the experimentally derived Sherwood numbers are less than theoretically predicted for Graetz numbers greater than 10. Furthermore, as expected, the rate of mass transfer does not depend on the Schmidt number.

At Graetz numbers less than 10, the Sherwood numbers determined experimentally are much lower than those predicted by the Lévêque solution. The Lévêque solution is a limiting case of the Graetz solution. It is valid for large Graetz numbers in the region where the concentration boundary layer is developing.[41] Once the concentration boundary layer is fully developed the Sherwood number will be a constant, independent of Graetz number. However, the results obtained here do not support this limit; rather the Sherwood number depends more strongly on Graetz number at low Graetz numbers.

In this analysis, it has been assumed that all the blood channels have the same thickness. In reality a distribution of channel thicknesses is likely. At higher flow rates this will have little effect on the overall mass transfer efficiency. However, at lower flow rates, this will lead to a lower gas transfer efficiency due to channelling and by passing by the liquid stream. In addition, contact between the screen in the blood channels and the membrane surface will continue to result in an overprediction of the mass transfer coefficient. Wickramasinghe *et al.*[17,45] modified Equation (2) to account for effect of polydisperse channel thicknesses on the rate of mass transfer. The dotted line in FIGURE 6 is the resulting prediction. As can be seen this equation is better able to predict the experimental results.

When we compare the BOs available today to the human lungs we see they are very different. The human lungs have a gas transfer surface area of 50–100 m^2, the capillary diameter in the alveolar wall is about 10 μm, the thickness of the blood–gas barrier is 0.5 μm, and pulmonary arterial pressure is about 2,000 Pa.[46] By providing slow flow in narrow capillaries, the gas transfer surface area is maximized and shear stress on the blood minimized. In the design of BOs, however, it is essential to minimize the priming volume in order to minimize the transfusion requirements. Additional clinical and economic considerations require that the membrane surface area is minimized. Clearly, the human lungs are very different to BOs.

In this work, the oxygenation and deoxygenation of water and glycerol/water mixtures have been investigated. This is much simpler than oxygenation of blood. Blood is a non-Newtonian fluid, whereas water and glycerol/water solutions are Newtonian. Furthermore, oxygen diffuses into the plasma and binds to hemoglobin, whereas in the liquid streams studied here, no binding of oxygen occurs. Thus, more sophisticated correlations will be required to predict the actual transfer rate of oxygen from the gas stream to blood. However, the dimensionless mass transfer correlations developed here may be used as a guide when designing better BOs.

CONCLUSIONS

Mass transfer correlations have been developed for flow across bundles of woven hollow fibers and in thin channels. For flow across bundles of woven fibers the results may be compared to analogous results for flow across tube banks in heat exchangers. For flow in thin channels we show that polydispersity in the channel thickness could lead to lower rates of mass transfer than for uniform channels.

ACKNOWLEDGMENTS

Financial support was provided by Cobe Cardiovascular, Arvada, Colorado, the Colorado Institute for Research in Biotechnology, and the National Science Foundation (CAREER Program Grant No. BES 9984095).

REFERENCES

1. GIBBON, JR., J.H. 1937. Artificial maintenance of circulation during experimental occlusion of pulmonary artery. Arch. Surg. **34:** 1105–1131.
2. GIBBON, JR., J.H. 1954. Application of a mechanical heart and lung apparatus to cardiac surgery. Minn. Med. **37:** 171–185.
3. KIRKLIN, J.W. & J.W. DUSHANE. 1955. Intracardiac surgery with the aid of a mechanical pump oxygenator system (Gibbon type): report of eight cases. Mayo. Clinic Proc. **30**(10): 201–206.
4. WEGNER, J.A. 1997. Oxygenator anatomy and function. J. Cardiothorac. Vascular Anesthes. **11**(3): 275–281.
5. CLARK, L.C., F. GOLLAN & V.P. GUPTA. 1950. The oxygenation of blood by gas dispersion. Science **111:** 85–87.
6. KLOFF, W.J. & R.R. BLAZER. 1955. Artificial coil lung. Trans. Am. Soc. Artif. Intern. Organs **1:** 39–42.
7. CLOWES, G.H.A., A.L. HOPKINS & W.E. NEVILLE. 1956. An artificial lung dependent upon diffusion of oxygen and carbon dioxide through plastic membranes. J. Thorac. Surg. **32:** 630–637.
8. VOORHEES, M.E. & B.F. BRIAN III. 1996. Blood gas exchange devices. Intl. Anesthes. Clinics **34**(2): 29–45.
9. BRAMSOM, M.L., J.J. OSBORN, F.B. MAIN, et al. 1965. A new disposable membrane oxygenator with integral heat exchange. J. Thorac. Cardiovasc. Surg. **50**(3): 391–400.
10. LANDÉ, A.J., L. EDWARDS, J.H. BLOCH, et al. 1970. Prolonged cardio-pulmonary support with a practical membrane oxygenator. Trans. Am. Soc. Artif. Int. Organs **16:** 352–356.
11. BODELL, B.R., L.R. HEAD, J.M. HEAD, et al. 1963. A capillary membrane oxygenator. J. Thorac. Cardiovasc. Surg. **46**(5): 639–650.
12. WILSON, R., D.J. SHEPLEY & E. LEWELLYN-THOMAS. 1965. A membrane oxygenator with a low priming volume for extracorporeal circulation. Can. J. Surg. **8:** 309–311.
13. DE FILIPPI, R.P., F.C. TOMPKINS, J.H. PORTER, et al. 1968. The capillary membrane blood oxygenator: in vitro and in vivo gas exchange measurements. Trans. Am. Soc. Artif. Int. Organs **14:** 236–241.
14. DORSON, W.J., E. BAKER & H. HULL. 1968. A shell and tube oxygenator. Trans. Am. Soc. Artif. Intern. Organs **14:** 242–249.
15. DUTTON, R.C., F.W. MATHER, S.N. WALKER, et al. 1971. Development and evaluation of a new hollow-fiber membrane oxygenator. Trans. Am. Soc. Artif. Intern. Organs **17:** 331–336.
16. ZINGG, W. 1967. Membrane oxygenator for infants. Trans. Am. Soc. Artif. Intern. Organs **13:** 334–340.
17. WICKRAMASINGHE, S.R., M.J. SEMMENS & E.L. CUSSLER. 1992. Mass transfer in various hollow fiber geometries. J. Membr. Sci. **69:** 235–250.
18. DANTOWITZ, P. & A. BORSANYI. 1969. A blood oxygenator with preformed, membrane lined channels. In Artificial Heart Program Proceedings. R. Hegyeli, Ed. Government Printing Office, Washington, DC.
19. LAUTIER, A., P. REY, J. BIZOT, et al. 1969. Comparison of gaseous transfers through synthetic membranes for oxygenators. Trans. Am. Soc. Artif. Intern. Organs **15:** 144–150.
20. GILLE, J.P., L. TRUDELL, M.T. SNIDER, et al. 1970. Capability of the microporous, membrane-lined, capillary oxygenator in hypercapnic dogs. Trans. Am. Soc. Artif. Intern. Organs **16:** 365–374.

21. MUSIC, K., W.J. HORGAN, R. RICHARDS, et al. 1982. Controversy concerning the use of membrane oxygenators. Proc. Am. Acad. Clin. Perf. **3:** 17–19.
22. MASSIMINO, R.J. & W.G. MAURER. 1983. Membrane versus bubbler. J. Extra-Corporeal Tech. **15:** 156–161.
23. KALSHOVEN, D., L. CAMERIENGO & J. DEARING. 1983. Membrane oxygenators: a few observations. J. Extra-Corporeal Tech. **15:** 41–44.
24. ELGAS, R.J. & T.M. GORDON, Inventors; Cobe Laboratories, Inc., Assignee. 1984. US Patent 4,451,562. Date of application: May 29.
25. IATRIDIS, A., T. CHAN & R. THOMPSON. 1985. Experimental and clinical trials with the extracorporeal hollow fiber lung. Proc. Am. Acad. Clin. Perf. **6:** 47–50.
26. BAURMEISTER, U., Inventor; Akzo NV, Assignee. 1992. US patent 5,143,312. Data of application: September 1.
27. BAURMEISTER, U., Inventor; Akzo NV, Assignee. 1990. US patent 4,940,617. Data of application: July 10.
28. GALLETTI, P.M. 1993. Cardiopulmonary bypass: a historical perspective. Artif. Organs **17**(8): 675–686.
29. ZYDNEY, A.L. 1985. Cross-flow Membrane Plasmapheresis: An Analysis of Flux and Hemolysis. Ph.D. Thesis, Massachusetts Institute of Technology, Cambridge, MA.
30. CUSSLER, E.L. 1984. Diffusion: Mass Transfer in Fluid Systems. Cambridge University Press, New York.
31. WICKRAMASINGHE, S.R., M.J. SEMMENS & E.L. CUSSLER. 1993. Hollow fiber modules made with hollow fiber fabric. J. Membr. Sci. **84:** 1–14.
32. SIRKAR, K.K. 1992. Other new membrane processes. *In* Membrane Handbook. W.S.W. Ho & K. K. Sirkar, Eds. Van Nostrand Reinhold, New York.
33. YANG, M.-C. & E.L. CUSSLER. 1986. Designing hollow fiber contactors. AIChE J. **32**(11): 1910–1916.
34. PRASAD, R. & K.K. SIRKAR. 1988. Dispersion free solvent extraction with microporous hollow fiber modules. AIChE J. **34**(2): 177–188.
35. QI, Z. & E.L. CUSSLER. 1985. Microporous hollow fibers for gas absorption: I. mass transfer in the liquid. J. Membr. Sci. **23:** 321–332.
36. QI, Z. & E.L. CUSSLER. 1985. Microporous hollow fibers for gas absorption: II. mass transfer across the membrane. J. Membr. Sci. **23:** 333–345.
37. KENFIELD, C.F., R. QIN, M.J. SEMMENS & E.L. CUSSLER. 1988. Cyanide recovery across hollow fiber gas membranes. Environ. Sci. Technol. **22**(10): 1151–1155.
38. YANG, M.-C. & E.L. CUSSLER. 1989. Artificial gills. J. Membr. Sci. **42:** 273–284.
39. WEAST, R.C. 1974. Handbook of Chemistry and Physics, 55th edit. CRC Press, Cleveland.
40. WILKE, C.R. & P.C. CHANG. 1955. Correlations of diffusion coefficients in dilute solutions. AIChE J. **1:** 264–270.
41. SHAH, R.K. & A.L. LONDON. 1974. Thermal boundary conditions and some solutions for laminar duct flow forced convection. J. Heat Transf. **96**(C2): 159–165.
42. ŽUKAUSKAS, A. & R. ULINSKAS. 1988. Heat Transfer in Tube Banks in Crossflow. Hemisphere Publishing Corporation, New York.
43. KREITH, F. & W.Z. BLACK. 1980. Basic Heat Transfer. Harper and Row, New York.
44. CATAPANO, G., H.D. PAPENFUSS, A. WODETZKI & U. BAURMEISTER. 2001. Mass and momentum transport in extra-luminal flow (ELF) membrane blood oxygenation. J. Membr. Sci. **184:** 123–135.
45. GOERKE, A.R., J. LEUNG & S.R. WICKRAMASINGHE. 2002. Mass and momentum transfer in blood oxygenators. Chem. Eng. Sci. **57**(11): 2035–2046.
46. WEST, J.B. 1985. Respiratory Physiology, 3rd edit. Williams and Wilkins, Baltimore.

Membrane Evaporative Cooling to 30°C or Less

1. Membrane Evaporative Cooling of Contained Water

SIDNEY LOEB

P.O. Box 41, Omer, Israel

ABSTRACT: Microporous hydrophobic membranes have been examined for possible use as containers in the evaporative cooling of water, particularly in desert climates. An experimental determination was made of the overall heat and mass transfer coefficients of these membranes while surmounting contained water and with air flowing over the surface of the membranes. Similar tests were made with water alone, that is, without a membrane. The coefficients were then used to compare the performance of existing (canvas water) coolers and membrane evaporative coolers under desert conditions. The performance of the membrane coolers was close enough to that of the canvas coolers that extensive investigation of various aspects of membrane evaporative cooling appears to be justified, particularly in view of the potential advantages of the latter over the existing evaporative cooling methods. For example, for cool storage of perishable goods in a desert climate, the membrane container might be uniquely qualified because of its low rate of water consumption compared to that of a canvas cooler.

KEYWORDS: evaporative water cooling; microporous hydrophobic membranes; membrane evaporative water cooling

NOMENCLATURE:

ABL	air boundary layer
amb	ambient
ΔW	humidity ratio driving force in water warming test, $W_{H_2O} - T_{WB,SU}$
DB	dry bulb
K	overall mass transfer coefficient of water vapor, $(gm/sec) \, m^{-2} W^{-1}$
M	measured property in water warming test
memb	membrane
MD	membrane distillation
MEC	membrane evaporative cooling
MH	microporous hydrophobic
PC	psychrometric chart
RH	relative humidity, $RH = 100 W/W_{H_2O}$ %
SU	air property near surface of membrane in water warming test
T	temperature, °C
t	time, seconds
U	overall heat transfer coefficient, $(kJ/sec) \, m^{-2} \deg^{-1}$
W	humidity ratio, grams water vapor per kilogram dry air (see FIG. 5)
$W_{DB,SU}$	humidity ratio of air close to surface of membrane
W_{H_2O}	equivalent humidity ratio of liquid water, expressed as humidity ratio of saturated air having same temperature as the water, gm water vapor per kg dry air
WB	wet bulb

Address for correspondence: Sidney Loeb, P.O. Box 41, Omer 84965, Israel.
sidloeb@bgumail.bgu.ac.il

INTRODUCTION

Small intermittently-filled evaporative water coolers are widely used by campers, particularly in desert climates. They are usually simple pouches of porous canvas. The pouch is originally filled with ambient temperature water that slowly permeates the canvas, during which time a fraction of it evaporates into the low humidity air, cooling the remaining water in the pouch. The liquid permeation rate through the canvas is controlled by the static head of the liquid water rather than the humidification demands of the air and, thus, the rate can be insufficient or excessive. If insufficient, the water cooling rate is too low. If excessive, liquid water is wasted. The wetness can be annoying if the cooler is carried on the person of the camper. The static head of water can also cause plugging of the porous canvas by particulate matter in the water. Air/water conjunction could lead to microorganism multiplication.

From the above discussion it is clear that the canvas cooler cannot be considered for a closely related application, extended storage of perishable goods, primarily because of the excessive losses of liquid water from this type of cooler.

It is believed that most of the drawbacks of evaporative cooling, as enumerated above, can be eliminated by *membrane evaporative water cooling*, described in the next section.

THE CONCEPT OF MEMBRANE EVAPORATIVE COOLING OF CONTAINED WATER

FIGURE 1 shows, conceptually, a membrane evaporative cooler. The liquid water is contained by a microporous hydrophobic (MH) membrane. Some water at the membrane/liquid water interface vaporizes and passes through the micropores and the air boundary layer (ABL) into the free air on the other side. The vaporization cools the adjacent bulk water in the container. The driving force for the process is essentially the difference between the vapor pressure of liquid water and the partial pressure of water vapor in the free air. However, the difference is expressed herein in terms of humidity ratios, $W_{H_2O} - W$. W is defined as grams of water vapor per gram of dry air in the free air phase. W_{H_2O} is the *equivalent* humidity ratio of liquid water, that is, of saturated air having the same temperature as the liquid water.

In principle, the membrane evaporative water cooler eliminates the faults of existing evaporative water coolers, viz: dripping, wastage, and corrosiveness of liquid water are minimized or eliminated; aerobic microorganism proliferation is prevented or minimized because the airstream and liquid water are physically separated by the membrane. Camping water coolers could be efficient, non-plugging devices for which water permeation is responsive only to evaporation requirements. Potable water containers made from a microporous hydrophobic membrane would be dry to the touch. Because of their good water conservation, membrane water coolers could be considered for extended storage of perishable goods at temperatures close to the wet bulb temperature.

To explore the possibilities of membrane evaporative cooling of contained water by microporous hydrophobic membranes, a rough comparison was made between the performance of canvas and membrane evaporative coolers. The comparison was

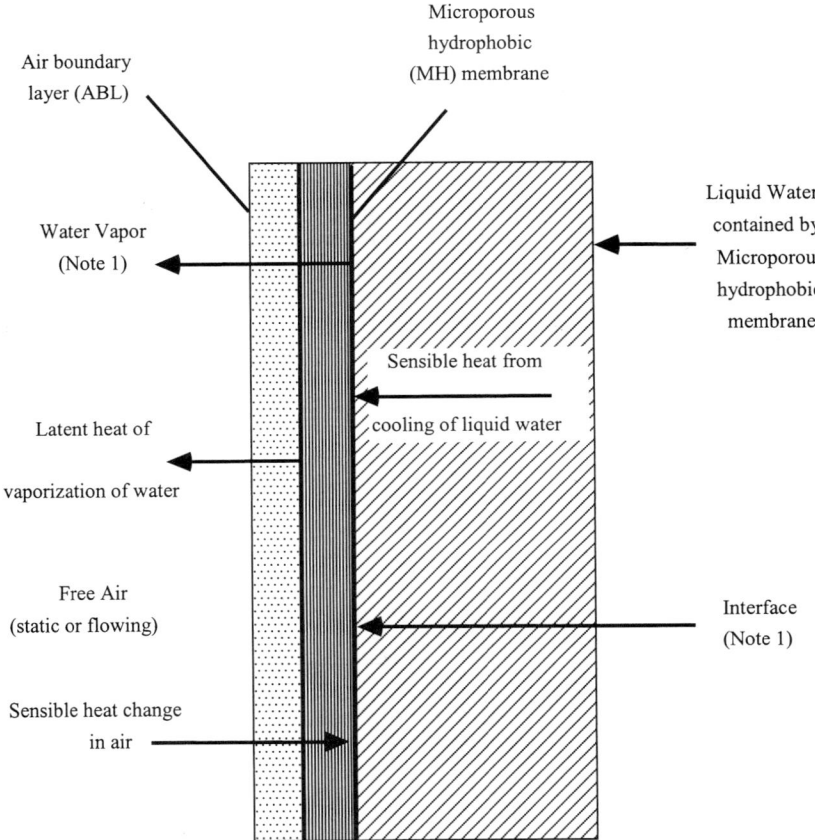

FIGURE 1. Membrane evaporative cooling of contained water. NOTE: Liquid water cannot enter micropores at interface because the static head of water is below breakthrough pressure; water vapor can enter.

based on a knowledge of the overall heat and mass transfer coefficients, U and K, in each type of cooler, these being obtained by evaporative experiments reported herein, both on microporous hydrophobic membranes over water and on water alone, that is, with no membrane.

EXPERIMENTAL

The performance of membrane evaporative water coolers depends on the overall coefficients mentioned above. These, in turn, are determined essentially by the individual coefficients of the membrane and the air boundary layer (ABL) shown in FIGURE 1. The value of the ABL coefficients is a function of air velocity and direction relative to the membrane.

The purpose of the experiments was primarily to determine the order of magnitude of the overall coefficients to be expected under conditions of evaporative water cooler operating conditions. The experimental apparatus is illustrated in FIGURE 2. Air was impinged vertically at about 3 m/sec on a microporous hydrophobic membrane surmounting the water in an insulated cup. Two tests were also made without any membrane surmounting the water. In all the tests, the impinging air was at a dry bulb (DB) temperature $T_{DB,SU}$, where SU refers to the air at the surface of the membrane. The temperature $T_{DB,SU}$ was slightly higher than $T_{DB,amb}$, the ambient temperature of the former being slightly higher because of the heating effect of the blower. The initial water temperature in the cup was either appreciably warmer than the wet bulb temperature, $T_{WB,SU}$, and the water was therefore cooled, or appreciably cooler and therefore warmed. The coefficients from the water warming test were more conservative and more consistent. Thus, these data are presented herein.

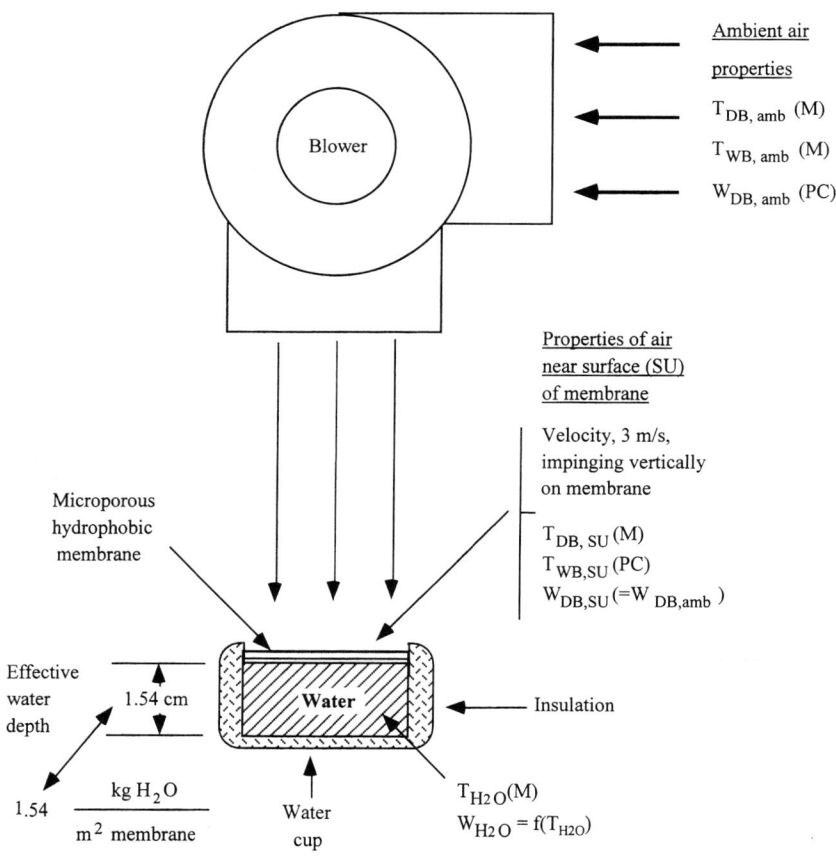

FIGURE 2. Apparatus for obtaining overall heat and mass transfer coefficients with air flowing over membrane surmounting contained water.

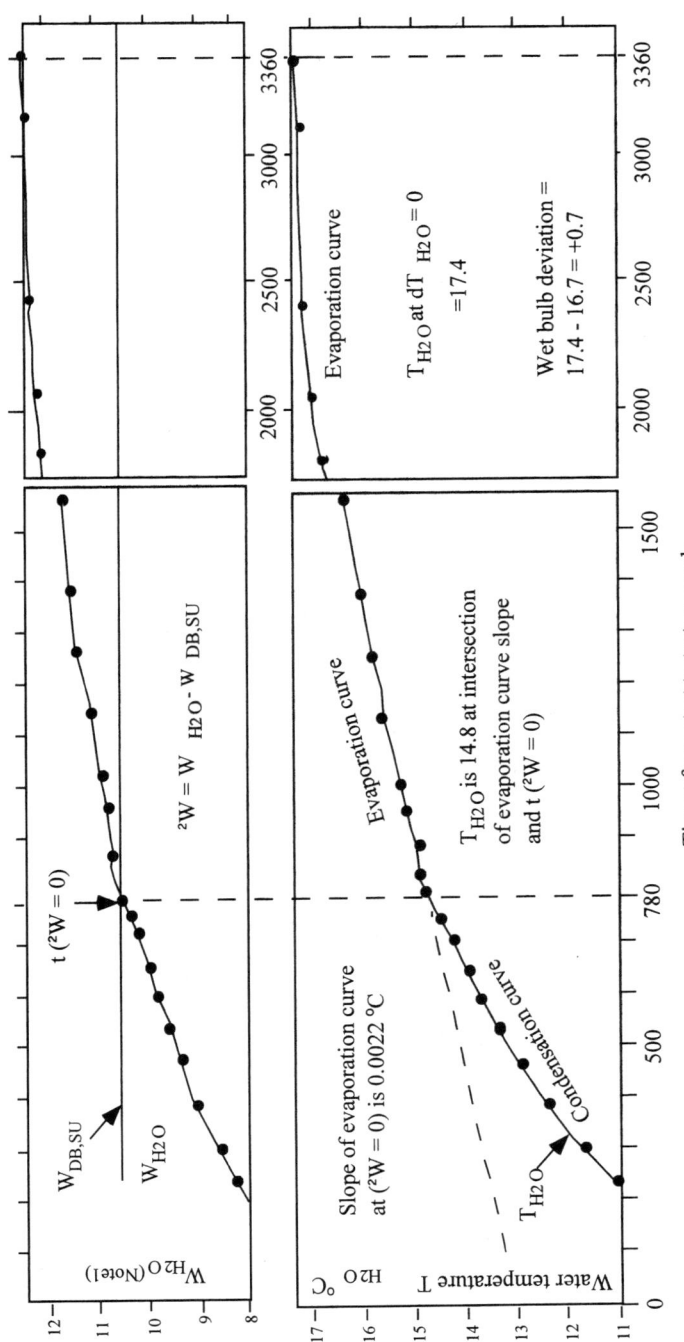

FIGURE 3. Water warming test of Accurel membrane. Air conditions: $T_{DB,amb} = 18.0$, $T_{WB,amb} = 15.9$, $T_{DB,SU} = 20.2$, $T_{WB,SU} = 16.7$, $W_{DB,amb} = W_{DB,SU} = 10.5$. W_{H_2O} is the equivalent humidity of liquid water, expressed as humidity ratio of saturated air having the same temperature, T_{H_2O}, as the water; units, gm water vapor/kg dry air.

TABLE 1. Microporous hydrophobic membranes tested

Manufacturer or Supplier	Designation	Composition	Thickness (microns)	Nominal Pore Size (microns)	Water Breakthrough Pressure (kg/cm^2)
Gelman[a]	TF200	polytetrafluoroethylene	135	0.2	4.9
Gelman[a]	H3101/single	polyurethane fluoroacrylate	73	0.2	4.75
Enka[b]	Accurel	polypropylene	155	0.4	—

[a]Gelman Sciences Technology, Ltd., formerly at Kiryat Weizmann, POB 2057, Rehovot, 76120 Israel. Present address: Gelman Sciences, Inc., 600 South Wagner Road, Ann Arbor, MI, USA.
[b]Enka AG, Oehder Strasse 28, Wuppertal 2, D-5600, Federal Republic of Germany. Enka is associated with Akzo, NV Netherlands. Sample obtained from E. Drioli, University of Napoli.

FIGURE 3 shows typical curves of water temperature, T_{H_2O}, and the equivalent water humidity ratio, W_{H_2O}, versus time, t, for a water warming rest with an Accurel membrane. Of special significance are the characteristics at times 780 and 3,360 seconds. The method of analysis of these results for calculating overall heat and mass transfer coefficients and other information is given in APPENDIX 1.

TABLE 2. Information gained from water warming tests

Designation		Overall Conductive Heat Transfer Coefficient U ($kJ/s\ m^2\ K$)	Overall Mass Transfer Coefficient K ($gm\ H_2O\ vapor\ sec\ m^2\ W$)	Wet Bulb Deviation (°C)
Water alone		0.0341	0.0304	0
		0.0445	0.0394	0
Average for H_2O		0.0393	0.0349	0
Microporous hydrophobic membranes over water	TF-200	0.0255	0.0152	+1.0
		0.0394	0.0180	+1.4
		0.0265	0.0148	+0.8
	H3101S	0.0322	0.0160	+1.1
		0.0392	0.0197	+1.0
	Accurel	0.0262	0.0149	+0.7
Average for MH membranes		0.0315	0.0164	+1.0

NOTE: $0.0164\ gm/sec\ m^2\ W = 1.01 \times 10^{-7}\ kg/sec\ m^2\ Pa = 1.01 \times 10^{-2}\ kg/sec\ m^2\ bar = 1.025 \times 10^{-2}\ kg/sec\ m^2\ atm = 1.35 \times 10^{-5}\ kg/sec\ m^2\ mmHg$.

SUMMARY OF EXPERIMENTAL INFORMATION GAINED FROM WATER WARMING TESTS

TABLE 1 gives information on relevant general properties of the three types of microporous hydrophobic membranes tested. TABLE 2 gives overall heat and mass transfer coefficients determined from the water warming tests on these three types. Coefficients are also given from warming tests on water alone, that is, with no membrane surmounting the water. In APPENDIX 2 a comparison is made between these coefficients and those obtained by other investigators in membrane distillation, which also uses microporous hydrophobic membranes.

In FIGURE 3, the water temperature, T_{H_2O}, was 17.4°C at the asymptote where $dT_{H_2O}/dt \to 0$. This may be considered the wet bulb temperature for the membrane–water system under test. The standard wet bulb temperature[a] of the air in contact with the membrane surface, $T_{WB,SU}$, was 16.7°C. The difference, 17.4 − 16.7 = 0.7°C, is called the *wet bulb deviation* and it averaged +1.0°C for all the microporous hydrophobic tests of TABLE 2. Thus, in membrane evaporative cooling of water, the lowest attainable temperature may be slightly higher than the standard wet bulb temperature of the contacting air.

A positive wet bulb deviation occurs with MH membranes because K/U is considerably less than in ordinary evaporative water cooling (U and K being the respective overall heat and mass transfer coefficients).

COMPARATIVE PERFORMANCE OF CANVAS AND MH MEMBRANE EVAPORATIVE WATER COOLERS

The overall transfer coefficients in TABLE 2 can be used to estimate the comparative performances of canvas and MH membrane coolers. The results are shown in FIGURE 4. The calculation methods leading to these curves are given in APPENDIX 3. (FIGURE 5, the psychrometric chart, was useful in calculations herein.)

The external conditions common to both of these materials are stated in FIGURE 4 and refer to the properties of warm desert air impinging on the pouch, inside of which is water, originally at 30°C. The ratio of water weight to permeation area is 20kg/m².

As can be seen in FIGURE 4 the canvas pouch is cooling water faster than the MH membrane pouch. Furthermore, the water in the canvas pouch ultimately cools to 16.1°C compared to 18.4°C for the membrane pouch, a difference discussed in the previous section.

As is shown in FIGURE 4, the overall mass transfer coefficient of the MH membrane is less than half that of the canvas container. The probable reason for this is

---

[a]*Standard wet bulb temperature.* Consider an air sample having a given humidity ratio, W, and a given fixed temperature. Assume that the air sample is in contact with liquid water alone (without a membrane). The liquid water will warm or cool to an equilibrium ($dT_{H_2O}/dT = 0$) temperature such that the latent heat lost from the liquid water by evaporation is equal to the sensible heat gained by conduction from the free air through the air boundary layer to the liquid water.

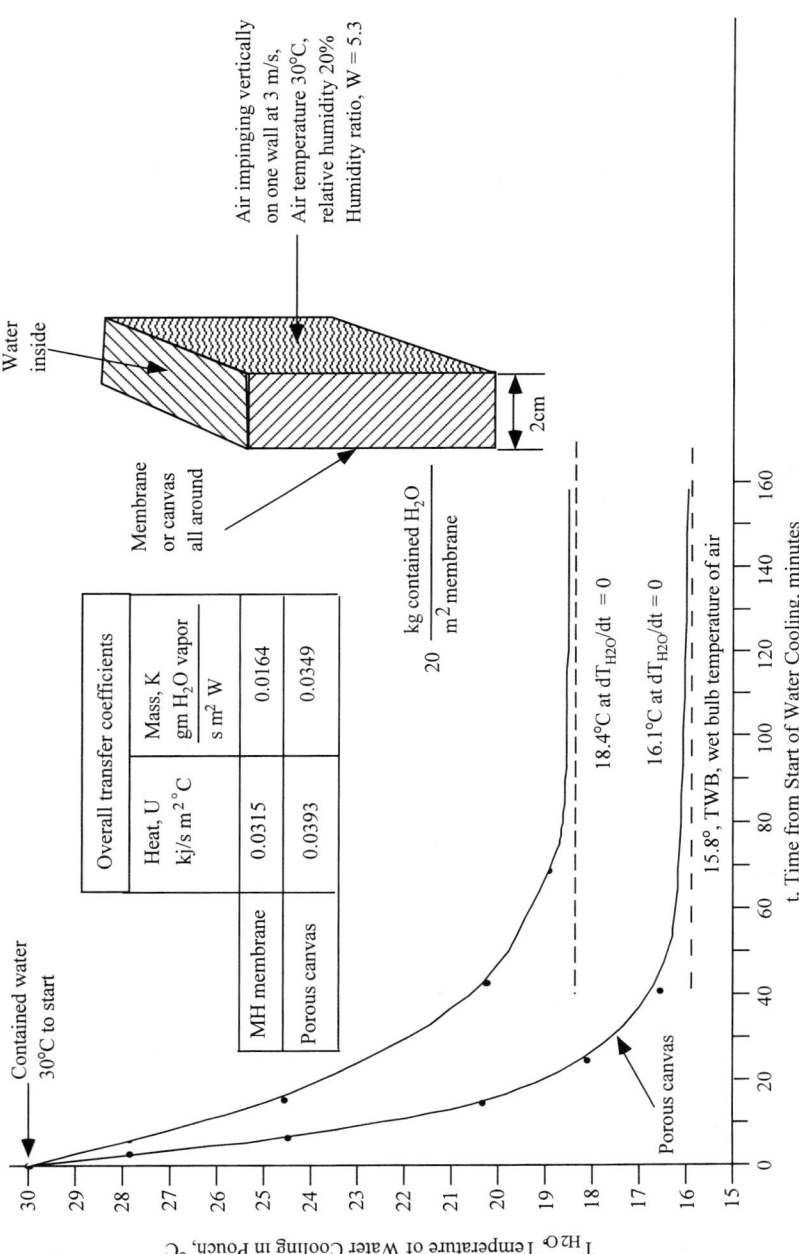

FIGURE 4. Estimated cooling characteristics of water contained in a canvas or MH membrane pouch with warm desert air impinging on pouch.

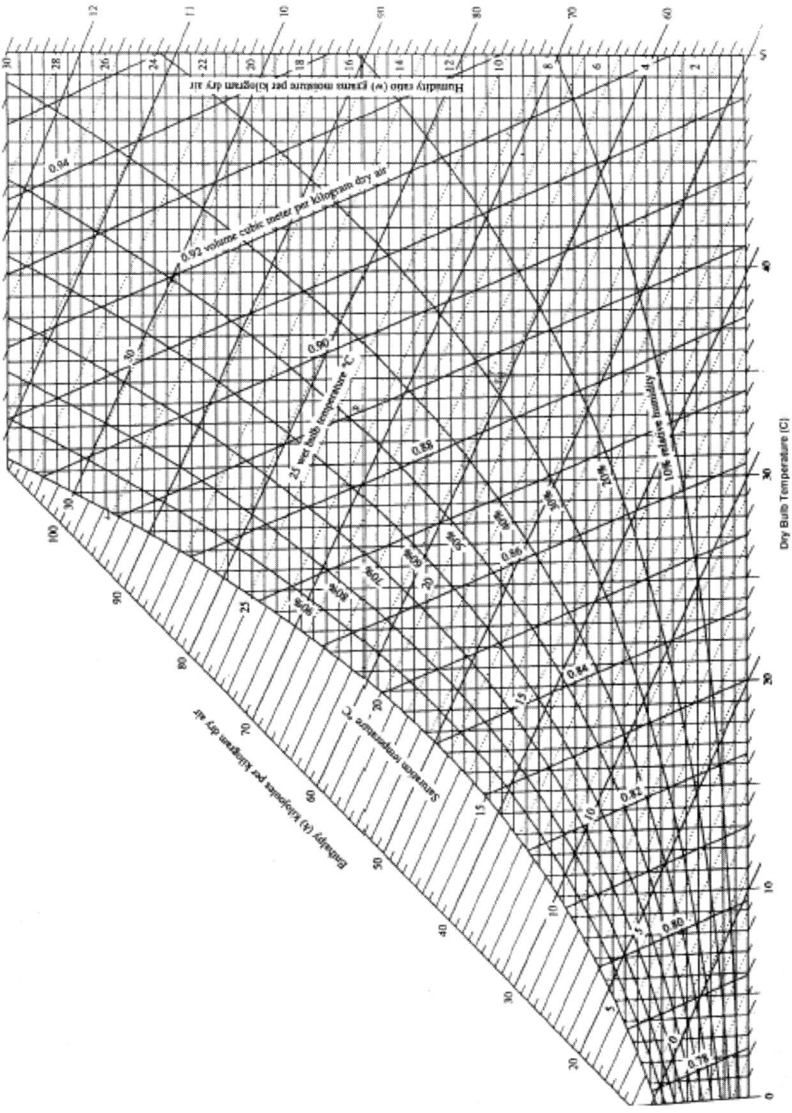

FIGURE 5. Psychrometric chart.

that the MH membrane has a considerable resistance to the passage of water vapor through its micropores.

The superior performance of the canvas membrane can also be understood qualitatively by reference to FIGURE 1. For the *membrane water cooler* case there are two resistances in series, the membrane and the air boundary layer. However, for the *canvas cooler* case there is only *one* resistance—the air boundary layer.

CONCLUSIONS

It is clear that the cooling performance of a membrane evaporative water cooler would not be quite as good as that of a canvas cooler nor as cheap. However the membrane cooler would be less susceptible to bacteria or viral contamination. Furthermore, the membrane water cooler would eliminate liquid water seepage through the container, water consumption being restricted to vapor transport through the membrane. Thus, the water in a given MH membrane container could be expected to last much longer than that in a canvas container before refilling. This raises the possibility that a membrane container could be used for extended storage of perishable goods at approximately the wet bulb temperature, a considerable advantage in dry climates.

A possible disadvantage of the MH membrane cooler is that if the contained water is evaporated close to dryness, there may be salt precipitation within the container.

APPENDIX 1:
ANALYSIS OF DATA FROM WATER WARMING TEST OF AN ACCUREL MEMBRANE

From consideration of the direction of the humidity ratio driving force,

$$\Delta W = W_{H_2O} - W_{DB,SU}$$

in the upper curve, W versus t, of FIGURE 3, it is possible to say that from zero to 780 seconds, at which time $\Delta W = 0$, water vapor is permeating from the air through the membrane and condensing in the liquid water. After time $t(\Delta W = 0)$ the vapor permeation direction is reversed, that is, liquid water is *evaporating*, then permeating through the membrane into the air. The lower curve, T_{H_2O} versus t, identically reflects conditions of water vapor permeation, first into and then out of the liquid water.

It is the evaporation section of the curves that is of interest to us because the water permeation direction is that in evaporative cooling. For the evaporation curve we can say qualitatively that the sensible heat change of the liquid water is equal to the conductive heat transfer into the water from the sensibly cooling air minus the latent heat loss from the water into the air, or in terms of heat fluxes, kJ/m²sec:

$$(15.4)(4.18)\left(\frac{dT_{H_2O}}{dt}\right) = U(T_{DB,SU} - T_{H_2O}) - K(W_{H_2O} - W_{DB,SU})(2.46) \quad (1.1)$$

where 15.4 kg/m^2 of water is in the water cup (effective water depth 1.54 cm); 4.18 kJ/kg°C is the specific heat of liquid water; dT_{H_2O}/dt °C/sec is the slope of the water heating curve; U is the overall heat transfer coefficient; K is the overall mass transfer coefficient of water vapor; and 2.46 kJ/g is the latent heat of vaporization of water. The other terms are already known or are defined in NOMENCLATURE.

Overall Heat Transfer Coefficient, U

At time $t(\Delta W = 0)$, which occurred at 780 seconds, the second term on the right of Equation (1.1) drops out, giving $(15.4)(4.18)(0.0022) = U(20.2 - 14.8)$ so that $U = 0.0262$ kJ/sec m^2 °C, where 0.0022, 20.2, and 14.8 can all be seen in FIGURE 3.

Overall Mass Transfer Coefficient, K

At time $t = 3,360$ seconds, $dT_{H_2O}/dt \approx 0$, thus eliminating the left-hand term of (1.1), giving $(0.0262)(20.2 - 17.4) = K(12.5 - 10.5)(2.46)$. Thus, $K = 0.0149$ gm water vapor/sec m^2 W, where 20.2, 17.4, 12.5, and 10.5 can all be seen in FIGURE 3. The water temperature of 17.4°C is essentially the wet bulb temperature for the membrane/air boundary layer combination.

These data are for an Accurel membrane. The average values for several membranes were, as shown in TABLE 2, $U = 0.0315$ and $K = 0.0164$.

APPENDIX 2:
COMPARISON OF COEFFICIENTS IN MEMBRANE EVAPORATIVE COOLING OF CONTAINED WATER AND MEMBRANE DISTILLATION

Membrane evaporative water cooling (MEC) is similar to membrane distillation (MD) in that both depend on the use of microporous hydrophobic membranes on the upstream side of which vaporization occurs and through which the vapor is transported.[1,2] However, in MD the downstream vapor is condensed either by direct contact with a *liquid coolant* or by a cold plate after the vapor has traveled through a *diffusion gap* of air.

Overall Heat Transfer Coefficients

The average overall heat transfer coefficient for MEC is 0.0315 kJ/sec m^2 °C (TABLE 2). Fane[3] obtained 0.392 kJ/sec m^2 °C for an MD system with liquid coolant. This suggests that the air boundary layer encountered in MEC is principally responsible for its low heat transfer coefficient. This view is supported by the data of Kubota[4] who obtained 0.04 kJ/sec m^2 °C with an MD system having a diffusion gap, that is, with an air boundary layer.

Overall Mass Transfer Coefficients

The average overall mass transfer coefficient for MEC in TABLE 2 is 1×10^{-2} kg/sec m^2 bar. This is near the lower end of values calculated in the literature for liquid coolant MD, values ranging from 0.42×10^{-2} to 9.6×10^{-2} kg/sec m^2 bar.[3,5,6,7] For diffusion gap MD the coefficients were 0.63×10^{-2} and 2.02×10^{-2} kg/sec m^2 bar, respectively.[7,8]

These data do not indicate a convincing difference in mass transfer coefficients between MEC and MD, whether the latter is with liquid coolant or diffusion gap. Perhaps this is because in both the cases the membrane itself may control mass flow due to the resistance to vapor passage through the micropores. This reason is cited in the section COMPARATIVE PERFORMANCE OF CANVAS AND MH MEMBRANE EVAPORATIVE WATER COOLERS as an explanation for the difference in mass transfer coefficients of the MH membranes and porous canvas.

APPENDIX 3:
CALCULATION OF COOLING CHARACTERISTICS OF WATER CONTAINED IN A MICROPOROUS HYDROPHOBIC MEMBRANE POUCH

The given conditions are tabulated in FIGURE 4. What must be calculated is the cooling of 30°C water as a function of time. First, we write Equation (1.1), from APPENDIX 1 as follows:

$$(20)(4.18)\left(\frac{dT_{H_2O}}{dt}\right) = U(T_{DB} - T_{H_2O}) - K(W_{H_2O} - W_{DB})2.46, \quad (1.1)$$

where 20 is now the ratio of contained water mass to membrane area, kg_{H_2O}/m^2 membrane, for a 2-cm pouch width, and other terms that have been defined.

Now we assume that

$$W_{H_2O} = 0.948 T_{H_2O} - 3.5. \quad (3.1)$$

This is a straight line correlation of W_{H_2O} and T_{H_2O} along the 100% relative humidity line of FIGURE 5 in the temperature region of about 20°C. We use the average U and K values in TABLE 2 for MH membranes. For the impinging air T_{DB} = 30°C, the humidity ratio at 20% humidity is 5.3 gm H_2O vapor/kg dry air, thus

$$(20)(4.18)\left(\frac{dT_{H_2O}}{dt}\right)$$
$$= (0.0315)(30 - T_{H_2O}) - (0.0164)(0.948 T_{H_2O} - 3.5 - 5.3)2.46. \quad (3.2)$$

This reduces to

$$\frac{dT_{H_2O}}{dt} = 0.0155 - 0.00084 T_{H_2O} \quad (3.3)$$

or

$$\frac{dT_{H_2O}}{0.0155 - 0.00084 T_{H_2O}} = dt. \quad (3.4)$$

This equation is integrated over the water temperature range 30 to 18.4°C, giving the curve of FIGURE 4. The temperature 18.4°C is the lowest possible water temperature, where $dT_{H_2O}/dt = 0$, that is, is the wet bulb temperature for the membrane/air boundary layer combination in this case.

REFERENCES

1. SARTI, G., C. GOSTOLI & S. MATULLI. 1985. Low energy cost desalination processes using hydrophobic membranes. Desalination **56:** 277–284.
2. GOSTOLI, C. 1993. Membrane distillation. Poster at ICOM 93, Heidelberg.
3. FANE, A., R. SCHOFIELD & C. FELL. 1987. The efficient use of energy in membrane distillation. Desalination **64:** 231–243.
4. KUBOTA, S., et al. 1988. Experiments on seawater desalination by membrane distillation. Desalination **69:** 19–26.
5. DRIOLI, E., Y. WU & V. CALABRO. 1987. Membrane distillation in the treatment of aqueous solutions. J. Memb. Sci. **33:** 277–284.
6. GORE, D. 1982. Gore-Tex membrane distillation. Presented at the Proc. 10th Annual Convention of the Water Supply Improvement Association, Honolulu, July 25–29.
7. SCHNEIDER, K., W. HOLZ & R. WOLLBECK. 1988. Membranes and modules for transmembrane distillation. J. Memb. Sci. **39:** 25–42.
8. KIMURA, S. & S. NAKAO. 1987. Transport phenomena in membrane distillation. J. Memb. Sci. **33:** 285–298.

Membrane Evaporative Cooling to 30°C or Less

2. Membrane Evaporative Air Cooling

SIDNEY LOEB

P.O. Box 41, Omer, Israel

ABSTRACT: Microporous hydrophobic membranes were examined for use in steady state membrane evaporative air cooling. The examination consisted of calculating membrane performance as a function of overall heat and mass transfer coefficients already obtained and reported in Part 1 (previous paper, this volume). This performance was compared with that obtained by similar calculations made on existing evaporative air coolers. It was found that the cooling performance of the membrane evaporative air cooler was not as good as that of the existing evaporative air cooler. This is to be expected since the existing cooler has only one resistance, the air boundary layer (ABL), whereas the membrane cooler has the ABL and the membrane resistance. However, the membrane air cooler has advantages, such as appreciably lower water consumption and operation under more sanitary conditions, that is, without intimate conjunction of flowing air and liquid water on solid surfaces.

KEYWORDS: evaporative air cooling; microporous hydrophobic membranes; membrane evaporative air cooling

NOMENCLATURE:

See also Part 1 (previous paper, this volume)
ABL	air boundary layer
EXC	existing air cooler (porous canvas)
f	final condition
i	initial condition
MC	membrane air cooler (microporous hydrophobic membrane)
m^2_{perm}	total permeation area of all capillary tubes, m^2
$m^2_{cs,avail}$	total available cross-sectional area for air flow
x^2	cross-sectional area of imaginary square just containing one capillary tube, m^2 (see APPENDIX 2)

GENERAL CONDITIONS FOR COMPARING PERFORMANCES OF EXISTING AND MEMBRANE EVAPORATIVE AIR COOLERS

Initial Air Properties

As shown in FIGURE 1, hot desert air is assumed to enter the apparatus (temperature 32°C, relative humidity 22%).

Address for correspondence: Sidney Loeb, P.O. Box 41, Omer 84965, Israel.
sidloeb@bgumail.bgu.ac.il

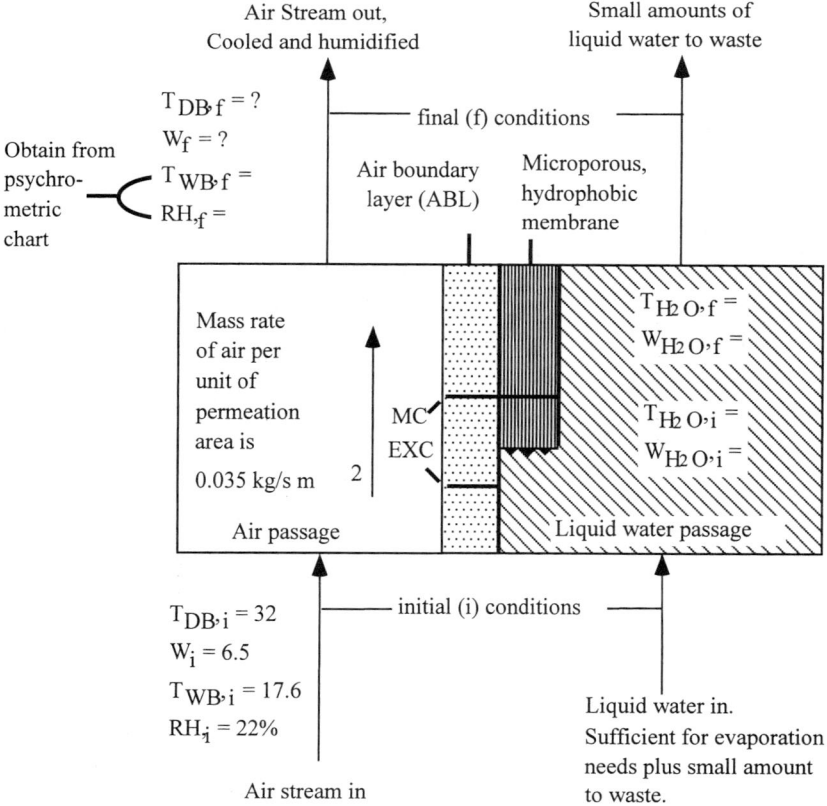

FIGURE 1. General conditions for comparing performances of evaporative air coolers. EXC, existing cooler (porous canvas); MC, membrane cooler (microporous hydrophobic).

Ratio of Mass Rate of Air Flow to Permeation Area, kg/sec m²

It was found by calculation that the liquid water temperature is close to the wet bulb temperature of the incoming air. Hence, if the above ratio of air rate to permeation area is inadequately low, the air may cool closely to its wet bulb temperature but at the expense of economic utility. If the ratio is too high the amount of air cooling will be insufficient. The calculations showed that the optimum value of this term is of the order of 0.035 kg/sec m² (see APPENDIX 1).

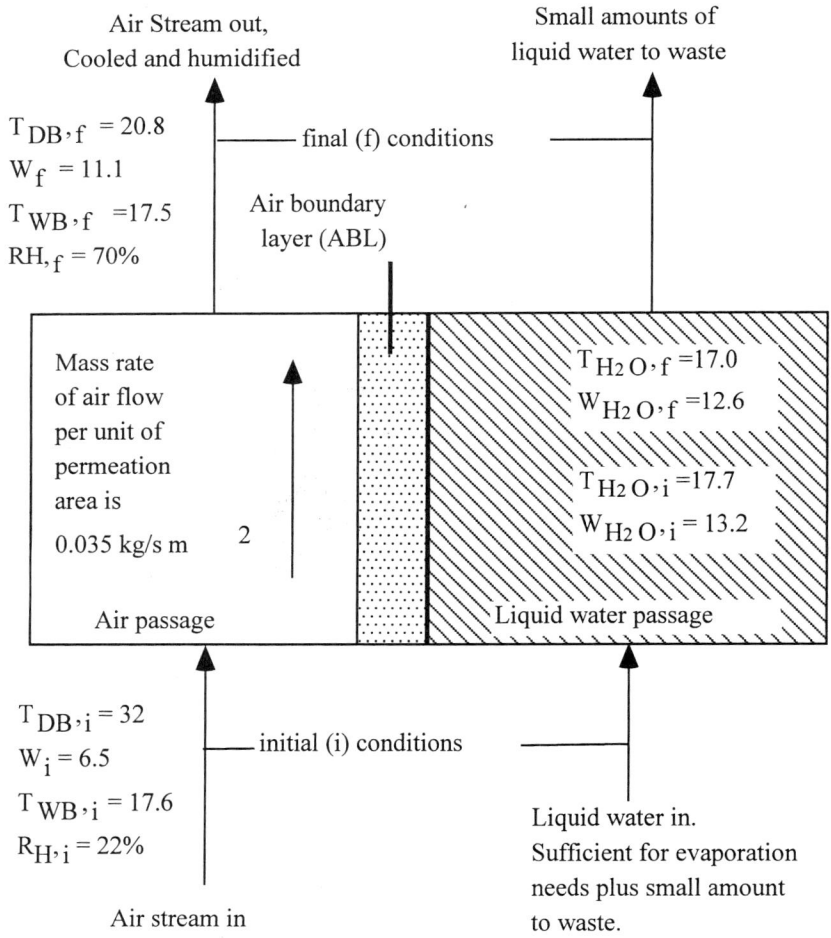

FIGURE 2. Existing evaporative air cooler. Calculated using the methods of APPENDIX 1. Overall transfer coefficients: heat, $U = 0.0393 \, kJ/sec \, m^2 \, °C$; mass, $K = 0.0349 \, gm \, H_2O$ vapor/sec m² W.

The Overall Heat and Mass Transfer Coefficients for Membrane Evaporative Air Cooling

These coefficients, shown in FIGURE 1, were determined in the first of these two papers (previous paper, this volume). The difference in performance cited herein between the existing and the membrane evaporative air cooler is due essentially to the differences in these overall transfer coefficients.

COMPARATIVE PERFORMANCE OF EXISTING AND MEMBRANE EVAPORATIVE AIR COOLERS

FIGURE 1 shows six terms to be evaluated. The procedure for this is shown in APPENDIX 1. The results are shown in FIGURES 2 and 3, which give the performance of existing and membrane evaporative air coolers, respectively. These data enable comparison between the two types of coolers.

The comparison is quantified in TABLE 1. According to Column 5 of this table, the air temperature decrease for the membrane air cooler was only $(7.8/11.2)(100) = 70\%$ that of the existing cooler. This ratio was the same for Column 6 and for Column 9. In this last column the criterion was kilowatts of cooling per square meter of permeation area. Thus, it is clear that the membrane evaporative air cooler does not function as well as the existing air cooler.

TENTATIVE DESIGN OF A MEMBRANE EVAPORATIVE AIR COOLER

The cooler is assumed to be a shell and capillary tube device, the flows occurring within a duct having a square cross-section with dimensions as shown in FIGURE 4. The air and water flows are parallel to the capillary tubes, that is, the long dimension of the duct. The water is supplied from within the capillary tubes and the air flows through the space between the tubes. The flow rates of air and water are not countercurrent, as is common in a heat exchanger, for example, because, as we have seen, the temperature of the water is virtually constant throughout the length of the flow path, that is, the length of the capillary tube.

Two questions arise from considerations of the optimum properties of the cooler.

Choice of Air Velocity

In the tests discussed in the water cooling paper, the air rate was 3 meters per second, impinging at 90° on the permeation surface. Because in the air cooling process, the air flow is parallel to the permeation surface, it is assumed herein that the transfer coefficients for cooling air is the same as those chosen for cooling water if the parallel flow rate of air is twice that in air cooling, that is, 6 meters per second.

The Influence of Capillary Tube Diameter on Performance

The shell cross-section area is $(0.5)^2 = 0.25\,\text{m}^2$. If the total cross-sectional area of all the capillary tubes is vanishingly small compared to the total cross-sectional area, the total air flow rate will be $(0.25)(6) = 1.50\,\text{m}^3/\text{sec}$.

In APPENDIX 2 calculations are shown for the influence of various capillary tube diameters on total volumetric air flow rate and other properties, with the results shown in TABLE 2. As can be seen the total volumetric air flow rate decreases slowly with increasing tube diameter, but at 0.005 meters approaches a limit for allowable tube diameter because of decreasing volumetric air flow rate. The required number of tubes diminishes with increasing tube diameter because the total permeation area of all the tubes is constant at about 50 square meters. FIGURE 4 shows the dimension of the tentative cooler and relevant data for the cooler if the tube outer diameter is 0.002 m.

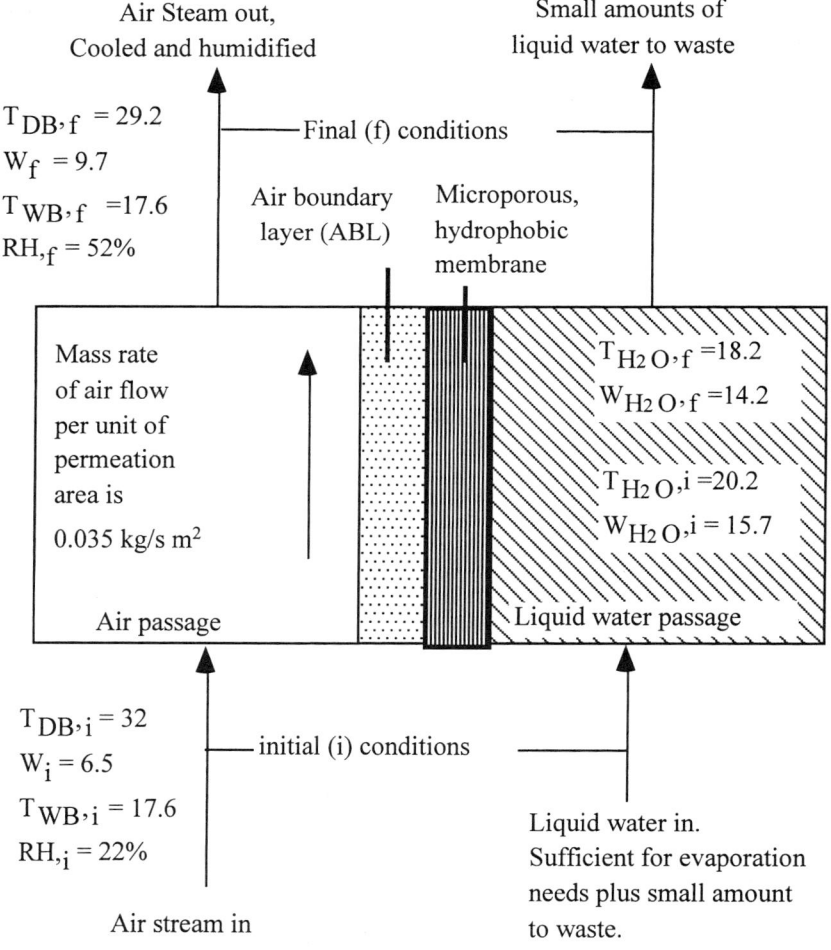

FIGURE 3. Membrane evaporative air cooler. Calculated using the methods of APPENDIX 1. Overall transfer coefficients: heat, $U = 0.0315 \, kJ/sec \, m^2 \, °C$; mass, $K = 0.0164 \, gm \, H_2O \, vapor/sec \, m^2 \, W$.

TABLE 1. Comparative performance of existing and membrane evaporative air coolers

Evaporative Air Cooler Type	Overall Transfer Coefficients		Air Temperature Decrease Initial dry bulb temperature, $T_{DB,i} = 32$ Initial wet bulb temperature, $T_{WB,i} = 17.6$ Max. possible temp. decrease, $32 - 17.6 = 14.4$			Relative Humidity, RH% Initial RH = 22%		Unit Refrigeration Rate kJ/sec m² = kilowatts/m²
	Heat, U	Mass, K	final air temperature, °C $T_{DB,f}$	air temp. decrease, °C $T_{DB,i} - T_{DB,f}$	percent of maximum possible temp. decrease	final relative humidity	relative humidity increase, RH_f	Notes 1 and 2
1	2	3	4	5	6	7	8	9
			FIGURES 2 and 3	32 − Col. 4	$\frac{(100)(\text{Col. 5})}{(14.4)}$	FIGURES 2 and 3	(Col. 7 − 22)	(0.035)(1.01)(Col. 5)
Existing (FIG. 2)	0.0393	0.0349	20.8	11.2	78	70	48	0.40
Membrane (FIG. 3)	0.0315	0.0164	24.2	7.8	54	52	30	0.28

NOTE 1: The value 0.035 is in FIGURES 1, 2, and 3; 1.01 kJ/kg°C is the specific heat of air.
NOTE 2: Tons of refrigeration per square meter of permeation area, Column 9/3.51.

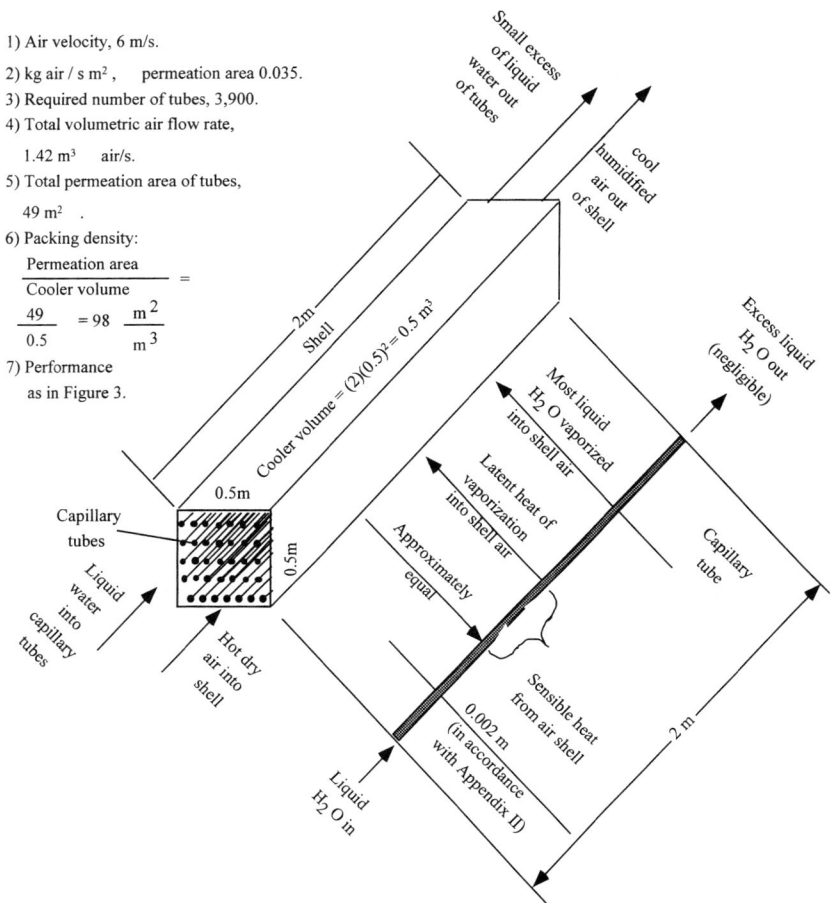

FIGURE 4. Characteristics of a (tentative) shell and capillary tube evaporative air cooler. Example: tube outer diameter, 0.002 m; see APPENDIX 2 and TABLE 2.

CONCLUSIONS

The conclusions are very similar to those of the first paper (previous paper, this volume), that on evaporative water cooling. A membrane evaporative air cooler would not work as well as an existing (canvas) air cooler, nor be as cheap. However, the membrane cooler would be less subject to viral or bacterial contamination because contact between air and liquid water would be eliminated.

Water consumption would tend to be less because it is virtually restricted to vapor transmission through the membrane, there being no liquid water seepage through it. However, a small amount of liquid water would have to be released on the downstream side of the unit to prevent precipitation of concentrated solutes inside the capillary tubes.

TABLE 2. Influence of capillary tube diameter on membrane evaporative air cooler of FIGURE 4[a,b]

Assumed Capillary Tube Diameter, m	Total Volumetric Air Flow Rate m³ air/sec	Required Number of Capillary Tubes	Total Permeation Area of all Capillary Tubes	Permeation Area Packing Density[c]
0	1.50	none	0	0
0.001	1.46	8,030	50.5	101
0.002	1.42	3,900	49	98
0.005	1.32	1,570	49	99

[a] Air velocity 6 m/sec.
[b] 0.035 kg air/sec m².
[c] Cooler dimensions of FIGURE 4. The cooler volume is 0.5 m³.

APPENDIX 1: CALCULATION OF AIR COOLER PERFORMANCE

Example: Existing Evaporative Air Cooler (Porous Canvas)

FIGURE 1 shows the given terms for the calculation. There are six unknown terms shown in the figure: $W_{H_2O,i}$, $W_{H_2O,f}$, $T_{H_2O,f}$, $T_{H_2O,i}$, $T_{DB,f}$ and W_f. (The terms $T_{WB,f}$ and RH_f are obtained from a psychrometric chart, FIGURE 5 of Part 1.) The subscripts i and f refer to initial and final conditions.

As can be seen, on the liquid water side the unknown terms are inside the cooler. This is because the externally added liquid water is scarcely more than sufficient for evaporation needs, so that changes, if any, in water temperature, T_{H_2O}, and water equivalent humidity ratio, W_{H_2O}, occur within the air cooler. In the following equations, the small overall sensible heat changes in the liquid water passing through the cooler are not considered.

The six equations for the six unknown terms are as follows:

$$W_{H_2O,i} = 0.948 T_{H_2O,i} - 3.53 \tag{1.1}$$

$$W_{H_2O,f} = 0.948 T_{H_2O,f} - 3.53. \tag{1.2}$$

Equations (1.1) and (1.2) are for straight lines representing conditions on the saturation curve of FIGURE 5 of Part 1 over the short range 15° to 25°.

$$K(W_{H_2O,i} - 6.5)2.46 = U(32 - T_{H_2O,i}) \tag{1.3}$$

$$K(W_{H_2O,f} - W_f)2.46 = U(T_{DB,f} - T_{H_2O,f}), \tag{1.4}$$

where 2.46 kJ/gm is the latent heat of vaporization of water, $W = 0.0393$ kJ/sec m² °C, $K = 0.0349$ gm H_2O vapor/sec m² W. K and U are the overall mass and heat transfer coefficients, respectively, those shown in FIGURE 4 of the previous paper.

Equations (1.3) and (1.4) state that the flux of latent heat in the water vapor permeating the membrane, kJ/sec m², is equal in magnitude and opposite in direction to the flux of sensible heat permeating the membrane due to the temperature difference between the air and the liquid water (see FIG. 4).

$$0.035(W_f - 6.5) = (W_{H_2O, f} - W_f + W_{H_2O, i} - 6.5)\frac{K}{2}. \tag{1.5}$$

According to (1.5), the mass flow rate of water vapor per unit area, $gm H_2O/sec\ m^2$, added to the independently variable flow rate of air passing longitudinally through the apparatus (FIG. 1) is equal to the average mass flux over the entire apparatus of water permeating perpendicularly through the membrane into the air. The 0.035 term ($kg air/sec m^2$) shown in FIGURE 2 was found to be about optimum in giving both adequate air flow rate and adequate air cooling simultaneously, a subject raised in the next section.

$$1.01(32 - T_{DB, f}) = 2.46(W_f - 6.5), \tag{1.6}$$

where $1.01 kJ/kg°C$ is the specific heat of air. Equation (1.6) states that the overall sensible cooling of a kilogram of air in the longitudinally-moving airstream is equal to the latent heat of vaporization of the water vapor finally added to this kilogram of air.

These six equations were solved for the existing (porous canvas) evaporative air cooler of FIGURE 2. From the values thus obtained for the final dry bulb temperature, $T_{DB,f}$, and final humidity ratio, W_f, both at the outlet air stream, the values of $T_{WB,f}$ and RH_f can be determined from the psychrometric chart (FIG. 5 of Part 1).

Similar performance calculations were made for the membrane evaporative air cooler, the principal difference being the values of 0.0315 and 0.0164 for the heat and mass transfer coefficients, respectively.

APPENDIX 2:
RELEVANT AIR COOLER PROPERTIES AS AFFECTED BY CAPILLARY TUBE DIAMETER (TABLE 2)

The selected air velocity in the shell and capillary tube cooler is 6 meters per second. We have also chosen 0.035 kilograms of air per second for each square meter of permeation area. Let m^2_{perm} denote the total permeation area of all the tubes. Then the velocity can be expressed as follows:

$$\frac{0.035}{1.19} \frac{m^2_{perm}}{m^2_{cs\ avail}} = 6,$$

where $1.19 kg air/m^3$ is the standard density of air and $m^2_{cs\ avail}$ is the total available cross-section area for air flow after taking the capillary tube cross-sectional area into account. The above relation may be written

$$\frac{m^2_{perm}}{m^2_{cs\ avail}} = 207. \tag{2.1}$$

Now consider a capillary tube having any given external diameter, say, 0.002 meters. The cross-section of the tube can be considered to be centered within a square having an area of x^2 meters. Then

$$m^2_{cs\ avail/tube} = x^2 - \frac{\pi}{4}(0.002)^2 \tag{2.2}$$

and

$$m^2_{\text{perm/tube}} = 2\pi(0.002), \quad (2.3)$$

where the cooler length is 2 meters.

Equations **(2.2)** and **(2.3)** are substituted in **(2.1)** to obtain $x = 0.008$ meters. Then, by **(2.2)** the unit available area, $m^2_{\text{cs avail/tube}}$, is 6.09×10^{-5} m²/tube. The volumetric air flow rate per tube is $(6)6.09 \times 10^{-5} = 36.5 \times 10^{-5}$ m³/sec tube.

The cross-sectional area of the shell is $(0.5)(0.5) = 0.25$ m². Thus, the total number of capillary tubes is $0.25/(0.008)^2 = 3{,}900$ tubes, where $(0.008)^2$ is the area of one square associated with a capillary tube.

The total volumetric air flow rate for the cooler is $(3{,}900)36.5 \times 10^{-5} = 1.42$ m³/sec. The total permeation area for all the tubes is, by **(2.3)**, $(3{,}900)(\pi)(0.002)(2) = 49$ m². The cooler volume is $(2)(0.5)(0.5) = 0.5$ m³. Thus, the air cooler packing density is $49/0.5 = 98$ m² of tube permeation area per cubic meter of cooler.

Index of Contributors

Abellán, M.J., 17–28
Abogrean, E.M., 85–96
Alberti, G., 208–225
Arcella, V., 226–244
Arnot, T.C., 411–419, 492–501

Bhanushali, D., 159–177
Bhardwaj, V., 318–328
Bhattacharyya, D., 159–177
Boerlage, S.F.E., 85–96
Borcherding, H., 470–479
Bottino, A., 29–38
Brinkmann, T., 306–317

Cao, B., 370–385
Capannelli, G., 29–38
Casciola, M., 208–225
Cheng, S., 245–255
Chmiel, H., 142–158, 178–193
Chua, H.C., 492–501
Ciabatti, I., 29–38, 53–64
Ciardelli, G., 29–38
Cowan, R.M., 453–469
Criscuoli, A., 1–16
Croce, F., 194–207

Daubert, I., 420–435
di Cesare, G., 208–225
Drioli, E., 1–16, 436–452

El-Azizi, I.M., 85–96
Engwall, E.E., 346–360

Field, R.W., 401–410
Fonade, C., 420–435, 480–491
Freitas dos Santos, L.M., 123–141
Fukuda, T., 256–266

Galjaard, G., 85–96
Garcia, J.D., 502–514
Ge, J.-J., 453–469
Ghielmi, A., 226–244
Giorno, L., 436–452
Goerke, A.R., 502–514
Goma, G., 420–435
Gordeyev, S.A., 318–328

Hägg, M.-B., 329–345
Han, B., 502–514
Han, S., 123–141
Henson, M.A., 370–385
Hicke, H.-G., 470–479
Ho, W.S.W., 97–122
Howell, J.A., 411–419, 492–501
Huuhilo, T., 39–52

Jorcke, D., 470–479

Kennedy, M.D., 85–96
Kovvali, A.S., 279–288
Kulprathipanja, S., 361–369

Lafforgue, C., 420–435
Li, N., 436–452
Li, N.N., ix
Lie, J.A., 329–345
Lindbråthen, A., 329–345
Liu, W., 411–419
Livingston, A., 123–141
Lobo, V., 401–410
Loeb, S., 515–527, 528–537

Ma, Y.H., 346–360
Macintosh, A., 318–328
Maranges, C., 420–435

Marcucci, M., 53–64
Mardilovich, I.P., 346–360
Matteucci, A., 53–64
Mavrov, V., 142–158, 178–193
McGregor, M.L., 453–469
Mercier-Bonin, M., 420–435, 480–491
Moretti, U., 1–16

Nair, D., 123–141
Noronha, M., 142–158
Nuortila-Jokinen, J., 39–52
Nyström, M., 39–52

Ohlrogge, K., 306–317
Ollis, D.F., 65–84
Ortiz, I., 17–28

Peeva, L., 123–141
Pellegrino, J., 289–305
Pica, M., 208–225

Qin, Y.-J., 453–469

Ranieri, L., 29–38

San Román, F., 17–28
Schippers, J.S., 85–96
Scrosati, B., 194–207
Sharpe, I.D., 318–328
Shilton, S.J., 318–328
Singh Luthra, S., 123–141
Sirkar, K.K., 279–288
Sun, B., 245–255, 267–278, 386–400
Sun, P.B., 267–278

Tommasi, G., 226–244
Trachtenberg, M.C., 453–469

Ulbricht, M., 470–479
Urtiaga, A., 17–28

Vernaglione, G., 53–64

Weber, R., 178–193
White, L.S., 123–141
Wickramasinghe, S.R., 502–514

Yamauchi, A., 256–266

Zou, J., 386–400

OHIO UNIVERSITY LIBRARY

Please return this book as soon as you have finished with it. In order to avoid a fine it must be returned by the latest date stamped below. All books are subject to recall after two weeks or immediately if needed for reserve.

DEC 1 5 2006

CF